STANDARD MODEL AND BEYOND

Proceedings of the XIII International School
of Theoretical Physics

STANDARD MODEL AND BEYOND

Szczyrk, September 19–26 1989
University of Silesia, Katowice

Edited by M. Zrałek and R. Mańka

NOVA SCIENCE PUBLISHERS, INC.

Nova Science Publishers, Inc.
6080 Jericho Turnpike, Suite 207
Commack, New York 11725

Book Production Manager: Janet Glanzman White
Graphics: Elenor Kallberg, Christopher Concannon, and Michael A. Masotti
Cover Design: Michael A. Masotti

Library of Congress Cataloging-in-Publication Data available upon request

ISBN 1-56072-027-1

Printed in the United States of America

LIST OF PARTICIPANTS

M.J. Abdelhafiz	University of Zagazig
S. Bilenky	Joint Institute for Nuclear Research, Dubna
K.S. Bjorkevoll	University of Bergen
D. Bödeker	University of Hannover
K. Bortel	University of Silesia
D. Böhm	University of Würzburg
P. Candelas	University of Texas
W. Cieślewicz	Polytechnic of Poznan
P. Christova	High Pedagogical Institute, Shoumen
D. Chruściński	University of Torun
M. Cierniak	University of Silesia
M. Czarnecki	Jagellonian University
H. Czyż	University of Silesia
G. D'Agostino	University of Roma
B.R. Greene	Harvard University
P.S. Green	University of Maryland
J.A. Grifols	Universitat Autonoma de Barcelona
J.L. Hewett	University of Wisconsin
T. Hübsch	University of Texas
C. Jarlskog	University of Stockholm

M. Ježabek	Jagellonian University
A. Kagan	University of Maryland
W.L. Kennedy	University of London
K. Kolodziej	University of Silesia
P. Krawczyk	University of Warsaw
D. Kotlorz	Institute of Nuclear Physics, Cracow
A. Kotlorz	Institute of Nuclear Physics, Cracow
A. Kalicińska	Medical Academy of Bydgoszcz
P. Korzybski	University of Toruń
M. Lynker	University of Texas
C.A. Lütken	NORDITA Copenhagen
Z. Lalak	University of Warsaw
Ly Sy Hoi	Institute of Nuclear Technology, Hanoi
M. Maśka	University of Silesia
M. Mańka	University of Silesia
I. Porchescu	Moldavian Academy of Science
Z. Pawełczyk	University of Warsaw
J. Polak	University of Silesia
T.G. Rizzo	University of Wisconsin
D. Spiesberger	University of Hamburg
D. Schildknecht	University of Bielefeld
J. Sładkowski	University of Silesia

J. Syska	University of Silesia
L. Turko	University of Wroclaw
S. Wojtyga	Polytechnic of Szczecin
B. Westwański	University of Silesia
P. Wegrzyn	Jagellonian University
E. Zipper	University of Silesia
M. Zrałek	University of Silesia

CONTENTS

FORWARD

During the last few particle physics theorists have been bombarding their experimental colleagues with new theoretical ideas. We have heard about grand unification, proton decay, neutrino and neutron oscillations, neutrino mass, supersymmetry, supergravity, strings, composite leptons and bosons, heavy leptons, fifth force, axions, Majorons etc. None of these very bright ideas have, up to now, been checked experimentally. On the other hand the Standard Model continues to give increasingly good results, at least for the presently available energies. It is really impressive how well the SU(3) ⊗ SU(2) ⊗ U(1) model describes all present high energy data including the very recent data from SLC and LEP. In spite of great success we still believe that the Standard Model is not the ultimate model of our physical world. An ideal theory for particle theorists is probably the one in which there are essentially no free parameters and an overall beautiful symmetry governs all its aspects. The Standard Model is far from achieving this goal. The number of free parameters is large (masses, mixing angles, coupling constants). There are also other unsolved questions like the strong CP problem, the triviality problem in the Higgs sector, the naturality problem etc. The efforts of physicists to understand the questions mentioned are usually referred to as "Beyond Standard Model Physics".

It is a common belief that the string model can give us a unified theory of particles and gravity. However, since the superstring theory may be naively built only in ten dimensions, the main task is to compactify the six internal dimensions and to obtain the observed four dimensional world. The compactification process is decisive for four dimensional field theory. However, in the result, the six dimensional background internal manifold will be hidden. if we require that supersymmetry survives in low energy we should assume that the internal manifold is the so called Calabi-Yau manifold.

In principle a given Calabi-Yau manifold should give us the Standard Model in four dimensions. Its fields are zero modes on the compact internal manifold. They carry topological and geometrical information about the manifold. The numbers of zero modes are topological invariants of the internal manifold and may be interpreted as a number of chiral families. The string theory gives us the pattern of Yukawa couplings of the particle field theory. However, there are a lot of Calabi-Yau manifolds and the question of how the string theory chooses one such peculiar manifold (a vacuum state) is still open. It seems to us that a physical mechanism which would be able to pin down one of such Calabi-Yau manifolds is desperately needed. Up to now there is no theoretical reason why the Standard Model group SU(3) ⊗ SU(2) ⊗ U(1) was chosen.

This year the Silesian School of Theoretical Physics was held at Szczyrk from 19–26 September. The main lecture courses were devoted to some aspects of particle physics phenomenology, in the first part, and to a few problems in superstring theory, in the second part. Sixty students of ten different nationalities spent eight days intensively studying the Standard Model and its generalizations. It is a pleasure for us to express our sincere gratitude to all those who contributed to the school: the lecturers who all came well prepared and who presented their written notes here, the chairmen of all sections and the students. Among the members of the Organizing Committee we would like to mention in particular H. Czyż and J. Sładkowski. Without their patient and effective work the School would not have been a success.

This School was supported in part by the University of Silesia and by the Polish Ministry of Education under Grant Project CPBP 01.03.

M. Zrałek and
R. Mańka

Part I: Particle Physics Phenomenology

ISBN 1-56072-027-1

PARTICLE PHYSICS PHENOMENOLOGY SUMMARIZED

J.A. Grifols

Grup de Física Teòrica, Universitat Autònoma de Barcelona, 08193 Bellaterra (Barcelona), Spain

The Silesian School of Theoretical Physics was devoted this year to the Standard Model and Beyond. It clearly splitted into two well differentiated parts. On the one side there were the lectures on pure string theory. On the other, a broad spectrum of phenomenological issues were discussed. I have been asked to summarize this second block of talks.

Rather the commenting each and everyone's talk, which I feel not competent to do and futhermore would be a nonsensical exercise since the reader himself can go over each individual contribution in these Proceedings, I shall generically and briefly survey the topics discussed in these lectures.

A large fraction of the lectures devoted to phenomenology were concerned with radiative corrections in the Standard Model. This obviously reflects the prominence that radiative corrections have for the interpretation of the experiments to come at the new machines, like LEP or HERA. Indeed, the experimental precision reached at these accelerators demands theory to match this accuracy and calculations must be done beyond the tree level. There is an impressive industry being developed around the radiative corrections of the electroweak model and for the first time after the advent of the $SU(2)_L \times U(1)$ theory it shall be possible to check in a systematic way the correctness of a renormalized quantum field theory. We are just now witnessing the start of this new era of Particle Physics with the impressive results on the mass and width of the Z boson from the first LEP run. The lectures at Szczyrk concerning the radiative corrections to a great variety of processes (see Proceedings) were an exponent of this extraordinary momentum of Particle Physics.

But there was more. There is general consensus in the high energy community that the Standard Model is just a low energy effective parametrization of a deeper underlying theory. There are at present too many free parameters in the Standard Model to be considered a satisfactory theory. The most popular completion of the Standard Model rests on the gauge unification paradigm and has led to superstring theory and to the so-called superstring inspired models. There has been a very exhaustive survey of the phenomenology associated to E_6 models (entertained in superstring inspired models) presented at this School. Indeed, the physics of extra Z' bosons new leptoquarks, etc has been throughly discussed.

Other aspects of new physics were also considered: fourth generations physics, compositeness related topics, mass hierarchy problems... . In the year where the 25 anniversary of the discovery of CP-violation is being celebrated, this question could not be absent in a School dedicated to the Standard Model and its extensions. We had in fact, and excellent review (with a quite nice historical hindsight) on CP-violation and its implications.

Finally, to close this very brief summary, we should mention that neutrino physics was also presented at this Silesian School. The phenomenology of neutrinos associated to old and new physics was throughly discussed.

Standard Model
and Beyond

ISBN 1-56072-027-1

CP VIOLATION IN THE STANDARD MODEL AND BEYOND

C. Jarlskog
Department of Physics, University of Stockholm, Stockholm, Sweden

Supported by the Swedish National Research Council

The Table of Contents of this talk is

First of all, I wish to thank the Organizers and specially Prof. Marek Zrałek for having invited me to lecture at this School.

This year, we have been celebrating the 25th anniversary of the discovery [1]of CP violation. Indeed, the discovery that nature does not respect the CP symmetry came as a great surprise to the physics community, in spite of the fact that one had got used to the idea that parity is not [2]conserved. Part of the reason was that although with the the fall of parity the C-symmetry also fell one could show that much of the developed formalism (for the neutral kaon system where CP was violated) could be "saved" by assuming that the product symmetry CP is a good symmetry. The antiparticle could then be identified with the CP-image of the particle. The CP "mirror" looked beautiful. In it the left and right were interchanged and the signs of the internal charges were reversed and that sufficed to turn a particle into its antiparticle. The discovery of CP violation meant that the CP mirror was cracked. It could not be used to define the antiparticle. One had now to rely on a more complex mirror, the CPT mirror, to get the antiparticle. An additional "mysterious" feature of CP violation was that it was "tiny". Parity and the C symmetry were manifestly maximally violated but apparently not! What could be the origin of such a tiny effect? Why did it only appear in the decays of the long-lived neutral kaon? Let me first give you a bit of history.

HISTORICAL COMMENTS

As soon as the physics community got over the CP-"shock" it tried to understand and explain it. In fact, in the early years of CP there was a great deal of theoretical as well as experimental interest in it. One can get a rough measure of that interest by consulting the proceedings of the biggest annual conference in particle physics and examining how much emphasize was put on CP violation and what was discussed. The result of such a "look" is the following. In the Proceedings of the Dubna Conference 1964 Fitch reported the discovery of the phenomenon and Treiman (who as you may know had been interested in CP for some time) made some comments about its origin. The proceedings of the following 10 conferences (1965-1974) show that CP violation quickly became a central theme and remained so during that period. Parallel as well as plenary sessions were devoted to it. Let me list the conferences and quote some of the relevant plenary talks:

Oxford (1965): Bell and Steinberger; Lee; Prentki

Berkeley (1966): Fitch; Lee

Heidelberg (1967): Okun and Rubbia

Vienna (1968): Cronin; Treiman

Lund (1969): Steinberger

Kiev (1970): Lobashev; Kobzarev

Amsterdam (1971): Winter

Chicago (1972): Rubbia

Aix-en-Provence (1973): Soergel

London (1974): Kleinknecht

During the above period (mostly in the first few years) the origin of CP violation was much debated. Was it due to strong interactions?, electromagnetic?, weak interactions?, superweak interactions? For example, one could argue that since α/π and ϵ were not so different the origin of CP violation is most probably electromagnetic. Some of the models proposed attributed CP violation to the existence of:

- a long-range (cosmic) force
- an octet of strongly interacting vector bosons
- new particles called the a-particles or chimerons
- mirror fermions
- leptoquarks
- gravitational dipole, etc.

That period was indeed a very exiting one for the theorists. On the experimental front, the proceedings of the conferences listed above witness that an enormous amount of work was done on improving the early results obtained in the neutral kaon system as well as on searches for CP violation in many other processes. During the above period it became gradually clear that there was no evidence for CP violation anywhere except in some decay modes of the K_L. Unfortunately, the situation has not changed since then and there is still no evidence for CP violation anywhere else. For a review of the theoretical as well the experimental situation of CP during those early years see the proceedings of the aforementioned conferences and the book [3]by Kabir.

By 1974 the experiments had become quite accurate and within the experimental uncertainties one had $\eta_{+-} = \eta_{00}$. Here η_{ij} denotes the ratio of the amplitudes $A(K_L \rightarrow \pi^i\pi^j)/A(K_S \rightarrow \pi^i\pi^j)$ Indeed all results were in agreement with Wolfenstein's superweak model. A substantial improvement of the experimental results appeared to be so difficult that the subject was no longer much debated at our largest annual conference. In fact, as far as I could see, during the period 1975 to 1983, not a single parallel session or plenary talk was devoted to CP violation at those conferences (they were held in Palermo (1975), Tbilisi (1976), Budapest (1977), Tokyo (1978), Geneve (1979), Madison (1980), Lisbon (1981), Paris (1982) and Brighton (1983)). But a year later, at the Leipzig Conference there was a notable change in the CP atmosphere which was undoubtedly caused by the promise of new measurements of η_{+-} and η_{00} at CERN and at Fermilab. Since then the interest in CP violation has grown more and more as witnessed by the proceedings of the conferences at Bari (1985), Berkeley (1986), Uppsala (1987) and München (1988). Let me also add that in 1987 Lepton-Photon Conference the NA31 Collaboration from CERN presented evidence for nonequality of η_{+-} and η_{00}, a result which would rule out the superweak model. By then **23 years** had passed since the original discovery of the CP violation. The message is very clear. We have to be very patient! Very recently, at the 1989 SLAC Lepton-Photon Conference as well as at the 1989 EPS Conference in Madrid, the CERN result on η_{+-} and η_{00} has been challenged by E731 Collaboration, doing experiments at Fermilab. The results read

$$Re\ \epsilon'/\epsilon\ = +(33 \pm 7\ \pm 8)10^{-4} \qquad NA31$$

$$Re\ \epsilon'/\epsilon\ = -(5\ \pm 14 \pm 6)10^{-4} \qquad E731$$

Here $\eta_{+-} = \epsilon + \epsilon'$ and $\eta_{00} = \epsilon - 2\epsilon'$ Thus the E731 result, if correct would imply a return to the "superweak era". However, at the very same conferences, evidence was presented for the transition $b \to u$ indicating that the element V_{bu}, in the Standard Three Family Model, is nonzero. If this result is further confirmed it would mean that none of the elements of the quark mixing matrix (in the 3 Family Model) can be zero and one generally would expect a nonvanishing value of $Re\ \epsilon'/\epsilon$.

Fortunately, the dark ages of CP violation, with no new experimental results, no plenary or parallel sessions, did not stop the theorists from working on various aspects of CP violation. There were several reasons for this lively activity. With the discovery that the gauge theories made sense (i.e., their being renormalizable and unitary) and the discovery of the weak neutral currents the CP problem had to be reinvestigated within the new framework. As the Standard Electroweak Model [4] with only two families of quarks and leptons could not incorporate CP violation the extension to 3 families was proposed by Kobayashi and Maskawa [5] The possibility of spontaneous CP violation was also investigated. For an early review see Ref. 6. Models for CP violation within the Standard Model, through the extension of the Higgs sector, were also proposed [7]. Some authors proposed that the origin of CP violation be sought beyond the Standard Model, for example in the framework of left-right symmetric models. For a recent review see, for example Ref. 8.

A second reason for a renewed interest in CP violation had to do with the discovery of new fundamental hadrons, charm and beauty, which called for an extension of the earlier works and promised new phenomena. Furthermore, the entrance of cosmology and the early universe into the lives of particle physicists triggered the study of the cosmological aspects of CP violation. For a recent review see, for example Ref. 9. To all this we should also add the discovery of the θ puzzle in QCD and its possible solutions on which a great deal of theoretical work was done during the "dark ages". For a very recent review see, for example, Ref. 10.

By far the most serious Model of CP violation in our time, i.e., 25 years after the discovery of the phenomenon, is the minimal version of the Standard Electroweak Model. Although the Standard Model is not expected to be the "final theory" it does seem to be a very good perhaps only "effective theory" and is highly successful in describing all relevant data.

THE ORIGIN OF CP VIOLATION IN THE STANDARD MODEL

The (minimal version of the) Standard Model says that, in general there will be CP violation provided there are (at least) three families of quarks and leptons. Although CP violation resides only in the hadronic sector of the Model the leptons are necessary for the consistency of the Model (anomaly cancellation).

The Lagrangian density of the Electroweak Model is, symbolically of the form

$$\begin{aligned} L &= L(f,G) + L(f,H) + L(G,H) + L(G) - V(H) \\ f &= \text{fermions (quarks, leptons)} \\ G &= \text{gauge bosons } (\vec{W} \text{ and } B) \\ H &= \text{the Higgs doublet } \phi \end{aligned} \tag{1}$$

Leaving aside the so-called strong CP problem all the terms in the above Lagrangian are, formaly, CP conserving (i.e., invariant under textbook CP transformations for fields) except the term $L(f,H)$ which describes the interactions of the massless-looking fermions with the Higgs doublet. The Higgs doublet interacts with quarks and leptons via scalar and pseudoscalar couplings. Due to presence of both such couplings CP is (formally) violated in the $L(f,H)$ sector. Under the spontaneous symmetry breaking, the field ϕ, which represents four scalars, undergoes a shift viz., $\phi \to \phi + v/\sqrt{2}$, where v is a real number called the vacuum expectation value of the field ϕ. Three out of the four scalar fields in ϕ are "eaten" by the gauge bosons (which thereby become massive). One of them, called the physical Higgs, ϕ_0 remains in the theory. The essential point is that what remains of $L(f, H)$ is

$$L(f,H) \xrightarrow{SSB} -\sum_{j,k=1}^{N} \{m_{jk}\, \overline{q_{jL}}\, q_{kR} + m'_{jk}\overline{q'_{jL}}q'_{kR} \;+ h.c.\}(1+\frac{1}{v}\phi_0) + leptonic\ terms \quad (2)$$

where the quantities m_{jk} and m'_{jk} are the so-called quark mass matrices for the up-type quarks (denoted by q_j) and the down-type quarks (denoted by q'_j) respectively. The Standard Model does not tell us what the elements of the mass matrices should be. They are all free parameters which eventually end up as quark masses and mixings. In fact the bulk of the free parameters of the Model come just from these mass matrices. The sum over j runs over the families, their number being N. The leptonic terms will in general have the same structure as the hadronic ones except that the first piece corresponding to the neutrino mass matrix will be absent because in the Minimal Electroweak Model there is no right-handed neutrino and thus the neutrino mass is automatically zero. The right-handed neutrino, if introduced into the theory, could be quite a remarkable object because it, unlike other fermions, would need no Higgs to "acquire" mass and thus the scale of its mass need not have anything to do with the scale of spontaneous symmetry breaking, v. The symmetry properties of the quark mass matrices are as follows. Consider, for example the up-kind sector. Suppressing the summation sign we have

$$\begin{aligned} m_{jk}\overline{q_{jL}}\, q_{kR} + \; h.c. &= \frac{1}{2}[\; m_{jk}\bar{q}_j(1+\gamma_5)q_k + m^*_{jk}\bar{q}_k(1-\gamma_5)q_j] \\ &= \frac{1}{2}\bar{q}\; [(m+m^\dagger) + (m-m^\dagger)\gamma_5]q \end{aligned}$$

Thus parity is conserved if the mass matrix is hermitian, because then the pseudoscalar terms, in the above equation, would be absent. Similarly, C-conjugation symmetry requires the mass matrix m to be a symmetric. Finally, under the CP operation

$$m_{jk}\overline{q_{jL}}\, q_{kR} + \; h.c. \xrightarrow{CP} \frac{1}{2}\bar{q}\; [(m+m^{\dagger})^{T} - (m-m^{\dagger})^{T}\gamma_5]q,$$

where, in addition, $\vec{x} \to -\vec{x}$. Thus CP invariance requires the mass matrices to be real

$$m \;=\; m^{*} \qquad m' \;=\; m'^{*} \tag{3}$$

In order to find the physical fields we must diagonalize the quark mass matrices, m and m' which is done via

$$U_L\; m\; U_R^{\dagger} = D \equiv Diag.(m_u, m_c, m_t, ...) \tag{4.a}$$

$$U_L'\; m'\; U_R'^{\dagger} = D' \equiv Diag.(m_d, m_s, m_b, ...) \tag{4.b}$$

Here the U_L and U_R are unitary matrices and the daggers on the matrices, on the RHS, are introduced for the sake of notational convenience. Furthermore, m_u, m_d, etc. denote the eigenvalues, of the mass matrices, i.e., the quark masses. Thus

$$U_L m U_R^{\dagger}\; U_R m^{\dagger} U_L^{\dagger} \;= U_L m m^{\dagger} U_L^{\dagger} \;= D^2 \;= Diag.(m_u^2, m_c^2, m_t^2, ...) \tag{5}$$

Thus we see that the matrix U_L diagonalizes the *hermitian* matrix $mm^{\dagger}$. Similarly, the matrix U_R diagonalizes the hermitian matrix $m^{\dagger}m$. Furthermore, there are analogous relations for the down-kind sector. The physical fields are

$$q_L^{phys} = U_L q_L \qquad q_L'^{phys} = U_L' q_L'$$

and their analog for the R-handed quarks. Eq. (2) rewritten in terms of the physical fields reads

$$L^{phys}(f,H) = -(1+\frac{\phi_0}{v})[m_u\bar{u}u + m_c\bar{c}c + ... + m_d\bar{d}d + m_s\bar{s}s + ...].$$

The essential feature of the above Lagrangian is that it conserves, seperately, P , C and thus also CP as well as T. This demonstrates that the apparent symmetry properties of a Lagrangian need not have anything to do with *physics*, if the Lagrangian in question is expressed in terms of unphysical fields. In rewriting the $L(f, H)$ in terms of the physical fields we find that there is no trace of the U and U' matrices left. The interactions of the Higgs with the fermions is automatically P and C conserving, as well as flavour conserving. In extensions of the Model these properties will, in general, not be valid.

What happens is that the original CP violation in the term $L(f, H)$ has now moved into a different sector of the theory, i.e., into the $L(f, G)$ sector which describes the interaction of the fermions with the gauge bosons. When one expresses the term $L(f, G)$ in terms of the physical fields several "miracles" happen. The neutral currents all conserve CP. In addition, there are no flavour changing neutral currents (the GIM mechanism [11]) because of two facts, (i) these terms are helicity conserving (i.e., they involve either the index L or R but not both) and (ii) the four matrices U and U' (with indices L or R) are unitary. The absence of flavour changing neutral currents means that there are no factors of the type $\bar{u}\gamma^{\mu}(a + b\gamma_5)c$, in the Lagrangian of the Standard Model. This absence of the flavour changing neutral currents is a very important feature of the Model. Note that the interactions of the physical vector bosons A (the photon) and the Z are then automatically flavour conserving because the photon and the Z are linear combinations of W^3 and B. As there are stringent limits, from experiment, on flavour changing neutral currents, the absence of such currents is one of the great successes of the Model. It should be noted that in the "extensions" of the Standard Model one generally encounters difficulties in this respect. However, in the charged current terms in $L(f, G)$ the up-type and the down-type quarks *do mix* simply because $W^{\pm}$ carry one unit of charge. Since the charged currents only involve left-handed quarks the matrices with the subscript R do not enter at all and the U with the index L gets multiplied by $U'^{\dagger}$ with the same index. But $UU'^{\dagger}$ need not be equal to unity. All we know is that it is a unitary matrix. Dropping the numerical factors and the coupling constant, the charge current terms in Eq. (1) are given by

$$X_C \equiv [W^1_\mu - iW^2_\mu]\ \overline{q_L}\ \gamma^\mu q'_L\ + h.c. = [W^1_\mu - iW^2_\mu]\ \overline{q_L^{phys}}\ \gamma^\mu U_L U'^{\dagger}_L q'^{phys}_L + h.c.$$

$$= [W^1_\mu - iW^2_\mu]\ \overline{q_L^{phys}}\ \gamma^\mu V q'^{phys}_L\ + h.c.$$

where $V \equiv U_L U'^{\dagger}_L$ is the so-called quark mixing matrix. It is, of course a unitary N by N matrix. CP conservation now requires V to be real. A general unitary N by N matrix has N^2 parameters $N(N-1)/2$ of these parameters may be taken as Euler angles which one introduces in dealing with rotations in N dimensional (Euclidean) space. The remaining parameters are called phases. However, $2N-1$ of these phases are not measurables because we have $2N$ quark fields. The phases of these quark fields may be chosen at will. Rephasing of the quark fields rephases the elements of the quark mixing matrix. However, only the phase differences appear there and there are $2N-1$ such quantities. Therefore, V has $N^2-(2N-1)=(N-1)^2$ parameters among which $N(N-1)/2$ are the rotation angles. Therefore, the number of phases in V is $(N-1)(N-2)/2$. For the case of 2 families there is only 1 rotation angle and no phases. The quark mixing matrix is then of the form

$$\begin{pmatrix} cos\theta & sin\theta \\ -sin\theta & cos\theta \end{pmatrix} \tag{6}$$

The angle θ is the Cabibbo angle [12]. It is amusing to note that, for the leptonic sector, the matrix (6) was written down [13] as early as in 1962, in connection with neutrino oscillations.

THE MINIMAL MODEL WITH 3 FAMILIES

Consider now the case of 3 families. The number of parameters in V is $(3-1)^2=4$, i.e., 3 rotation angles and one phase. The first example of such a matrix was given by Kobayashi and Maskawa [5]. One can show that there are 36 different looking parametrizations of the "KM-matrix" [14]. Of course, they are all equivalent to each other. In the original KM parmetrization with the angles θ_j, $j = 1, 2, 3$ and the phase δ eight conditions on these angles must be satisfied in order to get CP violation. These are

$$\theta_j \neq 0, \ \frac{\pi}{2} \qquad \delta \neq 0, \ \pi, \qquad j = 1, 2, 3 \tag{7}$$

Furthermore no two quarks with the same charge are allowed to be degenerate in mass because otherwise the phase can be removed and there will be no CP violation. Thus

$$m_u \neq m_c, \ m_c \neq m_t, \ m_t \neq m_u, \ m_d \neq m_s, \ m_s \neq m_b, \ m_b \neq m_d \tag{8}$$

It is interesting to note that the above 14 conditions, in Eqs. (7) and (8) are unified in a single relation, viz.,

$$det \ \mathbf{C} \neq 0 \tag{9}$$

where $\mathbf{C}$ is the commutator of the hermitian mass matrices (see Eq. (5))

$$\left[mm^\dagger, \ m'm'^\dagger\right] \equiv i\mathbf{C} \tag{10}$$

Here the quantities m and m' are the 3 by 3 quark mass matrices which were introduced before (see after Eq. (2)). Again the theory does not tell us what these mass matrices are. In fact their "determination", from first principles, is one of the most outstanding problems in particle physics. Although we can not determine the mass matrices, we can measure their eigenvalues (the 6 quark masses) and the four independent measurables of the quark mixing matrix. What is highly remarkable about the above commutator is that its determinant is given by (see Ref. [14] and references therein)

$$\begin{aligned} det \ \mathbf{C} = -2J \ (m_t^2 - m_c^2)(m_c^2 - m_u^2)(m_u^2 - m_t^2) \\ \times(m_b^2 - m_s^2)(m_s^2 - m_d^2)(m_d^2 - m_b^2) \end{aligned} \tag{11}$$

where J is an "invariant", related to CP violation. One can construct it quite simply for any parametrization. In the KM parametrization it is given by

$$J = sin^2\theta_1 sin\theta_2 sin\theta_3 cos\theta_1 cos\theta_2 cos\theta_3 sin\delta \tag{12}$$

Thus we see that the above 14 conditions are unified in the determinant. In other words **CP is violated if and only if** *det* $\mathbf{C} \neq 0$. This statement hardly needs a proof. If the determinant is nonzero then there is no way in which the phase can be removed since $sin\delta$ must be non-zero. For more details and references to the original papers see Ref. [14].

The invariant measurables of the quark mixing matrix, in the case of 3 families, are the moduli of its elements i.e., the quantities $|V_{\alpha j}|$, and the quantity J. The angle and phase independent parametrization is based on the single fact [14, 15] that even the quantity J, up to a sign, is a function of the moduli of the quark mixing matrix. In fact it is given by

$$4J^2 = -\lambda(a^2, b^2, c^2) \tag{13}$$

where the a, b, c are products of moduli of the elements of the quark mixing matrix. More precisely, they are the three sides of any of the 6 unitarity triangles [14, 15]. Let us first remind ourselves that the function λ is the triangular function sometimes also refered to as the Källén λ function. It is familiar from two body kinematics,

$$\lambda(x, y, z) = x^2 + y^2 + z^2 - 2xy - 2yz - 2zx \tag{14}$$

In fact all the six unitarity triangles have the same area and magnitude of J equals twice the area of any of these triangles (for a detailed discussion see Ref. 14). If CP would be conserved all the six triangles collapse to lines and their area vanishes.

A very useful phenomenological parametrization of the quark mixing matrix was suggested by Wolfenstein [16] and is much used in the literature.

BEYOND THE MINIMAL MODEL?

The Standard 3 Family Model tells us that it can accomodate CP violation but it can not predict the value of the magnitudes of the elements of the quark mixing matrix or the value of J. This is unsatisfactory. Either the Standard Model is just the low energy limit of a more fundamental theory (with more predictive power) or CP has nothing to do with the 3 Family Model.

Unfortunately, so far, essentially no progess has been made in finding the more fundamental theory (if it is there) with more predictive power. The extensions of the Standard Model considered are, so far, all even less predictive and much more ad hoc. In fact the Standard 3 Family Model has certain "charm" as follows:

A very essential point about the Standard Model is that, in its hadronic sector, going from one family to two leads to a new phenomenon, i.e., one gets flavour mixing. In other words, in general θ in Eq. (6) is non-zero. Going from two to three families again introduces a novel feature into the theory, i.e., in general CP is violated unless one of the 14 conditions mentioned before is not satisfied. However the transition from three to four is not known to lead to any new features. In fact, in the four family case there are six angles and three phases but these are of the same kind as before. Therefore, it is very hard to get excited about a "beyond" featuring a fourth family although one can not exclude the existence of more standard families. For a recent review of the 4 family extension of the Standard Model see, for example, Ref. 17.

What about other extensions of the Standard Model? Much work has been done on the Higgs mediated origin of CP violation. See, for example Refs. 6, 7 and 18. Another framework for CP violation has been the so called "Left-Right Symmetric Models". For a review of the present status of these models see, for example Ref. 8. An attractive feature of these models is that P and CP are assumed to be conserved "before" the process of spontaneous symmetry breaking takes place. However, the problem with these models is that they are not simpler or more predictive than the Standard Model. One can not even predict the mass of the most fundamental ingredient of the model, i. e., W_R. This is unsatisfactory. Remember that in the case of the Standard Model one had a prediction for M_W and M_Z and thus could look for them and find them.

Unfortunately, the present limits on the mass of the W_R indicate that the experimental test of the model may well be beyond us, at least in the near future.

Some years ago, supersymmetry was a popular model which, also in connection with CP violation. Unfortunately, however, supersymmetry had no predictive power. For example $Re\epsilon'/\epsilon$ could be positive or negative. As the lower limit on the mass of the supersymmetric particles is increasing, as a function of time, it looks less and less interesting. Indeed we have a great puzzle. We believe that there is a "beyond" but do not know what it is.

Observables

I shall be very brief in the written version of this Section as there is now a whole book [19]devoted to the subject. Hopefully, the reader will find the book pedagogical and useful. In 1964 there were only 3 quark degrees of freedom (u,d,s) and K^0 and $\overline{K^0}$ were the only known conjugate weakly decaying neutral mesons. Nowadays the situation is very different as there is another such system in the charm sector, D^0 and $\overline{D^0}$, and two more in the beauty sector. Indeed a great deal of theoretical effort has gone into investigating the CP properties of these new systems. One can, in principle repeat the old experiments for the new systems and, for example look for CP violating charge asymmetries. One can compare the decay rates of CP conjugate decay modes, etc. However, it is clear that one does not expect to see, in the near future, any CP violation anywhere else than in the K system if the minimal version of the Standard Model is correct. The branching ratios to the interesting final states are in general expected to be small which makes the CP violation in the amplitudes very difficult to measure. The analog of ϵ, i.e., mass matrix CP violation, for these systems is expected to be small. Furthermore, the lifetimes of the new states are much shorter than those of the kaons. The separation of the "long-lived" and the "short-lived" components appears very difficult.

For pedagogical reasons, let us go through the computation of the difference in the rates Γ and $\overline{\Gamma}$, where Γ and $\overline{\Gamma}$ denote respectively the rate for a process and its CP conjugate reaction. Such differences are signals of CP violation. Let me give a simple demonstration of how such an effect may come about. Consider the decay $(b\overline{u}) \rightarrow (c\overline{u}) + (d\overline{c})$, through the quark diagrams shown below:

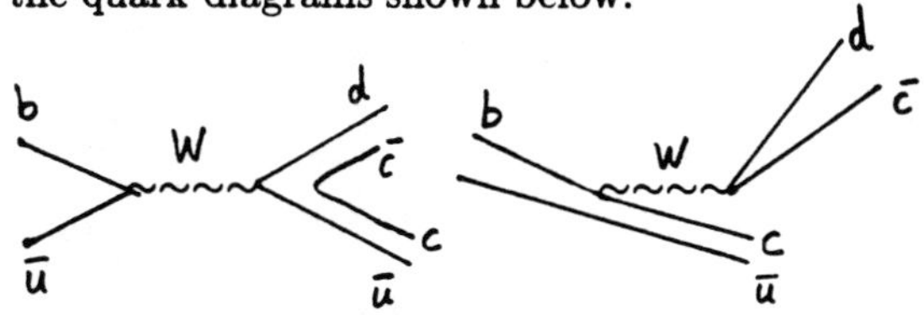

The amplitude for the process, shown above, is of the form

$$Amplitude \;=\; V_{ub}V^*_{ud}A_1e^{i\phi_1} \;+\; V_{cb}V^*_{cd}A_2e^{i\phi_2}$$

Here the A_j, j=1,2, are the weak amplitudes for the respective diagrams. These are real as we have explicitly factorized the (complex) weak coupling constants. The phases ϕ_j, j=1,2, are due to strong interactions. They are usually called the final state interaction phases. It is easy to show that to a good approximation all that the strong interactions do is to introduce these phases. When considering the CP conjugate process we draw the corresponding two diagrams by simply replacing each quark/antiquark, in the above diagrams, by its antiparticle. The corresponding amplitude is given by

$$\overline{Amplitude} \;=\; (V_{ub}V^*_{ud})^*A_1e^{i\phi_1} \;+\; (V_{cb}V^*_{cd})^*A_2e^{i\phi_2}$$

Note that in going from the process to its "antiprocess" we must complex conjugate the weak coupling constants. However, we are not allowed to complex conjugate the strong interaction factors (i.e., change ϕ to $-\phi$). A simple way to see this is by noting that the strong interaction factors come from loops. I shall soon give an example of such phases. Putting that issue aside, for the moment, from the above amplitudes we find the rates

$$Rate \sim \mid c_1 \mid^2 A_1^2 + \mid c_2 \mid^2 A_2^2 + 2A_1A_2Re\{c_1c_2^*e^{i(\phi_1-\phi_2)}\}$$

$$\overline{Rate} \sim \mid c_1 \mid^2 A_1^2 + \mid c_2 \mid^2 A_2^2 + 2A_1A_2Re\{c_2c_1^* e^{i(\phi_1-\phi_2)}\}$$

where $c_1 \equiv V_{ub}V_{ud}^*$ and $c_2 \equiv V_{cb}V_{cd}^*$. Using $Re(z_1z_2) = Rez_1Rez_2 - Imz_1Imz_2$, we find

$$CP\ asymmetry\ = \frac{(\Gamma - \overline{\Gamma})}{(\Gamma + \overline{\Gamma})} \sim Im(c_1c_2^*)\ sin(\phi_1 - \phi_2)$$

where a simple calculation shows that, as expected, $Im(c_1c^*{}_2) = J$. The above example illustrates the necessary requirements for obtaing CP violation effects. First of all we need to have more than one way of going from the initial state to the final state (as the terms 1 and 2 above), in order to get one or more interference terms. That is a necessary condition but not sufficient. In addition, the weak coupling constants must not be relatively real (i.e., $Im(c_1c_2^*)$ must be nonzero, in the above example). Further, two or more unequal final state interaction phases are needed ($\phi_1 \neq \phi_2$, in the above example). In the example we have been considering, the effect is going to be small because the first diagram is much suppressed. For present status of the expectations the reader may consult Ref. 19 and references therein.

Now, let me give an example of a phase which you are not allowed to flip, when going from a process to its antiprocess. Consider, for example, a four-fermion process mediated by the W, and its anti-reaction as shown below.

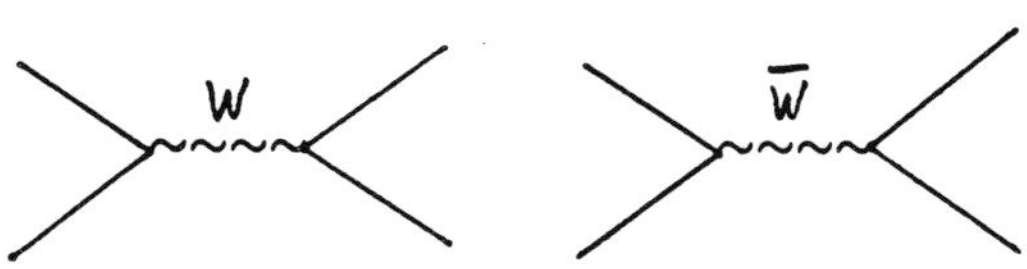

The propagator of the W is given by $(Q^2 - M_W^2 + iM_W\Gamma_W)^{-1}$. If one is naive, one may imagine that in the antiprocess, mediated by the $\overline{W}$, one should complex conjugate the propagator. But we know that we are not allowed to do so. The plus sign in front of the i is telling us that W decays (it does not grow in time). $\overline{W}$ must also decay and not grow, as required by unitarity. Further, by CPT, the lifetimes of W and $\overline{W}$ are equal. In other words, the propagator of the W and $\overline{W}$ look exactly the same. In fact the lines ∼∼ represent loops , due to quantum fluctuations, which give rise to the above phases.

OUTLOOK

I may have sounded very pessimistic throughout this article, as far as the observational aspects of CP violation and our chances of learning something new, in the near future, are concerned. Of course, we could be lucky and some unexpected discovery may reveal the nature of CP violation. Let us hope so.

With the recent coming into operation of the large electron-positron collider, LEP, at CERN you may wonder what do we expect to learn about CP violation from that lovely accelerator. After all the cannonical number of Z particles produced in one year of running is expected to be about 10^7 and the branching ratio of the mode $Z \to b\bar{b}$ is expected to be about 15 %

It is very easy to imagine many CP violating measurables in the transitions $Z \to f\bar{f}$, where $f\bar{f}$ represent a quark-antiquark pair. Consider the Lagrangian

$$L_{int} = \bar{\psi}_f \gamma^\mu (a_f \gamma_5 + v_f) \psi_f Z_\mu$$

Such a Lagrangian gives us, in addition to the decay rate for $Z \to f\bar{f}$, the following measurable quantities

$$\alpha = \frac{2Re(v_f a_f^*)}{| v_f |^2 + | a_f |^2}$$

$$\beta = \frac{2Im(v_f a_f^*)}{| v_f |^2 + | a_f |^2}$$

where α violates P and C but conserves CP while β conserves C but violates P and CP. Alas, in the Standard Model, the above vector and axial coupling constants are relatively real. Thus there is no CP violation in the above fundamental vertices.

Consider now the suppressed decays of the Z of the kind $Z \to f_1\bar{f}_2$, where the indices 1 and 2 denote two different flavours. In such processes there is no lowest order "fundamental" coupling, in the Standard Model, due to the absence of flavour changing neutral currents. However, diagrams such as

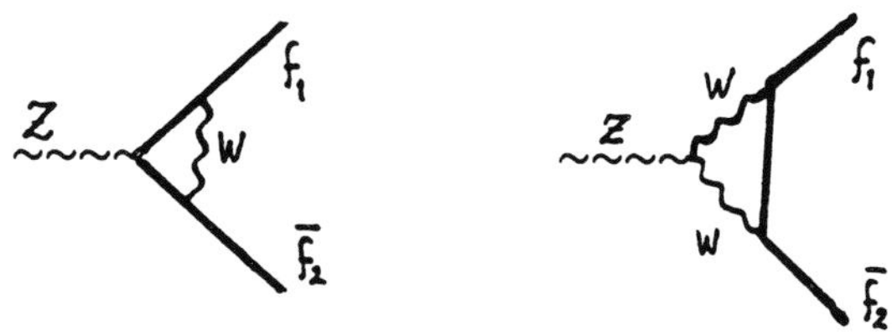

give CP violation. Thus, for example, the rates for conjugate decays are not equal, viz.,

$$\frac{\Gamma(Z \to s\bar{b}) \; - \; \Gamma(Z \to b\bar{s})}{\Gamma(Z \to s\bar{b}) \; + \; \Gamma(Z \to b\bar{s})} \sim J$$

Unfortunately, however, the individual rates of these flavour changing channels are minute, of the order of 10^{-7}. The situation for the leptons, in the Standard Model, is even more tragic. No CP violation is expected to be seen in the decays of the $Z \to l\bar{l}$.

CP impurity in the physical neutral pseudoscalar meson states is, as usual, expressed by

$$M_S = pM^0 + q\bar{M}^0$$

$$M_L = pM^0 - q\bar{M}^0$$

where M denotes a neutral pseudoscalar meson, $M = K, D, B$) and the indices S and L are the analogues of "short" and "long" in the neutral kaon system. However, for the heavy quarks both the long-lived and the short-lived states are quite short-lived and do not allow making beams of the long-lived components, etc. In the above relations, CP is violated if $| p | \neq | q |$. Now

$$< M_S \mid M_L > = | p |^2 - | q |^2 = 1 - 2 | q |^2$$

This "mass matrix" CP violation could exhibit itself, for example, in the same-sign dilepton production of the kind

$$e^+ + e^- \to B^0 + \bar{B}^0 + ... \to l^\pm + l^\pm + ...$$

One finds that the magnitude of the CP asymmetry is

$$\frac{N(l^+l^+) - N(l^-l^-)}{N(l^+l^+) + N(l^-l^-)} \approx 2 < B_S \mid B_L > \leq 10^{-3}$$

Again, the effect is too small to be seen at LEP, provided that the Standard Model is correct. For a more thourough discussion of the above points the reader may consult the recent book "CP Violation" [19].

Finally, a topic of great interest is the electric dipole moment of the neutron (EDM). Again, in the Standard Model the situation looks hopeless as the EDM of the neutron is expected to be smaller than $10^{-33}e - cm$, i.e., about 8 orders of magnitude below the present upper limit. For a recent review see Ref. [20]. However, the important point to keep in mind is that the neutron EDM is very sensitive to certain extensions of the Standard Model, specially where CP is due to Higgses. Then the EDM of the neutron tends to be large and often even larger than the present upper limit. This fact has been again and again stressed by many authors. For example, recently Weinberg [21]has found a new gluonic contribution of the kind $f_{ijk}\epsilon_{\mu\nu\lambda\sigma}G^i{}_{\mu\rho}G^j{}_{\rho\nu}G^k{}_{\lambda\sigma}$. It is not clear how big the contribution of this operator to the EDM of the neutron is going to be. But the point is not the specific value of such possible contributions. They are by far too uncertain, at the moment. The point is that perhaps the best hope of learning something new on CP violation in the near future may be in the EDM of the neutron. The present experimental limit on this quantity is

$$d_n = (-14 \pm 6)10^{-26} ecm \qquad (Leningrad\ Experiment)$$

$$d_n = (-3 \pm 5)10^{-26} ecm \qquad (Grenoble\ Experiment)$$

Since the predicted value, in the Standard Model, is much smaller than the present limit the observation of a relatively large EDM of the neutron would be very interesting.

REFERENCES

[1] J. H. Christenson, J. W. Cronin, V. L. Fitch and R. Turlay, Phys. Rev. Lett. 13, 138 (1964)

[2] T. D. Lee and C. N. Yang, Phys. Rev. 104, 254 (1956)

C. S. Wu, E. Ambler, R. W. Hayward, D. D. Hoppes and R. P. Hudson, Phys. Rev. 105, 1413 (1957);

R. L. Garwin, L. M. Lederman and M. Weinrich, *ibid*, p. 1415;

V. L. Telegdi and A. M. Friedman, *ibid*, p. 1681

[3] P. K. Kabir, The CP Puzzle (Academic Press, 1968)

[4] S. L. Glashow, Nucl. Phys. 22, 579 (1961);

S. Weinberg, Phys. Rev. Lett. 19, 1264 (1967);

A. Salam, *in* Elementary Particle Theory, Ed. N. Svartholm (Almqvist and Wiksell, 1968)

[5] M. Kobayashi and T. Maskawa, Prog. Theor. Phys. 49, 652 (1973)

[6] T. D. Lee, Phys. Reports 9C, 143 (1973)

[7] S. Weinberg, Phys. Rev. Lett. 37, 675 (1976);

P. Sikivie Phys. Lett. 65B, 141 (1976)

[8] R. N. Mohapatra in "CP Violation", Ed. C. Jarlskog (World Scientific, 1989), p. 384

[9] P. Langacker in "CP Violation", loc. cit., p. 552

[10] R. D. Peccei in "CP Violation", loc. cit., p. 503

[11] S. L. Glashow, J. Iliopoulos and L. Maiani, Phys. Rev. D2, 1285 (1970)

[12] N. Cabibbo, Phys. Rev. Lett. 10, 531 (1963);

M. Gell-Mann and M. Levy, Nuovo Cimento 16, 705 (1960);

M. Gell-Mann, Phys. Lett. 8, 214 (1964)

[13] Z. Maki, M. Nakagawa and S. Sakata, Proc. of 1962 Inter. Conf. (CERN) p. 663; Prog. Theor. Phys. 28, 870 (1962)

[14] C. Jarlskog in "CP Violation", loc. cit., p. 3

[15] See Ref. 14 and references thein. See also C. Jarlskog and R. Stora, Phys. Lett. B208, 268 (1988)

[16] L. Wolfenstein, Phys. Rev. Lett. 51, 1945 (1983) and these Proceedings

[17] A. Datta and E. A. Paschos in "CP Violation", loc. cit., p. 292

[18] I. I. Bigi and A. I. Sanda in "CP Violation", loc. cit., p. 362

[19] "CP Violation", Ed. C. Jarlskog, World Scientific Pub. Co. (1989)

[20] S. M. Barr and W. J. Marciano in "CP Violation", loc. cit., p. 455

[21] S. Weinberg, Preprint UTTG-30-89

Standard Model
and Beyond

ISBN 1-56072-027-1

PHENOMENOLOGY OF E_6 SUPERSTRING-INSPIRED MODELS

Thomas G. Rizzo

Physics Department, University of Wisconsin, Madison, WI 53706
and
Ames Laboratory and Department of Physics, Iowa State University
Ames, IA 50011 USA

1. ABSTRACT

In these lectures, we review the major low-energy phenomenological implications of E_6 superstring-inspired models emphasizing the new gauge bosons and exotic fermions present in such models.

2. INTRODUCTION

The rich phenomenology of E_6 theories has enjoyed a great popularity in recent years [1] due to the initial results from Superstring theories [2]. These results showed that the compactification of the ten-dimensional, anomaly-free, $E_8 \times E_8'$ string theory lead to an effective E_6 GUT with $N = 1$ SUSY in four dimensions below the Planck scale. The rich phenomenology of such models originates from various possible extensions of the Standard Model (SM) electroweak gauge group, the existence of several generations of vector-like exotic fermions, and an enrichment of the scalar sector beyond that of the SM. In these lectures we will concentrate on the production of and signals for these new gauge bosons and exotic fermions at e^+e^-, ep, and $\overset{(-)}{p}p$ colliders [3].

In these models, each chiral generation of fermions lies in the **27** of E_6 which contains the usual **16** of SO(10) (the SM fermion plus ν^c) as well as 11 new fields whose quantum numbers are listed in Table 1. Since the theory is supersymmetric, the corresponding scalar partners to these fermions are also present. Thus under SO(10), SU(5) we have

$$\mathbf{27} \Rightarrow (\mathbf{16},\, \mathbf{10}) + (\mathbf{16},\, \mathbf{5}^*) + (\mathbf{16},\, \mathbf{1}) + (\mathbf{10},\, \mathbf{5}) + (\mathbf{10},\, \mathbf{5}^*) + (\mathbf{1},\, \mathbf{1})\,. \tag{1}$$

In principle, an assignment ambiguity exists in that one could, *a priori*, assign the exotics to $(\mathbf{16},\, \mathbf{5}^*)$ and $(\mathbf{16},\, \mathbf{1})$ instead of $(\mathbf{10},\, \mathbf{5}^*)$ and $(\mathbf{1},\, \mathbf{1})$ without any effect on SM couplings. This ambiguity corresponds to the existence of an $SU(2)_I$ subgroup of E_6 which commutes with SU(5) and which can lead to interesting new physics for models with extended gauge groups.

Although their SM couplings are fixed by group theory, the exotic fields can have several possible Baryon (B) and lepton (L) number as well as R-parity (R) assignments. If only **27**'s are present, the most general renormalizable superpotential consistent with $SU(3)_c \times SU(2)_L \times U(1)_Y$ invariance is

$$\begin{aligned} W = &\lambda_1 H^c Q u^c + \lambda_2 H Q d^c + \lambda_3 H L e^c + \lambda_4 H^c H S^c + \lambda_5 h h^c S^c \\ &+ \lambda_6 h c^c e^c + \lambda_7 L h^c Q + \lambda_8 \nu^c h d^c \\ &+ \lambda_9 h Q Q + \lambda_{10} h^c u^c d^c \\ &+ \lambda_{11} H^c L \nu^c . \end{aligned} \tag{2}$$

Note that we have suppressed generation indices in writing W in this form. In the most commonly studied scenario, the scalar parts of the H, mH_1^c, and S^c (as well as ν^c in rank-6 model) superfields can obtain vev's and act as Higgs fields (this need not be the case, however). With these vev's, λ_1-λ_5 produce masses for e, u, d, N, E, and h, while $\lambda_{11} \neq 0$ produces a Dirac ν mass so that its value must be fined-tuned (*i.e.*, $\lambda_{11} \ll \lambda_3$) or absent. The terms λ_6-λ_8 and $\lambda_{9,10}$ cannot be present simultaneously as they lead to rapid proton decay so that at least one of these sets must be forced to vanish by some discrete or gauge symmetry. This leads to three possible B, L, and R assignments for the h fermion

			$B(h)$	$L(h)$	R
(i)	Leptoquark	$\lambda_9 = \lambda_{10} = 0$	1/3	1	−1
(ii)	Diquark	$\lambda_6 = \lambda_7 = \lambda_8 = 0$	−2/3	0	−1
(iii)	Quark	$\lambda_6 - \lambda_{10} = 0$	1/3	1	+

(3)

If the color-singlet exotic superpartners act as Higgs fields then they must also have $L = B = 0$ and have positive R-parity. Note a non-zero vev for ν^c (as required in rank-6 models) requires $L(\nu^c) = B(\nu^c) = 0$ and $R(\nu^c) = -1$. If, however, other fields are present besides the usual **27**'s then E and N can be just ordinary vector-like leptons. Since there are at least three generations of Higgs fields in W, it is always possible to work in a basis where only one generation gets non-zero vev's while the others are decoupled and are referred to as 'un-Higgs'.

3. NEW GAUGE BOSONS

3.1. Overview

In these string-inspired models the effective E_6 GUT breaks topologically at the Planck scale down to a rank-5 or rank-6 subgroup which may survive down to comparatively (?) low energies. The rank-5 model is unique: $SU(3)_C \times SU(2)_L \times U(1)_Y \times U(1)_\eta$ with the Z_η couplings from the new $U(1)_\eta$ factor fixed by group theory. In the rank-6 case there are many possibilities most of which break at a high (≥ 10 TeV) intermediate scale leaving an effective rank-5 group leading to two broad classes of models:

$$\begin{aligned} &(A) \quad SU(3)_C \times SU(2)_L \times U(1)_Y \times U(1)_\theta , \\ &(B) \quad SU(3)_C \times SU(2)_L \times SU(2) \times U(1) . \end{aligned} \tag{4}$$

IN (A) the $U(1)_\theta$ originates from the breaking $E_6 \to SO(10 \times U(1)_\psi \to SU(5) \times U(1)_\psi \to$ SM $\times\, U(1)_\theta$ where $Z_\theta = Z_\psi \cos\theta - Z_\chi \sin\theta$ remains light. θ is *a priori* a free parameter and the $U(1)_\eta$ rank-5 model can be studied as a sub-case of models of type (A) in that the Z_η couplings are recovered for $\theta = \sin^{-1}\sqrt{3/\gamma} \simeq 37.76°$. Models of this type are referred to as Effective Rank-5 Models (ER5M). There are several possible models type (B) but we limit ourselves to the case where the second SU(2) is $SU(2)_R$. Given the exotic fermion assignment ambiguity this leads to either an augmented version of the classic Left-Right Model (LRM) [4] or an interesting variant due to Ma [5] called the Alternative Left-Right Model [ALRM] both of which we will briefly discuss below. The couplings of the fermions in the ER5M and LRM are given in Table 2; for ALRM coupling we simply perform the interchange

$$\begin{bmatrix} \begin{pmatrix} \nu \\ e \end{pmatrix}_L \\ \\ d_L^c,\ \nu_L^c \end{bmatrix} \leftrightarrow \begin{bmatrix} \begin{pmatrix} \nu \\ E \end{pmatrix}_L \\ \\ h_L^c,\ S_L^c \end{bmatrix} . \tag{5}$$

In particular for Z, Z' in the ER5M we have ($x_W = \sin^2\theta_W$, $c_W = \cos\theta_W$)

$$\mathcal{L} = \frac{g}{c_W}(T_3 - x_W Q)Z + \frac{g}{c_W}\sqrt{\frac{5x_W}{3}}\,[Q_\psi\cos\theta - Q_\chi\sin\theta]\,Z' \tag{6}$$

while in the LRM and ALRM we have instead

$$\mathcal{L} = \frac{g}{c_W}(T_3 - x_W Q)Z + \frac{g}{c_W}(1-2x_W)^{-1/2}\,[x_W T_{3L} + (1-x_W)T_{3R} - x_W Q]\,Z' . \tag{6'}$$

Values of $T_{3L,R}$ and $Q_{\psi,\chi}$ are obtainable from Table 2. [Note that under the interchange (5) $\theta \to \theta' = \tan^{-1}\sqrt{15} - \theta$ so that $U(1)_\eta$ couplings remain unaltered.]

An important facet of Z' physics is its mixing with the SM Z which is a generic occurance for most models. Under $SU(2)_L$, the neutral fields in the **27** transform as two doublet members and two singlets. If the Z' couples to the charge Z' (via $\mathcal{L}' = g/c_W\, Q'Z'$) then we denote by Q_i' ($i = 1,2$) the changes of the two $SU(2)_L$ doublets with vev's $v_i/\sqrt{2}$. The Z-Z' mass matrix is then

$$\mathcal{M}^2 = \begin{pmatrix} M_Z^2 & \delta M^2 \\ \delta M^2 & M_{Z'}^2 \end{pmatrix} \tag{7}$$

with $M_Z^2 = \frac{1}{4}g^2v^2/c_W^2$ ($v^2 = v_1^2 + v_2^2$) and $\delta M^2/M_Z^2 = 2(Q_1'v_1^2 - Q_2'v_2^2)/v^2$ *etc.* Note M_Z is just the mass of the SM Z. To diagonalize $\mathcal{M}^2$ and bring us to the mass eigenstate basis we must rotate Z, Z' to $Z_{1,2}$ (which have physical masses $M_{1,2}$, respectfully), *i.e.*, $Z_1 = Z\cos\phi + Z'\sin\phi$, with Z_2 being the orthogonal combination. The angle ϕ is related to the entries and eigenvalues of $\mathcal{M}^2$ in a number of ways:

$$\tan 2\phi = \frac{2\delta M^2}{M_Z^2 - M_{Z'}^2}, \qquad \tan^2\phi = \frac{M_Z^2 - M_1^2}{M_2^2 - M_Z^2},$$
$$\delta M^2/M_Z^2 = -(M_2^2/M_Z^2 - 1)\tan\phi . \tag{8}$$

Note that $M_1 \le M_Z$, *i.e.*, mixing always lowers the mass of the SM Z. Equation (8) tells us many important things, *e.g.*, if I keep $\Delta m \equiv M_Z - M_1 > 0$ fixed, then $\phi \sim M_2^{-1}$ as M_2 gets large; this is the so-called Mass Constraint (MC) on ϕ. If one restricts the size of Δm then for a fixed $M_2(\phi)$ one obtains a bound on $\phi(M_2)$ from just the simple structure of $\mathcal{M}^2$. For example, if $\Delta m \le 1$ GeV, $M_1 = 91.17$ GeV and $M_2 = 300$ GeV one finds $|\phi| \le 0.045$. Figure 1 shows the minimum value of Δm as a function of M_2 for different $|\phi|$ values assuming $M_1 = 91.17$ GeV.

Stronger limits on ϕ are possible in E_6 models since we know something about the various entries in $\mathcal{M}^2$, *i.e.*, the Higgs structure of the models. Since $v_{1,2}^2/v^2 < 1$, combining (8) with the expression for δM^2 in terms of the Q'_L one obtains

$$2Q'_{\text{min}} \leq \left(\frac{M_2^2}{M_Z^2} - 1\right) \tan\phi \leq 2Q'_{\text{max}} \tag{9}$$

where $Q'_{\text{min(max)}}$ is the smallest (largest) of $-Q'_1$, Q_2. Equation (9) is referred to as the Higgs Constraint (HC) on ϕ and we see that $\phi \sim M_2^{-2}$ as M_2 gets large so that the HC becomes stronger than the MC for large M_2.

A last but important way to bound ϕ (as well as M_2) is to examine the modification the additional Z' produces in the low energy ($Q^2 \ll M_z^2$) neutral current Hamiltonian. In E_6 models, one has

$$H_{\text{NC}} = \frac{4G_F}{\sqrt{2}} \left(\frac{M_Z^2}{M_1^2} J_1^2 + \frac{M_Z^2}{M_2^2} J_2^2\right) \tag{10}$$

with $J_1 = J\cos\phi + J'\sin\phi$ being the current coupling to Z_1, *etc.* As $M_2 \to \infty$, $\phi \to 0$ so that $M_Z/M_1 = 1$ and the SM limit is recovered. There have been at least three detailed analyses of this type[6–8] which made use of all of the neutral current data available before 1988 as well as the UA1 and UA2 Collaboration measurements of the W and Z masses. Figure 2, from [6], shows the bounds on M_2 and ϕ for several different values of θ for the ER5M.

3.2. Z' Collider Searches

New gauge bosons can be search for directly (via production) or indirectly (by modifying SM predictions) at e^+e^-, ep, and hadron colliders. In hadron colliders, essentially only direct searches are possible with a new W or Z usually observed via its leptonic decay, *e.g.*, $q\bar{q} \to Z_2 \to \ell^+\ell^-$. Thus Z_2 search limits are particularly sensitive to the $u\bar{u}Z_2$ and $d\bar{d}Z_2$ couplings as well as the leptonic branching fractions of Z_2. Previous to 1989, the direct searches for Z_2 by the UA1 and UA2 Collaborations resulted in an approximately θ-independent limit of $\simeq 150$ GeV [9,10] assuming the exotics did not participate in the decay. Recently [11], the CDF Collaboration has announced new search limits for Z_2 which for, the ER5M, translates into the bounds in Fig. 3 [12]; for the LRM (ALRM) these search limits imply $M_2 > 374$ (468) GeV. These direct search limits are greatly improved over those from CERN and it is again assumed that the exotics do not participate in the decay of the Z_2.

One a new gauge boson is discovered a detailed study of its properties requires an e^+e^- collider. In order to discriminate new Z's from different models one needs measurements of the Z_2 width and branching fractions as well as forward-backward (A_{FB}) and left-right polarization (A_{LR}) asymmetries for the process $e^+e^- \to Z_2 \to f\bar{f}$. (To measure A_{LR}, polarized beams are necessary.)

Generally, the Z_2 width is dominated by the fermionic modes $Z_2 \to f\bar{f}$, *i.e.*, apart from QCD corrections ($\beta^2 \equiv 1 - 4m_f^2/M_2^2$)

$$\Gamma(Z_2 \to f\bar{f}) = \frac{g^2}{12\pi} M_2 c_f \beta \left[\frac{1}{2}(3 - \beta^2)v_f^2 + \beta^2 a_f^2\right], \tag{10}$$

where $c_f = 1$ for leptons and $= 3$ for quarks. Other modes, such as decays to SUSY particles, may also be present if kinematically allowed. Decays such as $Z_2 \to W^+W^-$ are also interesting since it probes the size of Z-Z' mixing:

$$\Gamma(Z_2 \to W^+W^-) = \frac{g^2 M_2}{192\,\pi c_W^2} \sin^2\phi \left(\frac{M_2}{M_Z}\right)^4 \left[1 + 20\frac{M_W^2}{M_w^2} + 12\frac{M_W^4}{M_2^4}\right]. \tag{11}$$

$\Gamma(Z_2 \to W^+W^-)$ can be sizeable since as $M_2/M_Z \to \infty$, $\sin^2\phi \sim (M_Z/M_2)^4$ from the HC; the LRM, for example, $\Gamma(Z_2 \to W^+W^-) \simeq \Gamma(Z_2 \to \ell^+\ell^-)$.

The asymmetries A_{FB} and A_{LR} for the reaction $e^+e^- \to Z_2 \to \bar{f}f$ (on resonance) are given by

$$A_{FB}^f = \frac{3v_e a_e v_f a_f}{(v_e^2 + a_e^2)(v_f^2 + a_f^2)},$$
$$A_{LR} = \frac{2v_e a_e}{(v_e^2 + a_e^2)}, \tag{12}$$

Note A_{LR} does not depend on the flavor of f; also note $A_{FB}^\ell = \frac{3}{4}A_{LR}^2$ if universality is assumed. The couplings of the ER5M are easy to pick out since they lead (independent of θ and neglecting QCD corrections) to the predictions

$$A_{FB}^u = 0\,, \qquad A_{FB}^d = -A_{FB}^\ell\,,$$
$$\Gamma(Z_2 \to d\bar{d}) = 3\Gamma(Z_2 \to \ell^+\ell^-)\,. \tag{13}$$

Neither the LRM nor the ALRM satisfy these conditions. Figures 4-7 summarize the branching fractions and asymmetries for the ER5M while the corresponding values for the LRM and the ALRM are found in Table 3.

Apart from direct Z_2 searches, both e^+e^- and ep colliders allow for indirect searches by looking for deviations from SM predictions. TRISTAN [13] has recently observed a deviation in $R = \sigma(e^+e^- \to \text{hadrons})/\sigma_{pt}$ (σ_{pt} being the photon mediated $e^+e^- \to \mu^+\mu^-$ cross section) of $\simeq$ 8-10% above the SM prediction for $\sqrt{s} \gtrsim 57$ GeV. Although a Z_2 can increase the value of R, if $M_2 \gtrsim 300$ GeV the increase can be at most $\simeq 1\%$ [12] so that such an explanation seems unlikely [14].

Z-Z' mixing can make modifications in the properties of the SM Z now being probed by the SLC and LEP. Such mixing can alter the Z width and branching fractions, the peak values of A_{LR} and A_{FB}, as well as the peak height/production cross section [1, 12]. Figures 8-10 show how $\phi \neq 0$ can modify the Z width, peak height and leptonic branching fraction in the various models under discussion. (Note that Z-Z' mixing can be sufficiently large as to hide almost a full extra neutrino in the Z width and peak height!)

LEP II and future e^+e^- colliders with larger values of $\sqrt{s}$ can search for Z_2 masses $> \sqrt{s}$ by looking at deviations in SM predictions for various cross sections and asymmetries. The limits so obtained are very model-dependent and are quite sensitive to the integrated luminosity and the flavor of the final state fermion as well as $\sqrt{s}$. Figures 11-13 show possible limits from LEP II while Figs. 14-16 show limits from a hypothetical $\sqrt{s} = 1$ TeV collider in both cases for the ER5M. Note that M_2 values in the 2-3$\sqrt{s}$ range can be probed in this way. These search limits compare favorably to what can be done at the SSC and upgraded Tevatron as shown in Figs. 17-19.

ep colliders, such as HERA, can search for new Z's by looking for deviations in the SM predictions for various asymmetries. At HERA ($\sqrt{s} = 314$ GeV) such searches are restricted to the range $M_2 \lesssim$ 300-400 GeV [15] which are already being probed at the Tevatron. A higher energy "LEP×LHC" collider with $\sqrt{s} =$ 1.4-1.8 TeV can search for new Z's beyond $\simeq 1$ TeV in mass.

A last possibility is very high precision measurement of $\overset{(-)}{\nu}_{\mu}e$ elastic scattering cross sections (which are free of QCD corrections) [16]. Z_2 masses as large as 800 GeV, competative with the Tevatron upgrades and LEP II, may be probed by this method.

3.3. W_R Collider Searches in the ALRM

Much of what has been said above about Z' searches at colliders applies to W_R as well except in the ALRM case. Here W_R carries $L \neq 0$ and $R = -1$ so that it cannot be produced in the usual Drell-Yan manner: $u\bar{d} \to W_R$. In fact W_R couples to $u\bar{h}$ and $S^c\bar{e}$ and so avoids the conventional constraints on the W_R mass from K_L-K_s mixing and polarized μ-decay. Since the Z_2 and W_R masses are related in this model by $M_2^2 \simeq (1-x_W)(1-2x_W)^{-1}M_{W_R}^2$, the preliminary CDF search limits on Z_2 in the ALRM force $M_{W_R} \gtrsim 390$ GeV. W_R production at hadron colliders proceeds via the process $gu \to W_R h$; for $M_{W_R} = 400$ GeV and $m_h = 50$ GeV the resulting production cross section at the Tevatron is quite small $\simeq 0.1$ pb [17]. The observation of W_R's in the ALRM must await higher energy machines such as the SSC or TeV e^+e^- colliders where the process $e^+e^- \to W_R^+W_R^-$ may be accessible.

4. EXOTIC FERMIONS

4.1. Overview

The properties and search limits for the E_6 exotic fermions has recently been reviewed by Hewett [18] and so we briefly summarize those results in this section.

What mass should we expect exotic fermions to have? One might expect that since the h and E obtain their masses from the same vev as does Z_2, h, E, and Z_2 would have comparable masses. This is only true, however, if the Yukawa couplings in the superpotential are of order g; by the same argument one might expect the electron and SM W to have comparable masses. If $m_{h,E}^2 \gg M_2^2$, then the model becomes strongly coupled sot that we may reasonably expect $m_{h,E} \lesssim$ a few M_2. Direct searches for new charged fermions at TRISTAN [19] strongly indicate $m_{h,E} \gtrsim 30$ GeV while Γ_Z measurements from Mark II [20] would also indicate $m_{h,E} \gtrsim 30$ GeV while Γ_Z measurements from Mark II [20] would also indicate $m_{h,E} \gtrsim 45$ GeV since their pair production would lead to unobserved events as well as increases in Γ_Z.

If a new fermion is produced at an e^+e^- collider one can make full use of cross section and asymmetry measurements to determine its electroweak quantum numbers [21]. For example, the production characteristics of h are very different from those of a sequential fourth-generation down-type quark, v. While v has chiral couplings to the Z, h is vector-like so that if Z_2 contributions are neglected the value of A_{FB} for h (and E) is zero. Unfortunately, the production of E and a corresponding sequential fourth generation charged lepton, L, would be quite difficult to distinguish at a hadron collider particularly if Z_2 contributions to the production cross section dominate.

We now turn to searches involving particular choices for the B, L, and R-parity assignments of the exotics.

4.2. Leptoquarks

Leptoquarks are color triplets carrying $L = 1$ and $B = 1/3$ resulting from the non-zero coupling λ_{6-8} in the superpotential (2). These couplings may in general be generation dependent so that a potential for the violation of universality exists. However, the imposition of discrete symmetries on W to avoid flavor-changing neutral current type interactions generally leads to a strong suppression of intergenerational Yukawa couplings. Since h has an odd R-parity it is actually not a leptoquark but a leptoquarkino (sleptoquark) and decays into SUSY particles via λ_{6-8}:

$$h \to d\tilde{\nu},\ \tilde{d}\nu,\ u\tilde{e},\ \tilde{u}e\,. \tag{14}$$

Depending on the mass spectrum, these SUSY particles may be real or virtual. The SUSY partner to h, $\tilde{h}$, is the true leptoquark with spin-0 and positive R-parity and decays via $\tilde{h} \to ue,\ d\nu$ with widths

$$\Gamma_{\tilde{h}} = \frac{(\lambda_6^2 + \lambda_8^2)}{16\,\pi}\, m_{\tilde{h}}\,, \qquad \Gamma_{\tilde{h}^c} = \frac{2\,\lambda_7^2}{16\,\pi}\, m_{\tilde{h}}\,, \tag{15}$$

neglecting $\tilde{h}$-$\tilde{h}^c$ mixing. Since λ_i are unknown, one either assumes $\lambda = \lambda_{6,7,8}$ and $\lambda^2 = Fe^2$ or one parameterizes the $\tilde{h}\ell eq$ vertex by $A + B\gamma_5$ and introduces $k = (A^2 + B^2)/e^2$ and $k' = 2Re(A^*B)/e^2$. Limits on h, $\tilde{h}$ production are then quoted in terms of F or k and k'.

Demanding that the cross section for $e^+e^- \to q\bar{q}$ not be significantly modified by the u-channel exchange of leptoquarks it is possible to rule out sizeable regions of the leptoquark coupling-mass (M) plane [22] as shown in Fig. 20 where $k = k'$ has been assumed. For $M \simeq 200$ GeV values of $k \simeq 1$ are allowed by larger values are excluded. Leptoquark exchange may provide an explanation [14] of the R increase observed by TRISTAN [13] if $k = 1$ and $M = 200$ GeV. Similarly, limits on M from data on atomic parity violation constrains [23] $M \gtrsim 300$ GeV for couplings $F \simeq 1$ if no other contributions (from Z_2, say) are present.

The leptoquark contribution to $e^+e^- \to q\bar{q}$ may provide a good indirect signal for leptoquarks at higher energy e^+e^- colliders. Figure 21 displays [24] contours of $\delta\sigma/\sigma$ in the leptoquark coupling-mass plane at LEP II, where $\delta\sigma$ represents the additional terms in the $e^+e^- \to u\bar{u}$ cross section due to leptoquark exchange and σ is the total cross section summed over all quark flavors. For $\lambda_{LQ} = 1.0,\ 0.5,\ 0.3$ ($\lambda_{LQ} = \lambda_6 = \lambda_7$) a sensitivity in the measurement of $\delta\sigma/\sigma$ of 2% leads to the leptoquark discovery limits $M \simeq 1.8,\ 0.9,\ 0.5$ TeV. This mechanism is clearly a very sensitive probe at LEP II. At LEP I energies, the cross section is dominated by the Z peak and is thus not sensitive to the virtual leptoquark exchange. Similarly, scalar leptoquarks can also indirectly contribute to the inverse subprocess $q\bar{q} \to e^+e^-$. However, this mechanism is also not as sensitive to the t-channel leptoquark exchange and is unlikely to provide a signal for leptoquark masses greater than that which can be produced directly. As an example, the limit [25]

$$M \lesssim 1.2\sqrt{F}\ \text{TeV} \tag{16}$$

is the best that can be achieved from examining $\tilde{h}$ contributions to Drell-Yan production at the LHC.

In e^+e^- annihilation, the spin 1/2 h-leptoquarkinos can only be pair produced (in order to conserve R-parity). This occurs through the usual s-channel γ, Z, and Z' exchange as well as by u-squark exchange in the t-channel via the Yukawa couplings λ_6 and λ_7. In this case the gauge contributions dominate the Yukawa interactions and the production rates are similar to those for any color triplet fermion pair. Schaile and Zerwas [26] have performed background simulations for $h\bar{h}$ production and found that appropriate cuts can result in a signal to background ratio of 10 and place a discovery limit at $M_h \simeq 0.9$ TeV at a $\sqrt{s} = 2$ TeV e^+e^- collider with an integrated luminosity of 10 fb^{-1}. The story is different for scalar leptoquark pair production. $e^+e^- \to \tilde{h}\bar{\tilde{h}}$ also receives contributions from γ, Z, and Z' s-channel diagrams as well as u-quark exchange in the t-channel due to the Yukawa interactions, but in this case the gauge contributions are suppressed by the usual $\beta^3 (= (1 - 4M^2/s)^{3/2})$ factor which is present for scalar particles. Assuming only the gauge interactions, Schaile and Zerwas [26] calculate a scalar leptoquark search limit of $M_{\tilde{h}} \simeq 0.6$ TeV for the same machine parameters as above. However, inclusion of the Yukawa interactions can significantly increase the $\tilde{h}$ pair production rates. Increases up to two orders of magnitude can be obtained for large values of the couplings, which could place the leptoquark discovery limit close to the beam energy.

Scalar leptoquarks can also be produced singly in e^+e^- collisions via the higher order process $e^+e^- \to e\bar{u}h$, which proceeds from a γ bremsstrahlung from the e^+ (or e^-) beam and through the consequent sub-process reaction $\gamma e^- \to \bar{u}h$. The resulting cross section [27], is obtained by use of the Effective γ Approximation, is proportional to the unknown Yukawa coupling, and is displayed in Fig. 22 at $\sqrt{s} = 2$ TeV for $k = 1.0$ and 0.1 (with k defined as above) and with a p_T cut on the associated jet of $p_T > 10$ GeV. The pair production cross section is also shown in the figure for comparison, where it is clear that the single production rate is competitive with that for pairs (even for small values of k) and will have a larger discovery reach. The signals for this process are 2 jets + $\not{p}_T$, or 2 jets + e with a small SM background arising from $\gamma\gamma$ events. This mechanism could greatly extend the scalar leptoquark discovery reach at e^+e^- colliders if the Yukawa couplings are not infinitesimally small.

Scalar leptoquarks have a non-zero forward-backward asymmetry in e^+e^- annihilation due to the existence of the u-channel Yukawa interactions which modify the angular distribution away from the standard $(1 - z^2)$ expected for s-channel scalar pair production. If the Yukawa couplings are large enough, they can lead to significant values of A_{FB} as are shown in Fig. 23. In this figure the maximum allowed values of k for each value of M as given in Fig. 20 have been assumed. Since 'ordinary' scalars (such as squarks) don't possess an A_{FB} (due to the $(1 - z^2)$ angular distribution), a non-zero measurement of A_{FB} would be a good signature for scalar leptoquarks and an indication of large Yukawa couplings.

ep colliders are especially well-suited for scalar leptoquark production and offer quite striking production signals. Direct production occurs through an s-channel resonance by $e + u \to \tilde{h}$ with a subsequent decay to eu or νd. The differential cross section for this process $(eu \to \tilde{h} \to eu,\ \nu d)$ is given by

$$\frac{d\sigma}{dx\,dy} = \frac{u(x, Q^2)}{32\pi} \frac{\lambda^4 \hat{s}}{(\hat{s} - M^2)^2 + M^2\Gamma^2} \tag{17}$$

where $u(x, Q^2)$ is the u-quark distribution function, $\hat{s} = xs$, $Q^2 = sxy$, the width Γ is given in Eq. (15), and $\lambda = \lambda_6 = \lambda_7$. Enormous event rates are obtained [28,29] and even though these rates are quite dependent on the strength of the coupling, 100 events/year are predicted at HERA for $M = 250$ GeV and $F = 0.01$. The s-channel resonance also yields distinctive peaks in the x-distribution, which have been shown [30] to lie 1-2 orders of magnitude above the ordinary electroweak background $eq \to eq$. Appropriate cuts in the y-distribution can also enhance the signal to background ratio since the scalar $\tilde{h}$ production is isotropic in y and the electroweak background is forward peaked. Discovery limits [29] for both decay channels are presented in the leptoquark coupling strength-mass plane in Fig. 29. Here the discovery criteria are a 5σ signal with $N_{\text{signal}} > 25$ events after a cut of $y > 0.5$ has been imposed. It is seen that even for $F = 0.01$, leptoquarks with mass up to ~ 250 GeV can be detected at HERA, and for larger values of F the discovery region approaches the kinematic limit of the machine, while the LHC $\times$ LEP I and II options dramatically increase the detectable region.

At hadron colliders the h-leptoquarkino can also only be pair produced with a cross section which is the same as that for quark pairs. An analysis of the potential signals from the SUSY decay chains and of possible backgrounds at the SSC is presented in Barger *et al.* [31], where the authors conclude that detection of a 500 GeV h leptoquarkino should be relatively straightforward. Pair production of the scalar partner, $\tilde{h}$, is similar, except for overall combinatorial factors which count the number of relative species, to that of any other color triplet scalar, such as squarks or technipions. We show the total production cross section [32] in Fig. 25a for only one generation of left- or right-handed leptoquarks. The signatures for scalar leptoquark pair production are 2 jets $+ \ell^+\ell^-$, 2 jets $+ \nu\bar{\nu}$, and 2 jets $+ \overset{(-)}{\nu} \ell^{\pm}$, where the most serious background arises for events of the third type. Estimated discovery limits [32] are placed at $M \simeq$ 40-60 GeV, 120 GeV, and 1.7 TeV at the CERN Collider, Tevatron, and SSC, respectively.

Single scalar leptoquark production at hadron colliders occurs via the parton level process $qg \to \tilde{h}\ell$ though the Yukawa interactions, where ℓ is a charged lepton or neutrino. The total cross section [32] for this mechanism is shown in Fig. 25b for the SSC and is directly proportional to the square of the Yukawa coupling; hence the production rates are very sensitive to the exact value of λ. This is in contrast to the case of pair production which is basically independent of λ. However the signatures for this process are quite distinct and are jet $+ \nu\bar{\nu}$, jet $+ \ell^+\ell^-$, and jet $+ \nu\ell^{\pm}$, where events of the first two types have the smallest SM backgrounds. Observable production rates are estimated [32] to occur up to $M \simeq$ 20-30 GeV and 80 GeV for the CERN Collider and Tevatron with $k = 0.1$, and for M in excess of 2 TeV at the SSC with $k = 1.0$. Clearly, this mechanism is only important for larger values of the Yukawa coupling.

4.3. Diquarks

Diquarks are color triplets with baryon number $B = -2/3$ and no lepton number, and are present in composite models as well as in E_6 theories. The diquark interactions are given in the superpotential terms involving the couplings λ_9 and λ_{10}. Here, h again has odd parity and is the fermionic superpartner, called diquarkino (or sdi-quark). The diquarkino has the tree level decays

$$h \to \bar{u}\bar{\tilde{d}}, \qquad h \to \bar{\tilde{u}}\bar{d} \tag{18}$$

where the relative mass spectrum again determines whether the u- or d-squarks are real or virtual. It is the scalar diquark (which has even R-parity) which couples to a $\bar{q}\bar{q}$ pair and decays via

$$\tilde{h} \to \bar{u}\bar{d} \tag{19}$$

with the partial widths (neglecting left-right mixing)

$$\lambda_{\tilde{h}} = \frac{4\lambda_9^2 M_{\tilde{h}}}{16}, \qquad \lambda_{\tilde{h}^c} = \frac{\lambda_{10}^2 M_{\tilde{h}}}{16\pi}. \tag{20}$$

s-channel pair production is the only production mechanism available at e^+e^- colliders for both the scalar diquark and the diquarkino. Schaile and Zerwas [24] repeated their Monte Carlo calculations for diquark production at a $\sqrt{s} = 2$ TeV collider, and assuming a 10 fb^{-1}

integrated luminosity found that 45-70 and 4-30 clean signal events/year could be obtained for the diquarkino and scalar diquark, respectively. These authors calculated search limits of $M_h \sim 0.9$ TeV and $M_{\tilde{h}} \sim 0.6$ TeV.

The production of scalar diquarks, both singly and in pairs, in ep collisions has recently been examined [33] using the Effective Photon Approximation. The pair production of diquarks proceeds through photon gluon fusion, $\gamma g \to \tilde{h}\bar{\tilde{h}}$, and is hence independent of the Yukawa coupling. The pair cross section depends only on the diquark mass and charge and is displayed in Fig. 26a for several different charge values. The corresponding signal is 4 jets with double peaking in the jet-jet invariant mass distribution and is relatively background free. Single diquark production occurs via photon quark fusion, $\gamma q \to \tilde{h}\bar{q}'$, and is directly dependent on the Yukawa couplings. The single cross section at HERA is presented in Fig. 26b for $k = 1.0$. For $k = 1.0$ the single production cross sections are larger than those for pair production for $Q = -1/3$ diquarks, but the single production rate decreases substantially for smaller values of k. The signal for single production is 3 (or 2) jets, with the jet from the associated anti-quark preferring to be at small angles. The background results from QCD processes and is potentially larger than that for diquark pairs since the associated antiquark jet produced along with the single diquark comes out a small angle relative to the beam. Although p_T cuts may decrease the QCD background, they will also substantially reduce the signal. Search limits, assuming 25 events constitutes a discovery, are then set at $\simeq 40$ GeV and 110 GeV for pair and single production, respectively, with $k = 1.0$.

h diquarkinos must be produced in pairs at hadron colliders at, again, the same rates as for quark pairs. Search limits of 500 GeV are expected [31] for diquarkinos at the SSC. Scalar diquark pair production is best characterized by 4 jet events with dijet mass bumps, provided that the jets don't coalesce, and hence has a tendency to be overwhelmed by QCD background at hadron colliders. A comparison [34] of the cross sections for 2, 3, and 4 jet final states from diquark $\tilde{h}\bar{\tilde{h}}$ production with the background QCD n jet cross sections is presented in Fig. 27 for the CERN Collider. It is evident that only the 4 jet diquark final states can compete with the QCD background, due to the QCD jet cross section behavior of $\sigma_{n+1} \sim \alpha_s \sigma_n$. These authors [34] compute that with appropriate cuts, 4 jet measurements could explore mass ranges up to $M_{\tilde{h}} \simeq 40$ GeV and $\simeq 70$ GeV at the CERN Collider and Tevatron, respectively, with 1 pb^{-1} of integrated luminosity.

The Yukawa couplings can also mediate single production of scalar diquarks in hadronic collisions via the direct subprocess s-channel resonance, $\bar{q}\bar{q} \to \bar{\tilde{h}}$ (or $qq \to \tilde{h}^c$). The total cross section is sensitive to the coupling strength and is shown [28] in Fig. 28a and 28b for the Tevatron and SSC, respectively, for various values of the parameter F. The experimental signature is a peak in the number of 2 jet events and is unfortunately overwhelmed at the SSC by the QCD 2 jet background by one to two orders of magnitude; even for large Yukawa couplings, it will be difficult to detect scalar diquarks from this mechanism.

4.3. h-Quarks and E-Leptons

In this case, the exotic fermions E and h have conventional baryon and lepton number assignments. This leads to a breakdown of the Glashow-Weinberg-Paschos [35] naturalness conditions implying the existence of flavor changing neutral currents and mixing between the E_6 exotic and SM fermions [1], such as h-d and E-e (as well as N-ν) mixing. In rank-6 models, where $\langle\tilde{\nu}^c\rangle \neq 0$, h-d mixing is also induced by the λ_8 term in the superpotential. These flavor changing couplings of the exotic to the ordinary fermions allow the exotic fermions to decay and provide signatures for their production.

Ordinary-exotic fermion mixing can lead to a variety of new phenomena, which I will only briefly mention here (for details see Ref. 1). The Kobayashi-Maskawa (KM) mixing matrix is no longer unitary; this leads to, amongst other things, an evasion of the usual GIM suppression of loop amplitudes, and right-handed charged currents in the leptonic sector. The non-unitarity of the KM matrix is also linked to the presence of FCNC couplings. The FCNC interactions (in the one generation limit) are

$$\frac{g}{2c_w}\left[-\frac{1}{2}\sin\theta_e^R\cos\theta_e^R\bar{e}\gamma_\mu(1+\gamma_5)E+\frac{1}{2}\sin\theta_d^L\cos\theta_d^L\bar{d}\gamma_\mu(1-\gamma_5)h\right]Z^\mu+\text{h.c.} \tag{21}$$

where θ_e^R and θ_d^L are the E-e and h-d mixing angles, respectively. This allows for the decays

$$h \to dZ\,, \qquad E \to eZ \tag{22}$$

with the Z being real or virtual depending on the masses of h and E. Similarly, the charged current (CC) sector contains couplings which mediate the decays

$$h \to uW\,, \qquad E \to \nu W\,. \tag{23}$$

Clearly, the ordinary-exotic mixing must be severely constrained in order to avoid conflicts with the experimental lack of FCNC. A systematic investigation of limits on this mixing imposed by data from high precision experiments on CC universality, W and Z masses, right-handed CC, and FC (and diagonal) NC has been performed by Langacker and London [36]. Allowing for all possible fermionic mixings to occur simultaneously and demanding consistency with the data, they have found constraints on the mixing angles, which are displayed in Fig. 29. Bounds on mixing between possible heavy fourth generation, mirror, and SM fermions are also shown in the figure for comparison. The parameter Λ labels the change in isospin associated with the mixing in the neutrino sector. As can be seen from the figure, the E_6 mixing is somewhat more constrained than the mixing in other models, with the general restriction $\sin^2\theta_i \lesssim 0.05$.

Pair production of h and E in e^+e^- annihilation proceeds through s-channel gauge boson exchange with event rates that are similar to those for SM quarks and leptons. Single exotic production, such as $e^+e^- \to d\bar{h} + h\bar{d}$ and $e^+e^- \to e\bar{E} + \bar{e}E$, can also occur via the ordinary-exotic mixing and Z, Z' exchange. However, the single production rates will be suppressed relative to the pair production rates due to the smallness of the mixing angles. Z and especially Z' decays could be a copious source of the exotic fermions (if the relative masses have favorable values). The single production mechanism could be observable on a Z or Z' resonance and would have striking production signatures.

Single exotic production can also occur in ep collisions via ordinary-exotic mixing [37]. The parton-level processes are $eq \to Eq$, $ed \to eh$, and $eu \to \nu_e h$ and result in 50-300 events/year at HERA for 150 GeV exotics, neglecting the mixing angle factors. Once the mixing angles are folded in, the production rates reduce dramatically.

The production of h-quark pairs at hadron colliders is the same as that of any heavy quark and has been thoroughly studied within the context of the SM [38] However, the FCNC decays lead to a unique signature of 2 jets $+\,\ell^+\ell^-\ell'^+\ell'^-$ which has little SM background and is completely reconstructible. At standard SSC luminosities, 67 events/year is predicted [39] for $M_h = 1$ TeV; hence h-quarks up to this mass could be easily detected in this manner. $E\bar{E}$ production in hadron collisions parallels that of heavy leptons and proceeds through the usual Drell-Yan process as well as gluon-gluon fusion. Here, the dominant gluon fusion diagram contains h-quarks in the triangle and Z' exchange. The integrated cross sections for these subprocesses are shown in Fig. 30 from Ref. 40. The FC couplings again allow for spectacular final states such as e^+e^-ZZ and $W^+W^-\not{p}_T$, and E-leptons with masses up to 1 TeV should be detectable at the SSC.

An interesting feature of the quark/lepton scenario for the exotics is the non-unitarity of the Kobayashi-Maskawa mixing matrix. Essentially, this non-unitarity results from the fact that whereas the usual SM $Q = -1/3$ isodoublet quarks each have exotic isosinglet partners, the SM $Q = 2/3$ quarks do not have such partners. In fact, one finds that $V^+V \neq VV^+$:

$$VV^+ = \begin{pmatrix} I & 0 \\ 0 & 0 \end{pmatrix}, \qquad V^+V = U\begin{pmatrix} I & 0 \\ 0 & 0 \end{pmatrix}, \tag{24}$$

where $I(0)$ is the $n\times n$ unit (null) matrix and U rotates the left-handed $Q = -1/3$ quarks to the mass eigenstate basis from the original weak basis. The fact that $V^+V \neq I$ is directly related tot he existence of FCNC in the Z and Z' couplings; for example, the off-diagonal couplings of left-handed $Q = -1/3$ quarks is directly proportional to the off-diagonal elements of V^+V.

To conclude this section, a summary of expected discovery limits from the various production mechanisms discussed above for the three distinct quantum number assignments of the exotic h is presented in Table 4. At e^+e^- colliders, the numbers inside the parenthesis for the cases of scalar leptoquarks and spin 1/2 h-quarks are the mass search limits for single production via the Yukawa interactions and ordinary-exotic mixing, respectively. These single production limits, which are highly dependent on the strength of the Yukawa coupling or mixing, are the maximum limits obtainable and, in reality, could be much lower if the respective interaction strength is weaker. The numbers listed outside the parenthesis are the limits from pair production and are model independent. The limits that we have quoted for the h-quark at ep and hadron colliders include only its semi-leptonic branching ratio, and not the other possible flavor changing decay modes which would distinguish it from a fourth generation, $Q = -1/3$, b'-quark. The inclusion of these FC decay modes would lower the h-quark search limits due to their smaller branching fractions. Scalar diquark production at the SSC has not been examined, but its detection will be difficult due to the overwhelming QCD background. If these exotic fermions are present at or slightly above the electroweak scale, they should be detectable at these present or future colliders.

ACKNOWLEDGEMENTS

The author would like to thank the Phenomenology Institute at the University of Wisconsin-Madison for its hospitality and use of its facilities. This research was supported in part by the University of Wisconsin Research Committee with funds granted by the Wisconsin Alumni Research Foundation, and in part by the U. S. Department of Energy under contract DE-AC02-76ER00881.

TABLE 1

Nomenclature and $SU(3)_C \times SU(2)_L \times U(1)_Y$ quantum numbers for the exotic fields in the **27** representation of E_6.

SO(10)	SU(5)		Color	T_{3L}	$Y/2$	Q
16	**10**	$Q \equiv \binom{u}{d}_L$	**3**	$\binom{1/2}{-1/2}$	1/6	$\binom{2/3}{1/3}$
		u^c_L	$\overline{\mathbf{3}}$	0	-2/3	-2/3
		e^c_L	**1**	0	1	1
	$\overline{\mathbf{5}}$	$L = \binom{\nu}{e}_L$	**1**	$\binom{1/2}{-1/2}$	-2/3	-2/3
		d^c_L	$\overline{\mathbf{3}}$	0	1/3	1/3
	1	ν^c_L	**1**	0	0	0
10	$\overline{\mathbf{5}}$	$H \equiv \binom{N}{E}_L$	**1**	$\binom{1/2}{-1/2}$	-1/2	$\binom{0}{-1}$
		h^c_L	**3**	0	1/3	1/3
	$\overline{\mathbf{5}}$	$H^c \equiv \binom{E}{N}^c_L$	**1**	$\binom{1/2}{-1/2}$	1/2	$\binom{1}{0}$
		h_L	**3**	0	-1/3	-1/3
1	**1**	S^c_L	**1**	0	0	0

TABLE 2

Quantum numbers for the fields in the **27** representation of E_6 under the groups $\mathrm{U}(1)_\eta$, $\mathrm{U}(1)_\psi$, $\mathrm{U}(1)_\chi$, $\mathrm{SU}(2)_R$, $\mathrm{U}(1)_V$, and $\mathrm{SU}(2)_I$, respectively.

	$2\sqrt{15}\ Q_\eta$	$2\sqrt{6} Q_\psi$	$2\sqrt{10}\ Q_\chi$	T_{3R}	$V/2$	T_{3I}
$Q \equiv \begin{pmatrix} u \\ d \end{pmatrix}_L$	2	1	-1	0	1/6	0
$L \equiv \begin{pmatrix} \nu \\ e \end{pmatrix}_L$	-1	1	3	0	-1/2	1/2
u_L^c	2	1	-1	-1/2	-1/6	0
d_L^c	-1	1	3	1/2	-1/6	1/2
e_L^c	2	1	-1	1/2	1/2	0
ν_L^c	5	1	-5	-1/2	1/2	1/2
$H \equiv \begin{pmatrix} N \\ E \end{pmatrix}_L$	-1	-2	-2	-1/2	0	-1/2
$H^c \equiv \begin{pmatrix} E^c \\ N^c \end{pmatrix}_L$	-4	-2	2	1/2	0	0
h_L	-4	-2	2	0	-1/3	0
h_L^c	-1	-2	-2	0	1/3	-1/2
S_L^c	5	4	0	0	0	-1/2

TABLE 3

A comparison of Z' properties in the LRM and ALRM including asymmetries and branching fractions in percent for $n_g = 3$ (0).

Asymmetry/Mode	LRM	ALRM
$A_{\rm LR}$	0.290	0.504
$A^{\mu}_{\rm FB}$	0.063	0.191
$A^{u}_{\rm FB}$	0.206	−0.358
$A^{d}_{\rm FB}$	0.211	−0.227
B		
e	1.27 (2.46)	3.29 (9.04)
u	5.62 (10.91)	5.62 (15.43)
d	9.84 (19.20)	0.75 (2.06)
ν_L	0.45 (0.87)	2.48 (6.80)
ν_R	5.04	0
h	1.20	10.29
E	4.96	2.93
N_L	2.48	0.45
S_R	0	5.04
N_R	2.48	2.48

TABLE 4

Summary of discovery limits for the three possible quantum number assignments of the exotic fermion h and its SUSY partner at various colliders.

		LEP I/SLC (GeV)	2 TeV CLIC (TeV)	HERA (GeV)	Tevatron (GeV)	SSC (TeV)
Leptoquark						
	h	44	0.9	40	120	0.5-1.0
	$\tilde{h}$	40(80)	0.6-1.2	300	100	1.7-2.0
Diquark						
	h	44	0.9	40	120	0.5-1.0
	$\tilde{h}$	40	0.6	110	70	"
Quark						
	h	44(80)	0.85	40-50	120	0.5-1.0
	$\tilde{h}$	40	0.8-0.9	40-80	120	0.5-1.0

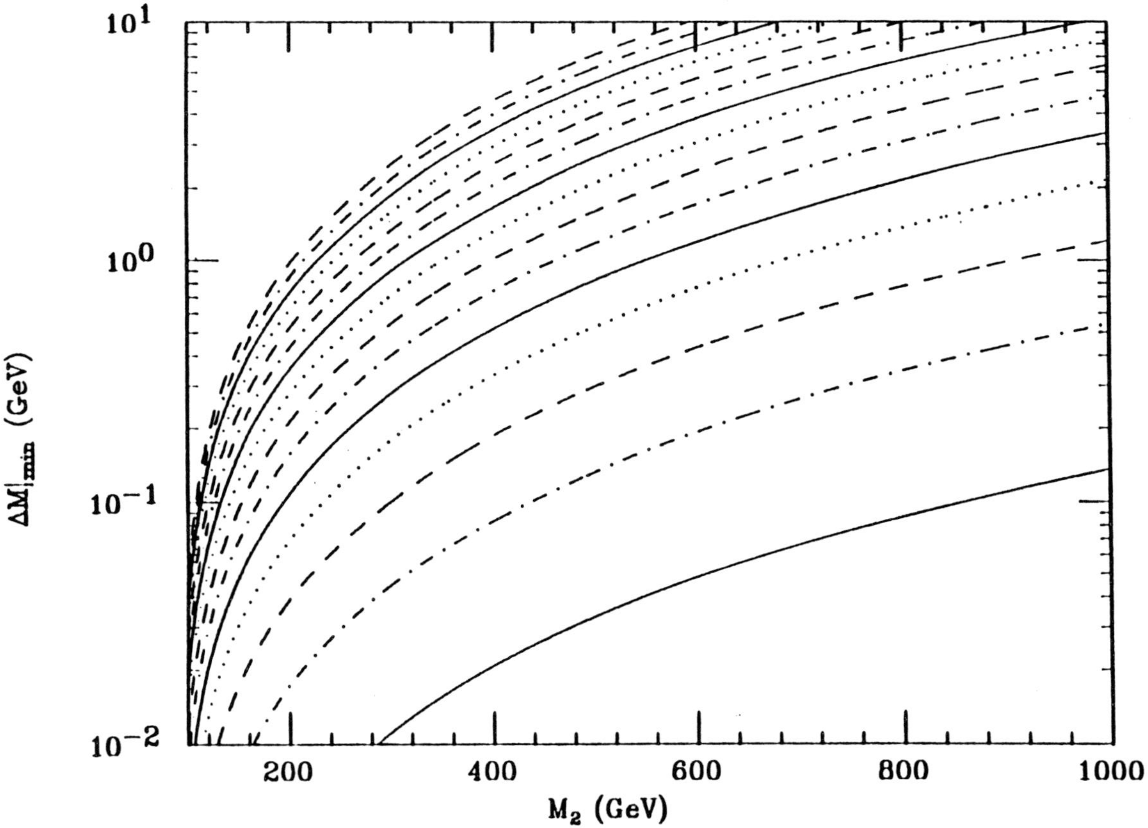

Fig. 1. Minimum value of $\Delta M = M_Z - M_1$ as a function of M_2 for different values of $|\phi|$. $M_1 = 91.17$ GeV has been assumed.

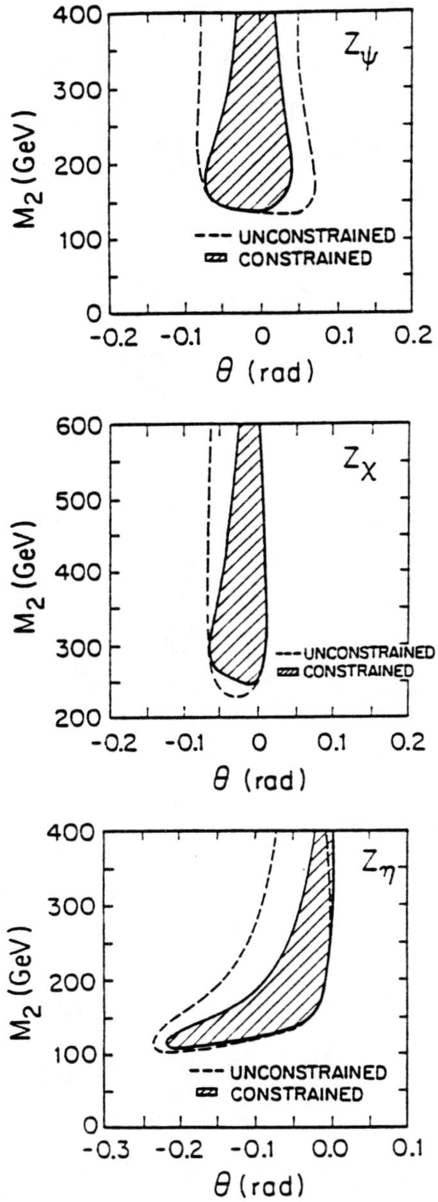

Fig. 2. Bounds from NC and CERN collider data on M_2 as a function of the Z-Z' mixing angle for constrained ($\rho = 1$) and unconstrained Higgs sectors for models (a) ψ, (b) χ, (c) η.

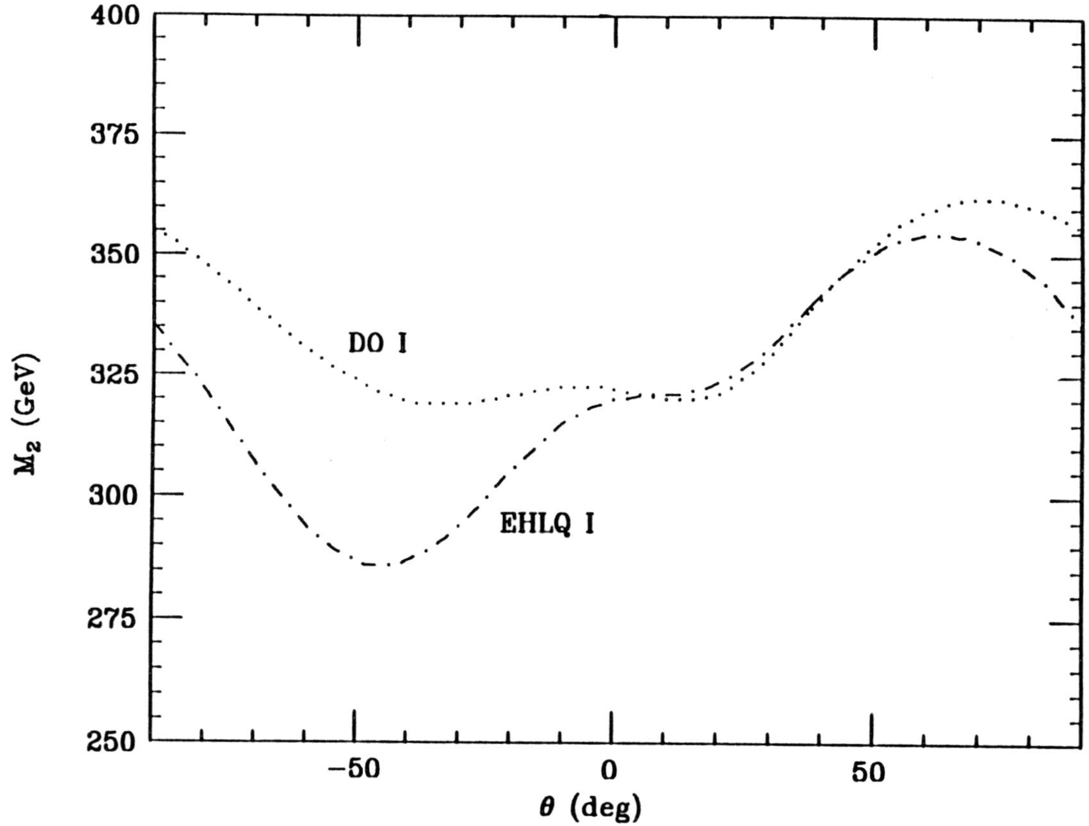

Fig. 3. Preliminary limits on the mass (M_2) of the Z_2 gauge boson as a function of θ in ER5M using either Duke and Owens (DO I) or EHLQ I parton distribution function.

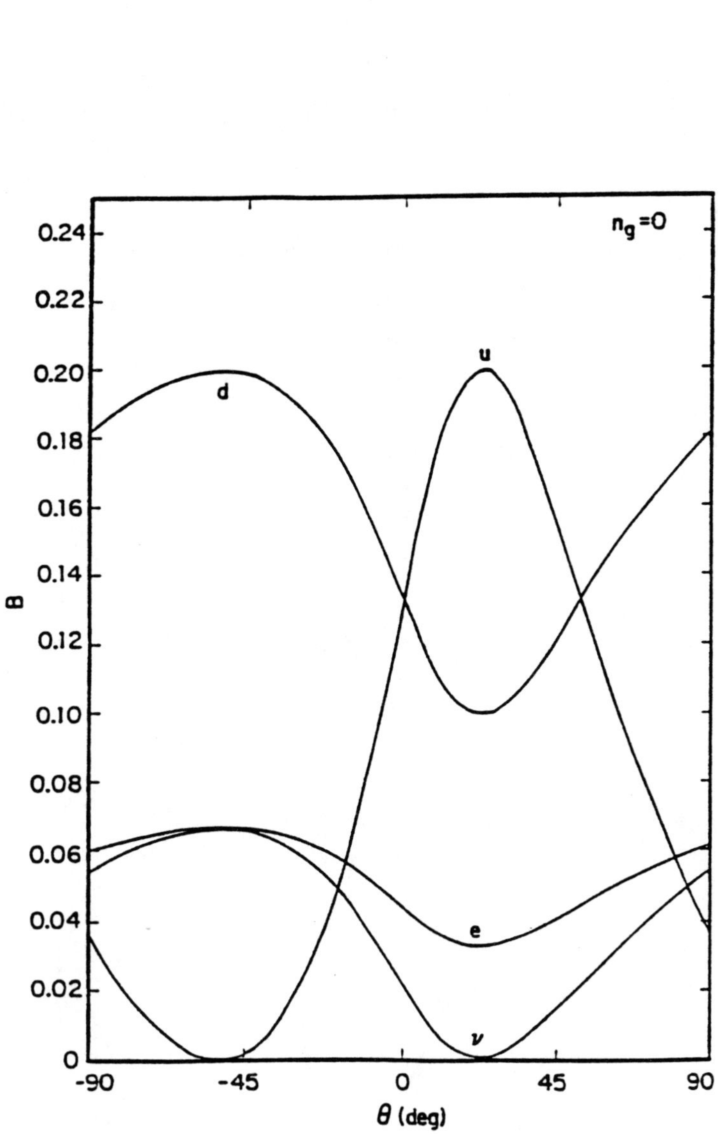

Fig. 4. Branching fractions for Z' in the ER5M for three ordinary massless generations of fermions assuming $n_g = 0$ as a function of θ.

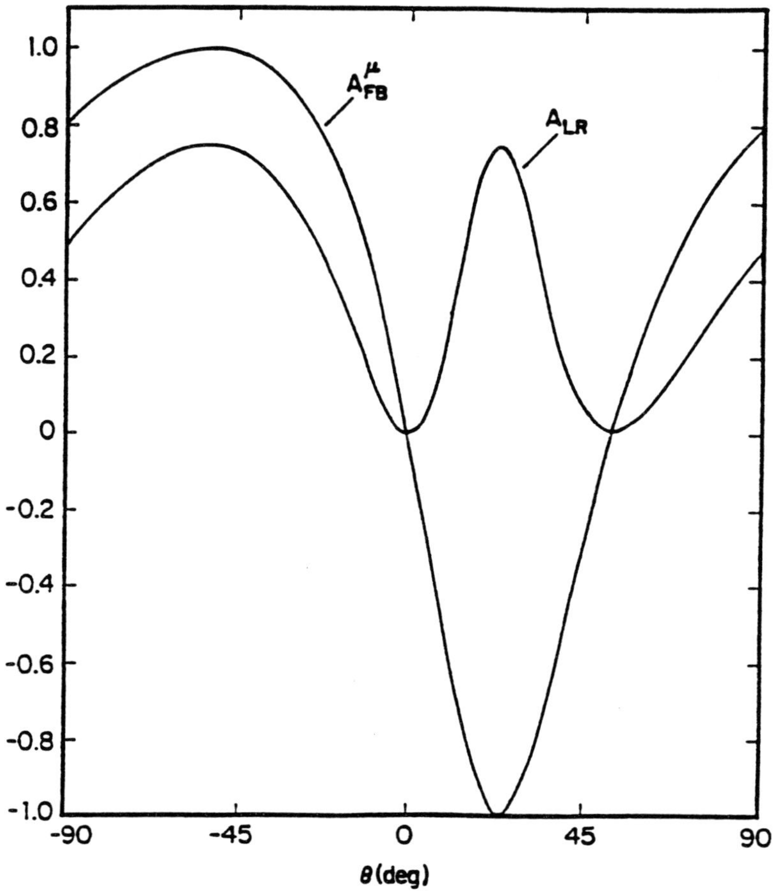

Fig. 5. Asymmetries on the Z' peak in the ER5M as a function of θ. Remember $A^{u}_{\mathrm{FB}} = 0$ and $A^{d}_{\mathrm{FB}} = -A^{\mu}_{\mathrm{FB}}$ on resonance in this model.

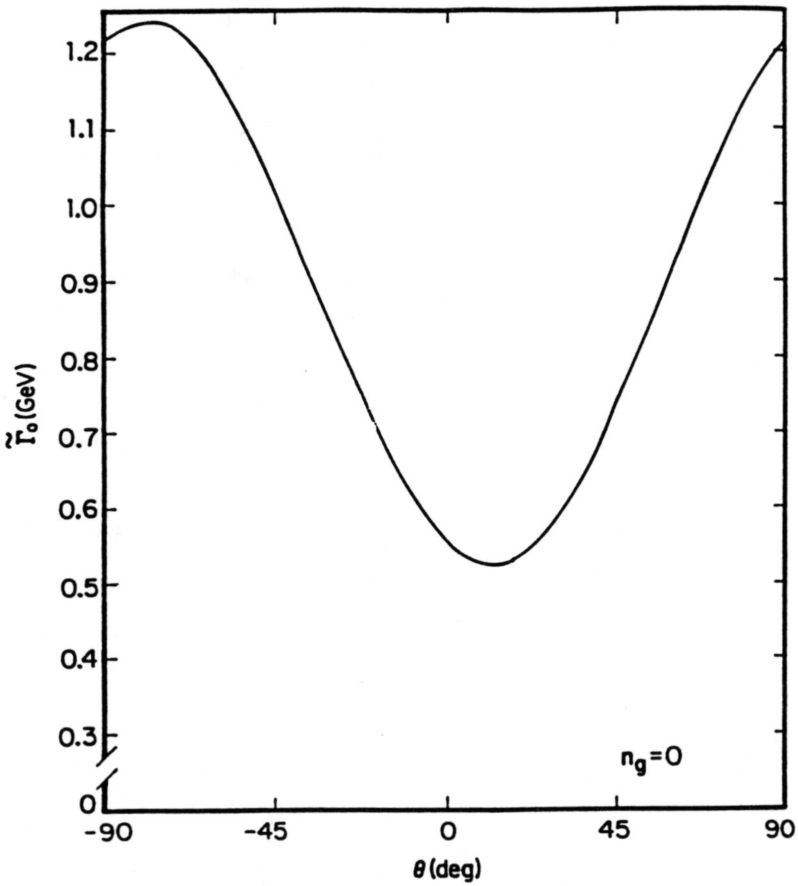

Fig. 6. Scaled width, $\tilde{\Gamma}_0 = \Gamma M_Z/M_{Z'}$, for Z' in the ER5M with $n_g = 0$ as a function of θ.

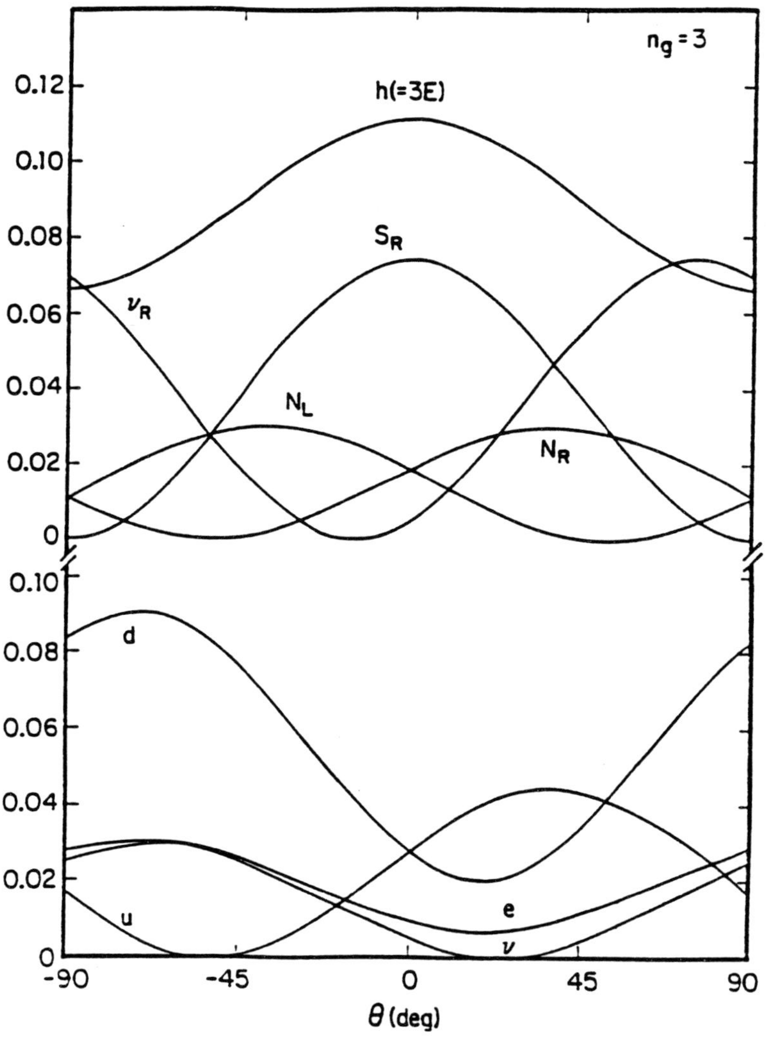

Fig. 7. Branching fractions for Z' in the ER5M for three ordinary plus three exotic ($n_g = 3$) massless generations of fermions as a function of θ.

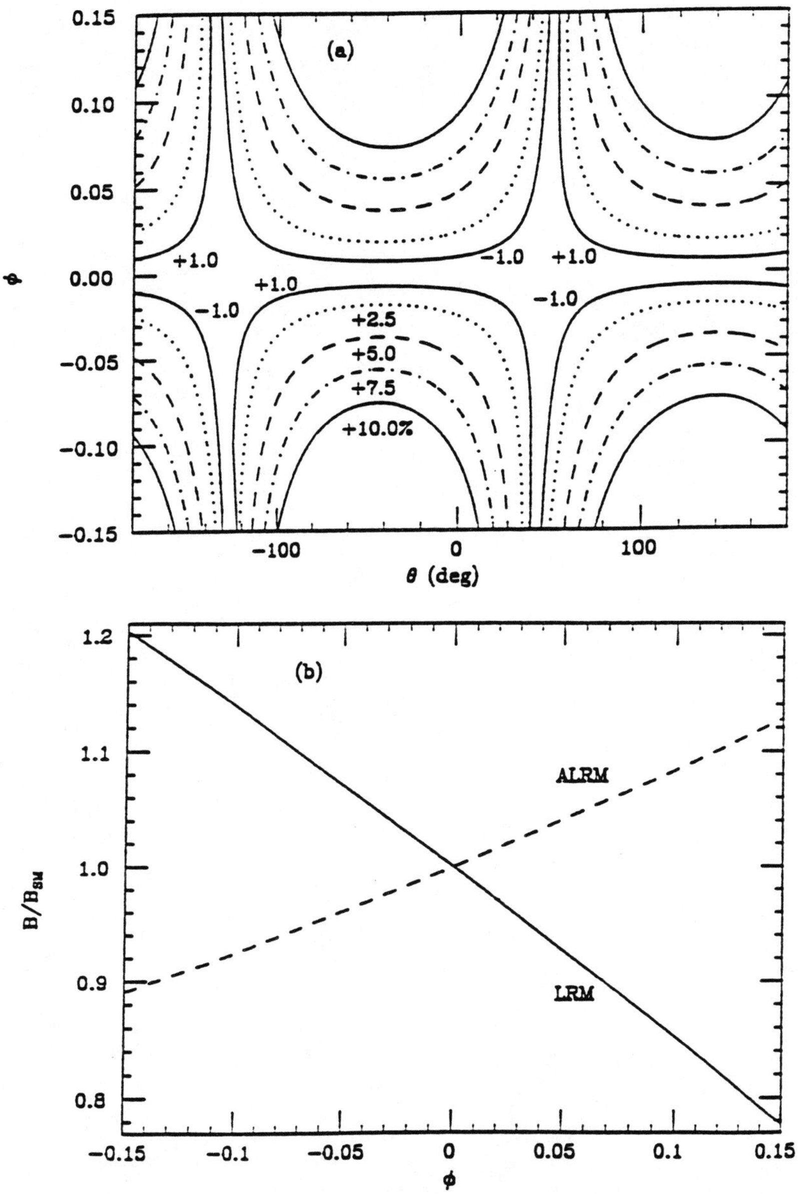

Fig. 8. Variations in the Z_1 leptonic branching fraction (B) due to Z-Z' mixing. (a) Percentage changes in B as a function of θ and ϕ for the ER5M using the same input as in Fig. 7 and using the same notation as in Fig. 7. (b) The ratio of B to its SM value as a function of ϕ for the LRM and ALRM.

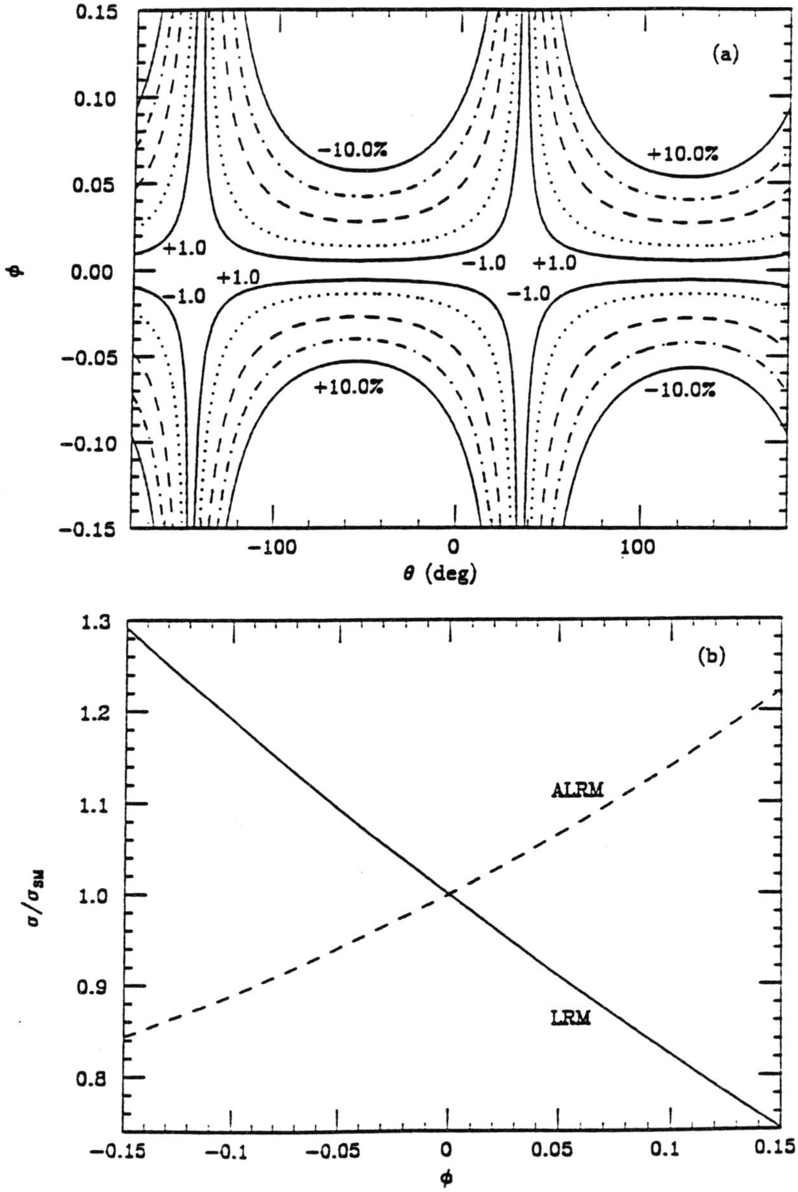

Fig. 9. Variations in the resonance peak cross section (σ) due to Z-Z' mixing. (a) Percentage change in σ as a function of θ and ϕ for the ER5M using the same input as in Fig. 7 and using the same notation as in Fig. 7. (b) The ratio of σ to its SM value as a function of ϕ for the LRM and ALRM.

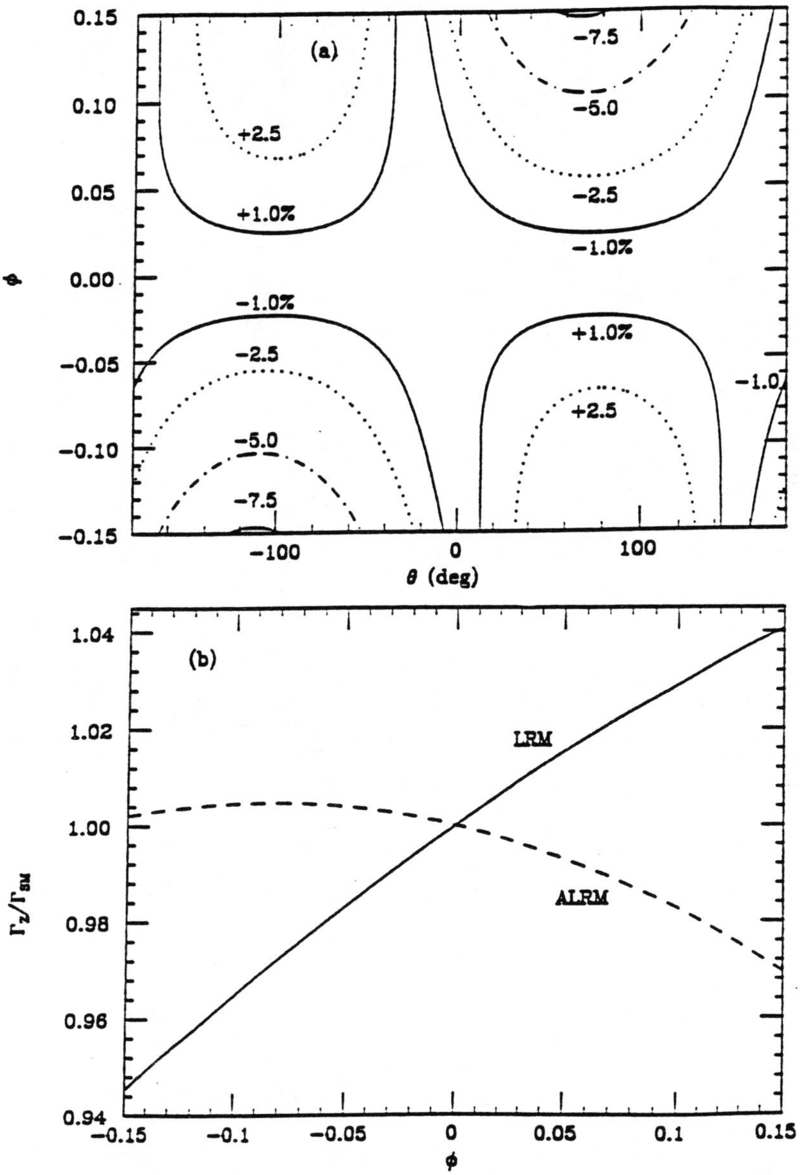

Fig. 10. Variations in the value of Γ_Z due to Z-Z' mixing. (a) Percentage changes in Γ_Z as a function of θ and ϕ for the ER5M assuming M_1 = 91.17 GeV, x_W = 0.2307, and α_s = 0.12. (b) The ratio of Γ_Z to its SM value as a function of ϕ for the LRM and ALRM with the same values of M_1, x_W, and α_s as in (a).

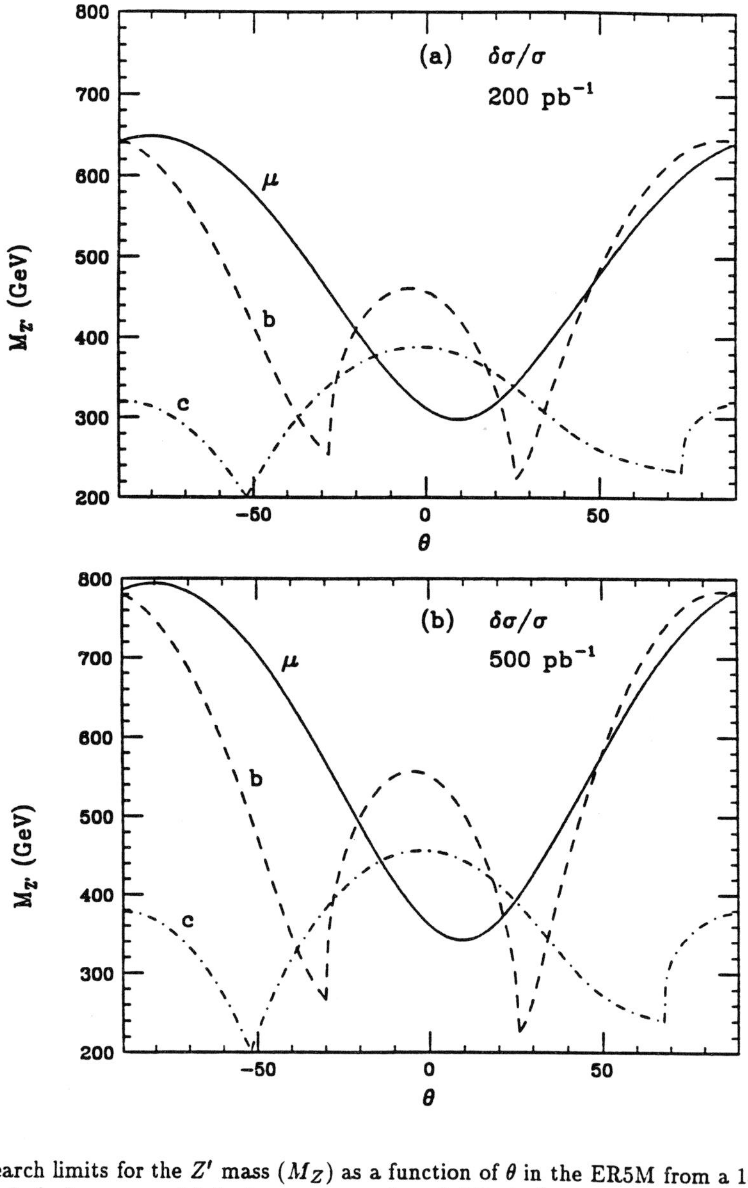

Fig. 11. Search limits for the Z' mass (M_Z) as a function of θ in the ER5M from a 1σ deviation in $\delta\sigma/\sigma$ for μ, c, and b final states assuming a) 200 pb^{-1} and b) 500 pb^{-1} of integrated luminosity at LEP II.

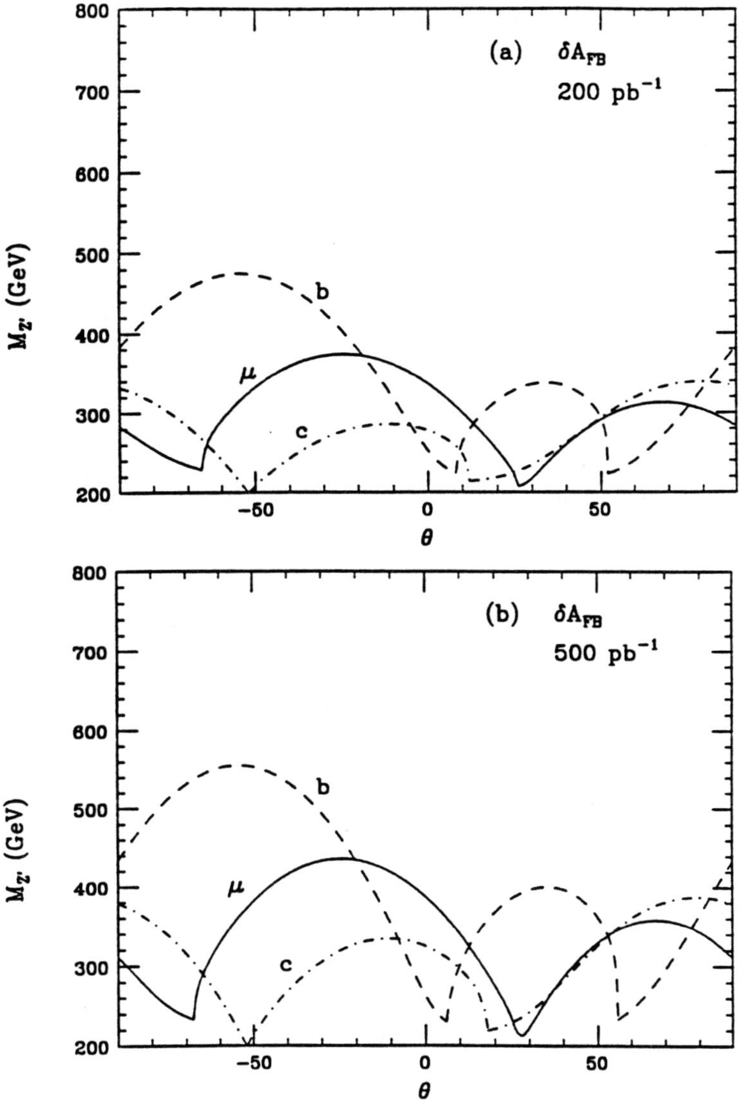

Fig. 12. $M_{Z'}$ search limits from 1σ deviations in δA_{FB} for the same process as in Fig. 11.

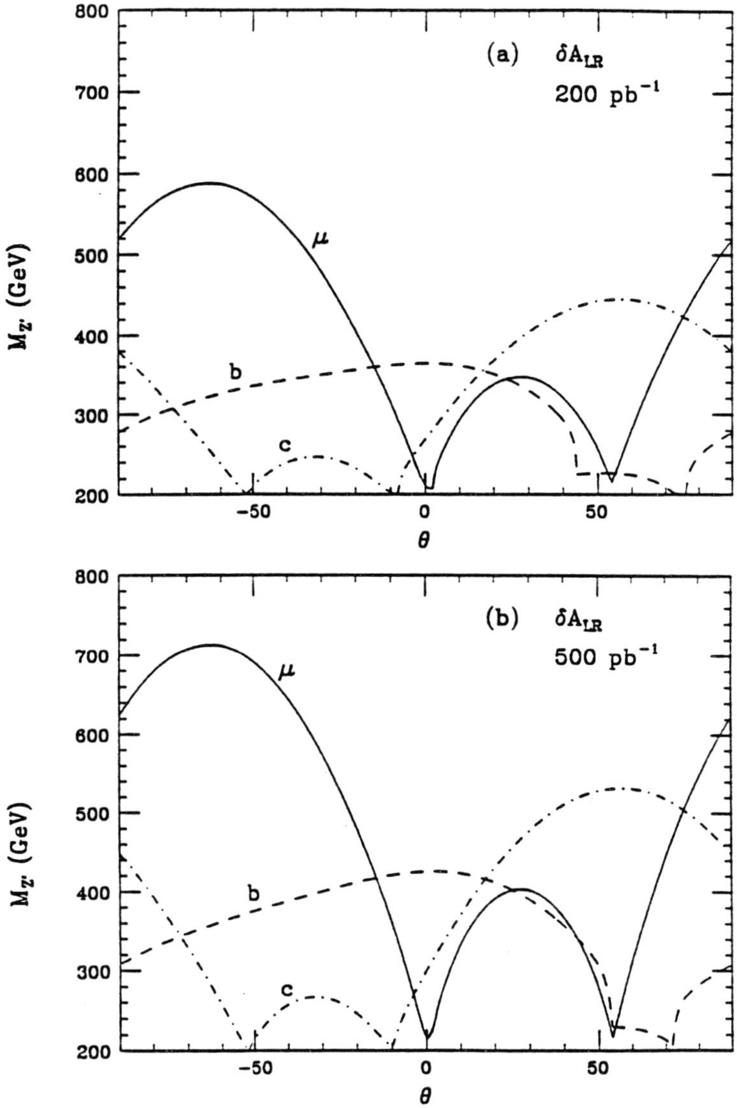

Fig. 13. Same as in Fig. 12 but for δA_{LR}.

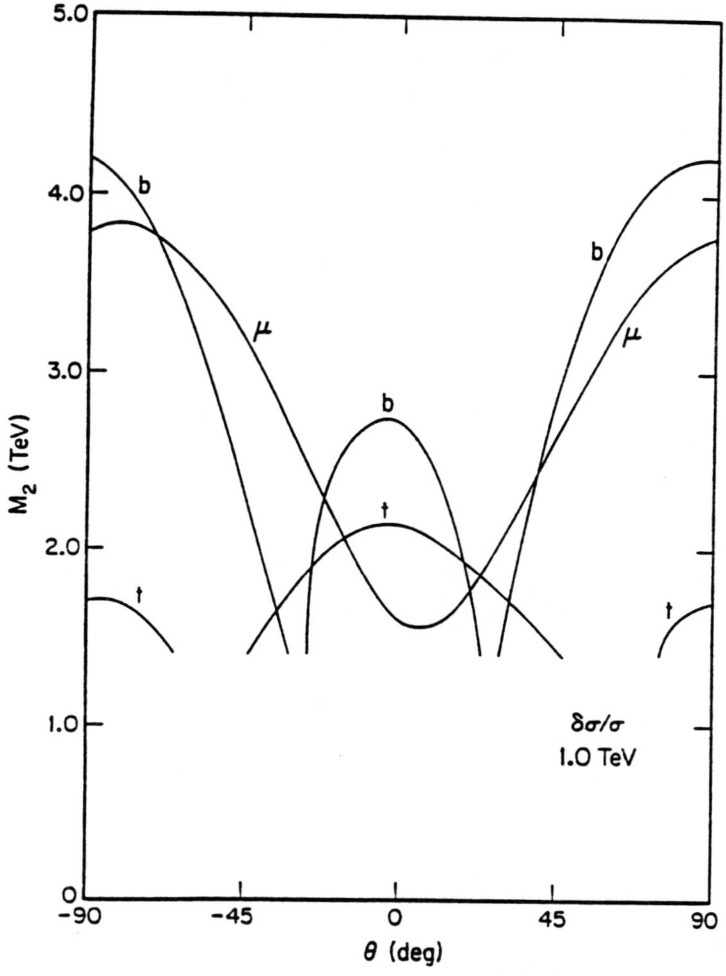

Fig. 14. Same as Fig. 11 but for $\sqrt{s} = 1$ TeV.

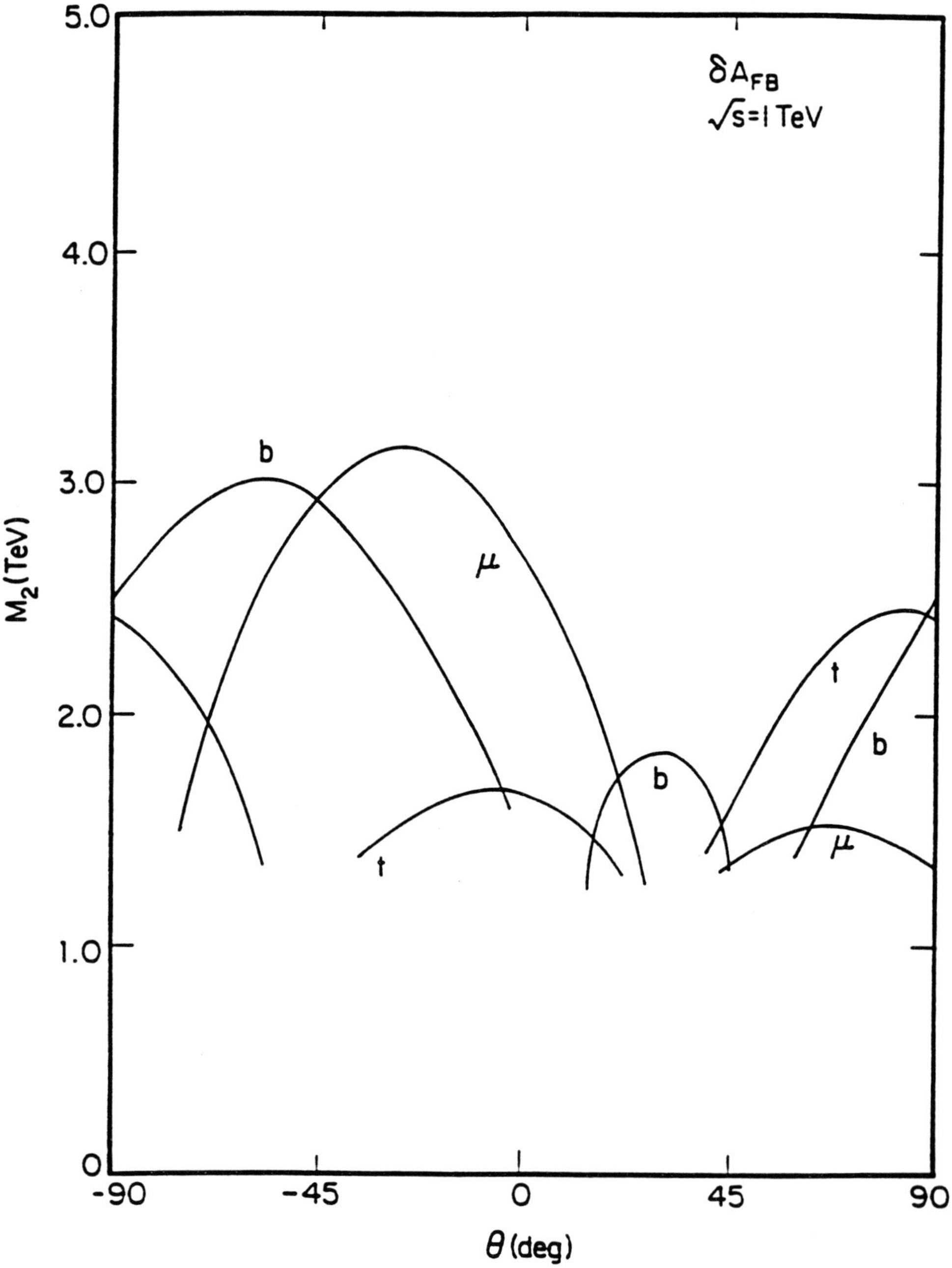

Fig. 15. Same as Fig. 12 but for $\sqrt{s} = 1$ TeV.

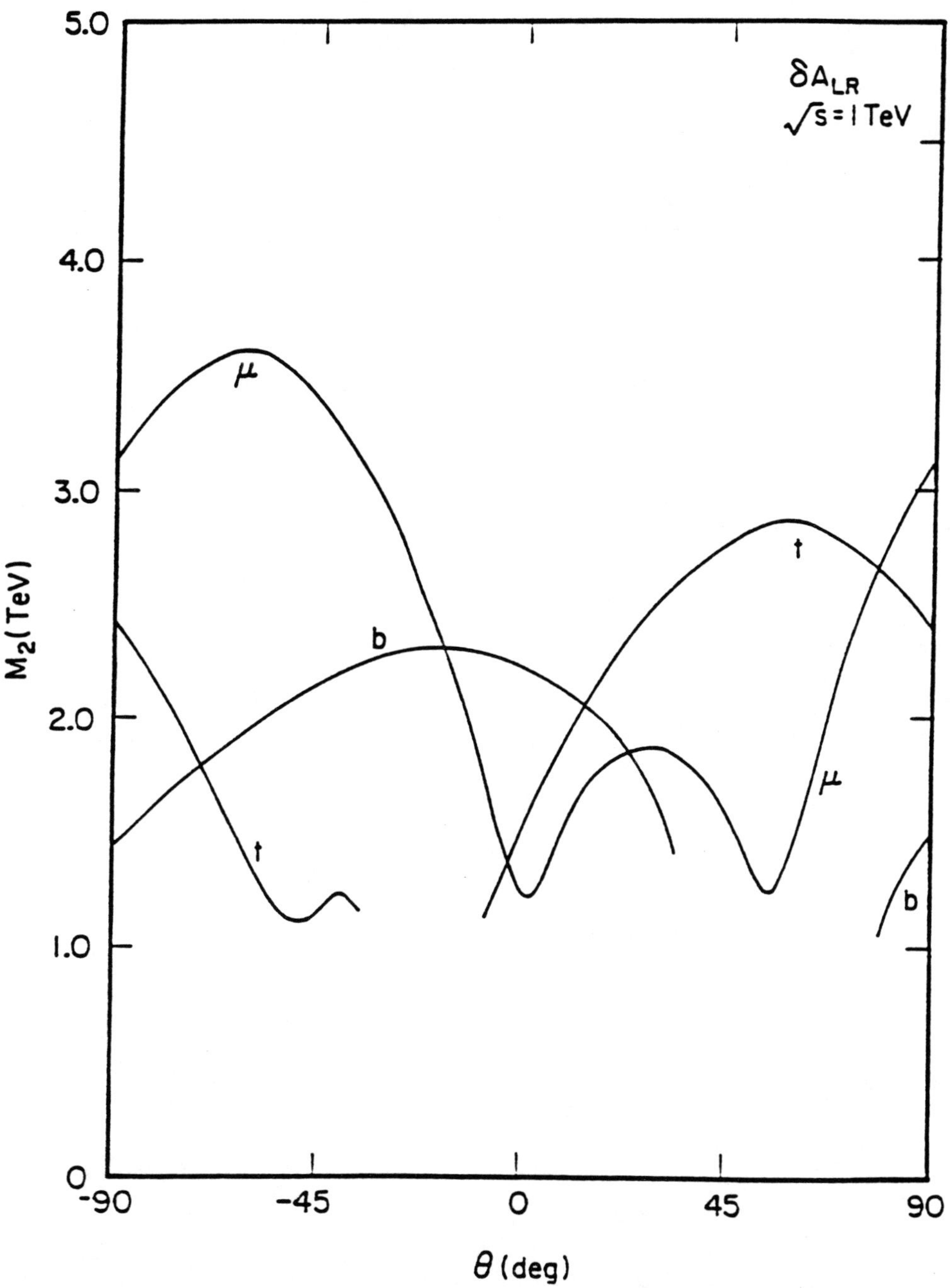

Fig. 16. Same as Fig. 13 but for $\sqrt{s} = 1$ TeV.

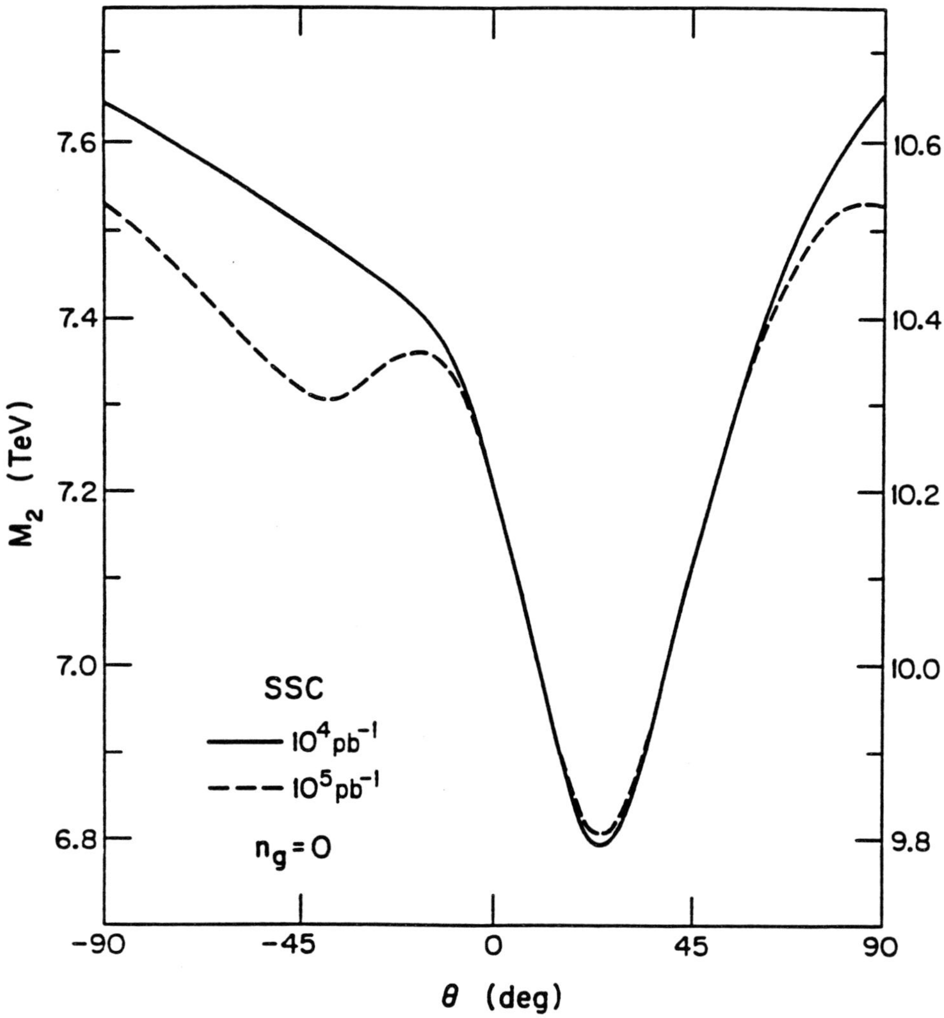

Fig. 17. Search limits for Z_2's originating from the ER5M as a function of θ at the SSC ($\sqrt{s}$ = 40 TeV).

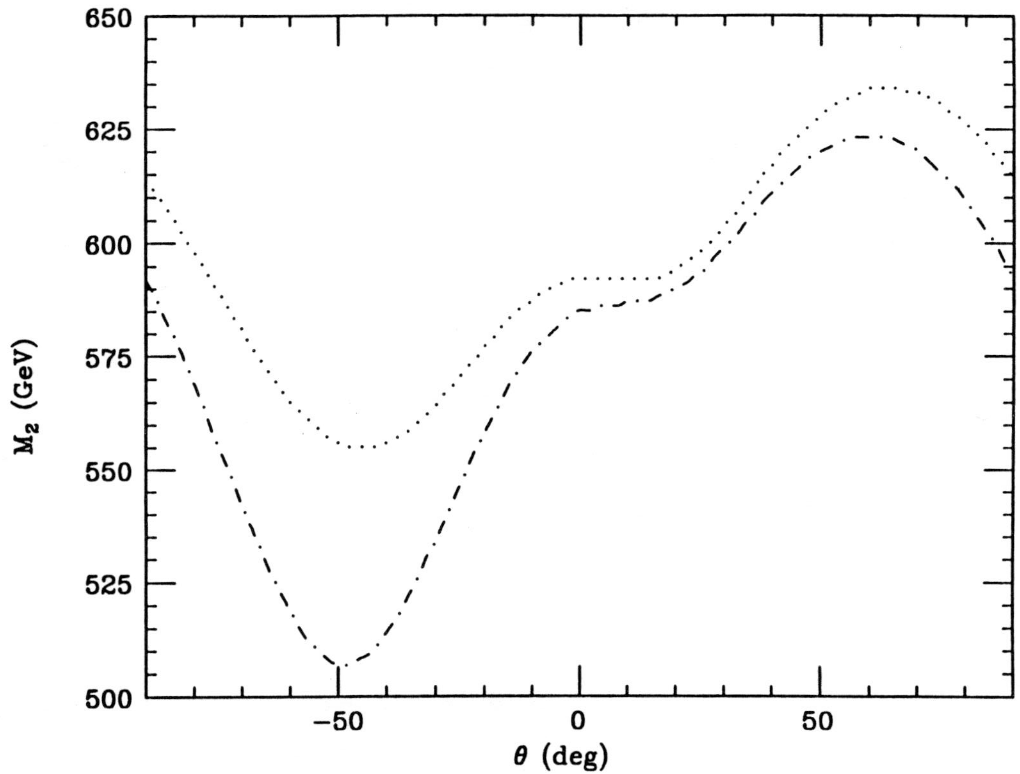

Fig. 18. Search limits for Z_2's in the ER5M as a function of θ at the upgraded Tevatron with $\sqrt{s} = 1.8$ TeV and an integrated luminosity of 100 pb^{-1}.

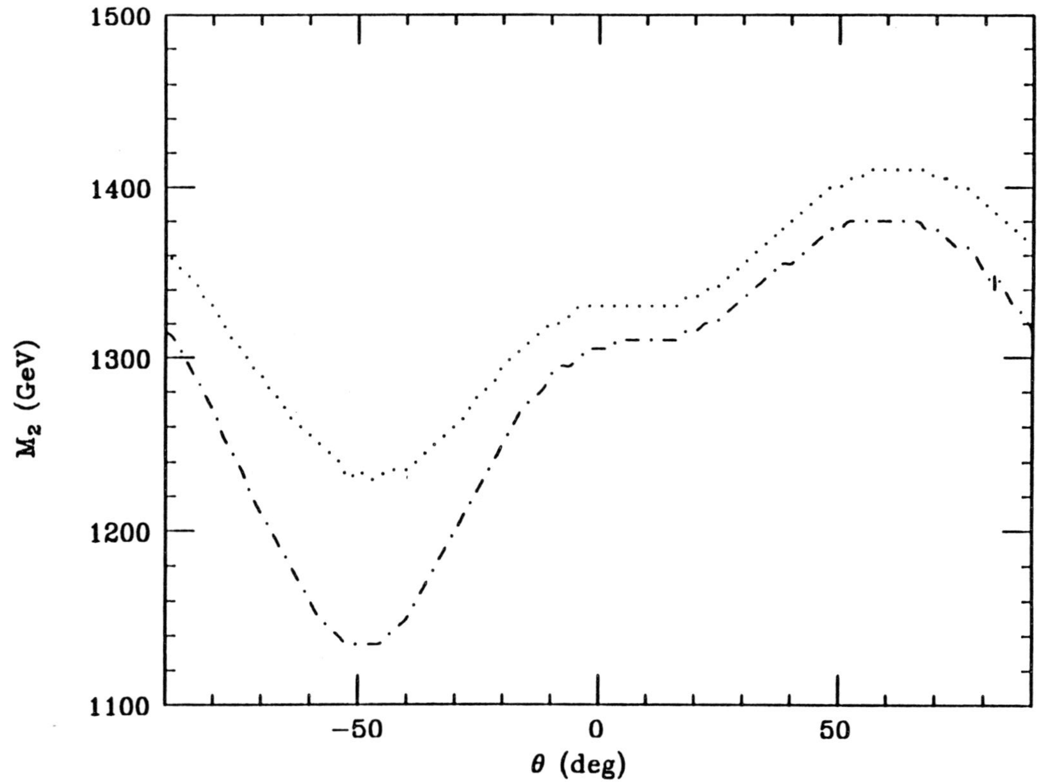

Fig. 19. Same as Fig. 18 but for $\sqrt{s} = 3.6$ TeV and an integrated luminosity of 1000 pb^{-1}.

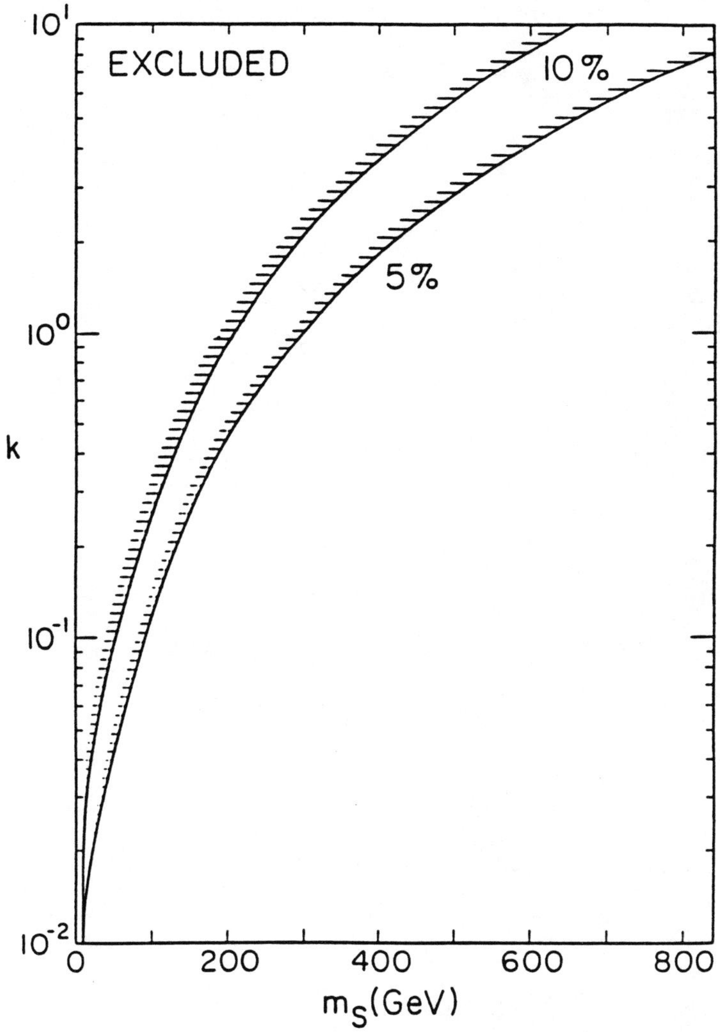

Fig. 20. Constraints on the scalar leptoquark mass as a function of the leptoquark coupling scale parameter, k.

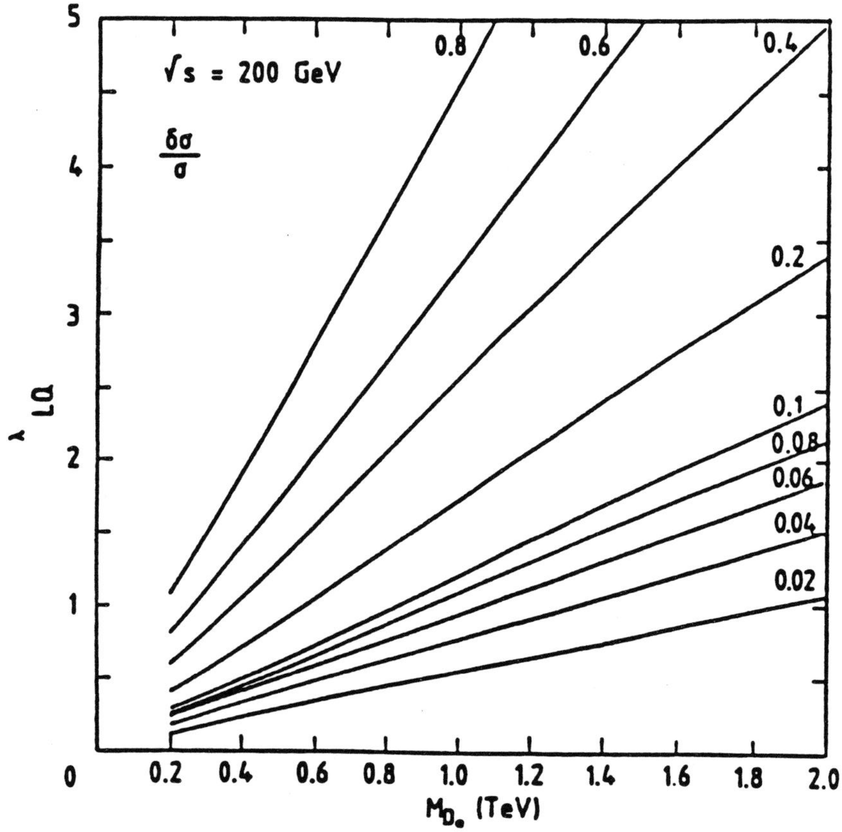

Fig. 21. Contours of $\delta\sigma/\sigma$ in the leptoquark coupling-mass plane for the process $e^+e^- \to q\bar{q}$ at LEP II.

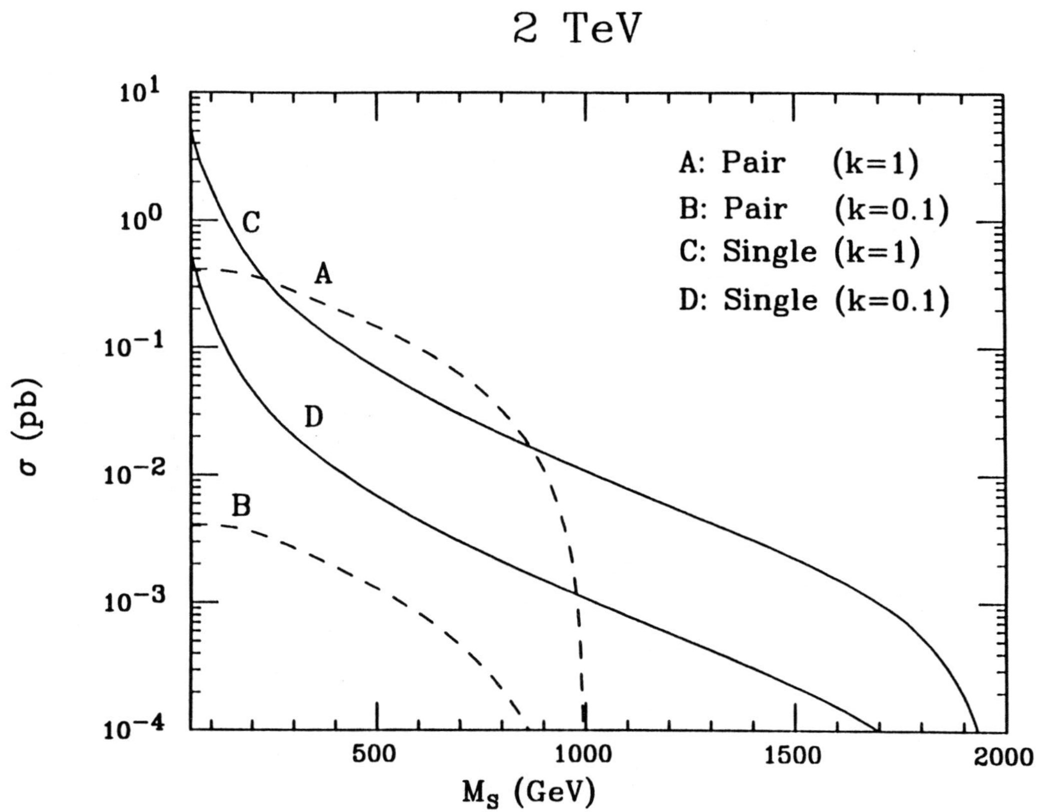

Fig. 22. Single (solid lines) and pair (dashed line) leptoquark production at a $\sqrt{s} = 2$ TeV e^+e^- collider for the values of k as indicated.

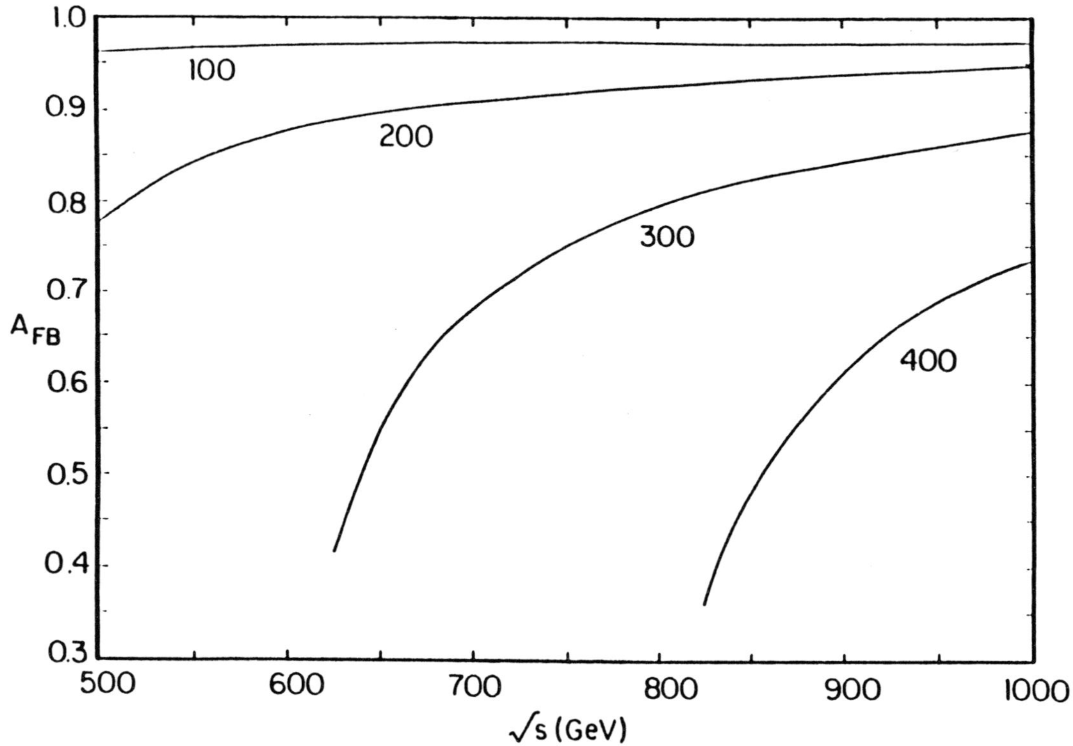

Fig. 23. A_{FB} versus $\sqrt{s}$ with $M_{\tilde{h}} = 100$ GeV ($k = 0.25$), $M_{\tilde{h}} = 200$ GeV ($k = 0.9$), $M_{\tilde{h}} = 300$ GeV ($k = 2.0$), and $M_{\tilde{h}} = 400$ GeV ($k = 3.6$).

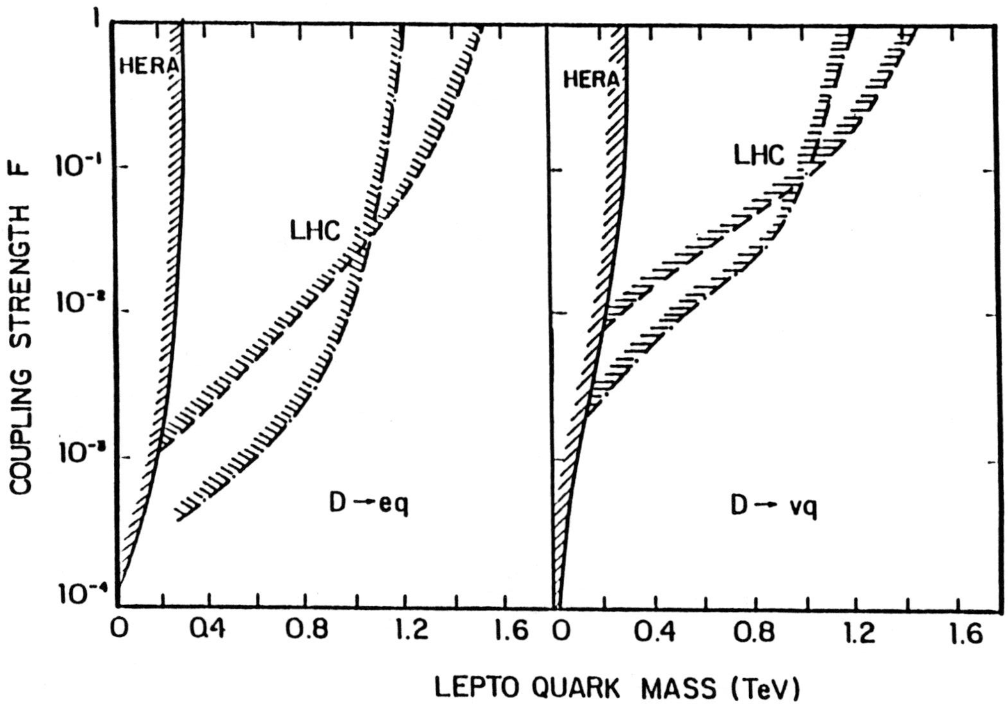

Fig. 24. A comparison of scalar leptoquark discovery limits at HERA and LEP × LHC (with $\sqrt{s} = 1.4$ or 1.8 GeV) in the leptoquark coupling-mass plane.

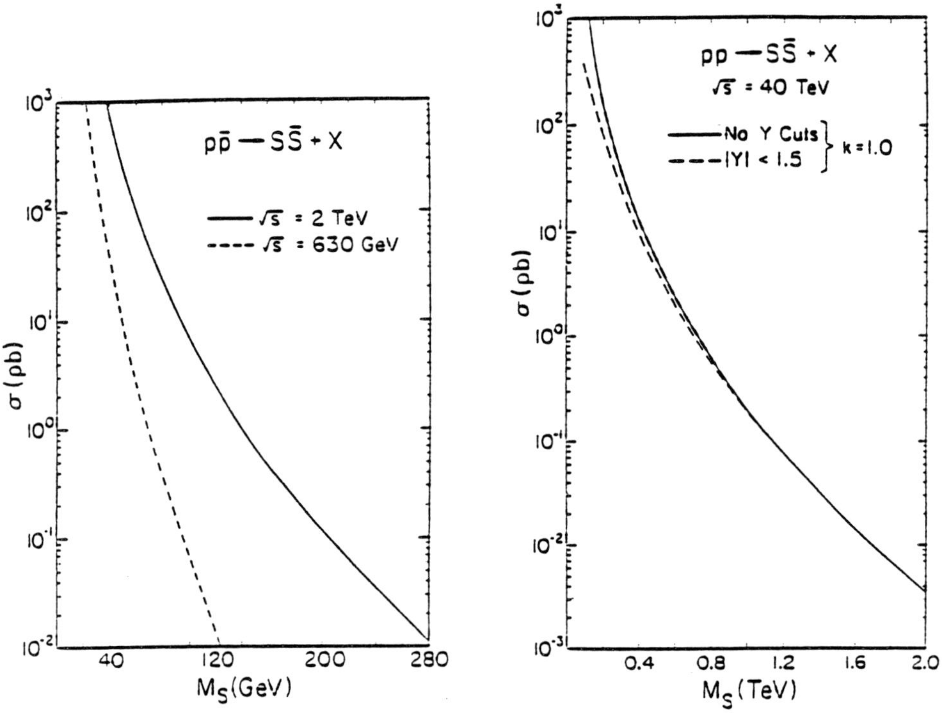

Fig. 25. a) Pair and b) single production of scalar leptoquarks at the SSC with rapidity cuts and values of k as indicated. (h is denoted as S.)

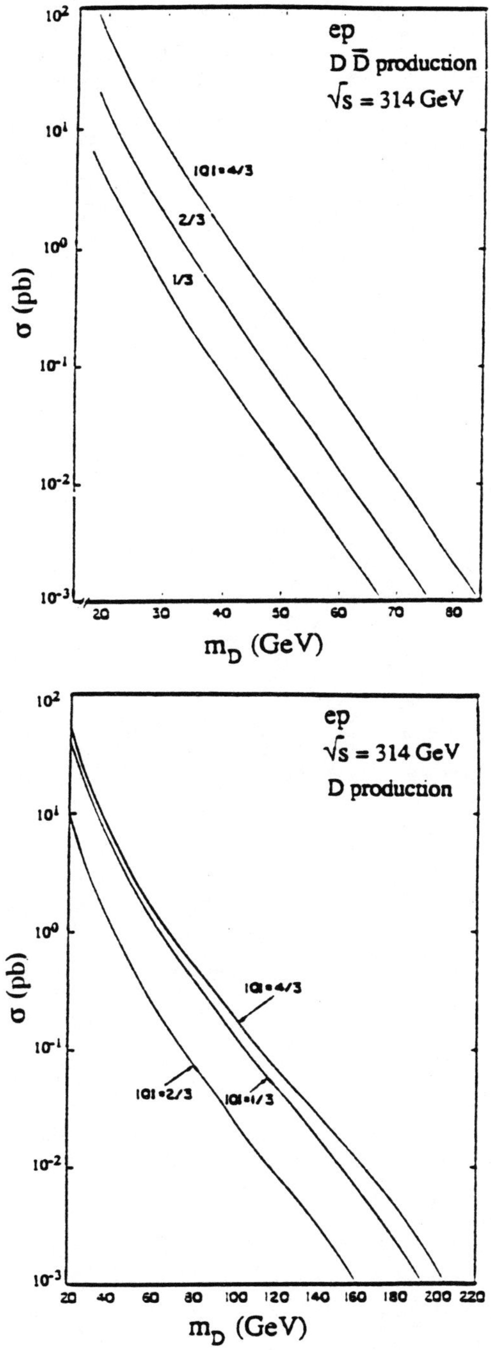

Fig. 26. a) Pair and b) single diquark production at HERA for various values of the diquark electric charge.

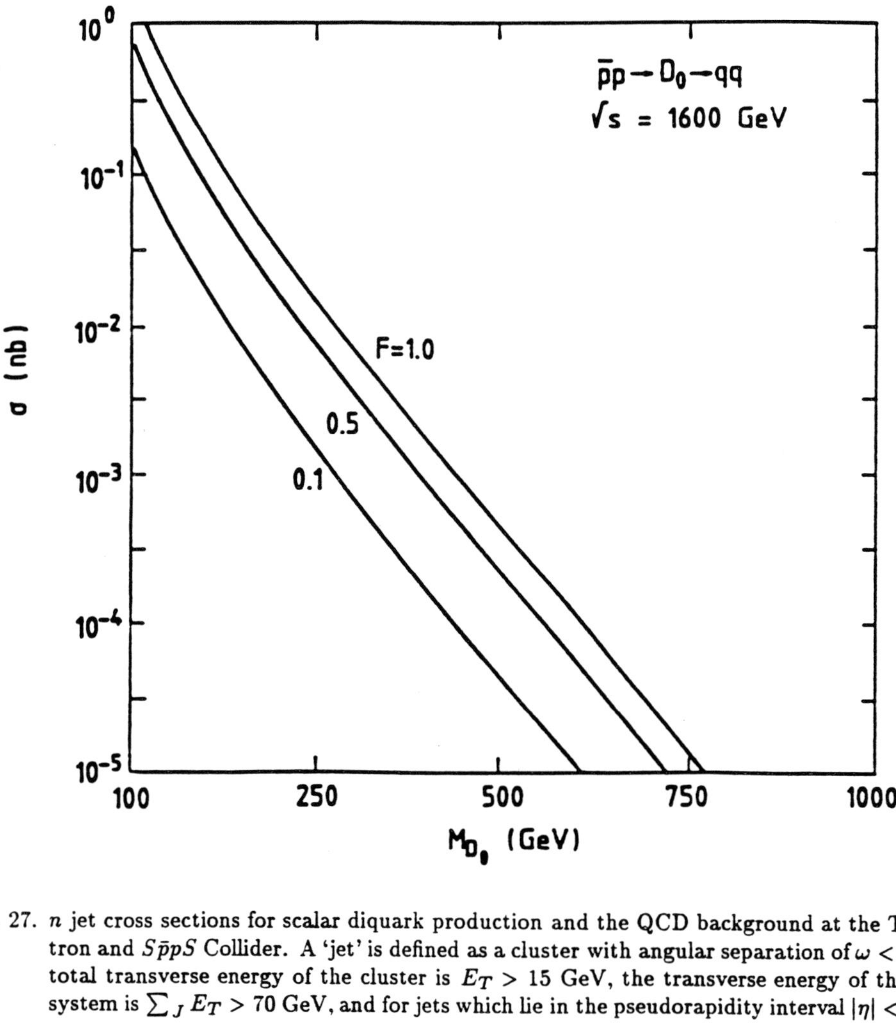

Fig. 27. n jet cross sections for scalar diquark production and the QCD background at the Tevatron and $Sp\bar{p}S$ Collider. A 'jet' is defined as a cluster with angular separation of $\omega < 50°$, total transverse energy of the cluster is $E_T > 15$ GeV, the transverse energy of the jet system is $\sum_J E_T > 70$ GeV, and for jets which lie in the pseudorapidity interval $|\eta| < 0.8$.

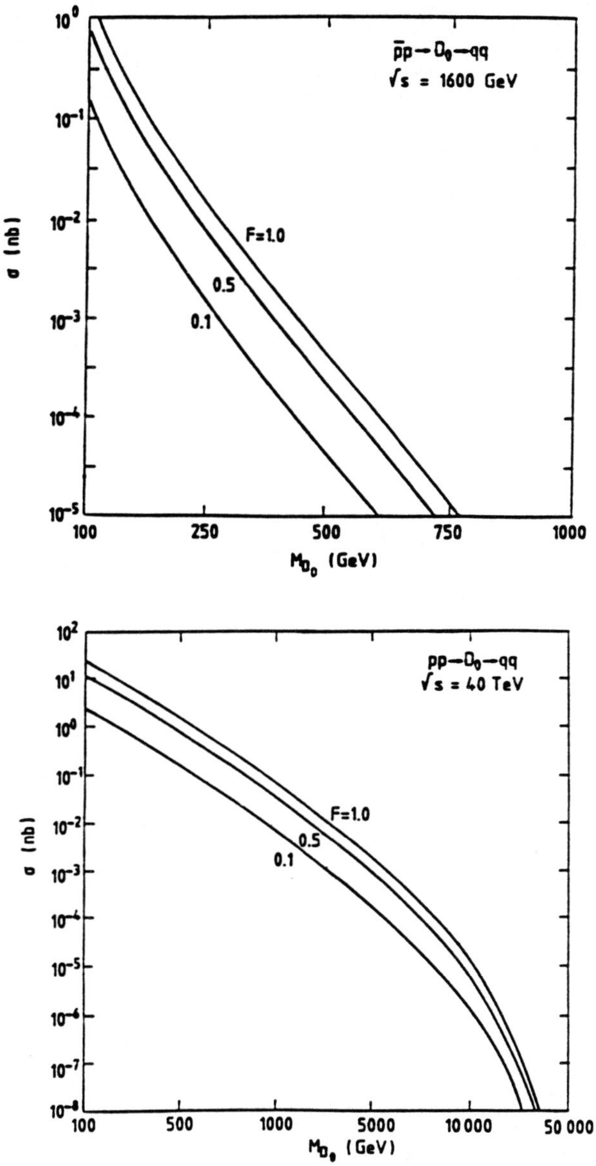

Fig. 28. Total cross section for the production of single scalar diquarks and subsequent decay into quark pairs at the a) Tevatron, and b) SSC for various values of F.

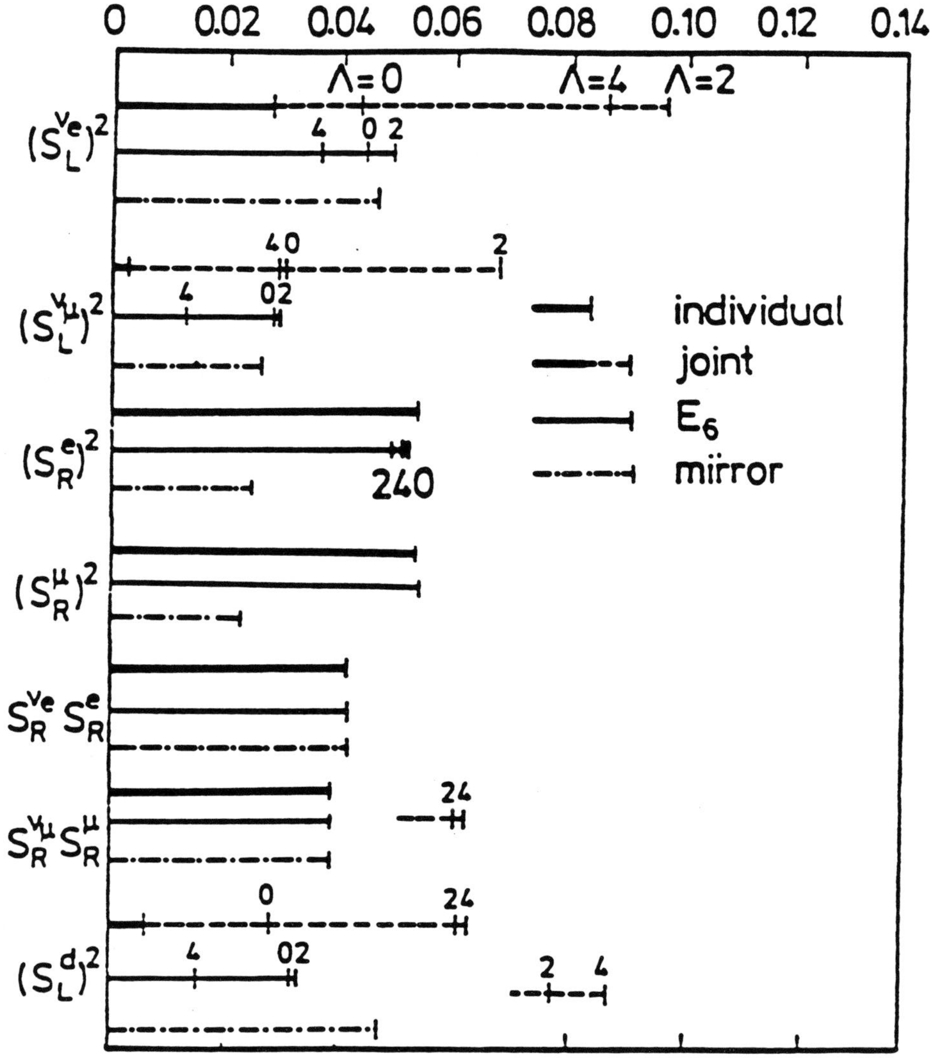

Fig. 29. 90% C.L. upper limits on mixing for the light left-handed fermions, for individual fits (heavy solid lines), joint fits to all mixings (dashed lines), E_6 models (solid lines), and the Hermitian mirror model (dot-dashed line).

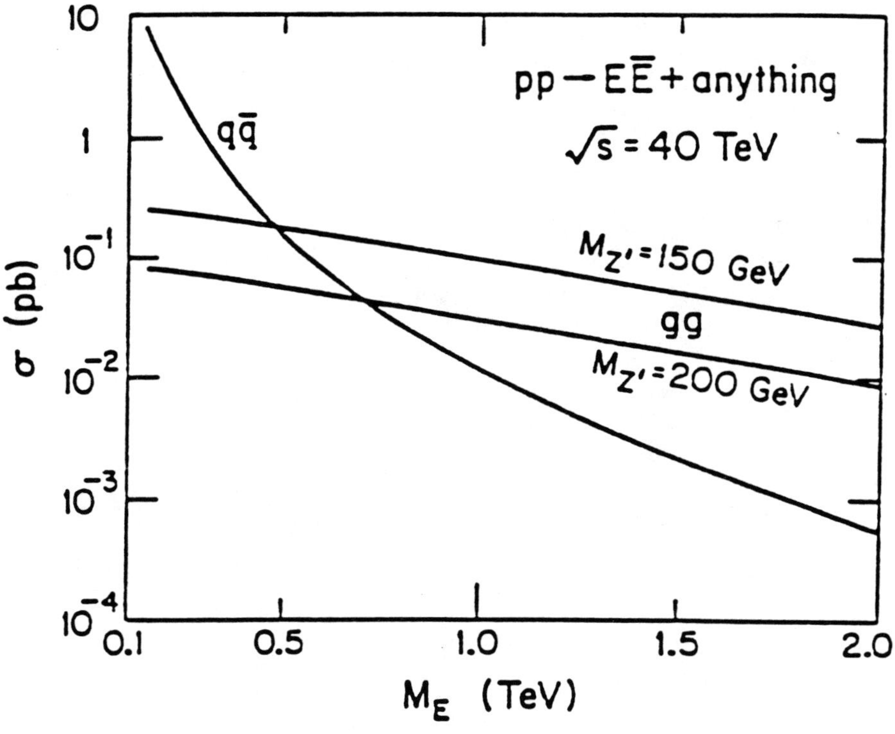

Fig. 30. Drell-Yan and gluon fusion contributions to $E\bar{E}$ production at the SSC.

REFERENCES

[1] For a general review of the phenomenology of E_6 theories, see J. L. Hewett and T. G. Rizzo, Univ. of Wisconsin Report MAD/PH/446 (1988), to appear in Phys. Rep.

[2] M. B. Green and J. H. Schwarz, Phys. Lett. 149B, (1984) 117; M. B. Green, J. H. Schwarz, and E. Witten, **Superstring Theory**, Cambridge Univ. Press, Cambridge, U. K., 1987.

[3] For a discussion of the physics of the scalar sector in these models, see Ref. 1 as well as J. F. Gunion *et al.*, Univ. of California-Davis preprint UCD-89-4 (1989).

[4] For a general survey of the LRM see R. N. Mohapatra, **Unification and Supersymmetry**, Springer-Verlag, New York, 1986.

[5] E. Ma, Phys. Rev. D36, (1987) 274; Mod. Phys. Lett. A3, (1988) 319; K. S. Babu, X.-G. He, and E. Ma, Phys. Rev. D36, (1987) 878; J. F. Gunion *et al.*, Int. J. Mod. Phys. A2, (1987) 1199; V. Barger and K. Whisnant, Int. J. Mod. Phys. A3, (1988) 879.

[6] U. Amaldi *et al.*, Phys. Rev. D36, (1987) 1385.

[7] G. Costa *et al.*, Nucl. Phys. B297, (1988) 244.

[8] J. Ellis and G. L. Fogli, Phys. Lett. 213B, (1988) 526.

[9] V. Barger, N. G. Deshpande, J. L. Rosner, and K. Whisnant, Phys. Rev. D35, (1987) 2893.

[10] J. Ellis, P. J. Franzini, and F. Zwirner, Phys. Lett. 202B, (1988) 417.

[11] S. Errede (CDF Collaboration), talk presented at TASI-1989, Boulder, CO, June 1989.

[12] V. Barger, J. L. Hewett, and T. G. Rizzo, University of Wisconsin-Madison preprint MAD/PH/514 (1989).

[13] A. Maki, talk presented at the 14th International Symposium on Lepton and Photon Interactions, Stanford, CA, August 1989; VENUS Collaboration, Phys. Rev. D39, (1989) 3524; AMY Collaboration, Phys. Lett. 218B, (1989) 499.

[14] For a discussion of these possibilities, see D. Zeppenfeld, University of Wisconsin-Madison preprint MAD/PH/512 (1989).

[15] S. Capstick and S. Godfrey, Phys. Rev. D35, (1987) 3351.

[16] S. Godfrey and W. J. Marciano, talk presented at the *BNL Neutrino Workshop*, Upton, NY, February (1987).

[17] J. F. Gunion *et al.*, Int. J. Mod. Phys. A2, (1987) 1199.

[18] J. L. Hewett, talk given at the *Beyond the Standard Model Conference*, Ames, IA, November 1988.

[19] See Ref. 13 for a recent review of the current experimental results.

[20] G. Feldman, talk presented at the *14th International Symposium on Lepton and Photon Interactions*, Stanford, CA, August 1989; Phys. Rev. Lett. 63, (1989) 724.

[21] T. G. Rizzo, Phys. Rev. D40, (1989) 1424.

[22] J. L. Hewett and T. G. Rizzo, Phys. Rev. D36, (1987) 3367.

[23] P. Langacker and R. Rückl, to be published.

[24] H. Dreiner *et al.*, Mod. Phys. Lett. A3, (1988) 443.

[25] R. Rückl and P. Zerwas, *Proceedings of the La Thuile Accelerator Workshop*, La Thuile, Italy, 1987, ed. J. H. Mulvay.

[26] D. Schaile and P. M. Zerwas, *Proceedings of the La Thuile Accelerator Workshop*, La Thuile, Italy, 1987, ed. J. H. Mulvay.

[27] J. L. Hewett and S. Pakvasa, Phys. Lett. 227B, (1989) 178.

[28] V. D. Angelopoulos *et al.*, Nucl. Phys. B292, (1987) 59.

[29] N. Harnew, Proceedings of the 1987 DESY Workshop on HERA Physics, Hamburg, Germany, October 1987; Proceedings of the La Thuile Accelerator Workshop, La Thuile, Italy, 1987, ed. J. H. Mulvay.

[30] J. F. Gunion and E. Ma, Phys. Lett. 195B, (1987) 257.

[31] V. Barger *et al.*, *1986 Summer Study on the Physics of the SSC*, Snowmass, CO, 1986, ed. R. Donaldson and J. N. Marx.

[32] J. L. Hewett and S. Pakvasa, Phys. Rev. D37, (1988) 3165.

[33] T. G. Rizzo, Z. Phys. C43, (1989) 223.

[34] E. N. Argyres *et al.*, Phys. Lett. 202B, (1988) 425.

[35] S. Glashow and S. Weinberg, Phys. Rev. D15, (1977) 1958; E. A. Paschos, Phys. Rev. D15, (1977) 1966.

[36] P. Langacker and D. London, Phys. Rev. D38, (1988) 886.

[37] J. L. Hewett, Phys. Lett. 196B, (1987) 223; T. G. Rizzo, Phys. Lett. 188B, (1987) 95.

[38] See, for example, E. Eichten *et al.*, Rev. Mod. Phys. 56, (1984) 579.

[39] T. G. Rizzo, Phys. Lett. 181B, (1986) 385.

[40] V. Barger and W.-Y. Keung, Phys. Rev. D34, (1986) 2902.

ISBN 1-56072-027-1

NEUTRINO MIXING

S.M. Bilenky
Joint Institute for Nuclear Research
Dubna, Russia

I. INTRODUCTION

Are neutrino masses different from zero? Are neutrinos with definite masses Dirac or truely neutral Majorana particles? Is there neutrino mixing? These fundamental problems are the most important problems of modern neutrino physics.

The mixing of massive neutrinos naturally arises in most models of Grand Unification and supersymmetric models. The discovery of the effects of nonzero neutrino masses would be very important for the construction of a theory beyond the standard theory.

In numerous laboratories all over the world many experiments in search of neutrino oscillations, neutrinoless double β-decay and other effect of nonzero neutrino masses have been done. Up to now no definite indications in favour of such effects have been obtained. However, the experiments that have been done allowed one to narrow considerably the region of the possible values of neutrino masses and mixing. New experiments aiming on the search for effects of nonzero neutrino masses are in progress now. There are no doubts that these experiments would allow one to make a new essential step in studing the fundamental properties of neutrinos.

In the beginning of these lectures we shall discuss the general phenomenological theory of the mixing of massive neutrinos. Further on we shall consider some physical consequences of neutrino mixing. Finally we shall present some data of the latest experiments.

Before presenting the theory of neutrino mixing I would like to remind you that in the standard theory of electroweak interactions the Lagrangians of the interaction of fundamental fermions (leptons and quarks) with W and Z have the form

$$\mathcal{L}^{W} = -\frac{g}{2\sqrt{2}} j_{\alpha}^{W} W^{\alpha} + h.c., \quad \mathcal{L}^{Z} = -\frac{g}{2\cos\theta_W} j_{\alpha}^{Z} Z_{\alpha}. \quad (1)$$

Here j_{α}^{W} and j_{α}^{Z} are the charged and neutral currents, g is the coupling constant and θ_W is Weinberg angle. For the charged current we have

$$j_{\alpha}^{W} = 2\left[\sum_{l=e,\mu,\tau} \bar{\nu}_{lL} \gamma_{\alpha} l_L + \bar{u}_L \gamma_{\alpha} d'_L + \bar{c}_L \gamma_{\alpha} s'_L + \bar{t}_L \gamma_{\alpha} b'_L\right]. \quad (2)$$

Here

$$d'_L = \sum_{q=d,s,b} V_{uq} q_L, \quad s'_L = \sum_{q=d,s,b} V_{cq} q_L, \quad b'_L = \sum_{q=d,s,b} V_{tq} q_L, \quad (3)$$

where V is a unitary matrix (the Cabibbo-Kobayashi-Maskawa matrix). For the neutrino's part of the neutral current we have

$$j_{\alpha}^{Z} = \sum_{l=e,\mu,\tau} \bar{\nu}_{lL} \gamma_{\alpha} \nu_{lL}. \quad (4)$$

It's well known that using expressions (1)-(4) all existing neutral current and charged current data can be described.

If neutrino fields enter <u>only</u> in the Lagrangians $\mathcal{L}^{W}$ and $\mathcal{L}^{Z}$ it means that, first, neutrinos are massless particles, and, second, the lepton numbers conservation law takes place. If we prescribe electron L_e, muon L_μ and taon L_τ lepton numbers to neutrinos, leptons and all other particles in accordance with Table I, then it follows from (1), (2) and (4) that the sum of the lepton numbers are separately conserved

$$\sum_i L_e^i = const, \quad \sum_i L_\mu^i = const, \quad \sum_i L_\tau^i = const. \quad (5)$$

Existing data agree with (5). The accuracies achieved can be seen from Table II where the upper limits of the relative probabilities of some forbidden by the lepton numbers conservation law processes are given.

Table I

Lepton numbers of particles

Particles	L_e	L_μ	L_τ
ν_e, e^-	1	0	0
ν_μ, μ^-	0	1	0
ν_τ, τ^-	0	0	1
Hadrons, γ, W, Z	0	0	0

As it is seen from Table II rather high accuracies in checking the conservation laws of the lepton numbers are achieved at present. Nevertheless the existing data do not exclude the possibility for violation of the law of lepton numbers conservation. In these lectures we shall consider the most probable mechanism for violation of the lepton number conservation law, namely neutrino mixing. Let us assume that the neutrino fields ν_ℓ in eqs. (2) and (4) are orthogonal superpositions of the left-handed components of the fields of neutrinos with definite masses

Table II

Upper limits for the relative probabilities of the processes that are forbidden by conservation of the lepton numbers

Process	Upper limit
$\mu^- \to e^- \gamma$	$5 \cdot 10^{-11}$
$\mu^- \to e^- e^+ e^-$	$1 \cdot 10^{-13}$
$\mu^- \to e^- \gamma \gamma$	$7 \cdot 10^{-11}$
$K^+ \to \pi^+ \mu^+ e^-$	$3 \cdot 10^{-10}$
$K_L \to \mu^\pm e^\mp$	$3 \cdot 10^{-10}$

$$\nu_{lL} = \sum_k U_{lk} \nu_{kL} . \tag{6}$$

Here ν_k is the field of neutrino with mass m_k and U is a unitary matrix. It is evident that in this case the electron, muon and taon lepton numbers are not conserved separately. For example, the decay $\mu \to e\gamma$ is possible. One of the diagrams of this process is shown on Fig. 1.

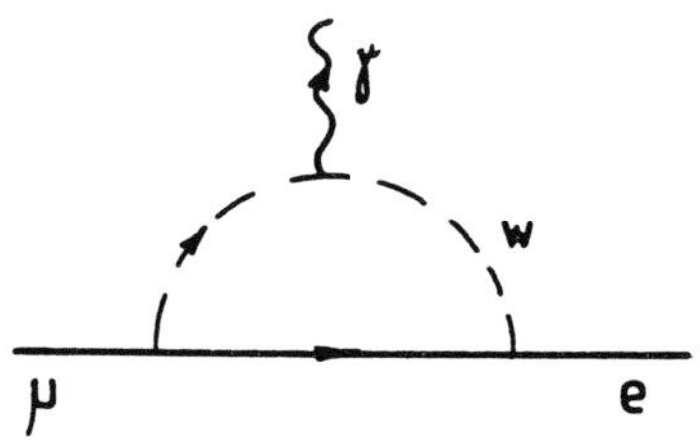

Fig. 1
Diagram of process $\mu \to e\gamma$.

For the relative probability of the decay $\mu \to e\gamma$ in case of neutrino mixing (6) we have [1]

$$\frac{\Gamma(\mu \to e\gamma)}{\Gamma(\mu \to e\nu_e \nu_\mu)} = \frac{3\alpha}{32\pi} \left| \sum_k U^*_{\mu k} U_{ek} \frac{m_k^2}{m_W^2} \right|^2 , \tag{7}$$

where α is the fine structure constant, m_W is the mass of W-boson. By substituting in (7) the experimental upper bounds on the neutrino masses, it is easy to see that upper bound of the relative probability of the decay $\mu \to e\gamma$ is much lower than the experimental one.

Expression (6) is analogous in form to expression (3). However, the mixing of neutrinos can differ in principal from the quark mixing. This is because of the quarks with electric charge different from zero are Dirac particles. Neutrinos with definite masses are electrically neutral particles and could be Dirac as well as Majorana particles. In the Dirac case, neutrinos with different masses possess conserved lepton number that differs neutrino from antineutrino. We shall turn now to the consideration of the phenomenological theory of neutrino mixing which takes into account all these possibilities.

II. PHENOMENOLOGICAL THEORY OF NEUTRINO MIXING

II.1. Mixing of Dirac Neutrinos (Dirac Mass Term)

If neutrino masses are different from zero this means that the neutrino mass term enters the Lagrangian of the system. Let us assume at the beginning that both left-handed current fields $\nu_{\ell L}$ and right-handed fields $\nu_{\ell R}$ enter the Lagrangian and that the neutrino mass term has the following form [2] *

$$\mathcal{L}^{D} = -\sum_{\ell',\ell} \bar{\nu}_{\ell' R} M_{\ell'\ell} \nu_{\ell L} + h.c. = -\bar{\nu}_R^{\,\prime} M \nu_L^{\,\prime} + h.c. \quad (7)$$

Here M is an arbitrary nxn complex matrix,

$$\nu_L^{\,\prime} = \begin{pmatrix} \nu_{eL} \\ \nu_{\mu L} \\ \nu_{\tau L} \\ \vdots \end{pmatrix}, \quad \nu_R^{\,\prime} = \begin{pmatrix} \nu_{eR} \\ \nu_{\mu R} \\ \nu_{\tau R} \\ \vdots \end{pmatrix}. \quad (8)$$

Let us reduce now the mass term (7) to the standard form. For this it is necessary to diagonalize the matrix M. An arbitrary complex matrix can be bringed to diagonal form with the help of the transformation **

$$M = V m U^{+}. \quad (9)$$

---

* We shall consider the general case of n types of neutrinos (n charged leptons $e, \mu, \tau, \ldots$).

** Below we shall give an easy proof of this statement. Let us consider the matrix MM^{+}. It is obviouse that this matrix is a hermitian one and all its eigenvalues are positive. We have

$$MM^{+} = V m^{2} V^{+}, \quad (10)$$

where $V^{+}V = 1$, and $m^{2}_{ik} = m^{2}_{k}\delta_{ik}$. Apparently

$$M = V m U^{+},$$

where $m_{ik} = \sqrt{m_k^2}\,\delta_{ik}$, and $U^{+} = m^{-1}V^{+}M$.
Further it is easy to proof that U is an unitary matrix. In fact from (10) it follows

$$U^{+}U = m^{-1}V^{+}MM^{+}Vm^{-1} = 1.$$

Here V and U are unitary nxn matrices, and $m_{ik} = m_k \delta_{ik}, m_k > 0$. Substituting (9) into (7), we have ,

$$\mathcal{L}^{D} = -\bar{\nu} m \nu = -\sum_{k=1}^{n} m_k \bar{\nu}_k \nu_k . \tag{11}$$

Here

$$\nu_L = U^+ \nu'_L , \tag{12}$$

$$\nu_R = V^+ \nu'_R , \quad \nu = \nu_L + \nu_R = \begin{pmatrix} \nu_1 \\ \nu_2 \\ \vdots \end{pmatrix} .$$

The expression (11) is the standard mass term. From (11) it follows that ν_k is the field of neutrino with mass m_k.

Apparently if in the Lagrangian of the system there is a term $\mathcal{L}^{D}$, electron, muon,... lepton numbers are not conserved. However, it is easy to see that the Lagrangian of the system is invariant under the global gauge transformations

$$\nu'_{\ell L}(x) = e^{i\Lambda} \nu_{\ell L}(x), \quad \nu'_{\ell R}(x) = e^{i\Lambda} \nu_{\ell R}(x), \quad \ell'(x) = e^{i\Lambda} \ell(x), \tag{13}$$

where Λ is a constant.
The invariance under the transformation (13) means that in the theory with mass term $\mathcal{L}^{D}$ the lepton number

$$L = L_e + L_\mu + L_\tau + \dots$$

identical for all types of neutrinos and leptons is conserved. From (12) and (13) it follows that the Lagrangian under consideration is invariant under the transformation

$$\nu'_k(x) = e^{i\Lambda} \nu_k(x), \quad \ell'(x) = e^{i\Lambda} \ell(x) \tag{14}$$

This means that neutrinos with definite masses (quanta of the fields ν_k) possess lepton number L and consequently are Dirac particles. The mass term $\mathcal{L}^{D}$ is called Dirac mass term.

Let us obtain now the relations between the current fields $\nu_{\ell L}$ and the fields of neutrinos with definite masses. For this purpose we multiply (12) by U, and we get

$$\nu'_L = U \nu_L .$$

In components we have

$$\nu_{\ell L} = \sum_{k=1}^{n} U_{\ell k} \nu_{kL} . \tag{15}$$

Thus if in the Lagrangian of the system there is a neutrino mass term (7), the n left-handed current fields $\nu_{\ell L}$ are orthogonal combinations of the n left-handed components of the fields of the massive neutrinos. Note that neutrino mixing (15) is analogous to quark mixing (3), and that the Dirac mass term (7) could arise in the standard model due to Higgs mechanism of mass generation.

II.2. Mixing of Majorana Neutrinos (Majorana Mass Term)

We assumed in II.1 that in the neutrino mass term both left-handed current fields $\nu_{\ell L}$ and right-handed sterile fields $\nu_{\ell R}$ enter. It is possible to construct a neutrino mass term only with the help of the left-handed current fields $\nu_{\ell L}$ [2]*. In fact it is easy to verify that the charge conjugated field $(\nu_{\ell L})^c = C\bar{\nu}_{\ell L}^{T}$ is a <u>right-handed</u> field (C is the matrix of charge conjugation, $C\gamma_\alpha^T C^{-1} = -\gamma_\alpha$). By means of $\nu_{\ell L}$ and $(\nu_{\ell L})^c$ the following mass term can be constructed

$$\mathcal{L}^M = -\frac{1}{2}\sum_{\ell',\ell} \overline{(\nu_{\ell' L})^c} M_{\ell'\ell}\, \nu_{\ell L} + h.c. = -\frac{1}{2}\overline{(\nu'_L)^c} M \nu'_L + h.c., \tag{16}$$

where M is a complex nxn matrix, and ν'_L is given by (8). It is easy to show that M must be a symmetric matrix.

Now let us reduce Lagrangian $\mathcal{L}^M$ to a diagonal form. For a complex <u>symmetrical</u> matrix we have

$$M = (U^+)^T m\, U^+ \tag{17}$$

where U is a unitary matrix and $m_{ik} = m_k \delta_{ik},\ m_k > 0$. By substituting (17) into (16) we get

$$\begin{aligned}\mathcal{L}^M &= -\frac{1}{2}\overline{(U^+\nu'_L)^c}\, m (U^+\nu'_L) + h.c. = \\ &= -\frac{1}{2}\bar{\chi} m \chi = -\frac{1}{2}\sum_{k=1}^{n} m_k \bar{\chi}_k \chi_k ,\end{aligned} \tag{18}$$

where

$$\chi = U^+\nu'_L + (U^+\nu'_L)^c = \begin{pmatrix}\chi_1\\ \chi_2\\ \vdots\end{pmatrix}. \tag{19}$$

---

* For the case of two types of neutrinos this was shown first in paper [3].

From (19) it follows that the field χ_k satisfies the Majorana condition

$$(\chi_k)^c = \chi_k \tag{20}$$

and consequently χ_k is a field of the truely neutral Majorana particles.

Now let us obtain the relations between the current fields $\nu_{\ell L}$ and the fields of neutrinos with definite masses. From (19) it follows

$$\nu'_L = U\chi_L .$$

In components we have

$$\nu_{\ell L} = \sum_{k=1}^{n} U_{\ell k}\chi_{kL} . \tag{21}$$

Thus, if the Lagrangian of the system contains a mass term $\mathcal{L}^M$, the left-handed fields $\nu_{\ell L}$ are orthogonal combinations of the left-handed components of the fields of the Majorana neutrinos with nonzero masses.

In conclusion we would like to make the following remark. From (16) it is obvious that the total Lagrangian of the system is not invariant under any global gauge transformations of the fields $\nu_{\ell L}$. This means that in the theory under consideration there are no conserved lepton numbers that would give us the possibility to distinguish neutrino and antineutrino. As a result after diagonalization of the mass term $\mathcal{L}^M$ we come to Majorana neutrinos with definite masses.

II.3. Mixing of Majorana Neutrinos (Dirac and Majorana Mass Term)

We shall consider here the most general neutrino mass term. In our disposal we have two columns ν'_L and $(\nu'_R)^c$ constructed from n left-handed components each and two columns $(\nu'_L)^c$ and ν'_R constructed from n right-handed components. Let us combine the left-handed and right-handed components in the columns

$$n_L = \begin{pmatrix} \nu'_L \\ (\nu'_R)^c \end{pmatrix} , \quad (n_L)^c = \begin{pmatrix} (\nu'_L)^c \\ \nu'_R \end{pmatrix} . \tag{22}$$

Apparently the most general neutrino mass term has the form [2]:

$$\mathcal{L}^{D,M} = -\frac{1}{2}\overline{(n_L)^c} M n_L + h.c. \tag{23}$$

Here M is a complex 2nx2n matrix. It is easy to prove that M is a symmetric matrix:

$$M^T = M.$$

Let us write down the matrix M in the form

$$M = \begin{pmatrix} M_L & M_D^T \\ M_D & M_R \end{pmatrix}, \tag{24}$$

where M_L and M_R are symmetric nxn matrices and M_D is an arbitrary complex matrix. From (22)-(24) we have

$$\mathcal{L}^{D,M} = -\bar{\nu}'_R M_D \nu'_L - \frac{1}{2}\overline{(\nu'_L)^c} M_L \nu'_L - \frac{1}{2}\bar{\nu}'_R M_R (\nu'_R)^c + h.c. \tag{25}$$

The first and the second terms in (25) are the Dirac ana Majorana mass terms, that we have considered in paragraphs II.1 and II.2. The third term in (25) has been constructed from the right-handed fields ν'_R and is a right-handed Majorana mass term. The mass term (25) is called Dirac and Majorana mass term.

The procedure of diagonalization of the mass term is similar to the one used for the Majorana mass term. For a symmetric matrix M we have

$$M = (U^+)^T m U^+, \tag{26}$$

where U is a unitary 2nx2n matrix, and $m_{ik} = m_k \delta_{ik}$, $m_k > 0$, i,k =1,2,... 2n. Substituting (26) into (23), we obtain

$$\mathcal{L}^{D,M} = -\frac{1}{2}\bar{\chi} m \chi = -\frac{1}{2}\sum_{k=1}^{2n} m_k \bar{\chi}_k \chi_k, \tag{27}$$

where

$$\chi = U^+ n_L + (U^+ n_L)^c = \begin{pmatrix} \chi_1 \\ \chi_2 \\ \vdots \\ \chi_{2n} \end{pmatrix}. \tag{28}$$

Apparently

$$\chi_k^c = \chi_k.$$

Thus χ_k is a field of a Majorana neutrino with mass m_k.

Further from (28) we have

$$n_L = U\chi_L .$$

From this relation it follows

$$\nu_{lL} = \sum_{k=1}^{2n} U_{lk}\chi_{kL} , \tag{29}$$

$$(\nu_{lR})^c = \sum U_{\bar{l}k}\chi_{kL} .$$

Thus if a Dirac and Majorana mass term enters the Lagrangian of the system, then the n current neutrino fields ν_{lL} are orthogonal combinations of 2n left-handed components of the fields of massive Majorana neutrinos. As it is seen from (29), in the case under consideration the n fields $(\nu_{lR})^c$ are orthogonal combinations of the same 2n left-handed components of the Majorana fields. From this important physical consequences follow. The fields $(\nu_{lR})^c$ do not enter the Lagrangian of the standard weak interaction. As a result left-handed antineutrinos $\bar{\nu}_{eL}$, $\bar{\nu}_{\mu L}$,..., that are quanta of the fields ν_{lR} , do not take part in the standard weak interaction (these neutrinos are called sterile ones). As we shall see later, when relations (29) hold, physical neutrinos (ν_e , ν_μ,...) may transform into sterile neutrinos (and vice versa).

II.4. The Simplest Case of Mixing of Two Fields of Majorana Neutrinos

It is very instructive to consider the Dirac and Majorana mass term in the simplest case of one neutrino type [2]. In this case we have

$$\mathcal{L}^{D,M} = -\frac{1}{2}\overline{(n_L)}^c M n_L + h.c., \tag{30}$$

where

$$n_L = \begin{pmatrix} \nu_L \\ (\nu_R)^c \end{pmatrix} , \quad M = \begin{pmatrix} m_L & m_D \\ m_D & m_R \end{pmatrix} .$$

We shall assume CP-invariance in the lepton sector. In this case $M=M^*$.

Let us carry out the standard procedure of diagonalization of the mass term. For this purpose we shall write down the matrix M in the form

$$M = \frac{1}{2} Sp M + M_0 , \tag{31}$$

where

$$M_0 = \begin{pmatrix} \frac{1}{2}(m_L - m_R) & m_D \\ m_D & -\frac{1}{2}(m_L - m_R) \end{pmatrix}.$$

The real symmetric matrix M_0 can be written in the form

$$M_0 = O m^\circ O^T, \tag{32}$$

where O is an orthogonal matrix, and $m^\circ_{ik} = m_k^\circ \delta_{ik}$, m_k° is k-th eigenvalue of the matrix M°. We have

$$m^\circ_{1,2} = \mp \sqrt{\frac{1}{4}(m_L - m_R)^2 + m_D^2} . \tag{33}$$

An orthogonal 2x2 matrix O has the form

$$O = \begin{pmatrix} \cos\theta & \sin\theta \\ -\sin\theta & \cos\theta \end{pmatrix}. \tag{34}$$

Let us substitute (33) and (34) into (32). For the angle θ we obtain the following relations

$$\sin 2\theta = \frac{m_D}{\sqrt{\frac{1}{4}(m_L - m_R)^2 + m_D^2}} , \tag{35}$$

$$\cos 2\theta = \frac{\frac{1}{2}(m_R - m_L)}{\sqrt{\frac{1}{4}(m_L - m_R)^2 + m_D^2}} .$$

Let us denote the eigenvalues of the matrix M by m_k'. We have

$$m'_{1,2} = \frac{1}{2}(m_L + m_R) \mp \sqrt{\frac{1}{4}(m_L - m_R)^2 + m_D^2} . \tag{36}$$

The quantities m_k' can be both positive and negative. Let us write down

$$m_k' = m_k \rho_k , \quad k = 1,2 \tag{37}$$

where $m_K = |m'_K|, \rho_K = \pm 1$. Futher we shall substitute (31), (32) and (36) into (30). For the mass term (30) we have

$$\mathcal{L}^{\mathcal{D},M} = -\frac{1}{2}\bar{\chi} m \chi = -\frac{1}{2}\sum_{K=1,2} m_K \bar{\chi}_K \chi_K, \qquad (38)$$

where

$$\chi = O^T n_L + \rho (O^T n_L)^c = \begin{pmatrix} \chi_1 \\ \chi_2 \end{pmatrix}. \qquad (39)$$

From (39) it is evident, that the field χ_K satisfies the following Majorana condition

$$\chi_K^c = \rho_K \chi_K. \qquad (40)$$

Thus in accordance with the general theory that we have discussed in paragraph II.3., the particles with definite masses are in the case we are considering here, two Majorana neutrinos. The fields ν_L and $(\nu_R)^c$ are connected with the left-handed components of the fields of Majorana neutrinos by the relations

$$\nu_L = \cos\theta \chi_{1L} + \sin\theta \chi_{2L}, \qquad (41)$$
$$(\nu_R)^c = -\sin\theta \chi_{1L} + \cos\theta \chi_{2L}.$$

Let us remark that the value $\eta_K = i\rho_K$ is the CP-parity of the Majorana neutrino with the mass m_K. The angle θ is the mixing angle. As it is seen from (34), $\theta = 0$ (no mixing) if $m_{\mathcal{D}}=0$. If $m_R = m_L$, the angle θ is equal to $\pi/4$ (maximal mixing). In the latter case neutrino mixing is analogous to the mixing of neutral kaons.

II.5. See-Saw Mechanism of Neutrino Mass Generation

From the data available it follows that neutrino masses (if any) are much smaller, than the masses of the other fundamental fermions (leptons and quarks). In this paragraph we shall discuss the popular mechanism, "explaining" the smallness of neutrino masses. Let us consider the Dirac and Majorana neutrino mass term in the case of one neutrino type (see II.4). We shall assume

$$m_L = 0, \quad m_R \gg m_{\mathcal{D}}. \qquad (42)$$

In this case from (35)-(37) we obtain

$$m_1 \simeq \frac{m_D^2}{m_R}, \; m_2 \simeq m_R, \; \theta \simeq \frac{m_D}{m_R}, \; \rho_1 = -1, \; \rho_2 = 1. \tag{43}$$

Further, from (41) we have

$$\nu_L = \chi_{1L} + \frac{m_D}{m_R}\chi_{2L},$$

$$(\nu_R)^c = -\frac{m_D}{m_R}\chi_{1L} + \chi_{2L}.$$

Thus, if conditions (42) are satisfied, particles with definite masses are light Majorana neutrino with mass $m_1 \simeq m_D^2/m_R$ and a heavy Majorana particle with mass $m_2 \simeq m_R$. The current left-handed field ν_L is a superposition of the fields χ_{1L} and χ_{2L}, the field χ_{2L} entering the mixture with coefficient $m_D/m_R \ll 1$.

This mechanism of generation of a small mass of current neutrino is called the see-saw mechanism [4]. Usually this mechanism is considered in the framework of Grand Unification models. In this case $m_R \simeq M \simeq 10^{14}$ GeV and the mass m_D is of an order of magnitude of the mass of a lepton or of a quark. For a small neutrino mass we have the see-saw formula

$$m_\nu \simeq \frac{m_q^2}{M} \quad \text{or} \quad m_\nu \simeq \frac{m_\ell^2}{M}. \tag{44}$$

III. PHYSICAL CONSEQUENCES OF THE HYPOTHESIS OF NEUTRINO MIXING

III.1. Oscillations of Neutrinos in Vacuum

Neutrino oscillations is one of the most important consequence of mixing of massive neutrinos (both Dirac and Majorana). Although neutrino oscillations were considered as early as at 1957-1958 (B.Pontecorvo [5]), the interest to this phenomena arose many years later, and was connected with the appearance of gauge theories. Many experiments searching for neutrino oscillations with different neutrino sources are going on at present (see the next paragraph).

We shall present here the phenomenological theory of neutrino oscillations in vacuum [2]. Let us consider a beam of neutrinos with momentum $\vec{p}$, that was formed as a result of a weak decays (of pions, kaons, β-decay of nuclei etc.). We shall assume $\vec{p}^{\,2} \gg m_k^2$ (m_k is the neutrino mass). If the current neutrino field $\nu_{\ell L}$ is a superposition of left-handed components of massive neutrinos (Dirac or Majorana)

$$\nu_{\ell L} = \sum_{k} U_{\ell k} \nu_{kL} \quad ,$$

for the state vector of the current neutrino we have

$$|\nu_{\ell}\rangle = \sum_{k} U^{*}_{\ell k} |\nu_{kL}\rangle \, . \tag{45}$$

Here $|\nu_{kL}\rangle$ is the state vector of a neutrino with momentum $\vec{p}$, mass m_k, and negative helicity.

In Eq.(45), we have assumed, that the state vectors of current neutrinos are <u>coherent</u> superpositions of state vectors of neutrinos with definite masses. This assumption is fulfilled if the differences of neutrino masses are sufficiently small.

Let us assume now, that in some weak decays neutrinos ν_ℓ with momentum $\vec{p}$ are produced. The state vector of the beam at the initial moment is given by expression (45). For the state vector after time t we have

$$|\nu_{\ell}\rangle_t = e^{-iH_0 t} |\nu_{\ell}\rangle \, , \tag{46}$$

where H_0 is the free Hamiltonian. The vectors $|\nu_{kL}\rangle$ are eigenvectors of H_0:

$$H_0 |\nu_{kL}\rangle = E_k |\nu_{kL}\rangle \, , \tag{47}$$

where $E_k = \sqrt{m_k^2 + \vec{p}^{\,2}}$.

Using (46) and (47), we obtain

$$|\nu_{\ell}\rangle_t = \sum_{k} |\nu_{kL}\rangle e^{-iE_k t} U^{*}_{\ell k} \, . \tag{48}$$

Thus, the state vector of the neutrino beam in the case of mixing will be a <u>sum</u> of stationary state vectors (and not a stationary state vector as in the case of absence of mixing). As a result, in a time t after the production of neutrinos, different vectors $|\nu_{kL}\rangle$ will acquire <u>different</u> phase factors. This is the principal quantum mechanical basis for the phenomenon of oscillations.

Neutrinos are detected via observation of weak processes. In such processes current neutrinos ν_e, ν_μ,... participate. Let us expand, consequently, the vector $|\nu_\ell\rangle_t$ over the total system, in which the vectors $|\nu_e\rangle$, $|\nu_\mu\rangle$... enter. First, we shall consider Dirac or Majorana mass terms. From (15) and (21) it is evident, that the state vectors of the current neutrinos form a total system. We have

$$|\nu_{kL}> = \sum_{l} |\nu_{l'}> U_{l'k}. \tag{49}$$

Substituting (49) into (48), we get

$$|\nu_l>_t = \sum_{l} |\nu_{l'}> a_{\nu_{l'};\nu_l}(t) \tag{50}$$

where

$$a_{\nu_{l'};\nu_l}(t) = \sum_{k=1}^{n} U_{l'k} e^{-iE_k t} U^*_{lk}. \tag{51}$$

Thus, if there is neutrino mixing, in time t after the production of neutrinos of a <u>definite</u> type, the state vector of the beam will be a superposition of the state vectors of neutrinos of <u>different</u> types. The amplitude of the transition $\nu_l \to \nu_{l'}$, after time t is given by the expression (51).

There exist an important physical difference between Dirac (Majorana) and Dirac and Majorana mass terms. In the latter case the state vectors of current neutrinos do not form a total system. The total system consists of n vectors of current ("active") neutrinos $|\nu_e>$, $|\nu_\mu>$, $|\nu_\tau>$, ... and n vectors of sterile left-handed antineutrinos $|\bar{\nu}_{eL}>$, $|\bar{\nu}_{\mu L}>$, $|\bar{\nu}_{\tau L}>$, ... From (29) we have

$$|\nu_{kL}> = \sum_{l'} |\nu_{l'}> U_{l'k} + \sum_{l'} |\bar{\nu}_{l'L}> U_{\bar{l}'k}. \tag{52}$$

By substituting (52) into (48), for the state vector we have

$$|\nu_l>_t = \sum_{l'} |\nu_{l'}> a_{\nu_{l'};\nu_l}(t) + \sum_{l'} |\bar{\nu}_{l'L}> a_{\bar{\nu}_{l'L};\nu_l}(t). \tag{53}$$

Here

$$\begin{aligned} a_{\nu_{l'};\nu_l}(t) &= \sum_{k=1}^{2n} U_{l'k} e^{-iE_k t} U^*_{lk}, \\ a_{\bar{\nu}_{l'L};\nu_l}(t) &= \sum_{k=1}^{2n} U_{\bar{l}'k} e^{-iE_k t} U^*_{lk} \end{aligned} \tag{54}$$

are, consequently, the transition amplitudes $\nu_l \to \nu_{l'}$ and $\nu_l \to \bar{\nu}_{l'L}$ for time t. So in the case of Dirac and Majorana mass term both transitions between active neutrinos $\nu_l \to \nu_{l'}$ and active and sterile ones $\nu_l \to \bar{\nu}_{l'L}$ become possible.

Now let us consider an antineutrino beam. With the help of (15) and (21), it is easy to demonstrate that in case of Dirac (Majorana) mass term the transition amplitude $\bar{\nu}_l \to \bar{\nu}_{l'}$ is given by

$$a_{\bar{\nu}_{l'};\bar{\nu}_l}(t) = \sum_{k=1}^{n} U_{l'k}^{*} e^{-iE_k t} U_{lk}. \tag{55}$$

For the Dirac and Majorana mass term the transition amplitudes $\bar{\nu}_l \to \bar{\nu}_{l'}$ and $\bar{\nu}_l \to \nu'_{l'R}$ are:

$$a_{\bar{\nu}_{l'};\bar{\nu}_l}(t) = \sum_{k=1}^{2n} U_{l'k}^{*} e^{-iE_k t} U_{lk}, \tag{56}$$

$$a_{\nu'_{l'R};\bar{\nu}_l}(t) = \sum_{k=1}^{2n} U_{\bar{l}'k}^{*} e^{-iE_k t} U_{lk}. \tag{57}$$

Now we shall obtain expressions for the transition probabilities. With the help of (51) and (54), for the probability of transition $\nu_l \to \nu_{l'}$ we have

$$P_{\nu_{l'};\nu_l}(R/p) = \delta_{l'l} - 2\,\mathrm{Re}\sum_{i<k} U_{l'i}^{*} U_{li} U_{l'k} U_{lk}^{*} \left(1 - e^{i\frac{1}{2}\Delta m_{ik}^{2}\frac{R}{p}}\right). \tag{58}$$

Here R is the distance between the detector and the source $(R \simeq t)$, $\Delta m_{ik}^2 = m_i^2 - m_k^2$, indexes i and k run from 1 to n for Dirac or Majorana mass terms and from 1 to 2n for Dirac and Majorana mass term. In order to obtain the probability of transition $\bar{\nu}_l \to \bar{\nu}_{l'}$, it is necessary to make the following replacement in (58): $U_{l'i}^{*} U_{li} U_{l'k} U_{lk}^{*} \to U_{l'i} U_{li}^{*} U_{l'k}^{*} U_{lk}$. Apparently, the expression for the probability of transition $\nu_l \to \bar{\nu}_{l'L}$ can be obtained from (58) by the substitution $U_{l'i}^{*} U_{l'k} \to U_{\bar{l}'i}^{*} U_{\bar{l}'k}$.

Thus, the probabilities of transition of the initial neutrinos into neutrinos of a different types are periodic functions of the variable R/p. From (58) it is seen also, that the probabilities $P_{\nu_{l'};\nu_l}$ depend on the mass square differences and on the elements of the mixing matrix.

It is easy to formulate the conditions, at which oscillations in neutrino beams are possible. If $\Delta m_{ik}^2 R/p \ll 1$ at all $i \neq k$, the second term in (58) vanishes and $P_{\nu_{l'};\nu_l} = \delta_{l'l}$ (oscillations are absent). In order that oscillations take place, it is necessary for the value p/R to be less or of the order of magnitude at least one of the neutrino mass square differences.

While comparing with experimental data, it is necessary to average the transition probabilities over neutrino spectrum, the region in which neutrinos are produced, the size of the detector etc. If $\Delta m^2_{ik} R/p \gg 1$ at all $i \neq k$, from (58) for averaged probabilities we have

$$\bar{P}_{\nu_{l'};\nu_l} = \bar{P}_{\bar{\nu}_{l'};\bar{\nu}_l} = \sum_k |U_{l'k}|^2 |U_{lk}|^2 . \tag{59}$$

Now let us obtain some general relations for the transition probabilities. Comparing (51) and (54) with (55) and (56), we have

$$P_{\nu_{l'};\nu_l} = P_{\bar{\nu}_l;\bar{\nu}_{l'}} \tag{60}$$

It is obvious, that this relation is a consequence of CPT-invariance. In the case of CP invariance in the lepton sector, we have

$$P_{\nu_{l'};\nu_l} = P_{\bar{\nu}_{l'};\bar{\nu}_l} . \tag{61}$$

Further, using the unitarity of the mixing matrix for the Dirac or Majorana mass terms we have

$$\sum_{l'} P_{\nu_{l'};\nu_l} = 1 \quad , \sum_{l'} P_{\bar{\nu}_{l'};\bar{\nu}_l} = 1 . \tag{62}$$

In the case of Dirac and Majorana mass term we get

$$\begin{aligned} &\sum_{l'} P_{\nu_{l'};\nu_l} + \sum P_{\bar{\nu}_{l'L};\nu_l} = 1 , \\ &\sum_{l'} P_{\bar{\nu}_{l'};\bar{\nu}_l} + \sum P_{\nu_{l'R};\bar{\nu}_l} = 1 . \end{aligned} \tag{63}$$

Finally, it is easy to show [2], that studying neutrino oscillations, it is impossible to distinguish the case of the Dirac mass term from the Majorana mass term.

Up to now we have not made any assumptions about neutrino masses. Supposing, neutrino masses satisfy the following conditions

$$m_1^2 \ll m_2^2 \ll \ldots \ll m_n^2 \tag{64}$$

These inequalities are true if neutrino masses satisfy the see-saw formula (44). Further,we shall assume that in the experiments with neutrinos from terrestrial sources (reactors, accelerators) only one mass squared difference could be relevant, i.e.

$$\begin{aligned} &(m_k^2 - m_1^2)\, R/p \ll 1 \quad , \quad k \neq n \\ &(m_n^2 - m_1^2)\, R/p \gtrsim 1 . \end{aligned} \tag{65}$$

In this case for the transition amplitude we have

$$\begin{aligned} a_{\nu_{l'};\nu_l} &= e^{-iE_1 t}\left(\sum U_{l'k}\left(e^{-i(E_k - E_1)t} - 1\right)U_{lk}^* + \delta_{l'l}\right) \simeq \\ &\simeq e^{-iE_1 t}\left[U_{l'n}U_{ln}^*\left(e^{-i\frac{1}{2}\Delta m^2 \frac{R}{p}} - 1\right) + \delta_{l'l}\right], \end{aligned} \tag{66}$$

where $\Delta m^2 = m_n^2 - m_1^2$.

From (66), it follows

$$P_{\nu_{l'};\nu_l} = P_{\bar\nu_{l'};\bar\nu_l} = 2|U_{l'n}|^2 |U_{ln}|^2 \left(1 - \cos\Delta m^2 \frac{R}{2p}\right), \quad l' \neq l \tag{67}$$

$$P_{\nu_l;\nu_l} = 1 - 2|U_{ln}|^2\left(1 - |U_{ln}|^2\right)\left(1 - \cos\Delta m^2 \frac{R}{2p}\right). \tag{68}$$

In the simplest case of the mixing of two neutrino types we have:

$$\begin{aligned} \nu_{lL} &= \cos\theta\, \nu_{1L} + \sin\theta\, \nu_{2L} \\ \nu_{l'L} &= -\sin\theta\, \nu_{1L} + \cos\theta\, \nu_{2L} . \end{aligned} \tag{69}$$

Here θ is the mixing angle, and l and $l'(l \neq l')$ equal to e and μ, μ and τ etc. The index k in (66) runs 1 and 2, and formula (67)-(68) are exact in this case. If we put $|U_{l2}|^2 = \sin^2\theta$ and $|U_{l'2}|^2 = \cos^2\theta$, from (67) and (68) for the transition probabilities we shall obtain the standard expressions

$$P_{\nu_{l'};\nu_l} = \frac{1}{2}\sin^2 2\theta\left(1 - \cos\Delta m^2 \frac{R}{p}\right), \tag{70}$$

$$P_{\nu_l;\nu_l} = 1 - \frac{1}{2}\sin^2 2\theta\left(1 - \cos\Delta m^2 \frac{R}{p}\right),$$

where $\Delta m^2 = m_2^2 - m_1^2$. As it is seen from (70), neutrino oscillations could be observed if the condition

$$\Delta m^2 \gtrsim \frac{p}{R} \tag{71}$$

fulfills.

Now let us return to the general case of n massive neutrinos. From (67) and (68) for the averaged probabilities we have

$$\bar{P}_{\nu_{l'};\nu_l} = 2|U_{l'n}|^2|U_{ln}|^2, \tag{72}$$

$$\bar{P}_{\nu_l;\nu_l} = 1 - 2|U_{ln}|^2(1 - |U_{ln}|^2). \tag{73}$$

The quantities $\bar{P}_{\nu_{l'};\nu_l}$ and $\bar{P}_{\nu_l;\nu_l}$ depend only on modulas squared of the elements of the n-th column. It allows us to get a number of relations between these quantities. In fact, from (73) we have

$$\frac{1}{2}\left(1 - \bar{P}_{\nu_e;\nu_e}\right) = |U_{en}|^2(1 - |U_{en}|^2). \tag{74}$$

On the other hand, from (72) it follows, that $|U_{en}|^2$ is connected with the transition probabilities by the relation

$$|U_{en}|^2 = \sqrt{\frac{\bar{P}_{\nu_e;\nu_\mu}\,\bar{P}_{\nu_\tau;\nu_e}}{2\,\bar{P}_{\nu_\tau;\nu_\mu}}} \tag{75}$$

It is obvious, that (74) and (75) is the relation between the quantities that could be measured experimentally (the probability $P_{\nu_e;\nu_e}$ can be measured in reactor experiments, $P_{\nu_e;\nu_\mu}$ can be measured in accelerator experiments etc.).

From (72) and (73) it also follows:

$$\frac{1}{2}(1-\bar{P}_{\nu_\mu;\nu_\mu}) = |U_{\mu n}|^2(1-|U_{\mu n}|^2), \tag{76}$$

$$|U_{\mu n}|^2 = \sqrt{\frac{\bar{P}_{\nu_e;\nu_\mu}\,\bar{P}_{\nu_\tau;\nu_\mu}}{2\,\bar{P}_{\nu_\tau;\nu_e}}} \tag{77}$$

Let us note, that relations (74)-(77) are valid for any number n of massive neutrinos.

The probability $\bar{P}_{\nu_\tau;\nu_e}$ that could be measured in experiments with high energy ν_e beams enters into the above given relations. Using (72) and (73), it is possible to obtain the relation that does not contain this probability. In fact, from (72) it follows

$$\bar{P}_{\nu_e;\nu_\mu} = 2|U_{en}|^2|U_{\mu n}|^2. \tag{78}$$

On the other hand using (73) we have

$$|U_{en}|^2 = \frac{1 \pm \sqrt{1-2(1-\bar{P}_{\nu_e;\nu_e})}}{2}, \tag{79}$$

$$|U_{\mu n}|^2 = \frac{1 \pm \sqrt{1-2(1-\bar{P}_{\nu_\mu;\nu_\mu})}}{2}. \tag{80}$$

Note, that in (79) and (80) both signs are possible. If the probability $\bar{P}_{\nu_e;\nu_\mu}$ coincides with one of the four values of the expression

$$\frac{1}{2}\left(1 \pm \sqrt{1-2(1-\bar{P}_{\nu_e;\nu_e})}\right)\left(1 \pm \sqrt{1-2(1-\bar{P}_{\nu_\mu;\nu_\mu})}\right)$$

this will be an argument in favour of the correctness of our assumption about neutrino mass hierarchy.

Finally, the last remark. From (73) it is easy to see that

$$\bar{P}_{\nu_e;\nu_e} \geq \frac{1}{2}, \quad \bar{P}_{\nu_\mu;\nu_\mu} \geq \frac{1}{2}. \tag{81}$$

The violation of this inequalities would mean that at least two differences of neutrino mass squared are relevant for neutrino oscillations.

III.2. Neutrinoless Double -Decay

In this paragraph we shall discuss briefly the neutrinoless double β -decay

$$(A,Z)\rightarrow(A,Z+2)+e^-+e^-. \tag{82}$$

It is obvious that the lepton charge L is not conserved in the process (82). This means that neutrinoless double β -decay is possible provided neutrinos with definite masses are Majorana particles. (Majorana or Dirac and Majorana mass terms).

In the standard theory the effective Hamiltonian of the β -decay has the following form

$$\mathcal{H}_I=\frac{G}{\sqrt{2}}2\left(\bar{e}_L\gamma_\alpha\nu_{eL}\right)j^\alpha+\quad h.c. \tag{83}$$

where G is the Fermi constant, j_α is the hadron current. In the theories considered here we have:

$$\nu_{eL}(x)=\sum_K U_{eK}\chi_{KL}(x). \tag{84}$$

Here $\chi_K(x)$ is the field of Majorana neutrino with the mass m_K. Apparently, neutrinoless double β-decay is the process in second order of perturbation theory in G. For the matrix element of the process we have

$$\langle p_1 p_2 f|S|i\rangle=\left(\frac{G}{\sqrt{2}}\right)^2\frac{1}{(2\pi)^3}\frac{1}{\sqrt{4p_{10}p_{20}}}\,2\Big[\int e^{-ip_1x_1-ip_2x_2}\cdot \bar{u}_L(p_1)\gamma_\alpha\overline{\nu_{eL}(x_1)\bar{\nu}^T_{eL}}(x_2)\gamma^T_\beta\bar{u}^T_L(p_2)\langle f|T(J^\alpha(x_1)J^\beta(x_2))|i\rangle dx_1dx_2 - (p_1\rightleftarrows p_2)\Big] \tag{85}$$

where p_1, and p_2 are the momenta of the final electrons, $|i\rangle$ and $|f\rangle$ are state vectors of initial and final nuclei. Further, from the Majorana condition it follows

$$\chi^T(x)=-\bar{\chi}(x)C \tag{86}$$

Using (86) we have

$$\overline{\nu_{eL}(x_1)\,\nu^{T}_{eL}}(x_2) = -\sum_k U^2_{ek}\frac{1+\gamma_5}{2}\overline{\chi_k(x_1)\,\bar{\chi}_k}(x_2)\frac{1+\gamma_5}{2}C =$$

$$= \sum_k U^2_{ek}\, m_k \frac{(-i)}{(2\pi)^4}\int e^{iq(x_1-x_2)}\frac{1}{q^2+m^2_k}\,dq\,\frac{1+\gamma_5}{2}C. \quad (87)$$

With the help of (85) and (87), we obtain the following expression for the matrix element of the process (82):

$$\langle p_1 p_2 f|S|i\rangle = 2\left(\frac{G}{\sqrt{2}}\right)^2\frac{1}{(2\pi)^3}\frac{1}{\sqrt{4p_{10}p_{20}}}\sum_k U^2_{ek} m_k \cdot \quad (88)$$

$$\cdot\, \bar{u}(p_1)\gamma_\alpha(1+\gamma_5)\gamma_\beta C\bar{u}^T(p_2)\int e^{-ip_1x_1-ip_2x_2}\langle f|T(J^\alpha(x_1)J^\beta(x_2))|i\rangle \cdot$$

$$\cdot\left(\frac{-i}{(2\pi)^4}\int e^{iq(x_1-x_2)}\frac{1}{q^2+m^2_k}\,dq\right)dx_1\,dx_2 .$$

We have considered the leptonic part of the matrix element of the process (82). The hadronic part of the matrix element can be calculated only approximately, and we shall not discuss it here (see review [6]).

We shall finish this paragraph by the following remarks:

1. From (88) it follows that the matrix element of the process (82) is equal to zero in the case of massless neutrinos. This corresponds to the well-known theorem of equivalence of the theories with massless Dirac and Majorana neutrinos.

2. For small neutrino masses ($\lesssim$ MeV), the term m^2_k in the neutrino propagator can be neglected. In this case the neutrino masses enter into the matrix element of the process in the form of the following factor

$$\langle m\rangle = \sum_k U^2_{ek} m_k . \quad (89)$$

3. If there is CP-invariance in the lepton sector, the mixing matrix satisfies the following condition

$$U_{ek} = U^*_{ek}\,\eta_k ,$$

where $\eta_k = -i\rho_k = \mp 1$, ρ_k being CP-parity of the Majorana neutrino with the mass m_k. In this case

$$\langle m \rangle = \sum_k |U_{ek}|^2 m_k \eta_k . \qquad (90)$$

Thus, the contributions to the quantity $\langle m \rangle$ of the neutrinos with different masses can compensate each other .

III.3. Tritium β-Decay and Neutrino Mass

As it is well-known, the precise measurement of the hard part of the spectrum of electrons from the decay

$$^3H \rightarrow {}^3He + e^- + \bar{\nu} \qquad (91)$$

could provide us with information about neutrino mass. The spectrum of electrons from the decay (91) is determined by the phase space and has the form

$$\frac{dN}{dE} = n(E) = C^2 p E p_\nu E_\nu F(E), \qquad (92)$$

where E, p (E_ν, p_ν) are the total energy and momentum of electrons (neutrinos), $F(E)$ is Fermi function, and C is a constant. Neglecting the recoil of ^{3}He, we have

$$E_\nu = Q - T \qquad (93)$$

where Q is the energy released (in the ^{3}H decay $Q \simeq 18{,}58$ keV) and T is the kinetic energy. In the simplest case of one massive neutrino, we have

$$p_\nu = \sqrt{(Q-T)^2 - m_\nu^2} \qquad (94)$$

From (92)-(94) we shall obtain the Kurie function

$$K(T) = \sqrt{\frac{n(E)}{pEF(E)}} = C\sqrt{(Q-T)\sqrt{(Q-T)^2 - m_\nu^2}}. \qquad (95)$$

For $m_\nu=0$ the function $K(T)$ is the straight line, crossing the abscissa in the point $T_{max} = Q$. For $m_\nu \neq 0$ the function $K(T)$ deviates from the straight line near the bound of the spectrum and crosses the abscissa in the point $T_{max} = Q - m_\nu$. Note that the measured electron spectrum is described by an integral which includes besides $n(E)$, spectrometer response function, the electron excitation function and other quantities. Thus, electron spectrum depends on five parameters ($m_\nu^2, Q, \ldots$).

IV. EXPERIMENTAL STATUS OF THE HYPOTHESIS OF NEUTRINO MIXING

Here we shall describe in brief the experimental situation with neutrino masses and mixing. The most sensitive method for testing neutrino mixing hypothesis is to search for neutrino oscillations [5]. This is due to the interference nature of oscillations. The phase difference of state vectors of neutrinos with definite masses is $(\Delta m^2/2p)R$ (Δm^2 is the difference of neutrino mass squared, p is the neutrino momentum, R is the distance between the detector and the source). Thus, the phase difference is proportional to the macroscopic parameter R and even for small Δm^2 can be of the order of magnitude or larger than 1.

The experiments in searching for neutrino oscillations are carried out on all neutrino facilities. There are two possible types of such experiments: inclusive experiments and exclusive ones. In inclusive experiments neutrinos of <u>the same type</u> ν_ℓ are detected at some distance from the source of ν_ℓ. In exclusive experiments neutrinos of a <u>different type</u> $\nu_{\ell'}$ ($\ell' \neq \ell$) are searched at some distance from the source of ν_ℓ.

Recently many experiments (both inclusive and exclusive) in search of neutrino oscillations have been carried out. No indications in favour of neutrino oscillations have been obtained up to now. The most sensitive experiments to the parameter Δm^2 are the reactor ones. As it is well-known, a reactor is a very intensive source of $\bar{\nu}_e$ of low energies (energy $\lesssim 10$ MeV, flux $\simeq 5\cdot 10^{20}$ $\bar{\nu}_e$/sec). Antineutrinos are detected through observation of the reaction $\bar{\nu}_e + p \to e^+ + n$ (experiments of the inclusive type). Up to now five groups have reported the results of their experiments: Gösgen (38m, 46m, 65m), Bugey (13,6 m, 18,3 m), Savanah River (18 m, 25 m), Rovno (18 m, 25 m), Kurchatov Institute (33 m, 92 m) (in brackets we indicated the distances between the detectors and the sources). These experiments considerably narrowed the allowed region of the possible values of the parameters Δm^2 and $\sin^2 2\theta$. (See, for example, [7]). Two groups in CERN and one group in Batavia have made inclusive experiments with ν_μ beam and two detectors at different distances from the sources. Many experiments of exclusive types have been carried out (searching for $\nu_\mu \to \nu_e$ ($\bar{\nu}_\mu \to \bar{\nu}_e$) and $\nu_\mu \to \nu_\tau$ transitions). The results of the analysis of these experiments can be found in [8]. The best limit of the parameter Δm^2 was obtained from the reactor experiments. From Gösgen's data it follows

$$\Delta m^2 < 0{,}019\ \mathrm{eV}^2 \quad (\sin^2 2\theta \simeq 1)$$
$$\sin^2 2\theta < 0{,}18 \quad (\Delta m^2 \gtrsim 5\ \mathrm{eV}^2)$$

Note, that at present new generation of reactor neutrino experiments are being prepared. The sensitivity of these experiments will be an order of magnitude higher than the previous one ($\Delta m^2 \simeq 5\cdot 10^{-3}$ eV^2).

In the recent years there was an essential progress in the experiments searching for neutrinoless double β-decay [9]. Some of the data obtained is presented in the Table III.

Table III

The data of the experiments searching for neutrinoless double β-decay

Group	Element	Lower limit of life-time
UCSB/LBL	^{76}Ge	$19 \cdot 10^{23}y$
ITEP/Yerevan	^{76}Ge	$13 \cdot 10^{23}y$
LBL	^{100}Mo	$1.3 \cdot 10^{22}y$
Baksan	^{136}Xe	$3.3 \cdot 10^{21}y$
UCI	^{82}Se	$1.1 \cdot 10^{22}y$

From Ge data it follows

$$|\langle m \rangle| < (1-2)\, eV.$$

Let us note, that new experiments searching for neutrinoless double β-decay are now carried out. An order of magnitude improvement is expected in sensitivity to the quantity $|\langle m \rangle|$.

Many experiments searching for nonzero neutrino mass effects by tritium method are performed all over the world. The latest results are presented in the Table IV.
As it is seen from the Table IV ITEP data have not been confirmed by the data of the other experiments.

In the conclusion we would like to stress that the problem of the neutrino mixing under discussion is very important for the future theory beyond the standard one. Up to now there were no unambiguous indications in favour of neutrino mass effects. This can only mean that the experiments performed up to now are not sensitive enough to see the effects. New progress in studying the problem of neutrino masses and mixing is expected in the nearest future.

Table IV

The data of the experiments measuring neutrino mass by tritium method

Group	Neutrino mass
ITEP	14 - 40 eV
Zurich	< 18 eV
Tokyo	< 11 eV
LANL	< 13,4 eV
Beijng	< 30 eV

REFERENCES

1 S.T.Petcov, Yad.Fiz. 25, (1977) 64 Sov.J.Nucl.Phys. 25, (1977) 340 .
S.M.Bilenky, S.T.Petcov and B.Pontecorvo, Phys.Lett. B67, (1977) 309.

2 See for example reviews: S.M.Bilenky and P.Pontecorvo, Phys.Rep. 41, (1978) 225;
F.Boehm and P.Vogel, Annu.Rev.Nucl.Part.Sci., 34, (1984) 125; S.M.Bilenky and S.T.Petcov, Rev.Mod.Phys. 59, N 3, Part 1, (1987), 671.

3 V.Gribov and B.Pontecorvo, Phys.Lett. B28, (1969) 463.

4 M.Gell-Mann, P.Ramond and R.Slansky, in Supergravity (Eds. P. van Nieuwenhizen and O.Freedman), North-Holland, Amsterdam (1979) p. 317; T.Yanagida in Proceeding of the Workshop on Unified Theories and Baryon Number in the Universe (Eds. A.Sawada and A.Sugamoto), KEK, Tsukuba (1979).

5 B.Pontecorvo, Zh.Eksp.Theor.Fiz. 33, (1957) 549; Zh.Eksp.Theor.Fiz. 34, (1958) 247.

6 M.Doi, T.Kotani, H.Nishiura and E.Takasugi, Osaka University Report, No. OS-GE 84-09 (1984). J.Vergados, Phys.Rep. 133, (1986) 1.

7 J.Bouchez in Proc. of the Intern. Conf. Neutrino - 88, 5 - 11 June, Boston (Medford), World Scientific, (1989) p. 28.

8 F.Vannucci in Proc. of the Intern. Conf. Neutrino-88, 5-11 June, Boston (Medford), World Scientific (1989) p.46.

9 F.T.Avignone in Proc. of the Intern. Conf. Neutrino-88, 5-11 June, Boston (Medford), World Scientific (1989) p.66.

ISBN 1-56072-027-1

THE ETHEREAL MATTER: NEUTRINOS AND HIGGS BOSONS

J.A. Grifols
Grup de Física Teòrica, Universitat Autònoma de Barcelona, 08193 Belleterra (Barcelona), Spain

INTRODUCTION

Neutrinos and Higgs bosons are essential ingredients of the Standard Model (SM). Both particles share a common feature: they interact very feebly with ordinary matter. However, while neutrinos do definitely exist, Higgs bosons have not been experimentally detected yet and might even not exist. In this respect, Higgs scalars would be literally very much like the aether of classical electromagnetism, *i.e.* a concept to be banned from the existing theoretical paradigm. Turning to the neutrino, almost all extensions of the SM contemplate the existence of right-handed (RH) neutrinos. But these particle degrees of freedom have not been found in the laboratory to this date and, again, they may or may not exist. If this latter possibility is true, RH-neutrinos will disappear from the Particle Physics Paradigm exactly as the aether did with the advent of Special Relativity.

These lectures are devoted to neutrino physics on the one hand and to Higgs bosons on the other. No attempt to be complete is intended. I simply picked a few issues in neutrino physics and I have expounded them. As to Higgs scalars this is even more so; I have selected a couple of miscellaneous topics and I have lectured on them. Besides the "ethereal" character of both particles remarked before, no obvious connection can be established between them and both sets of lectures are very much independent.

In the first part I discuss neutrino properties such as mass, charge, magnetic moment... . I also include a discussion of neutrino emission during gravitational collapse and present the implications of RH-neutrino (or photino) emission for extended theories like Supersymmetry, Left-Right Symmetry or E_6 models had these particles actually been emitted during the SN1987A collapse.

In the second part, which deals with Higgs physics, I examine lower mass limits for the SM Higgs boson based on nuclear scattering data. In a second section I consider extensions of the SM Higgs sector and discuss a theorem with phenomenological consequences. This part dedicated to the Higgs boson ends with a presentation of Higgs bosons in Supersymmetry.

1. NEUTRINO PHYSICS

1.1 NEUTRINO PROPERTIES

1a. Mass.

To this date there is no undisputed experimental evidence for a neutrino mass different from zero. However, there is no good theoretical reason for massless neutrinos either. Let us briefly review the status of the experimental limits on neutrino masses.

Tritium β-decay experiments

$$3H \rightarrow^3 He + e^- + \bar{\nu}_e$$

allow to place an upper limit to the electron neutrino mass of [1],

$$m_{\nu_e} \lesssim 18\ eV \tag{1}$$

The Q-value for the above transition is 18.6 keV and the study of the endpoint spectrum of the emitted electron provides the limit given above. This is simple kinematics: if neutrinos have a rest mass, electrons cannot carry all 18.6 keV away. Because the rates are very small at the endpoint, the experimental errors are large near the endpoint region and only an upper limit can be inferred. (An experimental group at ITEP claims that $m_{\nu_e} = 25 \pm {}^{6}_{5}\, eV$, but confirmation from other labs is lacking).

The mass of the muon neutrino has been bounded by considering two body pion decay data,

$$\pi \rightarrow \mu + \nu_\mu$$

The kinematics of this reaction is sensitive to a nonzero neutrino mass and thus a limit on it can be inferred. A group at the PSI obtains [2],

$$m_{\nu_\mu} \lesssim 250\ keV. \tag{2}$$

The ARGUS collaboration at DESY, by studying the 5 pion decay of the tau lepton

$$\tau \rightarrow 5\pi + \nu_\tau$$

gets the constraint

$$m_{\nu_\tau} \leq 35\ MeV \tag{3}$$

from a fit to the invariant mass spectrum about the tau mass [3].

Cosmology [4] allows also to put restrictions on the mass of neutrinos. Indeed, light stable neutrinos, if not massless, may contribute too much to the present energy density of the Universe. The number of photons per cubic cm at present is,

$$n_\gamma = \frac{2\xi(3)}{\pi^2} T^3 \tag{4}$$

For T=2.7 K this is about 400 photons/cm^3. If photons and neutrinos were at the same temperature then,

$$n_\nu = \frac{3}{4} n_\gamma \tag{5}$$

where the 3/4 factor comes from the different statistics obeyed by photons and neutrinos. However, their temperatures are not quite the same because between neutrino decoupling at about 10 MeV and photon decoupling at 1 eV there was electron-positron annihilation and the photon gas was reheated relative to the neutrino gas. The actual relation is

$$T_\nu = (4/11)^{1/3} T_\gamma \tag{6}$$

which just follows from entropy conservation at the e^+e^- annihilation stage (assumed to be reversible). Therefore, the actual number density of neutrinos is,

$$n_\nu = (3/4) \times (4/11) n_\gamma \tag{7}$$

which is about 100 neutrinos per cubic cm. Requiring the sum total of the rest mass associated to neutrinos not to exceed the critical energy density of the Universe one obtains,

$$\Sigma m_{\nu_i} \leq 65\ eV \tag{8}$$

1b.Electric charge.

Neutrinos are electrically neutral. This neutrality is experimentally sustained to a very high degree of accuracy. In fact,

$$|q_{\nu_e}| \leq 8 \times 10^{-20} e \tag{9}$$

where e is the electric charge of the positron. This limit follows from the assumption of charge conservation in neutron β-decay,

$$n \rightarrow p + e^- + \bar{\nu}_e$$

combined with the bound [5]

$$q_n \leq (-1.5 \pm 2.2) \times 10^{-20} e \tag{10}$$

obtained by studying the deflection of a neutron beam by a strong electric field and with the bound [6]

$$|q_p + q_e + q_n| \leq (0.8 \pm 0.8) \times 10^{-21} e \tag{11}$$

that was reached in an experiment with levitating steel balls. (This is a Millikan type experiment where gravity is counterbalanced by a magnetic force exerted on the balls and their stability is challenged by an electric field applied in the transverse direction.)

A direct bound on the electric charge of the neutrino can be derived from the recent supernova 1987 event [7]. Neutrinos emitted in a stellar collapse, when crossing the intergalactic and galactic magnetic fields would see - should they be charged - their trajectories lenghtened. As a consequence their bunching in time would spread beyond what was observed at the terrestrial detectors, should their charge exceed

$$|q_\nu| \leq 2 \times 10^{17} e. \tag{12}$$

This number is easily obtained from the curvature of the neutrino trajectory in a magnetic field B,

$$r = \frac{E(GeV)}{0.3B(Tesla)q_\nu} \tag{13}$$

and the implied time stretching,

$$\frac{\delta t}{t} = \frac{1}{12} \frac{(0.3Blq_\nu)^2}{E^2} \frac{\delta E}{E} \tag{14}$$

where l is the distance to the Large Magallanic Cloud (LMC) and E is the energy of the neutrino. Using as typical numbers,

$$\begin{array}{cc} \delta t \lesssim 5 \text{ sec} & l = 50\ Kpc \\ E \sim 15\ MeV & B \sim 1\mu G(\text{galactic field}) \\ \delta E/E \sim 1/2 & \end{array}$$

one gets the bound on q_ν given above (eq.(12)).

1c.Leptonic charge.

Baryon and lepton number play a very special role in Particle Physics. For all we know they are not associated to an interaction and hence there is no gauge particle linked to it. However, this notion might well be wrong. Not too long ago, the question concerning a fifth force was risen by a reanalysis of the Eotvos experiment. Even though the analysis turned out to be incorrect, it had the virtue of giving the whole subject a considerable momentum both theoretical and experimental.

The problem which concerns us now is the possible existence of a long-range force associated to lepton number [8].

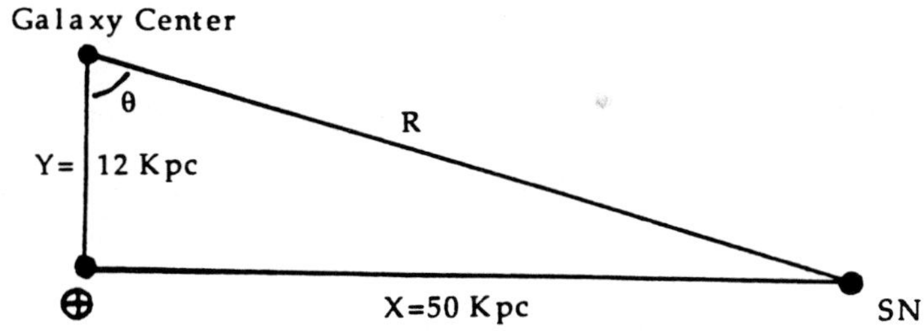

Fig. 1 Schematic geometry of the SN1987A event.

In february 1987 IMB and Kamioka detectors [9] were hit by electron antineutrinos from the Supernova explosion in the LMC.(The process is $\bar{\nu}_e + p \to e^+ + n.$). Antineutrinos carry opposite lepton quantum number to the electron and, should there be a long-range force associated to lepton number, these antineutrinos would experience an attraction towards the center of the Milky Way due to the huge lepton number of our galaxy (see fig.1).

In a Coulomb-like potential,

$$V = -K/r \tag{15}$$

the trajectory of a massless particle is

$$\frac{L^2 - K^2}{E}\frac{1}{r} = L\cos(\sqrt{1 - K^2/L^2}\ \theta + \theta_0) + K \tag{16}$$

where L is the angular momentum and E is the energy of the neutrino.

Perturbation theory can be used to compute the time delay δt induced by the curvature of the otherwise straight path,

$$\delta t = t - t_0 \tag{17}$$

where t is the time it takes for the $\bar{\nu}_e$ to travel from SN1987A to the earth and t_0 is the free-flight time.

The calculation gives,

$$\delta t = \frac{Y}{X}(R - Y - \theta X)\varepsilon^2 + ... \tag{18}$$

where $\varepsilon \equiv K/|\vec{p}|Y$.

Now,

$$K = \alpha_L \times \text{lepton number of the Milky Way} \simeq 6 \times 10^{68}\alpha_L \tag{19}$$

where α_L is the corresponding leptonic "fine structure constant" and we used a galaxy made mainly of hydrogen and with mass,

$$6 \times 10^{11} \text{solar masses}$$

The result for the time delay after using the actual values for the parameters is,

$$\delta t = 2.4 \times 10^{82} (\frac{\alpha_L}{|\vec{p}|(MeV)})^2 \tag{20}$$

Well understood SN dynamics tells us that thermal neutrinos are emitted over a period of 1 to 10 seconds. Hence, the inequality

$$10\,\text{sec} + |\delta t_{obs}| \geq |\delta t_{12}| \tag{21}$$

should hold for any two neutrino events. Here δt_{obs} is the time span betweeen two neutrino events as observed in the detector. δt_{12} is, on the other hand, given by

$$\delta t_{12} = 2.4 \times 10^{82} \alpha_L^2 (\frac{1}{|\vec{p_2}|^2} - \frac{1}{|\vec{p_1}|^2}) \tag{22}$$

Take, for instance, events # 3 and # 7 reported by the Kamioka group, whose parameters are explicitly

$$\begin{aligned} E_3 &= 7.5 \pm 2.0\ MeV \\ E_7 &= 35.4 \pm 8.0\ MeV \end{aligned} \tag{23}$$

then, the above inequality translates into [8]

$$\alpha_L \leq 1.7 \times 10^{-40} \tag{24}$$

We should mention that, although the stronger bound $\alpha_L \leq 10^{-47}$ is obtained from the analysis [10] of experiments where the equivalence of gravitational and inertial mass is tested, however, those experiments do not test the direct coupling of neutrinos to a fifth force.

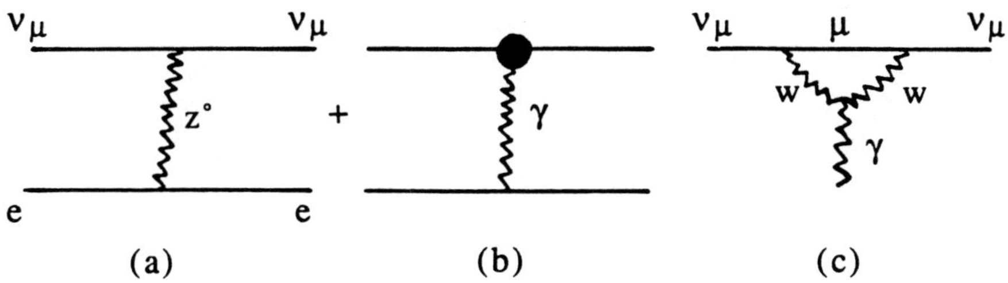

Fig. 2. (a, b) Amplitudes for neutrino-scattering (c) Contribution to the black blob.

1d.Charge radius.

Even though neutrinos are electrically neutral, they might have a "charge radius", *i.e.* the matrix element

$$< \nu | J^{\mu}_{e.m.} | \nu >= ieF_1(q^2)\bar{\nu}_L \gamma^{\mu} \nu_L \tag{25}$$

obeys

$$F_1(q^2 = 0) = 0 \tag{26}$$

but

$$F_1(q^2 \neq 0) \simeq q^2 \frac{\partial F_1}{\partial q^2} \equiv -\frac{1}{6} < r^2 > q^2 \neq 0 \tag{27}$$

an analysis of high energy $\nu_\mu(\bar{\nu}_\mu)$-electron scattering limits the charge radius of the (muon) neutrino [11],

$$< r^2 > \leq 10^{-32}\ cm^2 \tag{28}$$

This is obtained from considering the amplitudes in fig.2 and not permitting that the effects of the blob (charge radius) spoil the agreeement of the theoretical cross section with data.

The Standard Model (SM) gives radiative corrections to the tree level scattering cross section. In particular, diagram 2c contributes to the blob in fig. 2b. Numerically, these SM radiative corrections contribute about $10^{-34} - 10^{-33} cm^2$ to $< r^2 >$ which is still below the present sensitivity of experiments. We could, nonetheless, use present data to probe (sub) structure beyond the Fermi scale. Should compositeness of the building blocks of the SM, *i.e.* quarks/leptons and/or intermediate bosons, manifest itself at a higher energy scale, then there is no reason whatsoever for the 3- coupling in fig 2c to be exactly of the form it takes in the SM once this new energy threshold is crossed. Quite generally, the vertex can be parametrized as,

$$ieV^{\alpha\beta\gamma} = ieV_{SM}^{\alpha\beta\gamma} + ie\Delta V^{\alpha\beta\gamma} \tag{29}$$

where $\Delta V^{\alpha\beta\gamma}$ embodies the hypothesized departure from the gauge SM 3-vertex. It reads explicitly,

$$\begin{aligned}\Delta V^{\alpha\beta\gamma} =& g^{\alpha\beta}\frac{\lambda}{M_W^2}(p'\cdot q p^\gamma + p\cdot q p'^\gamma) + \frac{\lambda}{M_W^2}(p'^\gamma q^\alpha p^\beta - p^\gamma p'^\alpha q^\beta) - \\ & g^{\alpha\gamma}\left[(1-\chi)q^\beta + \frac{\lambda}{M_W^2}p'\cdot q p^\beta - \frac{\lambda}{M_W^2}p'\cdot p q^\beta\right] + \\ & g^{\beta\gamma}\left[(1-\chi)q^\alpha + \frac{\lambda}{M_W^2}p\cdot q p'^\alpha - \frac{\lambda}{M_W^2}p'\cdot p q^\alpha\right]\end{aligned} \tag{30}$$

which describes a $W^\pm$ with anomalous magnetic dipole

$$\mu_W = \frac{e}{2M_W}(1+\chi-\lambda) \tag{31}$$

and anomalous electric quadrupole

$$Q_W = -\frac{e}{M_W^2}(\chi+\lambda) \tag{32}$$

moments. The SM is recovered for $\chi = 1$ and $\lambda = 0$ for which evidently

$$\Delta V^{\alpha\beta\gamma} = 0 \tag{33}$$

We may evaluate the contribution to $< r^2 >$ coming from the W-loop diagram with the anomalous γWW - vertex $\Delta V^{\alpha\beta\gamma}$. One gets [11],

$$< r^2 >_{anom} = \frac{3}{32}\frac{\alpha}{\pi\sin^2\theta_W}\frac{1}{M_W^4}\Lambda^2(1-\chi) + \text{non leading terms in } \Lambda \tag{34}$$

The neutrino cross sections are [12],

$$\begin{aligned}\sigma(\nu_\mu e \to \nu_\mu e) =& \frac{2G_F^2 m_e E}{3\pi}\Big[\frac{3}{4} - 3s_W^2 + 4s_W^4 + < r^2 > M_W^2 s_W^2 \\ & (2 - \frac{16}{3}s_W^2) + \frac{16}{9} < r^2 >^2 M_W^4 s_W^4\Big]\end{aligned} \tag{35}$$

$$\begin{aligned}\sigma(\bar\nu_\mu e \to \bar\nu_\mu e) =& \frac{2G_F^2 m_e E}{3\pi}\Big[\frac{1}{4} - s_W^2 + 4s_W^4 + \frac{2}{3} < r^2 > M_W^2 \\ & (1 - 8s_W^2) + \frac{16}{9} < r^2 >^2 M_W^4 s_W^4\Big]\end{aligned} \tag{36}$$

The experimental constraints are,

$$1.5 \times 10^{-42} cm^2/GeV \le \frac{\sigma^{\nu}}{E} \le 2.3 \times 10^{-42} cm^2/GeV \tag{37}$$

and

$$1.2 \times 10^{-42} cm^2/GeV \le \frac{\sigma^{\bar{\nu}}}{E} \le 1.8 \times 10^{-42} cm^2/GeV \tag{38}$$

If we introduce $< r^2 >$ given in eq.(34) into eqs.(35) and (36) we obtain from the above constraints,

$$|1 - \chi| \le 0.3 \tag{39}$$

for an assumed compositeness scale Λ =1 TeV. (Recall that deviations from the SM are expected to take place around the 1 TeV energy scale.)

We see from the restriction in eq.(39) that the precision of present neutrino scattering experiments still allows much room for deviations from the SM structure to take place. Experiments in the future will test more tightly the correctness of the SM.

1e.Magnetic moment.

A massive Dirac neutrino admits a magnetic moment,

$$< \nu|J^{\mu}|\nu >_{m.d.m.} = \frac{\mu e}{2m} \sigma^{\mu\nu} q_{\nu} \bar{\nu}_L \nu_R + h.c. \tag{40}$$

Only upper limits for μ are known. The laboratory bounds are obtained from neutrino electron scattering data [13],

$$\begin{aligned} \mu_{\nu_\mu} &\le 10^{-9} \mu_B \\ \mu_{\nu_e} &\le 1.3 \times 10^{-9} \mu_B \end{aligned} \tag{41}$$

with $\mu_B = \frac{e}{2m}$ (Bohr magneton) by considering the effect on the cross sections of the neutrino magnetic coupling to the electromagnetic current. Better limits can be derived from astrophysics/cosmology [14]. In fact, nucleosynthesis of helium 4 tolerates at most 1 flavor with $\mu_\nu \ge 5 \times 10^{-11}$. The reasoning here is that a too large magnetic moment would imply that the additional right handed neutrino degrees of freedom were kept in equilibrium with the other species for too long a time as to not affect the primordial helium abundance in much the same way as extra light neutrinos would influence the expansion rate of the universe modifying the p/n ratio and as a consequence changing the helium abundance beyond observation. Also, SN1987A imposes the bound

$$\mu_\nu \le 2 \times 10^{-12} \mu_B \tag{42}$$

since otherwise RH neutrinos produced by magnetic transitions $\nu_L \rightarrow \nu_R$ in the core and escaping freely afterwards would carry off a too large a fraction of the energy generated in the stellar collapse. These RH-neutrinos would remain undetected [15].

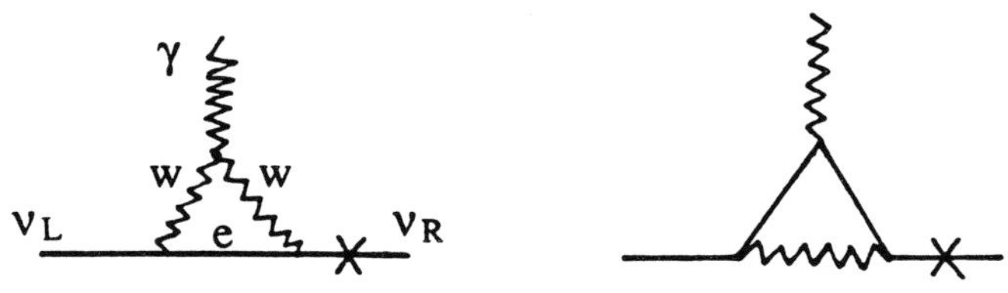

Fig. 3 Neutrino m.d.m. graph in SM.

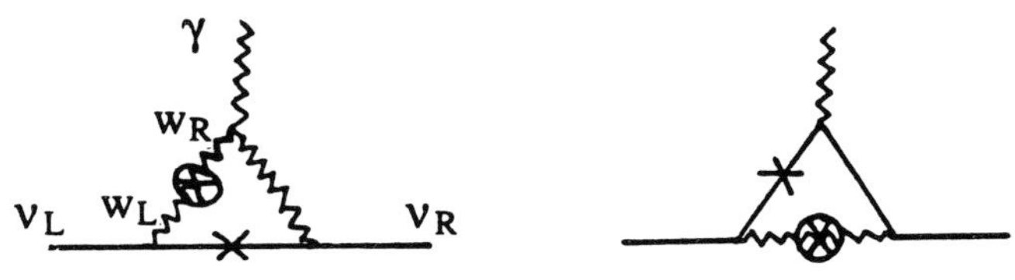

Fig. 4. Neutrino m.d.m. graph in LR.

The existence of a nonzero ν-magnetic moment is a physically relevant issue. For example, it has been advocated as a solution to the solar neutrino puzzle [16]. In this instance one would need $\mu \simeq 10^{-10}\mu_B$. The problem, however, is mainly theoretical. It is very difficult to produce a substantial magnetic moment (*e.g.* $10^{-10}\mu_B$) in current theoretical models [17]. Let us review this question in some of the most popular models.

Standard Model (with RH neutrino singlet). The diagrams which contribute to μ_ν are given in fig. 3. The necessary helicity flip is provided by the external neutrino and therefore the result is proportional to the neutrino mass [18],

$$\mu_\nu = \frac{3G_F m_e m_\nu}{4\sqrt{2}\pi^2} = 3.2 \times 10^{-19} \left(\frac{m_\nu}{1\ eV}\right) \tag{43}$$

which is obviously extremely small.

Left-Right Model. The gauge group is $SU(2)_L \times SU(2)_R \times U(1)$ and a detailed description of the model can be found in ref.[19]. The magnetic moment arises from the diagrams depicted in fig. 4 which involve both L-R charged weak boson mixing and a mass insertion in the internal fermion line. The result is,

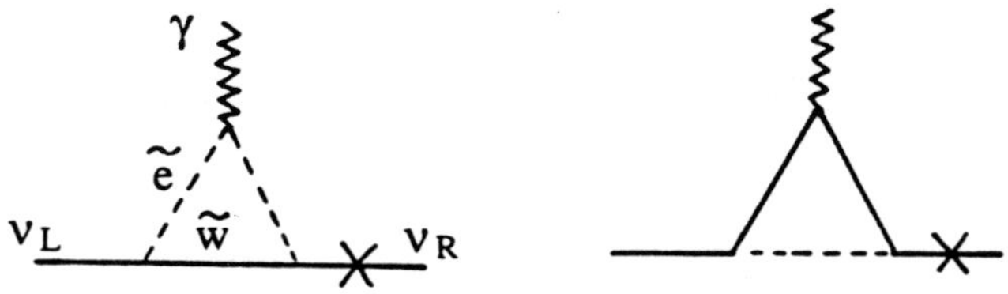

Fig. 5 Neutrino m.d.m. graph in SUSY.

$$\mu_{\nu_l} = \frac{G_F m_e m_l}{\sqrt{2}\pi^2} \sin 2\xi \tag{44}$$

and from the existing limits on LR mixing ($\sin 2\xi \leq 0.082$) we get

$$\mu_{\nu_l} \leq 1.7 \times 10^{-14} \mu_B \tag{45}$$

which is still too low.

Supersymmetric Model. Very complete reviews on SUSY can be found in the literature [20]. Here I only need to remind of the fact that many new particles are predicted and that some of them will contribute to the magnetic moment of the neutrino. In fig. 5 we give the relevant diagrams [21]. The chirality flip occurs in the external neutrino legs and in the Yukawa couplings. The calculation gives,

$$\mu_\nu = \frac{G_F m_e m_\nu}{\sqrt{2}\pi^2} \left(\frac{M_W}{M_{\tilde{W}}}\right)^2 \times f\left(\frac{M_{\tilde{e}}^2}{M_{\tilde{W}}^2}\right) \tag{46}$$

where the notation is obvious and $f(x) = (1-x)^{-2}(1 - x + x ln x)$. From the present limits on SUSY partners we have,

$$\mu_\nu \leq 10^{-19} \frac{m_\nu}{1\ eV} \tag{47}$$

which is at most of the same order of magnitude of the SM magnetic moment.

E_6 Models. In so called superstring inspired models each generation of matter particles is contained in a 27 representation of E_6. (We refer the reader to a very recent report on E_6 models for the details [22]).These 27 degrees of freedom decompose in the 15 degrees of freedom of the SM plus

$$\nu^c(1,1);\ H(1,2);\ \bar{H}(1,2);\ D(3,1);\ D^c(\bar{3},1);\ N(1,1) \tag{48}$$

where in parenthesis we give the $SU(3)_C$ and the $SU(2)_L$ quantum numbers. Worth noticing - as far as our problem is concerned - are the color triplet quark isosinglets

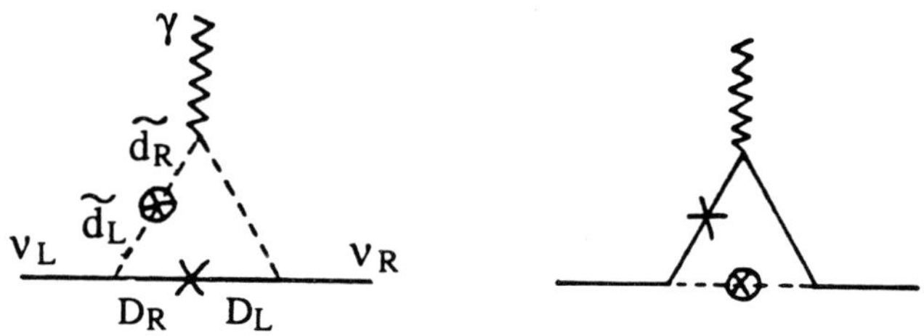

Fig. 6.Neutrino m.d.m. graph in E_6.

$D(3,1)$ and $D^c(\bar{3},1)$ and the right-handed neutrino $\nu^c(1,1)$. The new quarks participate in interactions with (s)leptons and (s)quarks of the type,

$$\lambda D^c L q + \lambda' D e^c u^c + \lambda'' D \nu^c d^c \tag{49}$$

where $L = \binom{\nu}{e}_L$ and $Q = \binom{u}{d}$ and the notation is very compact since the symbols above stand for supermultiplets. The Yukawa couplings λ and λ'' induce a magnetic dipole moment for the neutrino through the loops in fig. 6. The contribution from these diagrams to μ is [23],

$$\mu_\nu = \frac{\lambda\lambda'}{8\pi^2} \sin 2\chi_d \sum_{i=1,2} (-1)^i \frac{m_e m_D}{m_i^2 - m_D^2} \left[1 - \frac{m_i^2}{m_i^2 - m_D^2} log \frac{m_i^2}{m_D^2} \right] \mu_B \tag{50}$$

χ_d reflects the L-R mixing among squark eigenstates and m_D the fermion mass insertion, both necessary to provide the helicity flip characteristic of a magnetic transition. Phenomenological constraints on the relevant parameters imply,

$$\mu_\nu \leq 3 \times 10^{-13} \mu_B \tag{51}$$

still much too low to be relevant for the solar puzzle. Interfamily mixing, which we completely ignored, could eventually help to increase μ_ν .

1.2 Supernova 1987A : Neutrino emission

2a.Gravitational collapse.

In a SN event about 10^{53} erg gravitational binding energy are released (mostly) in form of neutrinos [24]. When nuclear fuels are exhausted the Fe-Ni core can no longer stand the gravitational pressure: electrons are captured (this is the neutronization process $e^- + p \to n + \nu$) and as a result the fermion pressure suddenly decreases and triggers the collapse of the star. This first neutronization burst carries at most 10%

of the total energy. The bulk of the 10^{53} erg is released as thermal neutrinos which diffuse their way out to the "neutrinosphere" from where they are thermally radiated. (Because of the large matter densities reached during the collapse the neutrinos follow a random-walk path scattering many times before leaving the star). The rate of energy depletion is $10^{52} - 10^{53}$ erg/s since the diffusion process lasts over a period of 1 to 10 seconds (a time scale much longer than the hydrodynamic collapse time). As inferred from the data from IMB and Kamiokande, the energy emitted in form of neutrinos by SN 1987A was $\geq 2 \times 10^{53}$ erg. The total gravitational energy is $\leq 4 \times 10^{53}$ erg in models of stellar collapse. As a consequence the difference of 2×10^{53} erg can be used to bind properties of Particle Physics (e.g. masses of exotic particles, the magnetic moment of the neutrino, axion couplings, etc).

2b.The physics of neutrino emission.

Neutrinos produced in the core of the star are trapped in the dense material ($8 \times 10^{14}\ g/cm^3$) and thus scatter very many times as they random-walk to the neutrinosphere. Neutrino cross-sections on matter depend quadratically on the neutrino energy [25] and hence the opacity of matter to neutrinos is also a function of the neutrino energy. The relevant parameter here is the mean-free-path of neutrinos in matter λ_ν and diffusion takes place when λ_ν is much smaller than the dimensions of the core. To define a neutrinosphere we need an energy averaged mean-free-path. Assuming local thermodynamic equilibrium one can write down a diffusion equation in the so called Rosseland approximation [26],

$$F(r) = -\frac{7\pi^2}{180}\bar{\lambda}_\nu T^3 \frac{\partial T}{\partial r} \tag{52}$$

where F(r) is the neutrino energy flux for a spherically symmetric star and where,

$$\bar{\lambda}_\nu \equiv \frac{\int dE \frac{1}{\sigma n} E^4 \frac{eE/T}{(1+e^{E/T})^2}}{\int dE E^4 \frac{e^{E/T}}{(1+e^{E/T})^2}} \tag{53}$$

σ is the microscopic neutrino cross-section, and n is the density of nucleon targets. Given a temperature profile T(r), a density of targets n(r), and the relevant cross section σ, one is able to compute the Rosseland mean-free-path as a function of the radial distance. We adopt for our calculations the profile introduced by Turner [27],

$$T(r) = \begin{cases} 15\ MeV\ c_1 \rho_{14}^{2/3} & \rho_{14} \geq 2 \\ 20\ MeV\ c_1 \rho_{14}^{1/3} & \rho_{14} \leq 2 \end{cases} \tag{54}$$

with $\rho_{14} = (r/1.5 \times 10^6 cm)^{-m}$ which is the core density in units of $10^{14}\ g/cm^3$, m=3-7, and c_1 is a parameter that takes into account uncertainties in the core adiabat.

The cross-section of ν_μ on nucleons is,

$$\sigma(\nu M) = \frac{1}{4}\sigma_0 (E/m_e)^2 \tag{55}$$

with $\sigma_0 \equiv 1.76\times10^{-44} cm^2$. (We restrict the discussion to ν_μ for simplicity; detailed calculations of neutrino transport support an approximate equipartition among neutrino species as far as energy flux is concerned [28].)

With all these inputs the Rosseland mean is [29],

$$\bar{\lambda}_\nu(r) = \frac{170}{c_1^2}\left(\frac{r}{1.5\times10^6 cm}\right)^{\frac{5m}{3}} cm \tag{56}$$

Physically the neutrinosphere corresponds to a distance R_ν for which $\bar{\lambda}_\nu$ is itself on the order of R_ν. So, we define

$$\bar{\lambda}_\nu(r = R_\nu) = \xi R_\nu \tag{57}$$

where ξ is a fudge factor of order 1. The solution to this equation is

$$\frac{R_\nu}{1.5\times10^6 cm} = (8.9\times10^3 c_1^2 \xi)^{3/5m-3} \tag{58}$$

The neutrino luminosity Q_ν (energy/time) is given by the Stefan-Boltzmann law for blackbody radiation from the surface R_ν at temperature $T_\nu = T(R_\nu)$, *i.e.* $Q_\nu = \frac{7\pi^3}{240}T_\nu^4 R_\nu^2$. Using T(r=$R_\nu$) and R_ν given above (eq.(54) and (58)) we get [29],

$$Q_\nu = \frac{2\times10^{54} c^{12m/5m-3}}{(8.9\times10^3\xi)^{4m-6/5m-3}} erg/s \tag{59}$$

The total neutrino luminosity is then,

$$Q_{TOT} \approx 6\times Q_\nu \tag{60}$$

where we took into account 3 generations of neutrinos and their antineutrinos. This Q_{TOT} is to be confronted with observation and should verify,

$$3\times10^{52} erg/s \le Q_{TOT} \le 3\times10^{53} erg/s \tag{61}$$

The curve in fig. 7 displays the set of points in the $T_\nu - R_\nu$ plane compatible with this last condition and where ξ departs from 1 by no more than an order of magnitude. The allowed neutrinosphere radius and neutrinosphere temperature are fully consistent with the conventional picture of SN 1987A. We took the isentropic solution $c_1 = 1$, but other reasonable values of c_1 are just fine. (Actually, the precise numerical values of ξ and c_1 are immaterial for our purposes). Also we took m=7 for this gives the best description of data.

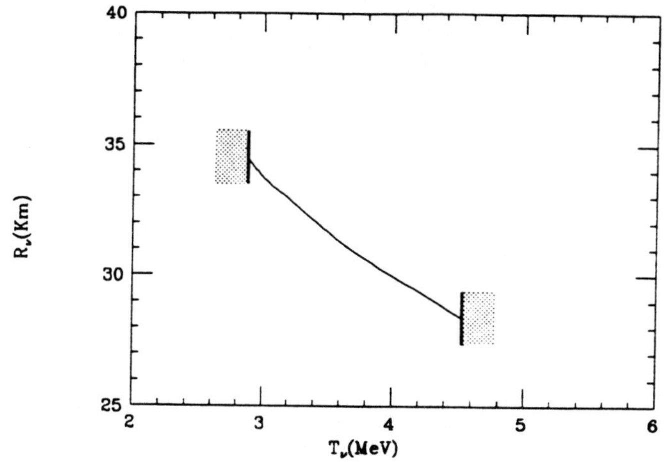

Fig. 7. Radius and temperature of the neutrinosphere.

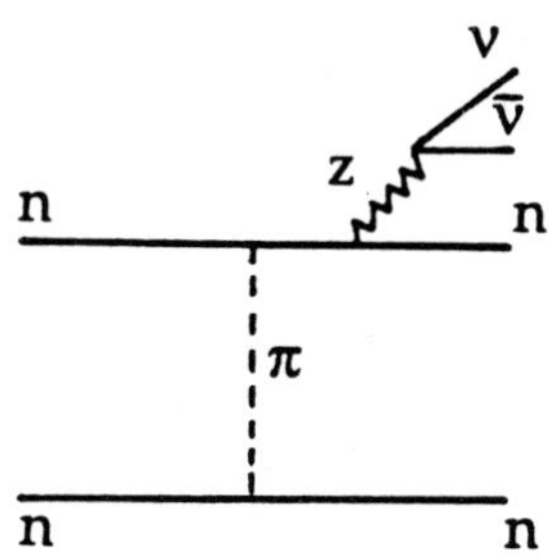

Fig. 8 Neutron-neutron bremsstrahlung.

2c.Production mechanisms.

The main microscopic source of neutrinos in a stellar collapse is nucleon-nucleon neutrino bremsstrahlung (e^+e^- annihilation is far less important) [30]. Let us consider for illustration neutron-neutron bremsstrahlung : $n + n \rightarrow n + n + \nu + \bar{\nu}$. (Should neutrinos stream out freely from the star - in actual fact they do not - then the following calculation would be a realistic luminosity calculation; here, it is only meant to illustrate the techniques used.)

Fig. 8 shows a representative diagram contributing to the bremsstrahlung process. The neutrino energy production rate (energy/volume/time) is given by

$$q_{nn} = \int \Pi_i^4 \frac{d^3 p_i}{(2\pi)^3 2Ei} \int \Pi_j^2 \frac{d^3 q_j}{(2\pi)^3 2Ej} \frac{1}{e^{(E_1-\mu)/T}+1} \frac{1}{e^{(E_2-\mu)/T}+1} \times$$
$$|M|^2 (2\pi)^4 \delta^4 (P_i - P_f)(E_1 + E_2) \qquad (62)$$

where $|M|^2$ is the matrix element squared for $nn \to nn\nu\bar{\nu}$, the kinematical notation is

$$q_j = (E_j, \vec{q_j}) \qquad P_i = p_i + p_2$$
$$p_i = (E_i, \vec{p_i}) \qquad P_f = q_1 + q_2 + p_3 + p_4$$

and μ is the chemical potential.

The high temperature in the core ($T_C \simeq 60\ MeV$) implies that neutrons are non degenerate (ND). Since, on the other hand, $T_C << m_n$ the nucleons are non relativistic (NR). In the ND and NR limit the statistical factors are just

$$e^{(\mu-m)/T} e^{-p_i^2/2mT} \qquad (63)$$

and the chemical potential is related to the density of neutrons through

$$e^{\mu-m/T} = \frac{\sqrt{2}\pi^{3/2}}{m^{5/2}} X_n \rho T^{-3/2} \qquad (64)$$

where X_n is the mass fraction of neutrons in the stellar core. In the one pion exchange approximation (OPE) the $nn \to nn\nu\bar{\nu}$ matrix element is [30]

$$|M|^2 = 256\ G_F^2 g_A^2 f^4 \left(\frac{m_n}{m\pi}\right)^4 \frac{E_1 E_2}{(E_1 + E_2)^2} \qquad (65)$$

with g_A the axial weak coupling and f being the pion-nucleon coupling constant. Then q_{nn} is [29],

$$q_{nn} = \frac{g^4 g_A^2}{M_W^4} X_n^2 \dot{\epsilon}_{NN} \qquad (66)$$

where

$$\dot{\epsilon}_{NN} \equiv \frac{16}{1155} \pi^{11/2} g^2 f^4 \left(\frac{m_n}{m_\pi}\right)^4 \frac{T^{11/2}}{m_n^{9/2}} \qquad (67)$$

Analogously, neutron-proton bremsstrahlung contributes to the neutrino production rate

$$q_{np} = 12\ \frac{g^4 g_A^2}{M_W^4} X_n X_p \dot{\epsilon}_{NN} \qquad (68)$$

A similar calculation can be done for the process e^+e^- [31] with the result

$$q_{ee} = \frac{g^4}{M_W^4}\dot{\epsilon}_{ee} \tag{69}$$

with

$$\dot{\epsilon}_{ee} = \frac{1}{480\pi^5}\mu^5 T^4 e^{-\mu/T} b(\mu/T) \tag{70}$$

where

$$b(y) \equiv \frac{5}{6} y^5 e^y (F_4^- F_5^+ + F_4^+ F_5^-); \quad F^{\pm}_{n(y)} = \int dx \frac{x^{n-1}}{1+e^{x\pm y}} \tag{71}$$

As stressed already before, these q's are not realistic neutrino luminosities because diffusion has been ignored. If neutrinos were not trapped in the stellar core, they would free-stream out of the star and the luminosity would indeed be

$$\frac{4}{3}\pi R^3 q\ .$$

But, in actual fact, they do not and luminosities are to be obtained from thermal radiation arguments. In general, however, when emission of light particles occurs in a SN explosion, two regimes must be considered: diffusion and free-streaming.

Diffusion. For interactions effectively on the order of or stronger than the weak interactions (characterized by G_F) the (hypothesized) light particles are trapped in the core exactly as ordinary neutrinos are.

Free-streaming. For interactions much weaker than the weak interactions the emitted particles stream out freely from the star. We shall now apply the techniques described so far to find bounds on masses of new particles predicted in current Extended Particle Physics Scenarios.

1.3 CONSEQUENCES OF ν_R EMISSION IN SN 1987A

3a.Left-Right Symmetry.

The model contains right-handed (RH) neutrinos. If light (cosmology requires light stable RH-neutrinos to be lighter than 2 KeV [32]), these RH-neutrinos should be emitted in a SN explosion. The neutral current interactions are described by the effective low energy hamiltonian [33],

$$H_N = \frac{1}{2}\Sigma_{i,j}\bar{f}\gamma^\mu n_i f \bar{f}'\gamma_\mu n_j f' M_{ij}^{-2} \tag{72}$$

with

$$n_{L,R} = g\left[T_3\frac{1}{2}(1\pm\gamma_5) - \frac{e^2}{g^2}Q\right]$$

and where $M_{i,j}^{-2}$ is the inverse mass matrix squared for the neutral gauge bosons except the one corresponding to the U(1) generator. From eq.(72) the effective $\bar{N}N\bar{\nu}_R\nu_R$ current-current interaction is,

$$\Sigma_{n,p}\frac{g^2}{2M_{W'}^2}v_N\bar{N}\gamma^\mu N\bar{\nu}_R\gamma_\mu\nu_R \tag{73}$$

with $v_p = \frac{1}{2} - 2\sin^2\theta_W$ and $v_N = -\frac{1}{2}$.

Diffusion will take place for a sufficiently low mass Z'. The opacity of matter to ν_R is governed by,

$$\begin{aligned}\sigma(\nu_R N \rightarrow \nu_R N) &\equiv \sigma(\nu_R p \rightarrow \nu_R p)X_p + \sigma(\nu_R n \rightarrow \nu_R n)X_n = \\ &= \frac{g^4}{2\pi}\frac{\cos^4\theta_W}{\cos^2 2\theta_W}E^2\left[\frac{1}{M_{Z'}}\right](X_n v_n^2 + X_p v_p^2)\end{aligned} \tag{74}$$

which allows us to calculate the Rosseland mean,

$$\bar{\lambda}_{\nu_R}(r) = \frac{6\times 10^{-7}}{c_1^2(\chi_n + 0.01\chi_p)}\left(\frac{r}{1.5\times 10^6 cm}\right)^{5m/3}[M_{Z'}(GeV)]^4\ cm \tag{75}$$

Then, the ν_R -sphere follows from,

$$\bar{\lambda}_{\nu_R}(r = R_{\nu_R}) = \xi R_{\nu_R} \tag{76}$$

Finally, blackbody radiation from this ν_R-sphere gives us the luminosity Q_{ν_R} we are after.

It is convenient to normalize

$$\frac{Q_{\nu_R}}{Q_{\nu_\mu}}$$

where Q_{ν_μ} is the ordinary muon-neutrino luminosity obtained in section 2b, to get rid of the uncertainties introduced by c_1 and ξ . Then, observational data imply,

$$\frac{Q_{\nu_R}}{Q_{\nu_\mu}} \leq 6 \tag{77}$$

(recall that Q(exotic) $\leq 2\times 10\ erg/\Delta t \leq Q_{TOT} \simeq 6\times Q_\nu$).From this condition we get the upper bound [29],

$$M_{Z'} \lesssim 190\ GeV \tag{78}$$

(having used $X_n = 0.7$, a preferred value in models of stellar collapse).

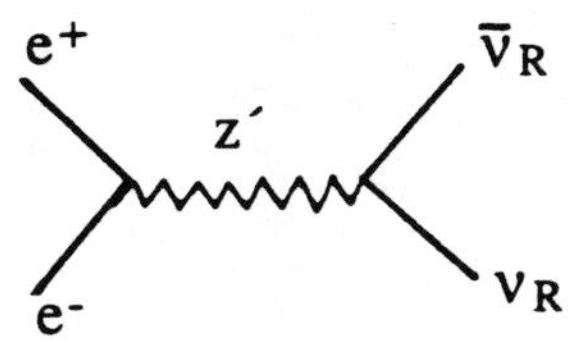

Fig. 9. RH-neutrino pair production process.

Free-streaming will occur for large enough Z' masses. The source of RH-neutrinos is e^+e^- annihilation (diagram in fig. 9) because nucleon-nucleon bremsstrahlung vanishes in the NR OPE approximation. This is due to the fact that in the NR OPE approximation only the axial component of the amplitude survives, as is explicitly shown in ref.[30]. But, as can be seen from eq.(73), the coupling $\bar{N}\gamma^\mu\gamma_5 N\bar{\nu}_R\gamma_\mu\nu_R$ is zero in the LR model. The rate at which the stellar heat bath transfers energy to the RH-neutrinos via e^+e^- annihilation is,

$$q_{ee}(\nu_R) = 64\, g^4 \left(\frac{\cos^2\theta_W}{\cos 2\theta_W}\right)^2 \left(-\frac{1}{2} + 2\sin^2\theta_W\right) \left(\frac{1}{M_{Z'}}\right)^4 \dot{\epsilon}_{ee} \tag{79}$$

The condition,

$$Q_{\nu_R} = \frac{4\pi}{3} R^3 q_{ee}(\nu_R) \lesssim 2\times 10^{53}\ erg/s \tag{80}$$

then implies [29]

$$M_{Z'} \gtrsim 1.2\ TeV$$

We used T=60 MeV, $\rho = 8\times 10^{14}\ g/cm^3$, and R=10 Km. To estimate the uncertainty in the bound we change ρ by a factor of 2 (ignorance of the equation of state at supernuclear densities is a major source of uncertainty). This amounts to varying T in the range 30 to 100 MeV. The bound worsens by a factor of 2.4, *i.e.*

$$M_{Z'} \gtrsim 0.5\ TeV \tag{81}$$

3b. E_6 Models.

Right-handed neutrinos are an ingredient of E_6 constructions (see eq.(48)). Light RH-neutrinos, should they exist, are emitted in a SN collapse. The lightness of ν_R is probably more natural here than in the LR-symmetry case (where a see-saw mechanism is viable).Majorana masses are difficult to implement in E_6. Dirac masses, on the contrary, emerge naturally from an E_6 invariant superpotential. E_6 models admit also extra Z' neutral gauge bosons. Their NC mediated interactions affect SN energy depletion. Therefore, constraints can be obtained from the restriction $Q_{\nu_R} \leq 2 \times 10^{53}\ erg/s$.

We consider explicitly two currently popular models,

A) Model of the abelian flux breaking type of ref.[34].

B) Intermediate scale Higgs breaking model of ref.[35].

Diffusion regime. The relevant cross-section that enters the opacity calculations is,

$$\sigma(\nu_R N \to \nu_R N) = \frac{g^4 \tan^4\theta_W}{8\pi}(c_n^2 X_n + c_p^2 X_p^2)\frac{1}{M_{Z'}^4}E^2 \tag{82}$$

where $c_n^2 \equiv V_n^2 + 3A_n^2$ and $c_p^2 \equiv V_p^2 + 3A_p^2$ with V_n, V_p and A_n, A_p given explicitly in table 1.

The by now familiar method gives,

$$M_{Z'} \leq 91\ GeV \tag{83a}$$

and

$$M_{Z'} \leq 81\ GeV \tag{83b}$$

($X_n = 0.7$ was used).

Table 1. Vector and axial couplings in eq. (82).

	$V_n(V_p)$	$A_n(A_p)$
model A	$\frac{4}{6}(\frac{5}{12})$	$\frac{5}{6}g_A(\frac{5}{4}g_A)$
model B	$\frac{5}{3}$	$\frac{5}{12}g_A(\frac{5}{12}g_A)$

Free-streaming regime. Nucleon-nucleon bremsstrahlung is the main source of RH-neutrinos. The power output is calculated to be,

$$q_{NN}(\nu_R) = 48\left(\frac{g\tan\theta_W}{M_{Z'}}\right)^4 X_n[A_n^2 X_n + 6(A_n^2 + A_p^2)X_p]\dot{\epsilon}_{NN} \tag{84}$$

The luminosity is,

$$Q_{\nu_R} = \frac{4}{3}\pi R^3 q_{NN}(\nu_R) \tag{85}$$

and the ensuing bounds are,

$$M_{Z'} \geq 7.9\ TeV \qquad \text{(Model A)} \tag{86a}$$

and

$$M_{Z'} \geq 5.0\ TeV \qquad \text{(Model B)} \tag{86b}$$

$(X_n = 0.7)$.

Again, the uncertainties in the core parameters relax these limits. Reliable, conservative limits are

$$M_{Z'} \geq 3.3\ TeV \quad \text{and} \quad M_{Z'} \geq 2.1\ TeV \tag{87}$$

$(X_n = 0.7)$.

3c. Supersymmetry.

From the strict phenomenological point of view photinos could be very light. Indeed, cosmology constrains photinos to weigh less than about 100 eV or else be heavier than 0.5 GeV [32]. No other experimental limits for the mass of photinos are known. We assume here that they are lighter than 1 MeV so that they can be abundantly produced in the course of stellar collapse.

Using the techniques presented so far we reach the conclusions [29],

$$M_{\tilde{q}} \lesssim 78\ GeV \tag{88}$$

or else

$$M_{\tilde{q}} \gtrsim 2.7\ GeV \qquad (X_n = 0.7) \tag{89}$$

A conservative statement, taking into account the uncertainties in the core parameters, is

$$M_{\tilde{q}} \gtrsim 1.1\ TeV \tag{90}$$

These results rely on

i) the opacity of matter to photino propagation is governed by elastic $\tilde{\gamma}$ N scattering via squark exchange [36],

$$\sigma(\tilde{\gamma}N) = \frac{2\pi\alpha^2}{27}\frac{E^2}{M_{\tilde{q}}^4}[25 + 9g_A^2 + 30g_A(X_p - X_n)] \tag{91}$$

and

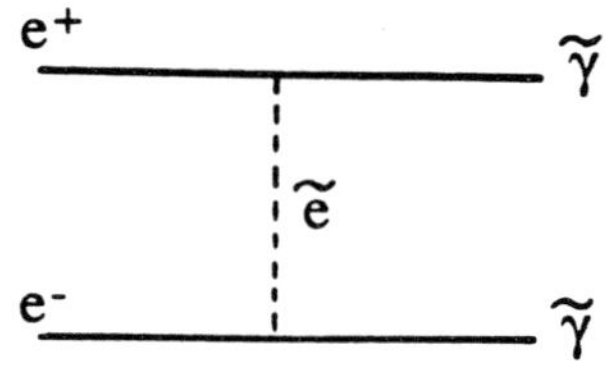

Fig. 10. Selectron exchange amplitude.

ii) the energy output rate associated to $NN \to NN\tilde{\gamma}\tilde{\gamma}$ processes is [29],

$$Q_{NN}(\tilde{\gamma}) = \frac{16}{81} e^4 X_n[(5 - 3g_A)^2 X_n + 12(25 + 9g_A^2)X_p] \left(\frac{1}{M_{\tilde{q}}}\right)^4 \dot{\epsilon}_{NN} \tag{92}$$

Since $M_{\tilde{q}} \leq$ 78 GeV is, of course, excluded by present accelerator data, the squarks are bound to be heavier than 1.1 TeV. This implies that photinos would free-stream once produced in the core of the star. This allows the determination of a bound for the selectron. In fact, photinos can be produced via the amplitude shown in fig. 10, where a selectron is exchanged in the t-channel. The energy emission rate is easily calculated to be,

$$q_{ee}(\tilde{\gamma}) = \frac{32e^4}{M_{\tilde{e}}^4} \dot{\epsilon}_{ee} \tag{93}$$

for extreme relativistic electrons and positrons. The implied luminosity should, as always, not exceed 2×10^{53} erg/s. The constraint [29, 36]

$$M_{\tilde{e}} \geq 2.0\ TeV \tag{94}$$

then follows. Again, from uncertainty considerations we get the more conservative bound,

$$M_{\tilde{e}} \geq 0.83\ TeV \tag{95}$$

Note.

All results obtained in subsections a,b, and c are summarized in table 2. Note that accelerator data rule out already, in all cases considered, the low mass end of the permitted windows and hence only the lower limits are reported on this table.

Table 2. Lower bounds to masses of particles of extended SM model scenarios.

Scenario	*Limit*
LR model	$M_{Z'} \geq 0.5\ TeV$
E_6 model A	$M_{Z'} \geq 3.3\ TeV$
E_6 model B	$M_{Z'} \geq 2.1\ TeV$
SUSY	$M_{\tilde{q}} \geq 1.1\ TeV$
SUSY	$M_{\tilde{e}} \geq 0.8\ TeV$

2. HIGGS PHYSICS MISCELLANY

2.1 Low mass limits on the Standard Model Higgs Boson

The mass of the Higgs boson is completely unknown. There is no theoretical input in the SM that determines its value. Actually, the mass of the Higgs scalar could lie anywhere in a wide range of masses extending from a few Mev's to about 1 TeV. Lower limits can be inferred from a variety of experimental data including X-ray transitions in muonic atoms, nuclear decays, nuclear scattering, meson decays, and other [37]. Here we shall consider explicitly neutron-nucleus scattering and derive a lower bound for the Higgs particle mass.

Scattering of neutrons on lead has been used to constrain the mass of the Higgs boson [38, 41]. The existence of a scalar particle implies the presence of a Yukawa potential

$$V_H = -A\frac{g^2}{4\pi}\frac{e^{-m_H r}}{r} \tag{96}$$

associated to its exchange, which adds to the strong nuclear potential. In the equation above A is the atomic number of the nucleus, m_H is the mass of the Higgs particle, and g is the Higgs-nucleon-nucleon coupling.

The Born amplitude associated to the interaction potential V_H is

$$f_h = \frac{-m_n}{2\pi}\int V_H e^{i\vec{q}\cdot\vec{r}} d^3r \tag{97}$$

where $\vec{q}$ is the momentum transfer. The integral can be calculated and gives,

$$f_H = 2m_n\frac{g^2}{4\pi}A\frac{1}{4m_nE(1-\cos\theta)+m_H^2} \tag{98}$$

with E being the kinetic energy of the incoming neutron and θ is the scattering angle.

On the other hand, data in the energy range 1 to 26 KeV can be parametrized as [39]

$$\frac{d\sigma}{d\pi} = \frac{\sigma_0}{4\pi}[1 + aE\cos\theta] \tag{99}$$

where $\frac{\sigma_0}{4\pi} = 100\ fm^2$ and $a = (1.91 \pm 0.42) \times 10^{-3} keV^{-1}$. The interference of the strong amplitude with f_H contributes to the constant a an amount

$$\Delta a = -16\frac{m^2}{\sqrt{\sigma_0/4\pi}}\frac{g^2}{4\pi}\frac{A}{m_H^4} \tag{100}$$

which modifies the angular dependence of the cross section. Requiring that,

$$|\Delta a| \leq 2 \times 10^{-3} keV$$

and using $g = 2.1 \times 10^{-3}$ (this is a recent determination of g by Cheng and by Barbieri and Curci [40]) leads to

$$m_H \geq 7\ GeV \tag{101}$$

To obtain this bound the angular dependence of the cross section was used. One could also use the energy dependence of the cross section [41]. A very recent measurement [42] of the neutron-lead scattering cross section gave,

$$\sigma(k) = \sigma_0 + a_1 k + a_2 k^2 \tag{102}$$

k is the wavevector of the incoming neutron and

$$\begin{aligned} \sigma_0 &= 11253(5)\ b \\ a_1 &= 0.60(51)\ bfm \\ a_2 &= -371(27)\ bfm^2 \end{aligned} \tag{103}$$

(fitted over the range $50\ eV \leq k \leq 20\ keV$). These data can be understood in terms of a simple hard-core potential and its effective range. Indeed, low energy nuclear scattering data can be parametrized as

$$k \cot \delta_0 = -\frac{1}{R} + \frac{1}{2} r_{eff} k^2 \tag{104}$$

where R is the scattering length, r_{eff} is the effective range, and δ_0 is the S-wave phase shift. Then, the cross section can be written as

$$\sigma \simeq 4\pi R^2 \left[1 - (kR)^2 (1 - \frac{r_{eff}}{R})\right] \tag{105}$$

which follows from eq.(104) and $\sigma = \frac{4\pi}{k^2} \sin^2 \delta_0$.

The scattering amplitude for our simple potential with hard core radius R is,

$$f_N = -R[1 - ikR - \frac{2}{3} k^2 R^2 + O(k^3 R^3)] \tag{106}$$

From this the effective range is $r_{eff} = \frac{2}{3} R$. On the other hand, from the data we extract

$$R = \sqrt{\sigma_0 / 4\pi} = 9.46\ fm \tag{107}$$

which means $r_{eff} = 6.3\ fm$ to be compared with the effective range extracted from experiment,

$$r_{eff} = 6.0 \pm 0.3\ fm \tag{108}$$

This value is in excellent agreement with the theoretical value found before.

The presence of the Higgs particle will change the potential felt by the neutron. Using the Distorted Wave Born Approximation (DWBA) [43] we write

$$f = f_N + f_H + \delta f \tag{109}$$

Here δf takes into account the distorting effect of the nuclear potential. Explicitly,

$$\delta f = -2\mu \int_0^\infty V_H(r)[R_0^2(r) - j_0^2(kr)]r^2 dr \tag{110}$$

with $R_0(r) = \theta(r-R)\frac{\sin k(r-R)}{kr}$ and $j_0(kr) = \frac{\sin kr}{kr}$.

f and δf can be expanded also in powers of k, for small k, exactly as was done for f_N

$$f_H(\theta) \simeq 2\mu \frac{Ag^2}{4\pi m_H^2}\left[1 - \frac{2k^2}{m_H^2}(1-\cos\theta) + O(\frac{k^4}{m_H^4})\right] \tag{111}$$

$$\delta f \simeq \frac{\mu A g^2}{2\pi}(I_0 + k^2 I_1) \tag{112}$$

with

$$I_0 \equiv \int_0^\infty dr e^{-m_H r} r^2 \left(\frac{e^{-m_H r}}{r+R} - \frac{1}{r}\right)$$

$$I_1 \equiv -\frac{2}{3}\int_0^\infty dr e^{-m_H r} r^4 \left(\frac{e^{-m_H r}}{r+R} - \frac{1}{r}\right)$$

Finally, we evaluate the total contribution to σ_0 and a_2 with R and m_H as free parameters (we take also $g = 2.1 \times 10^{-3}$) and require σ_0 and a_2 to agree with data at the (conservative) 2σ level. The result is shown in fig. 11. We see that the limit

$$m_H \geq 18 \; GeV \tag{113}$$

follows from this analysis [41].

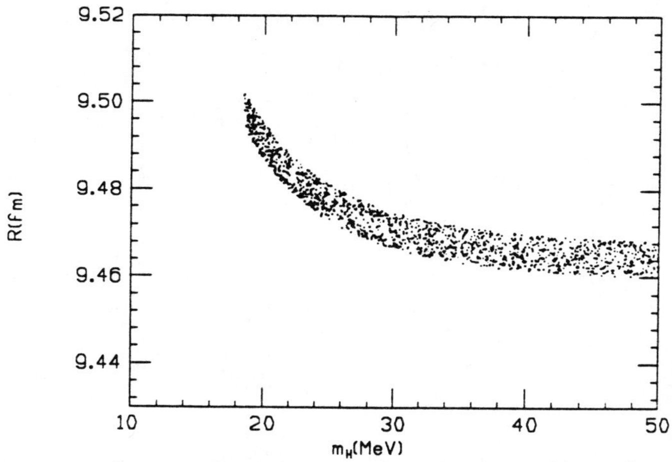

Fig. 11. Shaded area: allowed values of R and m_H.

2.2 Beyond the SM :Extended Higgs Sectors

2a.Model with 2 doublets.

The simplest extension of the $SU(2) \times U(1)$ Higgs sector consists in the inclusion of an extra scalar doublet,

$$\phi_1 = \begin{pmatrix} \phi_1^+ \\ \phi_1^0 \end{pmatrix} \text{ and } \phi_2 = \begin{pmatrix} \phi_2^+ \\ \phi_2^0 \end{pmatrix} \tag{114}$$

the v.e.v. $< \phi_{1,2}^0 >= v_{1,2}/\sqrt{2}$ breaks the gauge symmetry spontaneously and 5 scalar fields (2 charged, 3 neutral) remain physical.

The charged fields are

$$H^{\pm} = -\sin\beta\phi_1^{\pm} + \cos\beta\phi_1^{\pm} \tag{115}$$

where $\beta = \tan^{-1}\frac{v_2}{v_1}$.

The neutral fields are,

$$\begin{aligned} H_1^0 &= \sqrt{2}[(Re\phi_1^0 - v_1)\cos\alpha + (Re\phi_2^0 - v_2)\sin\alpha] \\ H_2^0 &= \sqrt{2}[-(Re\phi_1^0 - v_1)\sin\alpha + (Re\phi_2^0 - v_2)\cos\alpha] \\ H_3^0 &= \sqrt{2}[-Im\phi_1^0\sin\beta + Im\phi_2^0\cos\beta] \end{aligned} \tag{116}$$

where the mixing angle α depends explicitly on the parameters of the Higgs potential V(H).

The second doublet being just a replica of the first one, it is easy to understand that the masses of the gauge bosons are

$$\begin{aligned} M_Z^2 &= \frac{g^2 + g'^2}{4}(v_1^2 + v_2^2) \\ M_W^2 &= \frac{g^2}{4}(v_1^2 + v_2^2) \end{aligned} \tag{117}$$

and their ratio satisfies

$$\rho \equiv \frac{M_W^2}{M_Z^2 \cos^2 \theta_W} = 1 \tag{118}$$

automatically, just as in the SM.

2b.Doublet+Triplet Model.

It is useful to have another model to confront with the previous extension of the SM and with the SM itself. Let us consider the Higgs sector,

$$\phi = \begin{pmatrix} \phi^+ \\ \phi^0 \end{pmatrix} \text{ and } \chi = \begin{pmatrix} \chi^+ \\ \chi^0 \\ \chi^- \end{pmatrix}$$

After spontaneous symmetry breaking ($< \phi^0 >_0 = v_1$ and $< \chi^0 >_0 = v_2$) 4 physical states are left. Two of them are charged, the other two are neutral. These states are,

$$\begin{aligned} H^{\pm} &= \frac{-\sin\beta\phi^{\pm} + \frac{1}{\sqrt{2}}\cos\beta\chi^{\pm}}{(\sin^2\beta + \frac{1}{2}\cos^2\beta)^{1/2}} \\ H^0 &= \cos\alpha\frac{1}{\sqrt{2}}(\phi^0 + \phi^{0\dagger}) + \sin\alpha\chi^0 \\ H^0 &= \sin\alpha\frac{1}{\sqrt{2}}(\phi^0 + \phi^{0\dagger}) + \cos\alpha\chi^0 \end{aligned} \tag{119}$$

with $\beta = \tan^{-1}\sqrt{2}\,\frac{v_2}{v_1}$ and α depends on V(H).

Since the neutral χ^0 has Q=0 and Y=0 it does not couple to the Z boson. Therefore $< \chi^0 >$ contributes only to the charged $W^{\pm}$ boson mass,

$$\Delta M_W^2 = g^2 < \chi^0 >^2 \tag{120}$$

This piece introduces a shift away from 1 of the rho parameter,

$$\rho = 1 + \frac{4v_2^2}{v_1^2} \tag{121}$$

Phenomenology requires $\rho = 1$ within 1%. As a consequence $\frac{v_2}{v_1}$ must be small.

2c.A Theorem.

Is it possible to discriminate experimentally among the three different models (SM, 2 doublets, doublet+triplet)? In principle one could use the couplings of the Higgs scalars to the Z boson (and to the W boson) to distinguish among the models via the the decays of the intermediate bosons. The different branching ratios expected in the different models have been studied in ref. [44]. In table 3 we give a listing of the relevant couplings. On this table only a few entries take definite values. The other entries depend on the a priori undetermined parameters of the Higgs potential. The zero on the table is worth commenting. It is a particular case of the following theorem.

An SU(2)xU(1) model with two or more Higgs doublets has a vanishing tree level ZWH coupling [45].

Let us briefly sketch the proof. Suppose we have N scalar doublets

$$\begin{pmatrix}\phi^+\\ \phi^0\end{pmatrix}_1, \begin{pmatrix}\phi^+\\ \phi^0\end{pmatrix}_2 \cdots$$

with $<\phi_i^0>= v_i$.

Make now a rotation,

$$\phi_i' = \Sigma_j R_{ij}\phi_j$$

such that $<\phi_1'>\neq 0$ but $<\phi_{i\neq 1}'>= 0$. $(D_\mu\phi_j)^\dagger(D^\mu\phi_j)$ is invariant under R. The triple Z-W-Higgs vertices arise from the quartic coupling $ZW\phi_i'^0\phi_i'^\pm$ with the neutral scalar acquiring a vev. All of them vanish except the one corresponding to ϕ_1' which plays the role of the SM doublet. But, charged scalars in the SM are Goldstone bosons. Hence, charged scalars do not couple to ZW at tree level. The phenomenological consequences of this fact are clear. Recall that in the SM the ZZH vertex is very important because it provides the main mechanism for H^0 production in electron-positron collisions. Its charged counterpart $ZWH^\pm$, however, will not be a source of charged scalars (as one would a priori expect) in e^+e- annihilations. Therefore, the production of single charged Higgs particles is not as probable as the production of neutral ones. For further study of this and other issues related to table 3 we refer the reader to the literature.

Table 3. Coeficients of triple boson vertices (see ref. [44]).

Vertex	*SM*	*2 Doublets*	*Doublet + Triplet*
ZH^+H^-		$\frac{1}{2}\sec\theta_W(1-2\sin^2\theta_W)$	$\sec\theta_W\left[\cos^2\theta_W + O(\frac{v_2^2}{v_1^2})\right]$
$ZH_1^0H_2^0$		depends on V(H)	depends on V(H)
$WH^\pm H^0$		depends on V(H)	depends on V(H)
ZZH^0	$\frac{1}{2}\sec\theta_W$	depends on V(H)	depends on V(H)
WWH^0	1	depends on V(H)	depends on V(H)
$ZWH^\pm$		0	$\sqrt{2}\sec\theta_W\frac{\tan\beta}{1+2\tan^2\beta}$

2.3 Higgs Bosons in Supersymmetry

We turn now to the last topic of these lectures. What follows is a report on the work of Gunion and Haber in ref [46]. We urge the reader to consult ref. [46] for a much wider and detailed presentation of their results.

Supersymmetry (SUSY) provides a strong argument in favor of a two doublet Higgs sector. Furthermore, SUSY imposes severe constraints on the otherwise largely arbitrary Higgs potential. Let us discuss these points in the frame of the minimal supersymmetric 2-Higgs model.

The scalar particle spectrum of a 2 Higgs model was already discussed in a previous section. In the non-supersymmetric framework, the Higgs masses and α and β are unrelated, *i.e.* not fixed by theory. In the minimal SUSY model these parameters are quite constrained. Indeed, three independent parameters specify completly masses and mixings (and hence, couplings). In particular, the following relation among neutral boson masses holds,

$$m^2_{H_1^0,H_2^0} = \frac{1}{2}[m^2_{H_3^0} + m_Z^2 \pm \sqrt{(m^2_{H_3^0} + m^2)^2 - 4m_Z^2 m^2_{H_3^0}\cos_2^2\beta}] \tag{122}$$

We may take as the three independent parameters,

$$r_1 \equiv \frac{m^2_{H_1^0}}{m_Z^2}\,, \qquad r_2 \equiv \frac{m^2_{H_2^0}}{m_Z^2} \qquad \text{and } \epsilon = sign(v_2 - v_1)$$

Then, eq.(123) tells us immediately the following very important facts,

$$r_1 \geq 1 \qquad \text{and} \qquad r_2 \leq 1 \tag{123}$$

i.e. H_1^0 is heavier than the neutral Z boson and the scalar H_2^0 is lighter than the Z boson. The third neutral scalar particle can be either lighter or heavier than the Z boson,

$$r_3 \equiv \frac{m^2_{H^0_3}}{m^2_Z} = r_1 + r_2 - 1$$

Also, the relation

$$m^2_{H^\pm} = m^2_W + m^2_{H^0_3} \tag{124}$$

holds, which implies that the charged Higgs boson is always heavier than the W boson.

The angles α and β are, in terms of r_1 , r_2 , and ϵ,

$$\begin{aligned} \tan 2\beta &= \epsilon\sqrt{\frac{(r_1-1)(1-r_2)}{r_1 r_2}}, \quad 0 \leq \beta \leq \frac{\pi}{2} \\ \tan 2\alpha &= \tan 2\beta \frac{r_1+r_2}{r_3-1} \quad , \quad -\frac{\pi}{2} \leq \alpha \leq 0 \end{aligned} \tag{125}$$

Masses and mixings being specified, we can determine completely the Higgs couplings to gauge bosons, to quarks and to leptons, as well as to their SUSY partners. A few examples will suffice.

Couplings to gauge bosons:

$$\begin{aligned} H^0_1 WW/ZZ \quad &: -igm_{W/Z}\cos(\beta-\alpha)g^{\mu\nu} \times (1, \cos^{-1}\theta_W) \\ H^0_2 WW/ZZ \quad &: -igm_{W/Z}\sin(\beta-\alpha)g^{\mu\nu} \times (1, \cos^{-1}\theta_W) \end{aligned} \tag{126}$$

Couplings to quarks:

$$\begin{aligned} H^0_1 uu/dd \quad &: -ig\frac{m_{u/d}}{2m_W} \times (\frac{\sin\alpha}{\sin\beta}, \frac{\cos\alpha}{\cos\beta}) \\ H^0_2 uu/dd \quad &: -ig\frac{m_{u/d}}{2m_W} \times (\frac{\cos\alpha}{\sin\beta}, \frac{\sin\alpha}{\cos\beta}) \end{aligned} \tag{127}$$

etc.

Gunion and Haber [46] have done an exhaustive numerical analysis of the $H^0_{1,2,3}$ coupling strengths which we shall not make explicit here. We may summarize their results as follows. The light Higgs boson H^0_2 has production rates and a decay behavior similar to a light SM scalar. The heavy scalar particle H^0_1 on the contrary differs very much from a heavy SM Higgs boson. Indeed, its couplings to gauge bosons are highly suppressed and therefore WW/ZZ decay modes are never important (the main modes are SUSY modes; the width is rather narrow) and production cross-sections (via WW/ZZ fusion) are greatly reduced as compared to SM heavy Higgs boson production rates. The third Higgs scalar, H^0_1, cannot couple to WW/ZZ thus being also quite different from the SM Higgs particle. So, heavy SUSY Higgs physics is quite distinct from SM Higgs physics. In short, production and detection of heavy Higgs scalars in the SUSY scenario is harder.

ACKNOWLEDGEMENTS

I am grateful to the organizers of this School and in particular to Prof. M. Zrałek for the friendly hospitality shown to me in Szczyrk.

REFERENCES

1. M. Daniel, Proc. of XXIV Int. Conf. on High Energy Physics, Munich (1988).
2. R. Abela et al., *Phys. Lett.* **B** (1984) 43.
3. D. Wegener, Proc. of XXIV Int. Conf. on High Energy Physics, Munich (1988).
4. See *e.g.* J. A. Grifols, Proc. XIV Winter Meeting on Fundamental Physics, Sant Feliu de Guixols (1986) and references therein.
5. R. Gähler, J. Kalus and W. Mampe, *Phys. Lett.* **D25** (1982) 2887.
6. M. Marinelli and G. Morpurgo, *Phys. Lett.* **137B** (1984) 439.
7. G. Barbellini, and G. Cocconi, *Nature* **329** (1987) 21.
8. J. A. Grifols, E, Massó and S. Peris, *Phys. Lett.* **207B** (1988) 993.
9. K. Hirata, et al., *Phys. Rev. Lett.* **58** (1987) 1490.
   R. Bionta et al., *Phys. Rev. Lett.* **58** (1987) 1494.
10. L. B. Okun, *Yad. Fiz.* **10** (1969) 358.
11. A. Grau and J. A. Grifols, *Phys. Lett.* **166B** (1986) 233.
12. F. Bergsma et al., *Phys. Lett.* **147B** (1984) 481.
13. P. Vogel and J. Engel, *Phys. Rev.* **D39** (1989) 3378.
14. M. Fukugita and S. Yazaki, *Phys. Rev.* **D36** (1987) 3817.
15. R. Barbieri and R. N. Mohapatra, *Phys. Rev. Lett.* **61** (1988) 27
    J. Lattimer and J. Cooperstein, *Phys. Rev. Lett.* **61** (1988) 23
    M. B. Voloshin, *Phys. Lett* **219B** (1988) 360.
16. M. B. Voloshin and M. I. Vysotskii, *Sov, J. Nucl. Phys.* **44** (1986) 544.
17. M. J. Duncan, J. A. Grifols, A. Méndez and S. Uma Sankar, *Phys. Lett.* **B191** (1987) 304.
18. J. E. Kim, *Phys. Rev.* **D14** (1976) 3000
    B. W. Lee, R. E. Shrock, *Phys. Rev.* **D16** (1977) 1444
    W. J. Mariano and A. I. Sanda, *Phys. Lett.* **B67** (1977) 303.
19. For a recent review see R. N. Mohapatra, in Quarks, Leptons and Beyond, Proc. of the NATO Advanced Study Institute, Munich 1983, ed. H. Fritzsch et al. (NATO ASI Series B, vol. 122) Plenum, N.Y. 1985) p. 217.

20. H. E. Haber and G. L. Kane, *Phys. Rep.* **117C** (1985) 75.

21. S. N. Biswas, A. Goyal and J. N. Passi, *Phys. Rev.* **D28** (1983) 671.

22. J. L. Hewett and T. G. Rizzo, MAD/PH/446 (1988).

23. J. A. Grifols and J. Solà, *Phys. Lett.* **B182** (1986) 53.

24. See *e.g.* D. N. Schramm, Proc. 1987 Int. Symp. on Lepton and Photon Int. at high energies, eds. W. Bartel and R. Rückl (North-Holland, Amsterdam, 1988) p. 471.

25. D. L. Tubbs and D. N. Schramm, *Astrophys. J.* **201** (1975) 467.

26. S. L. Shapiro and S. Teukolski, Black Holes, White Dwarfs and Neutron Stars (Wiley, N. Y. 1983)
L. J. van den Horn and J. Cooperstein, *Astrophys. J.* **300** (1986) 142.

27. M. S. Turner, *Phys. Rev. Lett.* **60** (1988) 1797.

28. R. Mayle, J. R. Wilson and D. N. Schramm, *Nuovo Cimento* **9C** (1986) 443.

29. J. A. Grifols and E. Massó, *Nucl. Phys. B.* to be published.

30. B. L. Friman and D. V. Maxwell, *Astrophys. J.* **232** (1979) 451.

31. D. A. Dicus, *Phys. Rev.* **D6** (1972) 941.

32. Particle Data Group, M. Aguilar-Benitez et al., *Phys. Lett.* **B170** (1986) 1.

33. H. Georgi and S. Weinberg, *Phys. Rev.* **D17** (1978) 275.

34. E. Cohen et al., *Phys. Lett.* **B165** (1985) 76
J. Ellis et al., *Nucl. Phys.* **B227** (1986) 14; ibid *Mod. Phys. Lett.* **A2** (1986) 57.

35. F. del Aguila et al., *Nucl. Phys.* **B272** (1986) 413.

36. J. A. Grifols, E. Massó and S. Peris, *Phys. Lett.* **220** (1989) 591.

37. For a review see: R.N. Cahn, *Report on Progress in Physics* **52** (1989) 389 and J. F. Gunion, H. E. Haber, G. L. Kane and S. Dawson, The Higgs Hunter's guide, UCD-89-4.

38. R. Barbieri and T. E. O. Ericson, *Phys. Lett.* **B57** (1975) 276.

39. Y.A. Aleksandrov et al., *JETP Letters* **4** (1966) 134.

40. T. P. Cheng, *Phys. Rev.* **D38** (1988) 2869
R. Barbieri and G. Curci, IFUP-TH 35/88.

41. J. A. Grifols, E. Massó and S. Peris, *Phys. Rev. Lett.* **63** (1989) 1346.

42. S. Schmiedmeyer, H. Rauch and P. Riehs, *Phys. Rev. Lett.* **61** (1988) 1065.

43. L. I. Schiff, "Quantum Mechanics", McGraw-Hill (1968).

44. A. Grau, J. A. Grifols and N. Lupón, *Phys. Rev.* **D26** (1982) 324.

45. J. A. Grifols and A. Méndez, *Phys. Rev.* **D22** (1980) 1725.

46. J. G. Gunion and H. E. Haber, *Nucl. Phys.* **B272** (1986) 1; *ibid Nucl. Phys.* **B278** (1986) 449.

Standard Model
and Beyond

ISBN 1-56072-027-1

PREDICTING THE RELATIVE STRENGTHS OF THE FUNDAMENTAL COUPLING CONSTANTS VIA A STABILITY PRINCIPLE*

D. Schildknecht
Department of Theoretical Physics, University of Bielefeld
D-4800 Bielefeld 1, Germany

ABSTRACT

We propose a dynamical stability principle related to the interplay of electromagnetic and weak interactions and predict $sin^2\theta_W$ in good agreement with its empirical value. An attempt to apply the principle to strong interactions is also reported.

1. INTRODUCTION

The work I am going to talk about has been carried out in collaboration with Fernand Renard and is fully contained in two papers [1,2], which appeard in Phys. Lett. B. I will confine myself to giving a brief account of the essential results, and I refer to the mentioned papers for a more comprehensive one.

The attempts to understand the strengths of the different interactions observed in nature (as defined by their different symmetry properties), in recent years have mainly been concentrated on the GUT hypothesis [3]. On the basis of a simple underlying symmetry group the interaction strengths are conjectured to be identical, if studied at a sufficiently high energy scale, while differing from each other at the presently explored energies. More recently, a possible reduction of the number of coupling strengths appearing within a basic lagrangian has been studied [4] by exploiting consistency conditions on renormalization-group equations.

2. THE STABILITY PRINCIPLE

Our attempt to understand the observed ratios of the fundamental coupling constants, in particular the ratio of the electromagnetic and weak couplings, $sin^2\theta_W =$

---

* Partially supported by the program PROCOPE for French-German scientific collaboration
Presented at the Silesian Summer School for Particle Physics, Sczyrk (Poland), September 1989

$e^2/g_W^2 = \alpha/\alpha_W$, but also the ratio of the electromagnetic and strong couplings, α/α_s, is quite different from the above-mentioned attempts. Viewing the mixing phenomenon between electromagnetism and weak interactions as the essential link between a priori different interactions, we demand that the relative strength of these interactions, which determines the mixing parameter, is to fulfill a stability or minimization principle. The mixing phenomenon (obviously) appears in the neutral-current interactions and determines the properties of Z^0 propagation. Accordingly, it is required that Z^0 propagation (in particular the phenomenon of vacuum polarization) be stable under the relative variation of e and g_W, i.e., under variation of $s_W^2 \equiv sin^2\theta_W$. For additional comments on the intuitive background and the universality of our conjecture, let me refer to ref. 1.

Our principle of stability will first of all be formulated for Z^0 propagation in the high-energy limit [1] of $s \gg 4M_W^2, (M_Z + m_H)^2$, where all threshold effects due to the non-vanishing masses of the particles contributing to vacuum polarization are negligible. Subsequently, I will briefly mention a global (integral) form of our stability requirement, in which particle masses can explicitly be taken into account.

3. STABILITY OF Z^0 PROPAGATION AT ASYMPTOTIC ENERGIES

In principle, stability is to be required to all orders of perturbation theory. In practice, we will restrict ourselves to the one-loop order in the propagator, i.e., to the tree level in the Z^0 width. Accordingly, the imaginary part of the vacuum polarization function, $\Pi_Z(s)$, which determines the Z^0 width in the Z^0 propagator via

$$M_Z\Gamma_Z(s) = s\ Im\Pi_Z(s), \tag{1}$$

at asymptotic energies is given by [1,6]

$$Im\Pi_Z = \alpha_W \frac{c_{11}s_W^4 - 2c_{12}s_W^2 + c_{22}}{1 - s_W^2}, \tag{2}$$

$$(s \gg M_Z^2, 4M_W^2, (M_Z + m_H)^2),$$

where the short-hand notation

$$s_W^2 \equiv sin^2\theta_W(M) = \frac{\alpha(M^2)}{\alpha_W(M^2)}, \tag{3}$$

has been introduced for $sin^2\theta_W$ as measured at the (renormalization) scale $M \cong M_Z$. The constants c_{ij} in (2) contain fermion and boson contributions,

$$c_{ij} = c_{ij}^{f\bar{f}} + c_{ij}^{WW} + c_{ij}^{ZH}, \tag{4}$$

and are to be specified below. Noting that the limit of $s_W^2 \to 0$ in (2) yields the width of the unmixed neutral boson W^0, which is identical to the width of the charged $W^{\pm}$, we introduce the normalized reduced Z^0 width (in the limit of $s \to \infty$), which is given by

$$\lim_{s \to \infty} \left(\frac{M_Z \Gamma_Z(s)}{s} / \frac{M_W \Gamma_W(s)}{s} \right) = \frac{\frac{c_{11}}{c_{22}} s_W^4 - 2\frac{c_{12}}{c_{22}} s_W^2 + 1}{1 - s_W^2}. \qquad (5)$$

The stability principle [1], stating the vacuum-polarization effects in Z^0 propagation are to assume a stable minimum under the relative variation of e and g_W, now amounts to minimizing* (5) with respect to s_W^2. As a result, we obtain our basic formula [1] for s_W^2,

$$s_W^2 = 1 - \sqrt{1 - \frac{2c_{12} - c_{22}}{c_{11}}}, \qquad (6)$$

in terms of the coefficients c_{ij}, which are yet to be specified.

For the leptons and quarks, (ℓ, q), the vacuum-polarization coefficients are given by

$$\begin{aligned} c_{11}^{f\bar{f}} &= \frac{1}{3} \sum_i Q_i^2 = n_f \frac{8}{9}, \\ c_{22}^{f\bar{f}} &= \frac{1}{6} \sum_i \left(\frac{\tau_3^{(i)}}{2}\right)^2 = n_f \frac{1}{3}, \\ c_{12}^{f\bar{f}} &= \frac{1}{3} \sum_i Q_i \frac{\tau_3^{(i)}}{4} = n_f \frac{1}{3}, \end{aligned} \qquad (7)$$

where the sums over i run over all leptons and quarks to be taken into account when evaluating (4,6), and the equality on the right-hand side of (7) is valid for n_f complete families of leptons and quarks. The charges of the leptons and quarks and the weak isospin quantum numbers have been denoted by Q_i and $\tau_i = \pm 1$, respectively. If we neglect the bosonic contribution to vacuum polarization, we thus obtain from (6,7)

$$s_W^2 = 1 - \sqrt{1 - \frac{2\sum_i Q_i \frac{\tau_3(i)}{2} - \sum_i (\frac{\tau_3(i)}{2})^2}{2\sum_i Q_i^2}}, \qquad (8)$$

which yields

$$s_W^2 = 1 - \sqrt{\frac{5}{8}} = 0.2094 \cong 0.21, \qquad \text{(leptons and quarks)}, \qquad (9)$$

when evaluated for (an arbitrary number of) n_f complete families of leptons and quarks.

In order to appreciate the significance of our prediction (8,9) following from the stability principle, it is useful to quote a few values obtained for s_W^2, if (8) is evaluated for the case of incomplete families of leptons or quarks. In particular, in a world of n_f

---

* Note that the real part of the vacuum polarization function, $Re\Pi_Z(s)$, is then stabilized automatically, as it is related [1] to $Im\Pi_Z(s)$ via a dispersion relation.

families of only leptons (and no quarks) one would have $s_W^2 = 0.13$, while for quarks only (and no leptons) one finds $s_W^2 = 0.26$. The result (9) is changed only slightly, if $n_f = 3$ families are present, but no top-quark,

$$s_W^2 = 0.20 \quad (n_f = 3, \text{ no top} - \text{quark}), \tag{10}$$

while adding N neutrinos without corresponding charged leptons to the $n_f = 3$ complete families, decreases s_W^2 slightly to

$$s_W^2 = 1 - \sqrt{\frac{5}{8} - \frac{N}{64}} = \begin{cases} 0.20 & \text{for N} = 1, \\ 0.18 & \text{for N} = 3. \end{cases} \tag{11}$$

When evaluating the bosonic vacuum polarization, in the case of W^+W^- pairs, care must be taken to include the "universal" parts (independent of the properties of the particles coupled to the Z^0 propagator) from box and vertex diagrams to insure decent high-energy behaviour of the Z^0 width in the Z^0 propagator. A careful analysis yields [2]

$$c_{11}^{WW} = \frac{1}{4} \;, \quad c_{12}^{WW} = \frac{11}{24} \;; \quad c_{22}^{WW} = \frac{11}{16} \;, \tag{12}$$

while for the ZH loop one finds

$$c_{11}^{ZH} = c_{12}^{ZH} = 0 \;, \quad c_{22}^{ZH} = \frac{1}{48}. \tag{13}$$

Inclucing these contributions when evaluating (4,6) then yields a correction of approximately 10 % to the value given in (9),

$$s_W^2 = 0.235 \;, \quad (\text{leptons, quarks, W}^+\text{W}^- \text{ and ZH}), \tag{14}$$

in good agreement with the empirical value of [5]

$$s_W^2 = 0.230 \pm 0.005. \tag{15}$$

Fig. 1 shows the minimum of $s_W^2 \cong 0.23$ in a plot of the ratio of the reduced widths given by (5) as a function of s_W^2.

In summary: The requirement of stability of Z^0 propagation under variation of s_W^2 in the limit of $s \to \infty$ implies an empirically highly successful prediction for $s_W^2 \equiv sin^2\theta_W(M)$.

4. THE GLOBAL (INTEGRAL) FORM OF THE STABILITY PRINCIPLE

If the Z^0 propagator, including vacuum polarization, is considered at energies comparable to the masses of the particles causing the vacuum polarization, the function $Im\Pi_Z$, given in (2) for the case of $s \to \infty$, becomes dependent on s, $Im\Pi_Z = Im\Pi_Z(s)$. A global (integral) quantity, defined in terms of the Z^0 propagator, has obviously to be

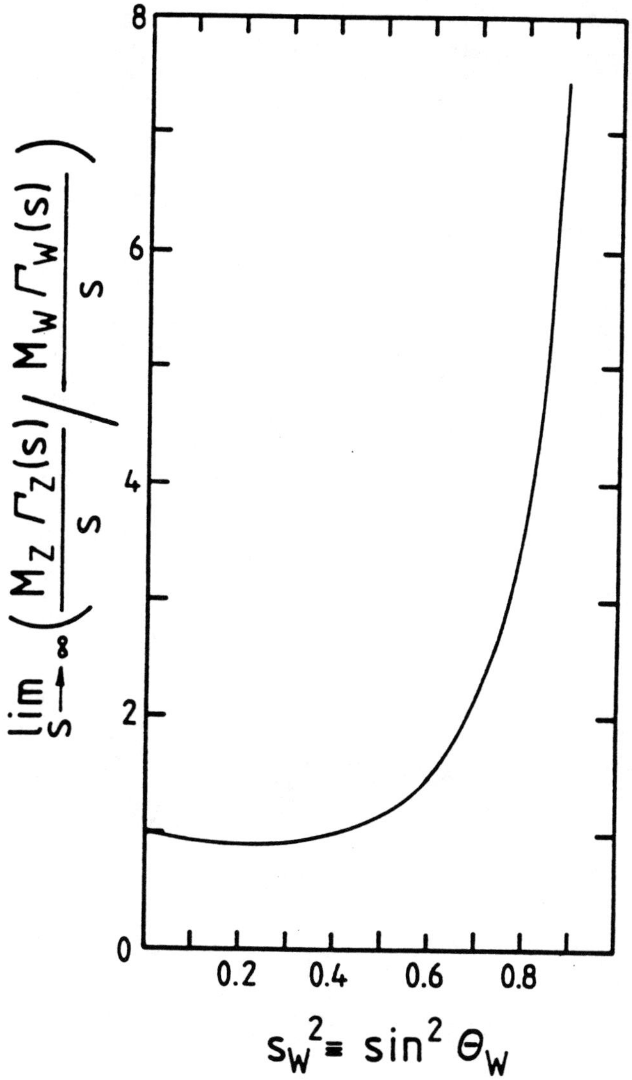

Figure 1: The ratio of the reduced widths of the Z^0 and the $W^{\pm}$ in the limit of $s \to \infty$ as a function of $sin^2\theta_W$.

introduced as a basis for a minimization principle. Let us thus consider the quantity [2]

$$a_Z(s_W^2) = [\frac{M_Z^2}{\pi} \int_{-\infty}^{+\infty} \frac{ds}{(s - M_Z^2)^2(1 + (s - M_Z^2)G_Z(s))^2 + M_Z^2\Gamma_Z^2(s)}]^{-1}, \qquad (16)$$

defined via an integral over the Z^0 mass distribution determined by the Z^0 propagator*. Normalizing $a_Z(s_W^2)$ by its value for $s_W^2 \to 0$, by introducing the "normalized global absorption strength",

$$A_Z(s_W^2) = \frac{a_Z(s_W^2)}{a_Z(s_W^2 \to 0)}, \tag{17}$$

we formulate the global stability requirement by demanding $A_Z(s_W^2)$ to assume a stable minimum under variation of s_W^2,

$$\frac{d}{ds_W^2} A_Z(s_W^2) = 0. \tag{18}$$

The physical meaning of (18) is clarified by first of all noting that the evaluation of (16) in the narrow width Breit-Wigner approximation ($\Gamma_Z = const.$) yields the reduced Z^0 width, $a_Z(s_W^2) = \Gamma_Z/M_Z$. We expect that the equality between $a_Z(s_W^2)$ and the reduced Z^0 width will remain true in an approximate sense,

$$a_Z(s_W^2) \approx \frac{\Gamma_Z(M_Z^2)}{M_Z}, \tag{19}$$

even when the s-dependence of $\Gamma_Z(s)$ is fully taken into account and all thresholds are included. The connection of (18) with the minimization requirement of section 3, formulated for $s \to \infty$, is established by evaluating (16,17) in the limit of massless particles withing the vacuum-polarization loops. Inserting (2), which in the present context is assumed to be valid for all $s \geqq 0$, into (16), one finds

$$A_Z(s_W^2) = \frac{\frac{c_{11}^{f\bar{f}}}{c_{22}^{ff}} s_W^4 - 2\frac{c_{12}^{f\bar{f}}}{c_{22}^{ff}} s_W^2 + 1}{1 - s_W^2}. \tag{20}$$

The expression (20) for $A_Z(s_W^2)$ in the massless limit thus coincides with the limit of $s \to \infty$ of the reduced Z^0 width, (5), apart from the suppression of the bosons in (20), which originates from the presence of the Z^0 mass, which sets the scale in (16). The requirement that the global normalized absorption strength (16,17) be minimal under variation of s_W^2, in view of the asymptotic and massless cases (5) and (20), thus amounts to the appropriate generalization from a local ($s \to \infty$) to a global quantity, which allows one to properly evaluate the effects caused by the non-vanishing masses of the particles contributing to vacuum polarization.

Due to the dependence of $A_Z(s_W^2)$ on M_Z, where

$$M_Z^2 = \frac{\pi}{\sqrt{2}} \frac{\alpha}{G_F s_W^2 (1 - s_W^2)(1 - \Delta r)}, \tag{21}$$

---

* $G_Z(s)$ denotes the finite width correction related to $Im\Pi_Z(s)$ via dispersion relations [1,2].

and on $\Gamma_Z(s)$, which for leptons and quarks is given by

$$M_Z \Gamma_{Z \to \Sigma f\bar{f}} = s \frac{\alpha_W}{1 - s_W^2} \frac{1}{3 \cdot 16} \sum_{f,\bar{f}} (1 - \frac{4m_f^2}{s})^{1/2} \cdot \left[(v_f^2 + a_f^2)(1 + 2\frac{m_f^2}{s}) - 6a_f^2 \frac{m_f^2}{s}\right], \tag{22}$$

the value of s_W^2 which minimizes $A_Z(s_W^2)$ will depend on α, G_F and the masses of the leptons and quarks, in contrast to the simple expression (6) for s_W^2, which only contained the quantum numbers of these particles. The notation [2] in (21) and (22) is completely standard [6], Δr denoting the vacuum-polarization correction to the Z^0 mass, and v_f and a_f the vector and axial-vector coupling constants. We refrain from giving the $Z^0 \to W^+W^-$ and $Z^0 \to WH$ widths [2], as these yield negligible contributions to $A_Z(s_W^2)$ anyway, the dominating ones arising from the region of $s \cong M_Z^2$ in the integrand, and we refer to ref. 2 for further details.

The numerical evaluation of the minimization of $A_Z(s_W^2)$ yields

$$s_W^2 \cong 0.20 \qquad (\text{for } \mathrm{m_t} \gtrsim 50\mathrm{GeV}) \tag{23}$$

unless additional Z^0 decay modes, such as $Z^0 \to t\bar{t}, b'\bar{b}', \tilde{\nu}_s\bar{\tilde{\nu}}_s, ...$ are present, in which case one finds

$$s_W^2 \cong 0.23. \tag{24}$$

The upper limit on the mass of the top quark from the UA2 and CDF experiments is given by $m_t \gtrsim 80GeV$, while LEP data also do not indicate Z^0 decays into hitherto unknown particles, such as supersymmetric ones. A straightforward minimization of $A_Z(s_W^2)$ thus leaves us with the result given by (23), which is somewhat low when compared with the empirical one given in (15). Further theoretical work on the global formulation of the stability principle will be required, in particular higher electroweak orders are to be more fully taken into account.

5. THE RELATIVE STRENGTH OF ELECTROMAGNETIC AND STRONG INTERACTIONS

Evidently, the case of strong interactions (QCD), i.e. the question of the relative magnitude of the quark-gluon coupling constant and the electromagnetic one, at first sight is of completely different nature than the case of electroweak interactions. In contrast to the electroweak case, there is indeed no a priori mixing between the photon and the gluon.

As a consequence of the coupling of the photon to quarks and the formation of hadronic bound states, mixing of the photon reappears, however, at the bound-state level as a mixing of the photon with the neutral hadronic vector-meson states, ρ^0, ω, ϕ, etc. The $\gamma\rho^0(\omega, \phi, J/\psi, ...)$ mixing phenomenon leads to a successful description of many features of photon-hadron interactions within the framework of an effective lagrangian, such as high-energy photoproduction, shadowing in photon reactions with complex nuclei, $\rho^0\omega$ mixing, etc. The $\gamma\rho^0$-mixing parameter, the quantity analogous to s_W in the $\gamma\rho^0$ case is given by

$$\lambda_\rho = e/g_{\rho\pi\pi}, \tag{25}$$

where the $\rho\pi\pi$ coupling, $g_{\rho\pi\pi}$, is in turn related to the QCD coupling constant α_s. A prediction of $g_{\rho\pi\pi}$ thus indirectly leads to a prediction for α_s. Empirically we have

$$\lambda_\rho^2 \cong 0.004, \tag{26}$$

which is almost two orders of magnitude smaller than $s_W^2 \cong 0.23$. While from our present point of view* the phenomenon of $\gamma\rho^0$ mixing is by far not as fundamental as γW^3 mixing, it is in fact tempting to apply the stability principle also to ρ^0 propagation.

We require [1] that the value of e be such that the disturbance of the short-range ρ^3 interaction by the photon is to lead to a stable minimum of the reduced ρ^0 width under variation of e relative to $g_{\rho\pi\pi}(\alpha_s)$.

Referring to ref. 1 for somewhat more details, we simply have to write down the ρ^0-vacuum-polarization function which in complete analogy to (2) is given by

$$Im\Pi_\rho = \frac{g_{\rho\pi\pi}^2}{4\pi}\frac{c_{11}\lambda_\rho^4 - 2\bar{c}_{12}\lambda_\rho^2 + \bar{c}_{22}}{1-\lambda_\rho^2} \tag{27}$$

While

$$c_{11} = n_F\frac{8}{9} = \begin{cases} 16/9, & \mathrm{n}_F = 2, \\ 8/3, & \mathrm{n}_F = 3, \end{cases} \tag{28}$$

as in the electroweak case due to the photon-quark interaction, the value for $\bar{c}_{12} = \bar{c}_{22}$ is now dramatically different,

$$\bar{c}_{12} = \bar{c}_{22} = \frac{1}{12}, \tag{29}$$

from the electroweak case, where $c_{12}^{f\bar{f}} = c_{22}^{f\bar{f}} = 1$. The difference is due to the spin 0, isospin 1 nature of the pion.

The stability principle now yields

$$\lambda_\rho^2 = 1 - \sqrt{1 - \frac{2\bar{c}_{12} - \bar{c}_{22}}{c_{11}}} = \begin{cases} 0.016, & (\mathrm{n}_F = 3), \\ 0.024, & (\mathrm{n}_F = 2). \end{cases} \tag{30}$$

to be compared with the empirical value $\lambda_\rho^2 \cong 0.004$ given in (26). The result (30) is to be considered quite encouraging in view of the crudeness of the approximations

* This point of view would have to be changed dramatically, if the weak bosons would eventually turn out to be of composite nature. One should remember that also ρ^0, ω, ϕ were once conjectured to be elementary gauge bosons [7].

involved, e.g., the neglect of other channels, like $\pi\omega, \pi A_1, K\overline{K}$, etc. apart from the $\rho^0 \rightarrow \pi\pi$ decay, the neglect of hadronic form factors, etc.

6. CONCLUSIONS

We propose to look at the relative strengths of the different interactions observed in nature from a basically dynamical as opposed to a primarily group-theoretical point of view. The proposed stability postulate on Z^0 propagation leads to an empirically successful relationship between $sin^2\theta_W$ and the quantum numbers, weak isospin and electric charge, of the leptons and quarks. In other words, it leads to an interesting consistency condition between the relative magnitude of the interaction strengths and the spectrum of particles. This is also seen explicitly by the application of our principle to the "electrostrong" case, $\lambda_\rho^2 = e^2/g_{\rho\pi\pi}^2(\alpha_s)$, where the different spectrum of contributing states leads to a value of λ_ρ^2 much smaller than the corresponding quantity in the electroweak case, $\lambda_\rho^2 \ll sin^2\theta_W$, which is quite in line with empirical observation.

ACKNOWLEDGEMENT

The author thanks M. Zrałek for a very pleasant stay at the Sczyrk summer school.

REFERENCES

[1] F.M. Renard and D. Schildknecht, Phys. Lett. B 219, (1989) 481.
[2] F.M. Renard and D. Schildknecht, Phys. Lett. B 225, (1989) 443.
[3] H. Georgi, H. Quinn and S. Weinberg, Phys. Rev. Lett. 33, (1974) 451.
[4] W. Zimmermann, Comm. Math. Phys. 97, (1985) 211;
J. Kubo, K. Sibold and W. Zimmermann, Nucl. Phys. B 259, (1985) 331.
[5] e.g., U. Amaldi et al., Phys. Rev. D 36, (1987) 1385.
[6] G. Gounaris and D. Schildknecht, Z. Phys. C 40, (1988) 447;
Z. Phys. C 42, (1989) 107.
[7] J.J. Sakurai, Ann. of Nucl. Phys. 1 (1960).

ISBN 1-56072-027-1

RADIATIVE CORRECTIONS TO DEEP INELASTIC *eP* SCATTERING AT HERA

H. Spiesberger*

II. Institut für Theoretische Physik der Universität Hamburg, Luruper Chaussee 149, 2000 Hamburg 50, Germany

Abstract

We give a review of the $\mathcal{O}(\alpha)$ electroweak radiative corrections to neutral and charged current processes in deep inelastic electron proton scattering at HERA. These corrections include 1−loop and 1−photon bremsstrahlung contributions. Cross sections as well as charge and polarization asymmetries and NC/CC ratios are discussed.

Lecture given at the 13th International School of Theoretical Physics in Szczyrk, Poland, Sept., 19th - 26th, 1989

*Supported by Bundesministerium für Forschung und Technologie, 05 4HH 92P/3, Bonn, FRG

1. Introduction

Deep inelastic lepton nucleon scattering has been since long an important particle reaction that provided insight into the structure of matter and interactions. In the past fixed target experiments at SLAC and CERN have reached a momentum transfer Q^2 of up to 500 GeV^2. HERA will be the first storage ring for electrons and protons. It will significantly enlarge the range of energy and momentum transfer that is accessible to experiment. The center of mass energy will reach 314 GeV ($S \simeq 10^5\, GeV^2$ for an electron energy of $E_e = 30\, GeV$ and a proton energy of $E_P = 800\, GeV$). Events with Q^2 of up to several $10^4\, GeV^2$ and with x down to 10^{-4} will easily be observed. Moreover, both electrons and positrons can be accelerated and longitudinal polarization of up to a theoretical limit of 80% will probably be available.

Important topics of physics investigations [1] will be the measurement of the shape in Bjorken x of the structure functions, of their scaling violations, and the determination of the running of α_S as well as the search for tests of the electroweak standard model. In addition, HERA will open a large searching ground for any kind of new physics. The perspectives on HERA physics have been discussed *e.g.* in [2]. Deviations from the standard model - be it new symmetries like supersymmetry or extensions of the standard model gauge group, or new particle degrees of freedom - will in most cases show up as small effects. A precise knowledge of the standard model predictions is therefore indispensible.

In the standard theory based on the gauge group $SU(3)_C \times SU(2)_L \times U(1)_Y$ the deep inelastic scattering is described by independent 4-fermion scattering processes between the lepton and the quarks inside the nucleon. In this partonic picture the cross section for electron proton scattering reads

$$d\sigma(eP \to eX) = \sum_f \int dx' q_f(x') d\hat{\sigma}(eq_f \to eX) \qquad (1)$$

where $q_f(x')$ are the probability distributions of the quark with flavor f inside the proton and $d\hat{\sigma}$ are electron parton subprocess cross sections rescaled by x' in their center of mass energy. This relation will get corrections from the strong interactions. In leading logarithmic order their effect can be respected by using Q^2 dependent distributions $q_f(x', Q^2)$. In this report we do not discuss these QCD corrections but we are interested in the corrections of electroweak origin to the basic electron quark cross section.

The calculation of the electroweak radiative corrections presented here is performed in the on-mass-shell scheme which uses the particle masses M_W, M_Z of the vector bosons besides the Higgs boson mass M_H and the fermion masses m_f together with the fine structure constant α as basic parameters. A detailed discussion of the renormalization procedure and the basic building blocks needed in the calculation of electroweak 1−loop corrections can be found in [3][1].

At low $Q^2 \ll M_Z^2$ electromagnetic corrections are the only important ones. For the pure photon exchange they have been calculated by Mo and Tsai [4] and by Akhundov et al. [5]. There is an extensive literature concerning the radiative corrections to charged current deep inelastic neutrino scattering at low Q^2 [6]. In this context de Rújula et al. [7] have discussed the appearance of quark mass singularities and their absorption into the structure functions. For neutral current scattering this has recently been done also in [8].

At large energies weak interaction effects come into play but still photonic corrections to the lepton line are the dominating ones. These corrections have been calculated in the leading logarithmic approximation in [9,10] and recently also by [11]. They can most conveniently be described in the structure function formalism where also the summation of the leading logarithms to all orders can easily be performed. This has first been discussed by Consoli and Greco in [12] and later in [13].

The complete $\mathcal{O}(\alpha)$ electromagnetic and weak corrections [14] have been calculated by Bardin et al. [15,16,17] and by Böhm and Spiesberger [18,19]. The calculations of both groups differ by the use of the gauge (unitary in [15,16,17] versus 't Hooft-Feynman in [18,19]) and in the details of the methods used for the numerical integration over the photon phase space. In both calculations the photon is treated as totally inclusive. The comparison of both calculations show now agreement of the numerical results within less than 1% in the whole x, y range ($0.01 \leq x, y \leq 0.99$). This makes us confident in that the corrections are now known with sufficient accuracy. In this paper we present mainly results of [18,19] but discuss also those of [16,17].

2. Lowest Order Formulas

In order to define our notations we repeat here the Born-level expressions for the NC cross section. We use as usual the kinematical variables

$$\begin{aligned} Q^2 &= -(p-p')^2, \\ x &= \frac{Q^2}{2P(p-p')}, \\ y &= \frac{P(p-p')}{Pp} = \frac{Q^2}{xS} \end{aligned} \tag{2}$$

where $S = (p+P)^2$ and the meaning of the momenta is indicated in fig. 1. For scattering of electrons or positrons with degree of longitudinal polarization P_L the

[1] See also the contribution of M. Böhm in these proceedings

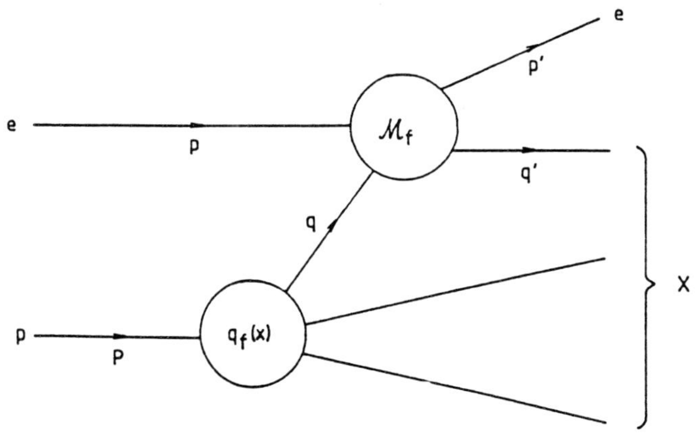

Figure 1: Notations for deep inelastic lepton proton scattering in the parton model.

cross section reads:

$$\left.\frac{d^2\sigma}{dx dy}\right|_{e^\pm P \to e^\pm X} = \frac{2\pi\alpha^2}{xyQ^2}\left(\bar{\sigma}_0^\pm + P_L\bar{\sigma}_L^\pm\right), \tag{3}$$

and the reduced cross sections $\bar{\sigma}_i^\pm$ can be expressed in terms of structure functions:

$$\begin{aligned} \bar{\sigma}_0^\pm &= F_2^0[1+(1-y)^2] \pm xF_3^0[1-(1-y)^2], \\ \bar{\sigma}_L^\pm &= F_2^L[1+(1-y)^2] \mp xF_3^L[1-(1-y)^2], \end{aligned} \tag{4}$$

with

$$\begin{aligned} F_2^{0,L} &= \sum_f A_{0,L}^f(Q^2)x\left(q_f(x)+\bar{q}_f(x)\right), \\ xF_3^{0,L} &= \sum_f B_{0,L}^f(Q^2)x\left(q_f(x)-\bar{q}_f(x)\right), \end{aligned} \tag{5}$$

and

$$\begin{aligned} A_0^f &= Q_f^2 - 2v_eQ_fv_f\,\chi_Z(Q^2) + (v_e^2+a_e^2)(v_f^2+a_f^2)\left(\chi_Z(Q^2)\right)^2, \\ A_L^f &= \quad - 2a_eQ_fv_f\,\chi_Z(Q^2) + 2v_ea_e(v_f^2+a_f^2)\left(\chi_Z(Q^2)\right)^2, \\ B_0^f &= \quad - 2a_eQ_fa_f\,\chi_Z(Q^2) + 2v_ea_e2v_fa_f\left(\chi_Z(Q^2)\right)^2, \\ B_L^f &= \quad - 2v_eQ_fa_f\,\chi_Z(Q^2) + (v_e^2+a_e^2)2v_fa_f\left(\chi_Z(Q^2)\right)^2. \end{aligned} \tag{6}$$

In (6) one recognizes the different contributions of the pure γ exchange (where only $A_0^f = Q_f^2$ is nonzero and therefore the cross section does not depend on the charge and polarization of the leptons), the γZ interference and the pure Z exchange which contains the reduced Z propagator

$$\chi(Q^2) = \frac{Q^2}{Q^2+M_Z^2}. \tag{7}$$

v_f and a_f are the vector and axial vector coupling constants of the fermions to the Z given by their charge Q_f and isospin I_3^f:

$$v_f = \frac{2I_3^f - 4s_W^2 Q_f}{4s_W c_W}, \quad a_f = \frac{I_3^f}{2s_W c_W}. \tag{8}$$

The weak mixing angle is determined by the gauge boson masses:

$$c_W = \frac{M_W}{M_Z}, \quad s_W^2 = 1 - c_W^2. \tag{9}$$

One could also use a parametrization in terms of the μ decay constant G_μ using the realtion [20]

$$G_\mu = \frac{\pi\alpha}{\sqrt{2}\, s_W^2 M_W^2} \frac{1}{1-\Delta r}. \tag{10}$$

In lowest order one has $\Delta r = 0$. Inclusion of radiative corrections in this parametrization leads to the modified on-mass-shell scheme (MOMS). We will discuss it in some more detail below.

3. Radiative Corrections to the Neutral Current

The $\mathcal{O}(\alpha)$ corrections to the NC cross section are described by virtual 1-loop diagrams and real 1-photon bremsstrahlung. They are shown in figs. 2 and 3. We are not going to present the complete set of formulas for the $\mathcal{O}(\alpha)$ radiative corrections. They can be found *e.g.* in [14]. Instead we only discuss some important features:

- For a better understanding we need a classification of the various contributions to the radiative corrections . For this we look again at the cross section formula. The Born cross section for eP scattering is a sum over quark flavors and over the type of the exchanged boson:

$$\left.\frac{d^2\sigma}{dxdy}\right|_{eP\to eX} = \sum_f q_f(x) \sum_B \left.\frac{d^2\sigma}{dxdy}\right|^B_{eq_f\to eq_f}, \quad B = \gamma,\ Int,\ Z. \tag{11}$$

  Each contribution in (11) gets different corrections. They depend on both the external quark line and on the type of the exchanged boson. But there are certain parts of the radiative corrections that do not depend on f or B or only on one of these indices. For example one could write for the $\mathcal{O}(\alpha)$-corrected cross section

$$\left.\frac{d^2\sigma}{dxdy}\right|_{eP\to eX} = (1+\delta_0)\sum_f q_f(x)(1+\delta_f)\sum_B(1+\delta_B+\delta_{f,B}) \left.\frac{d^2\sigma}{dxdy}\right|^B_{eq_f\to eq_f}. \tag{12}$$

  By this a classification of the various parts of the corrections with respect to their dependence on the theoretical input can be achieved: δ_0 does depend

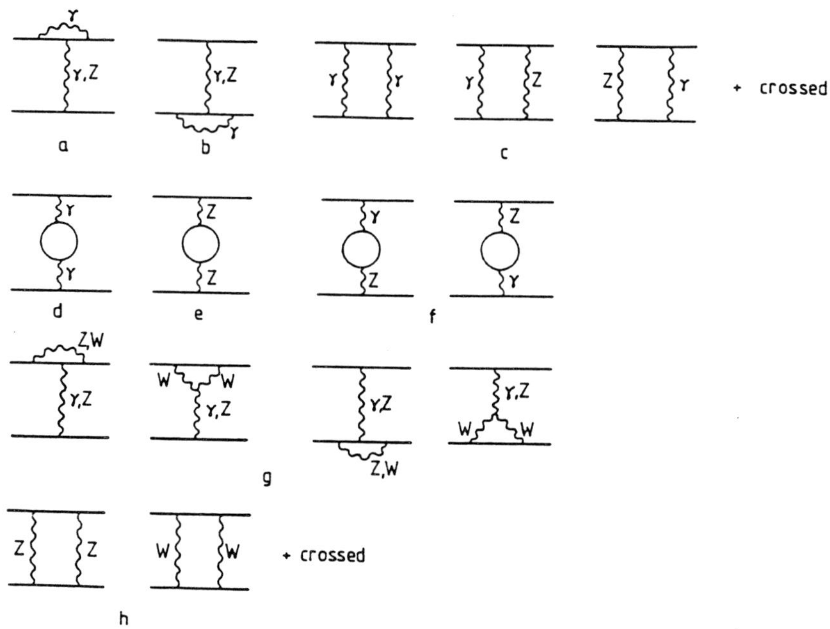

Figure 2: 1-loop diagrams for neutral current lepton quark scattering.

neither on the quark content of the proton nor on whether the photon or the Z mediates the interaction. δ_0 could therefore be denoted as the universal part of the corrections. It is given by the photonic corrections at the lepton line (soft and virtual parts). δ_f contains the analogous photonic corrections at the quark line, whereas δ_B contains the self energy contributions together with the weak leptonic vertex corrections. Finally, $\delta_{f,B}$ arises from all other weak corrections (quark vertex, box diagrams) and from the photonic lepton-quark interference (Bremsstrahlung and boxes with at least one photon).

The hard Bremsstrahlung does not fit into this classification because it results

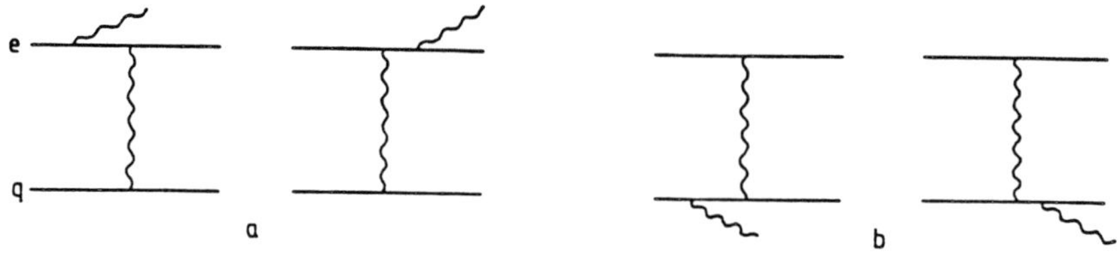

Figure 3: 1-photon bremsstrahlung diagrams for neutral current lepton quark scattering.

from a convolution of radiative cross sections with the parton distribution functions. This means that also the leptonic corrections do depend on the proton structure - although not dramatically. The separation of corrections in this way is therefore only of limited practical use.

Instead of this we separate the corrections according to gauge invariant subsets of diagrams, namely into *i*) the leptonic corrections that consist of the diagrams containing an additional photon attached to the lepton line (*i.e.* the photonic lepton vertex[2] fig. 2a, and the photon emission from the lepton line, fig. 3a, *ii*) the hadronic corrections described by diagrams with an additional photon at the quark line analogous to *i*) (diagrams of figs. 2b, 3b). This part is proportional to the square of the quark charge. *iii*) The lepton-hadron interference part consisting of the $\gamma\gamma$ and γZ box diagrams (fig. 2c) and the interference of the bremsstrahlung diagrams figs. 3a with figs. 3b. This contribution is proportional to the quark charge. Finally we have *iv*) the purely weak corrections consisting of all the other diagrams that do not contain an additional photon. This part is IR finite and contains the diagonal γ and Z self energies (figs. 2d,e), the γZ mixing (fig. 2f), the weak lepton and quark vertex corrections (fig. 2g), and the boxes with two heavy gauge bosons (fig. 2h).

- The self energy diagrams (fig. 2d-f) contain loop diagrams that are built with all particle degrees of freedom that couple to the gauge bosons. Therefore they contain information on the whole theory. From them there arises a dependence on the top mass, the Higgs mass, and on the masses and couplings of eventually existing other unknown particles. Fig. 4 contains as an example the list of diagrams that contribute to the Z self energy.

- Each contribution to the $\mathcal{O}(\alpha)$ corrected matrix element can be written in the current $\times$ current form:

$$\delta\mathcal{M}_i = i\frac{\alpha}{\pi}\; J_{lept} \;\times\; J_{quark}\; D_i \times \frac{\alpha}{2\pi} F_i(S,x,y), \tag{13}$$

where

$$\begin{aligned} J_{lept} &= \bar{u}(p')\gamma_\mu(G_V^{e,i} - G_A^{e,i}\gamma_5)u(p), \\ J_{quark} &= \bar{u}(q')\gamma_\mu(G_V^{f,i} - G_A^{f,i}\gamma_5)u(q), \end{aligned} \tag{14}$$

and

$$D_i = 1/Q^2, \quad \text{or} \quad 1/(Q^2 + M_Z^2). \tag{15}$$

[2]The photonic vertex corrections are always combined with the self energies of the external fermion lines

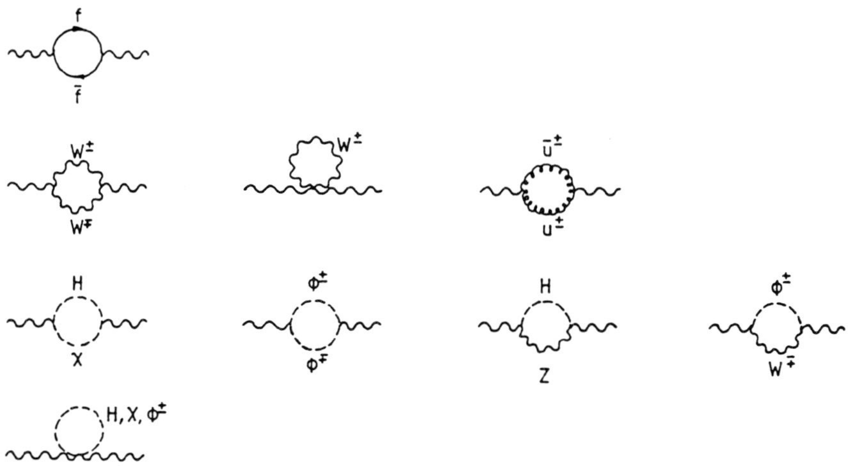

Figure 4: 1-loop Z self energy diagrams.

F_i are scalar form factors that depend in general on the invariants S, x, and y. The vertex corrections and self energies are described by form factors that depend only on Q^2.

- The self energies are dominated by the fermion loops. This contribution is sometimes refered to as a QED part. It can partly be accounted for by the use of the running fine structure constant

$$\alpha(Q^2) = \frac{\alpha(0)}{1 - \Pi^\gamma(Q^2)}. \tag{16}$$

$\Pi^\gamma = \hat{\Sigma}^\gamma/Q^2$ is the vacuum polarization derived from the renormalized photon self energy $\hat{\Sigma}^\gamma$ and $\alpha(0) = 1/137.036$. Its hadronic part is connected with the total hadronic cross section in e^+e^- annihilation [21]. At $Q^2 \simeq M_Z^2$ its value is $\Pi^\gamma \simeq 0.06$. The prescription (16) together with the inclusion of the leptonic QED corrections (discussed below) gives the $\mathcal{O}(\alpha)$ corrected cross section already with a precision of a few %. The self energy correction to the γ exchange is treated correctly in this way. By the use of (16) also the Z self energy is respected approximately. The appropriate factor in this case contains the Z self energy in the combination $\Pi^Z(Q^2) = \hat{\Sigma}^Z(-Q^2)/(Q^2 + M_Z^2)$ and

$$\frac{\alpha(0)}{1 - \Pi^Z(Q^2)} \simeq \alpha(Q^2). \tag{17}$$

But this is a good approximation only if Q^2 is not small and if the top quark is not too heavy. Instead of this, it is a better approximation to absorb the Z self energy by normalizing the Z exchange part with help of the μ decay constant. In lowest order one has the relation

$$\frac{\alpha}{s_W^2(1-s_W^2)} = \frac{\sqrt{2}M_Z^2 G_\mu}{\pi}. \tag{18}$$

This factor appears in front of the Z exchange contribution to the NC cross section. If 1-loop corrections are consistently included, one has to take into account also the corrections to this relation, *i.e.* the corrections to the μ decay, denoted usually by Δr. Then one has

$$\frac{\alpha}{s_W^2(1-s_W^2)}\frac{1}{1-\Pi^Z} = \frac{\sqrt{2}M_Z^2 G_\mu}{\pi}\frac{1-\Delta r}{1-\Pi^Z}. \tag{19}$$

Δr is dominated by the W self energy and the approximate relation

$$1-\Delta r \simeq 1-\Pi^Z \tag{20}$$

holds also for a heavy top. The modified on-mass-shell scheme based on the relation (19) leads to smaller numerical values for the weak corrections. They are shown in fig. 5b. Fig. 5a contains for comparison the results in the on-mass-shell scheme where $M_W = 82\,GeV$ and $M_Z = 93\,GeV$ are fixed input parameters. Both figures are taken from [16]. Fig. 5a contains also results of [18] (dashed curve). The difference between the two results is due to the resummation of higher order terms by including the photon self energy in accordance with (16) whereas [18] contains the pure $\mathcal{O}(\alpha)$ corrections only (linearized in Π^γ). The dashed-dotted curve results after linearizing the results of [16] to $\mathcal{O}(\alpha)$. The agreement between the two calculations is perfect up to $y = 0.7$ and still satisfactory for larger y.

3.1 Bremsstrahlung

The virtual photonic corrections contain IR divergences. In our calculation they have been regularized by the use of a small finite photon mass λ. After the inclusion of real photonic corrections these singularities cancel. The resulting cross section then depends on the maximal energy of the emitted photon K_{max}. Because of the complexity of the calculations, an experimental cutoff for this energy is usually imposed in a Monte Carlo treatment. It is more easy to calculate the integrals over the phase space for the momentum of the emitted photon up to the kinematically allowed maximum. K_{max} is then determined by the kinematical variables S, x, and y and due to the fact that the HERA laboratory system is not equal to the center of mass system also on the angle of the photon momentum with respect to the beam axis. The final result

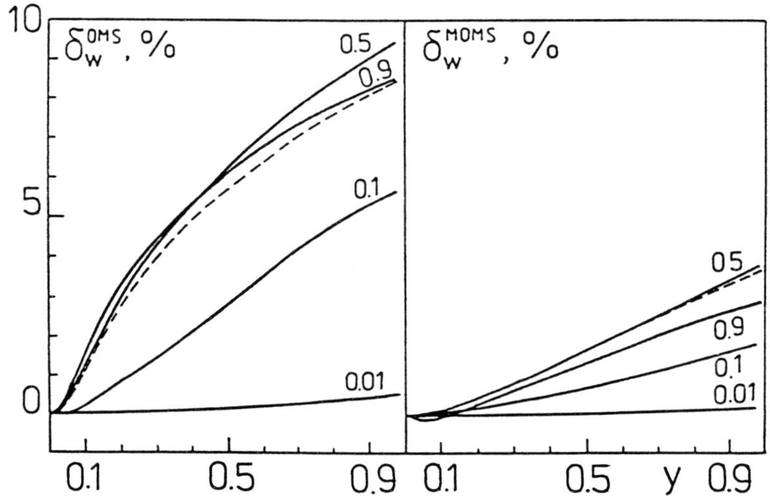

Figure 5: Weak 1-loop corrections for electron proton scattering. Full curves from [16], dashed from [18], dashed-dotted: in α linearized version of [16]. $M_W = 82\,GeV$, $M_Z = 93\,GeV$, $M_H = 100\,GeV$, $m_t = 30\,GeV$, $S = 10^5\,GeV^2$, quark distribution functions of [22].

for the $\mathcal{O}(\alpha)$ corrected cross section (if also no angle cutoff is used in the integration over the solid angle) depends only on S. x, and y.

In order to get a correct finite result, the real Bremsstrahlung has to be calculated using the same regularization of the IR divergence as for the virtual ones. To simplify the calculation, the phase space is cut into a soft and a hard region and the λ-regularization is only performed in the soft region where the IR singularity is located. The soft region is chosen such that the photon momentum can be neglected if compared to the momenta of the other particles. The resulting soft photon approximation is much simpler to calculate. The integral over the hard region is finite and the limit $\lambda \to 0$ can be taken before the integration is performed. The separation of the soft and hard region is arbitrary. The final result must not depend on the description for this separation. In our calculation it is described by a limit on the photon energy ΔE in the HERA lab-system. The verification of the independence of the results on ΔE is an important check of the correctness of the calculations.

The diagrams of figs. 3a, b describe the Bremsstrahlung for the neutral current process. The corresponding matrix element squared has been given in [18]. It contains

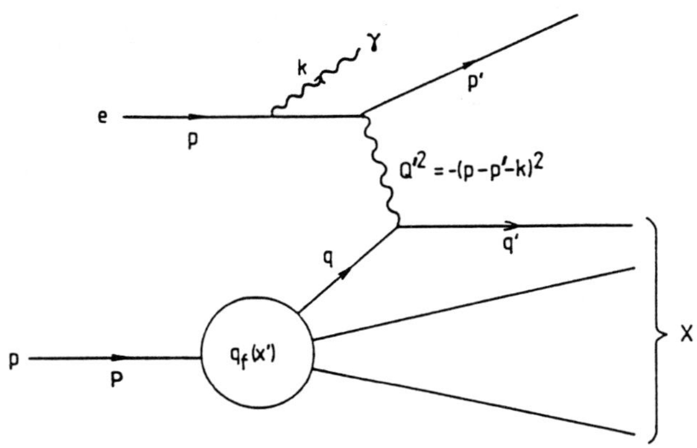

Figure 6: Notations for the bremsstrahlung in lepton quark scattering.

terms of the form

$$\frac{m_i^2}{(kp_i)^2} \quad \text{and} \quad \frac{p_i p_j}{(kp_i)(kp_j)}, \tag{21}$$

where k is the photon momentum and p_i denote the momenta of the fermions with mass m_i. Upon integration over the photon energy these factors lead to the IR divergence. The angular integration is of the form

$$\int_{-1}^{1} \frac{dz}{E - |p|z}$$

which leads to numerically large logarithms

$$\ln \frac{m_i^2}{4E_i^2}.$$

They arise from the configuration where the photon is emitted parallel to one of the fermion momenta (collinear logarithms).

The presence of an additional photon in the final state forces us to review the definition of the kinematical variables (see fig. 6). We want to calculate a correction to the differential cross section $d^2\sigma/dxdy$ where x and y are still defined from the measured momentum of the final state lepton according to (2). In order that the emission of a photon is kinematically allowed, x can now no more be interpreted as being the fraction of the energy of the incoming quark with respect to the proton energy. We introduce for this energy ratio the new variable x' defined by

$$x' = \frac{(p+q)^2}{S}, \tag{22}$$

The kinematical lower limit of x'

$$x'_{min} = x + \frac{2(m_e^2 + m_q^2)}{yS} \tag{23}$$

is reached in the soft photon limit.

The correction to the differential cross section can be written in the form

$$\delta\left(\left.\frac{d^2\sigma}{dxdy}\right|_{eP\to eX}\right) \simeq \sum_f \int_{x'_{min}}^1 dx' q_f(x') \int d^{(3)}PS \left|\mathcal{M}^f(x')\right|^2. \tag{24}$$

$d^{(3)}PS$ is the 3-particle phase space $d^3k/2k_0\delta(q'^2 - m_q^2)$ which can be written in terms of two further dimensionless Lorentz-invariant variables:

$$d^{(3)}PS \simeq \int_{y'_{min}}^{y'_{max}} dy' \int_{\tau_{min}}^{\tau_{max}} \frac{d\tau}{\sqrt{(\tau_{max}-\tau)(\tau-\tau_{min})}}, \tag{25}$$

where

$$y' = \frac{Q'^2}{x'S}, \quad Q'^2 = -(q-q')^2, \tag{26}$$

and

$$\tau = \frac{(p'+q')^2}{(p+q)^2}. \tag{27}$$

The integration limits depend in a complicated way on the fermion masses and the soft/hard separation ΔE. For vanishing m_i and ΔE they are

$$\begin{aligned} &x'_{min} \simeq x, \\ &y'_{min} \simeq 0, \quad y'_{max} \simeq y, \\ &\tau_{min,max} \simeq (1-x')(1+y'-y) + \frac{1}{y}\left(\sqrt{(1-y)(x'-x)(y-y')} \pm \sqrt{xy'}\right)^2. \end{aligned} \tag{28}$$

The integration can only partly be done analytically because of the appearance of the quark distribution functions which depend in a complicated way on the integration variable x' and even on y' if non-scaling distribution functions are used. The results of a numerical calculation where the τ integration is done analytically but the x' and y' integration numerically are shown in figs. 7, 9-10. In order to have as much as possible separated parts that can be compared among different groups, we show separately the leptonic, hadronic, and lepton-hadron interference parts each for the pure photon and Z exchange and the photon Z interference

$$\delta_i(B) = \frac{\left.\frac{d^2\sigma}{dxdy}\right|_i^{1-loop}(B)}{\left.\frac{d^2\sigma}{dxdy}\right|_i^{Born}(B)} - 1, \tag{29}$$

with $i = lep, qua, int$ and $B = \gamma,\ \gamma Z,\ Z$.

- **Radiation from the lepton line.** The leptonic corrections are shown in figs. 7a-c. They show a very pronounced increase to large values at small x and large y and can become also large but negative for small y and large x. This behaviour can be understood from the following observations:

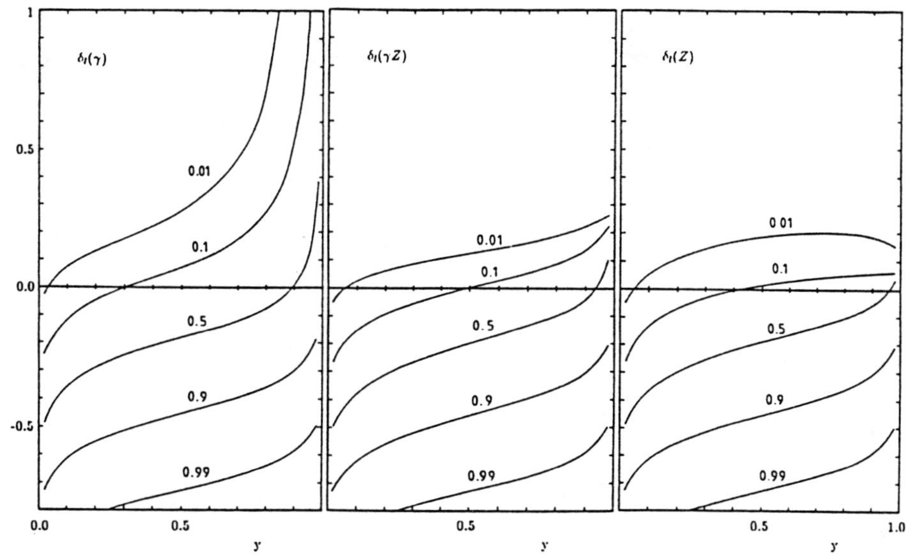

Figure 7: Leptonic QED corrections for electron proton scattering. Parameters as in fig. 5.

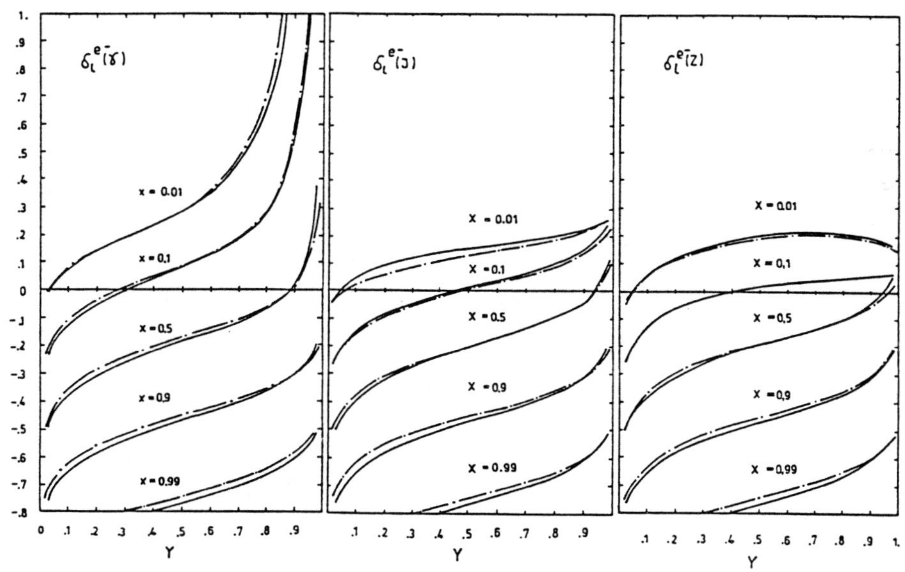

Figure 8: Leptonic corrections to lepton proton scattering in the leading logarithmic approximation (full line, [11]) in comparison with the exact results of [16] (dashed-dotted line). Parameters as in fig. 5.

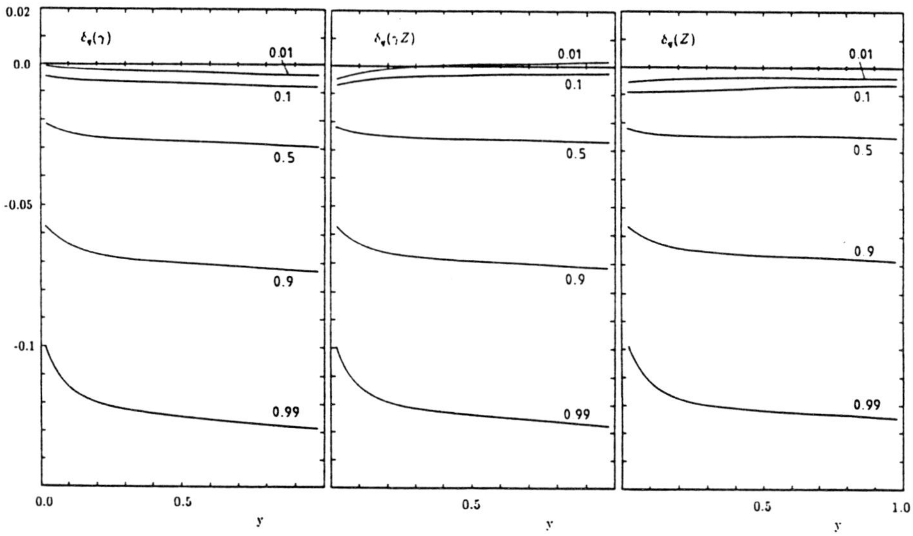

Figure 9: Hadronic QED corrections for electron proton scattering. Parameters as in fig. 5.

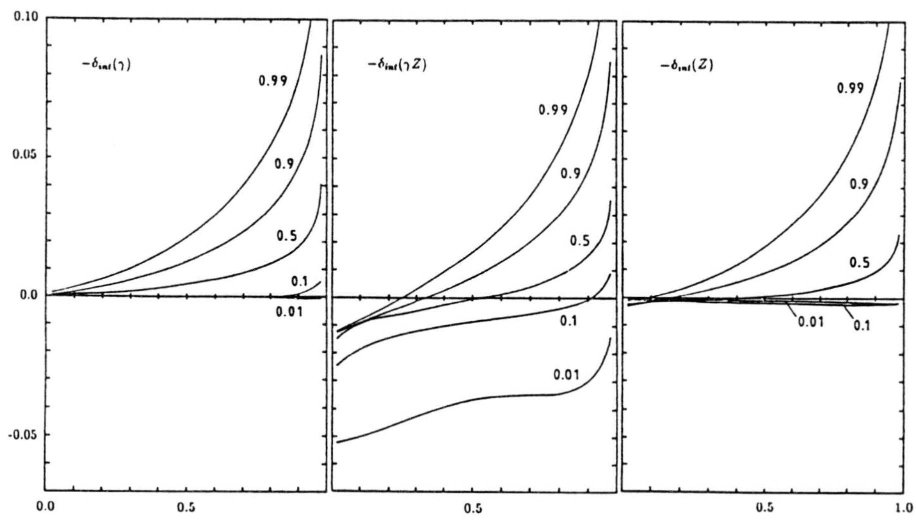

Figure 10: Lepton-hadron QED interference corrections for electron proton scattering. Parameters as in fig. 5.

i) The order of magnitude is determined by the factor

$$\frac{\alpha}{\pi} \ln \frac{m_e^2}{Q^2} \simeq (0.24\%) \times (-25) \simeq \mathcal{O}(10\,\%).$$

This number is multiplied by logarithms of ratios of the maximal photon energy and the center of mass energy which can also become large of the order of 10 for small x and large y.

ii) The emission of an energetic photon can shift the value of the momentum transfer seen from the quark line to very small values (see fig. 6). x and y and consequently $Q^2 = xyS$ are determined from the momentum of the outgoing electron. But $Q'^2 = (p - p' - k)^2$ is also determined by the momentum of the emitted photon and $Q'^2 \neq Q^2$ in general and Q'^2 can be very small compared to Q^2 if the emitted photon takes away a large energy. Thus the bremsstrahlung contribution is enhanced through the photon propagator $1/Q'^4$. This effect is similar to the radiative tail effect above the peak of a resonant cross section. It is responsible for the large increase of the corrections at large y. In principle its magnitude can be reduced by applying suitable cuts, *e.g.* to the energy of the emitted photon E_γ. If $E_\gamma \leq 10\,GeV$ is required the corrections remain below 20 % [9].

These statements are supported by the results shown in figs 7b, c. These figures contain the leptonic corrections to the γZ interference and the Z exchange parts of the NC cross section. In the first case the photon propagator comes in only with the smaller power $1/Q'^2$ and in the second case there is only the Z propagator $1/(Q'^2 + M_Z^2)$ which cannot lead to the strong increase of the corrections. The raise of the corrections with y at large x is due to the increasing phase space volume alone.

iii) At small values of y and large x the photon phase space volume shrinks as is seen from the approximate integration limits given above:

$$\int d^{(3)}PS \simeq \int_x^1 dx' \int_0^y dy'.$$

Therefore for $x \to 1$ and $y \to 0$ the virtual and soft real corrections dominate and lead to large negative contributions.

The purely leptonic corrections discussed up to now can be calculated with an astonishingly well precision in the collinear approximation. This technique isolates regions in the phase space where large logarithmic contributions $\simeq \alpha/\pi \times \ln Q^2/m_e^2$ can arise. These correspond to configurations where the momentum of the photon (either emitted or virtually exchanged) can become parallel to one of the fermion momenta. Fig. 8 shows the results of a calculation by Blümlein [11] (full line). These figures contain also the complete calculation for the corresponding contributions of Bardin et al. [16] (dashed line) which show good agreement with the results in figs. 7a-c.

- **Radiation from the quark line.** The numerical results for the corrections $\delta_{qua}(B)$ arising from the diagrams fig. 2b and 3b are shown in figs. 9a-c. They do not receive large positive contributions because the emission of a photon from the quark line cannot shift Q^2. Moreover they are numerically smaller because of the smaller quark charges and the larger values used for the quark masses ($m_f = 30\,MeV$).

  Due to the Kinoshita-Lee-Nauenberg theorem the sum of virtual and real photonic corrections is independent of logarithms of the final quark mass. But this class of corrections depends logarithmically on the masses of the incoming quarks.

  However, this does not pose a serious problem because the appearance of quark mass singularities is an artifact of a naive interpretation of the parton distribution functions as "bare" ones [7]. The mass singularities contained in δ_{qua} can be factorized from the cross section and it is therefore possible to absorb them into the definition of the distribution functions:

$$\begin{aligned} \left.\frac{d^2\sigma}{dxdy}\right|_{eP\to eX} &= \sum_f \left.\frac{d^2\sigma}{dxdy}\right|_{eq_f\to eq_f} q_f^0(x)\left(1+\delta_{qua}(Q^2,m_f)\right) \\ &= \sum_f \left.\frac{d^2\sigma}{dxdy}\right|_{eq_f\to eq_f} \bar{q}_f(x,Q^2). \end{aligned} \tag{30}$$

  In fact, any experiment can only measure the "dressed" distribution functions $\bar{q}_f$. If this measurement was performed at a certain scale Q_0^2 the predictions for the deep inelastic cross section at a different momentum transfer Q^2 in terms of the dressed distribution functions $\bar{q}_f$ reads

$$\left.\frac{d^2\sigma}{dxdy}\right|_{eP\to eX} = \sum_f \left.\frac{d^2\sigma}{dxdy}\right|_{eq_f\to eq_f} \bar{q}_f(x)\left(1+\delta_{qua}(Q^2,m_f)-\delta_{qua}(Q_0^2,m_f)\right). \tag{31}$$

  The correction now does no more contain mass singularities but is of the form

$$\frac{\alpha}{\pi}\log\frac{Q^2}{Q_0^2}.$$

  Thus the only effect of the photonic quark line corrections is to introduce an additional Q^2 dependence. This can be described in complete analogy to the

Q^2 dependence arising from gluonic corrections, *e.g.* with help of the Altarelli-Parisi equations. Including the term of QED origin they read

$$\begin{aligned} \frac{d\bar{q}_f}{dt}(x,t) = & \frac{\alpha_S}{2\pi}\int_x^1 \frac{dy}{y}\left(\sum_f' \bar{q}_{f'}(y,t)\cdot P_{q_f q_{f'}}\left(\frac{x}{y}\right) + G(y,t)\cdot P_{q_f G}\left(\frac{x}{y}\right)\right) \\ & + \frac{\alpha}{2\pi}\int_x^1 \frac{dy}{y}\bar{q}_f(y,t)\cdot P^{\gamma}_{q_f q_f}\left(\frac{x}{y}\right)\cdot Q_f^2 \\ & + \mathcal{O}(\alpha_S^2) + \mathcal{O}(\alpha_S\alpha). \end{aligned} \tag{32}$$

The Altarelli-Parisi kernels in 1−loop order

$$P^{\gamma}_{qq} = \frac{1+z^2}{(1-z)_+} + \frac{3}{2}\delta(z-1) \tag{33}$$

differ from the QCD kernels only by a group factor. We have solved these equations numerically and found corrections $(q_f^0(x) - \bar{q}_f(x,Q^2))/q_f^0(x)$ at most of the order of magnitude of 2% in the range of x and Q^2 accessible at HERA. [8]. In addition to these leading logarithmic corrections there are also non-logarithmic contributions. But they are small and can be neglected if one aims not more than an accuracy at the 1 % level.

The results in fig. 9 were obtained using the parametrizations for the quark distribution functions of [22] as bare distributions, *i.e.* they contain explicitly the quark mass logarithms. The prescription (32) requires parton distributions that were determined from experiment without subtracting the hadronic logarithmic corrections from the data.

It is important to note that the corrections from radiation at the lepton line and at the quark line are gauge invariant by themselves. Only this allows a separate treatment of them.

- **The lepton-hadron interference** of the diagrams in fig. 3a with fig. 3b and the box contributions of fig. 2c are shown in figs. 10. They do not contain a logarithmic dependence neither on the lepton nor on the quark mass and therefore remain numerically small except at extreme values for x and y. Technically they are described by more complicated formulas than the purely leptonic and hadronic parts. The corresponding expressions contain logarithms and dilogarithms (Spence functions) of arguments that depend on both Q^2 and x. These corrections cannot be absorbed by a redefinition of the quark distribution functions but have to be included as explicit corrections to the cross section if it is required by the requested accuracy. In a first approach to physics analysis at HERA, if one contents oneself with a precision at a few percent level, these corrections can be neglected.

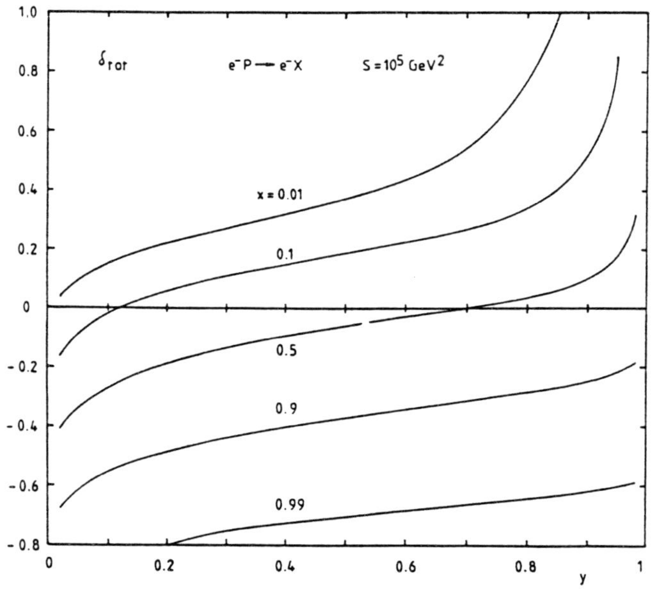

Figure 11: Complete $\mathcal{O}(\alpha)$ corrections to neutral current electron proton scattering. Parameters as in fig. 5.

- Finally we present in fig. 11 the results for the complete $\mathcal{O}(\alpha)$ corrections to $\left.\frac{d^2\sigma}{dxdy}\right|_{eP\to eX}$ including both photonic and weak 1-loop corrections in the on-mass-shell scheme. We mention that to the level of 5% accuracy the complete result is obtained already when only the leptonic QED corrections together with the photon and Z self energies are included.

4. Radiative Corrections to the Charged Current

4.1 Lowest Order Cross Section

The formulas for the charged current differential cross section for

$$e^-P \to \nu_e X, \qquad e^+P \to \bar{\nu}_e X \tag{34}$$

are simpler than in the NC case because here only one diagram appears in lowest order and because the CC interaction is purely left-handed. The Born result reads

$$\left.\frac{d^2\sigma}{dxdy}\right|_{eP\to\nu X} = \frac{\pi\alpha^2}{2s_W^2}\frac{xS}{(Q^2+M_W^2)^2}\left\{q+(1-y)^2\bar{q}\right\}, \tag{35}$$

where

$$q = \begin{cases} u+c & \text{for } e^-, \\ d+s & \text{for } e^+, \end{cases} \qquad \text{and } \bar{q} = \begin{cases} \bar{d}+\bar{s} & \text{for } e^-, \\ \bar{u}+\bar{c} & \text{for } e^+. \end{cases} \tag{36}$$

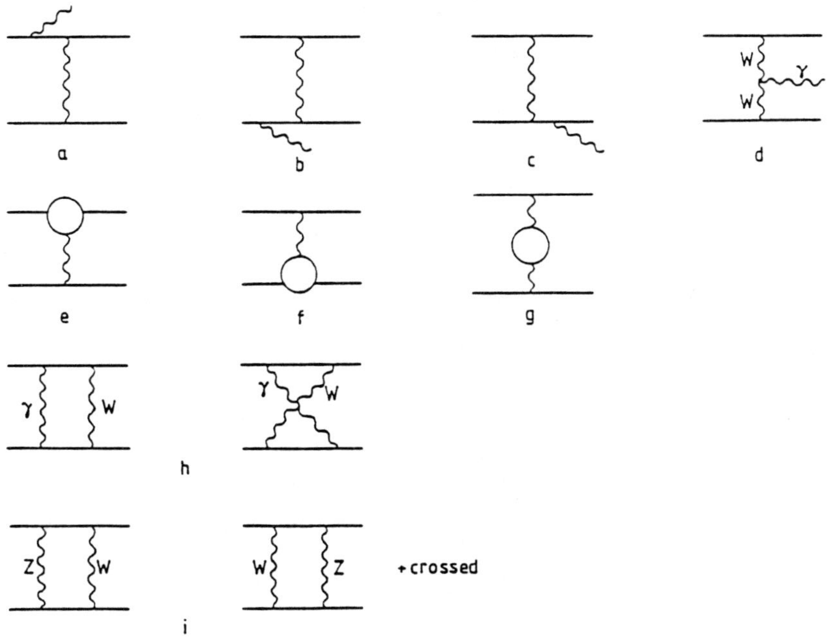

Figure 12: Feynman diagrams contributing to the $\mathcal{O}(\alpha)$ radiative corrections for charged current scattering.

As already discussed in connection with the Z exchange in the NC process one can also normalize the cross section in terms of G_μ with help of the relation (10). Then again the cross section does not receive large radiative corrections associated with the renormalization of the electric charge.

4.2 Radiative Corrections

The 1-loop diagrams of fig. 12 lead to corrections that can be collected together into a single factor $(1 + \delta_{1-loop})$. The corresponding formulas can be found in [17,19].

Compared to the NC cross section the real bremsstrahlung has however a more complicated structure. This is due to the diagram 12d which contains the non-abelian γWW vertex. Due to the presence of this diagram there is no obvious separation of the radiative corrections into gauge invariant classes that could be termed as leptonic, quarkonic or lepton-quark interference parts. By inspection of the expression for the modulus squared of the matrix element for $eq \rightarrow \nu q'\gamma$ [19] one can see that for example the logarithms of the lepton mass arise not only from diagram 12a but also from the interference of this diagram with 12b - d. In [17] a possible way to separate the cross section into different gauge invariant pieces was proposed. For this the relations between the fermion charges and the charge of the W

$$|Q_e| = |Q_W|, \quad Q_a = Q_c + Q_u, \cdots \tag{37}$$

was used to express the cross section in terms of the electron charge and the charge of the incoming quark. The bremsstrahlung contribution then has the form

$$|\mathcal{M}|^2_{eq\to\nu q'\gamma} \sim (Q_e^2 I_{lep} + Q_e Q_q I_{int} + Q_q^2 I_{qua}). \tag{38}$$

After integration over the photon phase space and adding the virtual photonic corrections separated in the same way one can see that the collinear logarithms of the lepton mass are all contained in the term proportional to Q_e^2, those of the incoming quark in the term proportional to Q_q^2 and the contribution proportional to $Q_e Q_q$ contains no mass logarithms at all. The quark mass singularities appear in the same universal form as in the NC case. It is therefore possible to absorb them into the distribution functions in a process independent way.

For the calculation of the real bremsstrahlung we applied a technique different from that in the NC case. The cross section was rewritten in the reference system where the outgoing quark and the photon are at rest: $\vec{k} + \vec{q}\,' = 0$. In this reference system the phase space is independent of the photon emission angle:

$$\int dx' \int \frac{d^3k}{2k_0} \delta(q'^2 - m_{q'}^2) = \frac{1}{4} \int dx' \frac{k_0}{p_0 + q_0 - p_0'} d\cos\theta_\gamma d\phi_\gamma \tag{39}$$

with $\mu^2 = m_e^2 + m_q^2 + m_{q'}^2$. The IR singularity appears now in the x' integration at its lower limit $x'_{min} = x + \mu^2/yS$. After substitution for new variables $u = \log(x' - x - \mu^2/yS)$ and $w = \log(p_0/|\vec{p}| - \cos\theta_\gamma)$ to absorb the rapidly changing factors, the integration can easily be done with help of a standard Monte Carlo algorithm [23]. As checks of the calculation we have in addition to the independence of the final result on the soft/hard separation ΔE the independence on the mass of the final quark.

In fig. 13 we show the results for the complete electromagnetic and weak 1-loop and 1-γ bremsstrahlung corrections to charged current electron and positron scattering. These results are obtained in the on-mass-shell scheme where the cross section is parametrized in terms of M_W and M_Z. They contain the quark mass singularities explicitly and use the quark distribution functions of [22]. The results show a similarity to those for the Z exchange contribution in the NC cross section. The large negative corrections at large x and small y are due to the tight phase space in this region, whereas at large y and small x the corrections become positive, however without getting huge because in the CC case there is no hard radiative tail from the low Q^2 region.

The corrections shown in fig. 13 are bigger by about 14% compared to those derived in the modified on-mass-shell scheme [17]. A comparison of separate parts of the

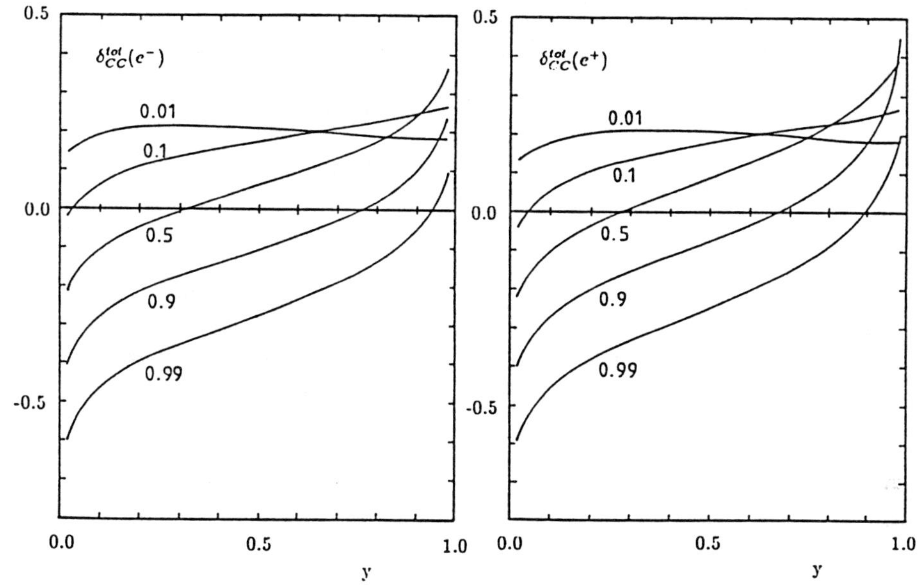

Figure 13: Complete $\mathcal{O}(\alpha)$ radiative corrections to charged current electron and positron quark scattering. Parameters as in fig. 5.

photonic corrections obtained by ourselves and of Bardin et al. show perfect agreement at the level of 1% for the leptonic corrections and even better at the level of 0.1% for the hadronic and interference parts.

5. Electroweak Physics at HERA

As a consequence of the high Q^2 that is reachable in deep inelastic scattering at HERA, the cross sections receive sizable contributions from the pure weak interaction mediated by Z or W exchange. We want now to discuss the question whether and to what extent HERA could be a tool to measure the parameters of the weak sector of the electroweak standard model.

In principle one could use the cross sections themselves, *i.e.* the event rates, to derive the weak parameters. They enter into the cross section formulas first of all in the normalization factors, namely

$$\frac{\pi\alpha^2}{2s_W^2c_W^2}\frac{1}{Q^2+M_Z^2}=\frac{\alpha G_F}{\sqrt{2}}\frac{M_Z^2}{Q^2+M_Z^2}\qquad \text{NC, Z exchange} \tag{40}$$

and

$$\frac{\pi\alpha}{2s_W^2}\frac{1}{Q^2+M_W^2}=\frac{G_F}{\sqrt{2}}\frac{M_W^2}{Q^2+M_W^2}\qquad \text{CC, W exchange.} \tag{41}$$

The advantage of using the cross sections is the high integrated luminosity and as a consequence of this the small statistical errors. But in turn this also means, that the event rates are sensitive to systematic errors in the luminosity. In contrast to this, cross section ratios are much less sensitive to a variation in the luminosity but suffer from limited statistics. From the discussion of the radiative corrections it seems preferable to use cross section ratios because the cross sections receive large corrections of purely electromagnetic origin. These photonic corrections cancel to a large extent in ratios and also beyond the Born level the various cross section ratios are more sensitive to the weak part of the input than to the pure QED contributions.

From the NC cross sections for polarized electrons and positrons one can build the following asymmetries:

$$A_{\pm} = \frac{d\sigma(e_L^{\pm}) - d\sigma(e_R^{\pm})}{d\sigma(e_L^{\pm}) + d\sigma(e_R^{\pm})},$$
$$B_{\pm} = \frac{d\sigma(e_L^{\pm}) - d\sigma(e_R^{\mp})}{d\sigma(e_L^{\pm}) + d\sigma(e_R^{\mp})}, \tag{42}$$
$$C_{L,R} = \frac{d\sigma(e_{L,R}^{-}) - d\sigma(e_{L,R}^{+})}{d\sigma(e_{L,R}^{-}) + d\sigma(e_{L,R}^{+})}.$$

Including the CC cross section one has also the possibility to measure

$$R_{\pm} = \frac{d\sigma(e^{\pm} \to e^{\pm}X)}{d\sigma(e^{\pm} \to \nu_e X)}. \tag{43}$$

In the following we discuss the case where for $d\sigma$ the twofold differential cross sections $d^2\sigma/dxdy$ with maximal degree of longitudinal polarization $P_L = 1.0$ are used but these ratios could also be formed with (partly) integrated cross sections which could lead probably to a gain in statistical accuracy. In figs. 14 and 15 we show the complete $\mathcal{O}(\alpha)$ corrections to A_- and B_+ for $x = 0.2$ as a function of y. The corrections are small. A_- is influenced practically only by the purely weak corrections. The corrections to the asymmetry B_+ is dominated by the photonic lepton-quark interference contribution (including the $\gamma\gamma$ box). These corrections are odd under charge conjugation.

The figs. 14 and 15 contain results for three values of the W mass. From them one can deduce that the asymmetry A_- is sensitive to M_W (resp. s_W^2) and suitable for a determination of the weak mixing angle at HERA. This is also the case for the other asymmetries and $R_{\pm}$. The mixed charge-polarization asymmetry B_+ is exceptional because it is almost independent of M_W. B_+ is therefore an interesting observable that could eventually signal non-standard physics if s_W^2 is known with good accuracy from other sources.

Via the loop contributions to the Z and W self energies the corrections to these asymmetries and the ratios $R_{\pm}$ depend on the values for the top quark mass and the

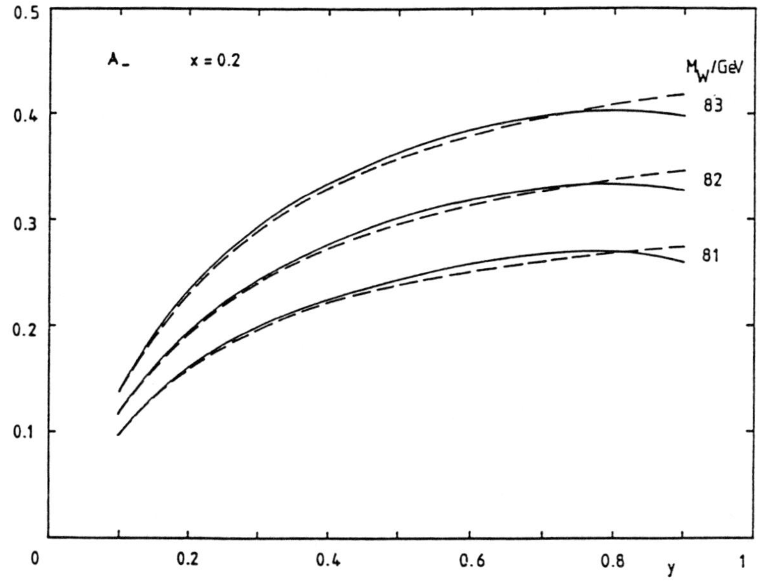

Figure 14: Born result (dashed line) and 1-loop corrected result (full line) for $A_-(x = 0.2,\ P_L = 1.0)$ for $M_W = 81,\ 82$, and $83\,GeV$. Other parameters as in fig. 5.

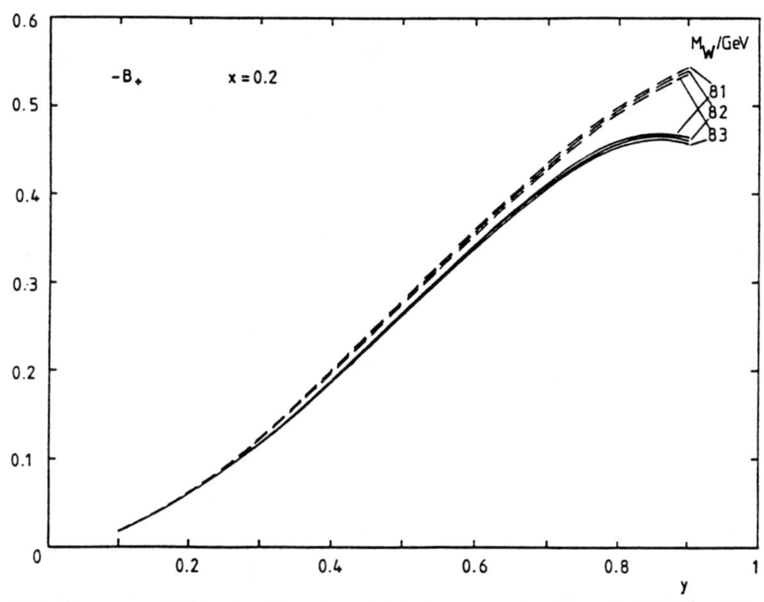

Figure 15: Born result (dashed line) and 1-loop corrected result (full line) for $B_+(x = 0.2,\ P_L = 1.0)$ for $M_W = 81,\ 82$, and $83\,GeV$. Other parameters as in fig. 5.

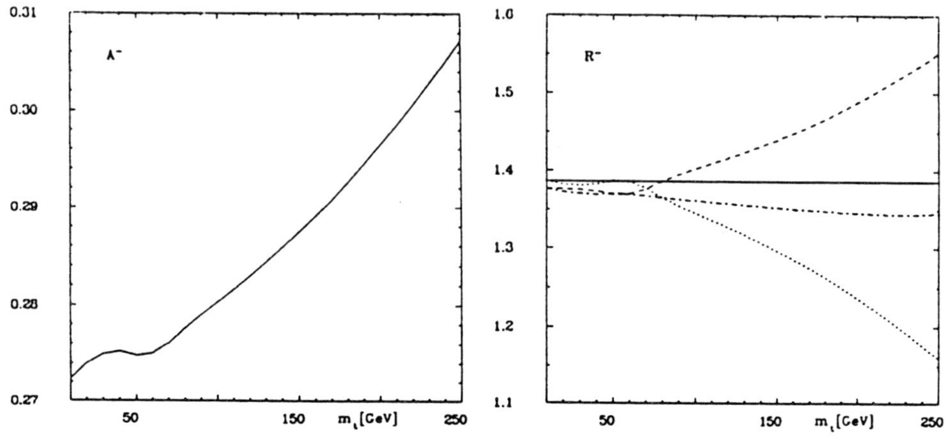

Figure 16: Top mass dependence of A_- (a) and R_- (b) for $x = y = 0.3$, $M_H = 100\,GeV$. A_- is calculated with fixed M_Z and G_μ, R_- for fixed M_Z and G_μ (dashed-dotted) and for fixed M_Z and M_W (dashed). The Born value is indicated by the full line. The dotted line includes only the effect of Δr in (19) but not the genuine radiative corrections.

Higgs mass. This dependence for R_- and A_- is shown in fig. 16 for the two schemes where else M_W and M_Z or G_μ and M_Z are used as fixed input parameters. Due to the cancellation between Δr and the Z, resp. W self energy the corrections are smaller in the second case and show a much weaker dependence on m_t and M_H. If the polarization asymmetry A_- could be measured with an accuracy of 1% then one could derive a value for m_t with an accuracy of $\pm 40\,GeV$. One can not expect to improve the sensitivity on m_t by just using a parametrization in terms of fixed M_W and M_Z as it seems to be suggested from fig. 16b. In fact, when HERA has gone into operation, M_W will still be known only with a moderate accuracy and the lines in fig. 16b is after inclusion of the error on M_W broadened to a band. The error on m_t derived from this will certainly be of the same magnitude as in the scheme with fixed M_Z and G_μ. It would be nice to be able to derive from HERA an additional information on the parameters of the weak sector. For a preparation of such an analysis it would be helpful to repeat the study of [24] where the top mass dependence from Δr was included but not that from the genuine 1−loop corrections.

6. Conclusion

In conclusion we have presented an overview over the $\mathcal{O}(\alpha)$ electroweak radiative corrections to the NC and CC differential cross sections for deep inelastic lepton proton scattering at HERA. These corrections can become large in general. They are dominated by purely photonic corrections associated with radiation (virtual or real) from the lepton line. Large corrections connected with radiation from the quark line can be absorbed into the parton distribution functions and induce only a small additional Q^2 dependence. The leptonic QED corrections together with the vacuum polarization (*e.g.* included by using the running fine structure constant) give the complete corrections with an accuracy of a few %. A good agreement at the %-level between results of different groups has now been achieved.

The value of the $\mathcal{O}(\alpha)$ corrections makes it probable that higher order corrections will also be important. A part of them is the soft photon corrections that can be obtained by exponentiation [25,16]. This leads to a positive contribution of the order of a few % for large x and small y but negligible corrections for small x and large y. The latter kinematical region will probably be influenced by higher order leading logarithmic corrections of the form $\mathcal{O}(\alpha \log(m_e^2/Q^2))^n$. These have not yet been worked out explicitly for the deep inelastic cross sections at HERA energies.

In addition to this a further correction will come from the radiative tail of elastic eP scattering. For a treatment where only the momentum of the outgoing electron is measured but the rest of the final state is taken totally inclusive, this process will also contribute. The effect has been calculated in [26] where it was shown that for $x \leq 0.1$ and $y \geq 0.7$ it will be non-negligible.

Many important questions can not be answered on the basis of the existing calculations where the radiated photon is treated totally inclusive. Among them is the problem of what effects will experimental cuts have for the magnitude of the radiative corrections . Also the study of combined effects of electroweak radiative corrections and QCD radiative corrections or the influence of electroweak radiative corrections on the measurement of the structure functions is difficult to treat in the existing framework of the calculations. For the investigation of these problems an event generator that includes electroweak radiative effects would be a very helpful tool. Such a Monte Carlo program is presently being developed at DESY.

Acknowledgment

I want to thank M. Böhm, J. Blümlein, W. Hollik, T. Riemann, and G. Schuler for helpful discussions. I also thank M. Zrałek and his colleagues for the perfect organization of the conference in Sczcyrk.

References

[1] Proceedings of the HERA Workshop, Hamburg 1987, ed. R. D. Peccei, Vol. 1 and 2.

[2] G. Schuler, DESY preprint DESY 89-075.

[3] M. Böhm, W. Hollik, H. Spiesberger, Fortschr. Phys. 34 (1986) 687.

[4] L. W. Mo, and Y. S. Tsai, Rev. Mod. Phys. 41 (1969) 205;
Y. S. Tsai, SLAC-PUB-848 (1971).

[5] A. A. Akhundov et al., Sov. J. Nucl. Phys. 26 (1977) 660.

[6] J. Kiskis, Phys. Rev. D 8 (1973) 2129; W. J. Marciano, A. Sirlin, Phys. Rev. D 22(1980) 2695, Nucl. Phys. B 189 (1981) 442; J. F. Wheater, C. H. Llewellyn-Smith, Nucl. Phys. B 208 (1982) 27;

[7] A. de Rújula, R. Petronzio and A. Savoy-Navarro, Nucl. Phys. B 154 (1979) 394.

[8] J. Kripfganz, H. Perlt, Z. Phys. C 41 (1988) 319.

[9] J. Kripfganz, H. J. Möhring, Z. Phys. C 38 (1988) 653.

[10] W. Beenakker, F. A. Berends, W. L. van Neerven, Proceedings of the Ringberg Workshop on "Radiative Corrections for e^+e^- Collisions", Ringberg 1989, ed. J. H. Kühn, p.3.

[11] J. Blümlein, Berlin-Zeuthen preprint PHE 89-8, PHE89-10, and Proceedings of the International Conference "Frontiers in Elementary Particle Physics", Kazimierz 1989, eds. Z. Adjuk and S. Pokorski.

[12] M. Consoli, M. Greco, Nucl. Phys. B 186 (1981) 519.

[13] E. A. Kuraev, N. P. Merenkov, V. S. Fadin, Sov. J. Nucl. Phys. 47 (1988) 1009.

[14] Proceedings of the HERA Workshop, Hamburg 1987, ed. R. D. Peccei, Vol. 2, p. 577.

[15] D. Yu. Bardin, O. M. Fedorenko, N. M. Shumeiko, J. Phys. G: Nucl. Phys. 7 (1981) 1331.

[16] D. Yu. Bardin, C. Burdik, P. Ch. Cristova, T. Riemann, Dubna preprint E2-87-595, E2-88-682, and Z. Phys. C 42 (1989) 679.

[17] D. Yu. Bardin, C. Burdik, P. Ch. Cristova, T. Riemann, Z. Phys. C 44 (1989) 149.

[18] M. Böhm, H. Spiesberger, Nucl. Phys. B 294 (1987) 1081.

[19] M. Böhm, H. Spiesberger, Nucl. Phys. B 304 (1987) 749.

[20] A. Sirlin, Phys. Rev. D 22 (1980) 971.

[21] H. Burkhardt, F. Jegerlehner, G. Penso, C. Verzegnassi, in: "Polarization at LEP", CERN 88-06, eds. G. Alexander et al., Vol. 1, p. 145.

[22] D. W. Duke, F. F. Owens, Phys. Rev. D 30 (1984) 49.

[23] S. de Jong, J. Vermaseren, 'AXO user manual', NIKHEF-H, Amsterdam, 1987.

[24] J. Blümlein, M. Klein, T. Riemann, in: Proceedings of the HERA Workshop, Hamburg 1987, ed. R. D. Peccei, Vol. 2, p. 687.

[25] D. R. Yennie, S. C. Frautschi, H. Suura, Ann. of Physics 13 (1961) 379.

[26] A. A. Akhundov, D. Yu. Bardin, C. Burdik, P. Ch. Christova, L. V. Kalinovskaja, Dubna preprint E2-89-405 (1989).

ISBN 1-56072-027-1

RADIATIVE CORRECTION TO THE NEUTRINO COUNTING REACTION

Masataka Igarashi

Department of Phyics, Tokai University, Hiratsuka, Kanagawa 259-12, Japan

Abstract

I review the topics related with the neutrino counting reaction $e^+e^- \rightarrow \nu\bar{\nu}\gamma$. First I discuss the validity of various treatments of the lowest order cross section by stressing the possibility that the reaction may be used for the study of γ-W-W gauge coupling in the energy region higher than M_Z. The reaction becomes directly sizable and feasible around the Z-boson peak due to the resonance nature of it. The latter property in turn necessitates the inclusion of the radiative correction to it. The known QED radiative correction to the reaction mediated by Z-boson is shown to be extended to that to the case mediated by the Z-boson and by the W-boson in the point interaction approximation. This corrected cross section is numerically compared with the cross section due to the main background, i.e., the radiative Bhabha scattering under realistic conditions. Finally I point out possible difficulties in calculating the $SU(2) \times U(1)$ correction to it.

§ 1. Introduction

Since the early 80's, a theory group at KEK has made efforts to calculate the higher order corrections to e^+e^- annihilation processes before the construction of TRISTAN. The purpose was to calculate the corrections by ourselves to understand the details of the results for theoretical and experimental application of them. We have calculated the corrections in the $SU(2) \times U(1)$ standard theory to important e^+e^- annihilation processes in the energy range of TRISTAN to LEP II[1]. Among them I here talk about the radiative correction to the so-called neutrino counting reaction $e^+e^- \rightarrow \nu\bar{\nu}\gamma$ and the related topics.

The number of neutrino species (N_ν) is very important not only in particle physics but also in cosmology. In particle physics the knowledge of it is directly related with the number of generations with almost massless neutrinos and, though depending on the processes studying it, may give a clue to possible new physics. In cosmology it is crucial to the primordial abundance of He^4, which limits N_ν less than or equal to four[2]. Conversely, if N_ν is determined it gives another new constraint on this branch of cosmology. Among several methods of determining it, it is considered that the most promising method is that using the neutrino counting reaction.

In section 2, I review the methods of determining N_ν. Thereafter I concentrate on the neutrino counting reaction (NCR). In section 3, I give the Born cross section and explain its usefulness for studying new physics. In section 4 I give the calculational method and the result of QED virtual corrections to NCR including the corrections due to real photon emission process. The numerical results are compared in section 5 with the cross section of the main background, i.e., that of the radiative Bhabha scattering. In section 6, I explain our renormalization scheme of the $SU(2) \times U(1)$ standard theory and report our recent efforts toward the calculation of the $SU(2) \times U(1)$ correction to NCR. The summary is given in the last section.

§ 2. How to determine N_ν

Decay width measurement. A pair of neutrino and antineutrino is produced by the Z-boson in s-channel or by the W-boson in t-channel. Thus it is a pure weak process. In order to determine N_ν by decay of a particle, one must search particle with a large branching ratio into $\nu\bar{\nu}$ or particle with a large number of decay events for the case that there are other decay modes mediated by stronger force.

The first candidate is the decay of charged kaon into charged pion with missing neutrino pairs. Although this process is second order in weak interaction, there is possibility of measuring it since the main decay modes of kaon are first order in weak interaction and the accumulated number of decay events is copious. According to recent calculations[3], its branching ratio is given by

$$BR(K^+ \to \pi^+\nu\bar{\nu}) \sim N_\nu \times 10^{-10}. \tag{2.1}$$

On the other hand the available data[4] give only the upper limit

$$BR(K^+ \to \pi^+\nu\bar{\nu})_{exp} < 1.4 \times 10^{-7} \tag{2.2}$$

which correspond to a limit on $N\nu$ more than one thousand. This result is out of our interest; but the experiment is in progress [3].

The second candidate is the ψ family ,of which data are accumulated very much in comparison with other particles. An example is the decay

$$\psi' \to \psi\pi^+\pi^- \tag{2.3}$$

associated with the decay

$$\psi \to \nu\bar{\nu}. \tag{2.4}$$

The latter mode proceeds through weak interaction while ψ mainly decays through electromagnetic interaction, e.g.,

$$\psi \to l\bar{l}, \tag{2.5}$$

where $l = e$ or μ. Actually this couple of decays is proposed for the determination of the neutral weak coupling constants of charmed quark[5]. The order of the ratio of Eq.(2.4) to Eq. (2.5) can be easily estimated and it is given by

$$\begin{aligned}\frac{\Gamma(\psi \to \nu\bar{\nu})}{\Gamma(\psi \to l\bar{l})} &= \left(\frac{ratio\ of}{propagators} \times \frac{ratio\ of\ cc}{with\ quark} \times \frac{ratio\ of\ cc}{with\ lepton}\right)^2 \\ &= \left(\frac{m_\psi^2}{M_Z^2 - m_\psi^2}\frac{3}{2}\frac{-5M_Z^2 + 8M_W^2}{12M_W\sqrt{M_Z^2 - M_W^2}}\frac{M_Z^2}{4M_W\sqrt{M_Z^2 - M_W^2}}\right)^2 \\ &\sim 5 \times 10^{-8}.\end{aligned} \tag{2.6}$$

Since the branching ratio of ψ' into ψ with a pion pair is one half and that of ψ into a charged lepton pair is 7%, the final result is given by

$$BR(\psi' \to \psi(\nu\bar{\nu})\pi^+\pi^-) \sim 2 \times 10^{-9}. \tag{2.7}$$

This is well below the present data.

The third candidate is the Z-boson. It is well-known that its peak is very high compared with the background and thus it can give very clean data. Another good feature of Z-boson is that, among the decays of it, a kind of democracy is realized; i.e., all the main decay modes proceed via the neutral weak interaction. Thus the branching ratio into a neutrino pair reaches to 7% which is comparable to that into charged fermion pair. However there is a difficulty in the precise determination of N_ν from the decay widths. Because the visible decay widths must be subtracted from the total decay width and this subtraction may lead to a large accumulated error. For the methods of subtraction and the estimate of errors, please consult in the forthcoming book on LEP Physics[6]. I personally think that the determination of N_ν and the related physics must be studied by combining the data on the pole and the information from NCR.

Measurement of scattering cross section. As you know the UA1 and UA2 data on $p\bar{p}$ have given information on N_ν with the condition on the top quark mass. I am not in a situation discussing them in detail. Please consult in other reports[7,8].

Another scattering process for measuring N_ν is just the NCR $e^+e^- \rightarrow \nu\bar{\nu}\gamma$. It was proposed by Ma and Okada in the standard theory[9]. Because of an extra factor α, its cross section is rather small compared with other important annihilation processes. Nevertheless it becomes sizable even in the energy region off the Z-boson peak if reasonable integrated luminosity is obtained[10]. This is mainly due to the fact that, in the e^+e^- annihilation, the incident energy is in most cases fully converted to a pair of final particles, the data are thus clean and the signal can be eliminated from the background.

The original proposal was to study N_ν by making use of the resonance nature of the Z-boson peak. There the experiment is of course possible and feasible. This feasibility was studied by taking into account the realistic backgrounds in ref. [11]. The resonance nature of Z-boson enhances the signal of this reaction. But just this resonance nature requires the inclusion of radiative correction to it just as in the case of extremely narrow resonance like J/ψ. This is our topic.

§ 3. The Born cross secton

The lowest order process is described by the five Feynman diagrams in Fig. 3.1 in the approximation neglecting electron mass. The gauge invariance with respect to the emitted real photon is valid in the

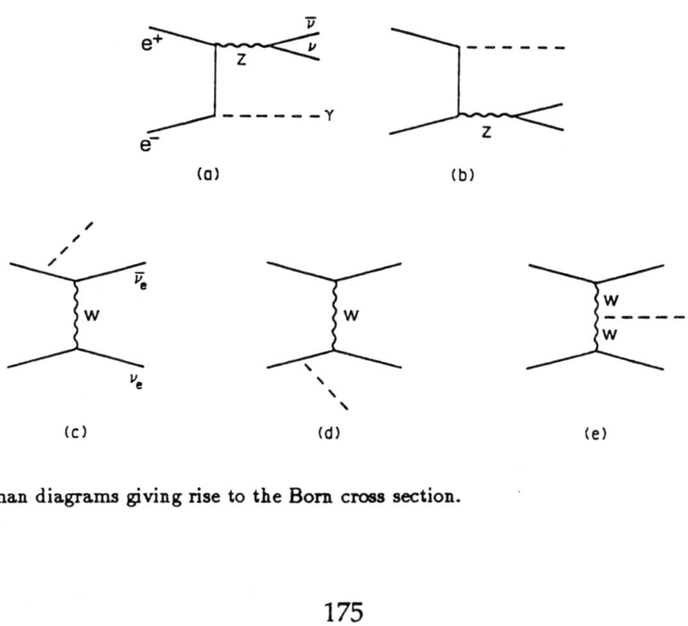

Fig. 3.1. Feynman diagrams giving rise to the Born cross section.

sets Fig. 3.1a and b and Fig. 3.1c, d and e, respectively. If the electron mass is kept, other diagrams mediated by unphysical Higgs particles must be added to Fig. 3.1c, d and e. However, since the photon is detected, the electron mass in the propagator can be neglected and the electron mass in the numerator can be of course neglected.

Although the Born cross section is the lowest order cross section by definition, there are a few physical approximations in treating it. If one stick to the determination of N_ν around the Z-boson peak, the t-channel contribution of W-boson may be neglected against the s-channel Z-boson contribution. This treatment is denoted as Z. The electron neutrino can be produced by Z-boson in a pair and also converted from the electron or positron by the W-boson. These two amplitudes can interfere with each other. It is dubious that the interference is neglected from the outset. Much better way is to include the interference by treating the t-channel W-boson as an infinitely heavy object. Since this is a point interaction approximation for the W-boson interaction, we call this treatment PIA. In this case Fig. 3.1e makes no contribution. The original calculation was made in this approximation. The result is as follows.

$$\frac{d^2\sigma^Z}{dxdy} = 8\alpha^3\frac{N_\nu}{3}\frac{(f^2+g^2)(F^2+G^2)}{[M_Z^2-s(1-x)]^2+M_Z^2\Gamma_Z^2}\frac{s(1-x)}{x(1-y^2)}\left[\left(1-\frac{1}{2}x\right)^2+\frac{1}{4}x^2y^2\right], \tag{3.1}$$

$$\begin{aligned}\frac{d^2\sigma^{PIA}}{dxdy} &= \frac{d^2\sigma^Z}{dxdy}\\ &+\frac{\alpha^3}{6}\frac{M_Z^4}{M_W^4}\frac{1}{(M_Z^2-M_W^2)}\left[\left(1-2\frac{M_W^2}{M_Z^2}\right)\frac{M_Z^4-M_Z^2s(1-x)}{[M_Z^2-s(1-x)]^2+M_Z^2\Gamma_Z^2}+1\right]\\ &\times\frac{s(1-x)}{x(1-y^2)}\left[\left(1-\frac{1}{2}x\right)^2+\frac{1}{4}x^2y^2\right],\end{aligned} \tag{3.2}$$

where the coupling constants are

$$\begin{aligned} f &= \frac{M_Z}{4\sqrt{M_Z^2-M_W^2}}\frac{M_Z}{M_W}\left(3-4\frac{M_W^2}{M_Z^2}\right),\\ g = F = -G &= \frac{M_Z}{4\sqrt{M_Z^2-M_W^2}}\frac{M_Z}{M_W}\end{aligned} \tag{3.3}$$

and

$$x = k/E = 2\times k/\sqrt{s}, \quad y=\cos\theta \tag{3.4}$$

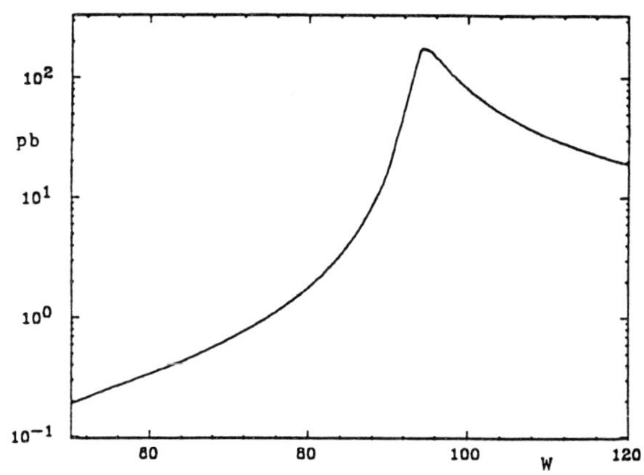

Fig. 3.2. The total Born cross section integrated over $k = xE \geq 1GeV$ and $0.96 \geq |y|$.

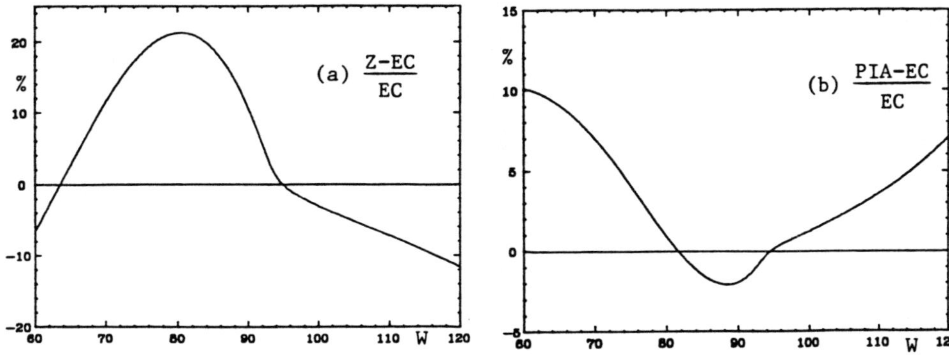

Fig. 3.3. Comparison of various treatments of the lowest order cross section; these are obtained under the same conditions as in Fig. 3.2.

with k and θ being the photon energy and the scattering angle between the incident positron and the photon, respectively.

The third treatment is the exact calculation of the Born cross section, which is denoted as EC. There are various methods of obtaining the results in EC; the problem is how to perform the phase space integral when the integral variables are contained in the denominators. The most easiest way is to use the Monte Carlo multiple integration program. Another method is to make use of the recurrence formula which is obtained by dividing the invariant scalars in the numerator of the integrand by the W-boson propagators[12]. The third method is to analytically perform the phase space integral of the square of matrix element[13]. In this case we done it in a specific frame where the positron momentum $\mathbf{p}_+$ is along the z-axis and the photon momentum $\mathbf{k}$ lies in the x-z plane. We used these three methods to get the answer; these results are in agreement with each other within four digids.

First we show the total cross section in Fig. 3.2; the parameters are taken as

$$\begin{aligned} N_\nu &= 3, \\ M_W &= 81.0Gev, \\ M_Z &= 92.4GeV, \\ \Gamma_Z &= 2.7GeV. \end{aligned} \tag{3.5}$$

It steeply increases to the Z-boson peak and thereafter decreases gradually. The Z-boson peak is caught when the invariant mass of $\nu\bar{\nu}$ system is equal to the Z-boson mass. Because of the photon tagging the Z-boson peak is shifted to a position slightly higher than the input mass. Next we compare Z and PIA against EC in Fig. 3.3. Within 2% the former approximation is valid in the energy region $\sqrt{s} = 96$ to $102GeV$ and the latter in 80 to $108GeV$. The validity of the approximation Z means that NCR is proportional to N_ν independent of the contribution of W-boson. In order to study the W-boson contribution, we should make experiments in the range $\sqrt{s}$ lower than 80 or higher than $110GeV$. The precise data in these regions may give the information on weakly coupling particles with electron number[12] and/or on the magnitude of the gauge coupling of γ-W-W[14]. The latter type of experiment in higher energy region is promising and quite interesting, because there are arguments that the gauge couplings could deviate from those in the standard theory[15].

§ 4. QED correction to neutrino counting reaction

First we consider the QED virtual corrections to NCR in the approximation Z[16,17]. They are given by the diagrams in Fig. 4.1. It should be noted that, since the $SU(2) \times U(1)$ standard theory is anomaly free, the contribution of Z-γ-γ and Z-Z-γ triangle is negligible[18]. The electron self-energy correction inFig. 4.1a can be found in any book of QED. The vertex correction in Fig. 4.1b is obtained in the book by Akhiezer and Berestetskii[19]. For another vertex correction in Fig. 4.1c, there is no way of renormalization in the framework of QED because the axial-vector vertex is not contained in it. Therefore we followed a convention taken in literatures[20,21] that the counterterm to the axial-vector vertex is the same as that to the vector vertex. This assumption leads to a finite term different from that calculated in the $SU(2) \times U(1)$ theory; but the coefficients of mass-singular terms are correctly obtained[22].

The remaining task is to evaluate the box contribution in Fig. 4.1d. Thereby one must tackle the following type of integrals.

$$[J, J_\mu, J_{\mu\nu}, J_{\lambda\mu\nu}] = \frac{1}{i\pi^2}\int d^4l \frac{1}{(+)(-)(3)(0)}[l, l_\mu, l_{\mu\nu}, l_{\lambda\mu\nu}], \tag{4.1}$$

$$\begin{aligned} (+) &= l^2 + 2(l\cdot p_+) & p_3 &= p_- - k \\ (-) &= l^2 + 2(l\cdot p_-) & t &= (p_+ - k)^2 \\ (3) &= l^2 + 2(l\cdot p_3) + u & u &= (p_- - k)^2 \\ (0) &= l^2 & Q^2 &= s - t - u \end{aligned} \tag{4.2}$$

Covariant method. The covariance dictates that these tensors after the integral over l must be the tensors composed of $p_{i\mu} = p_{+\mu}, p_{-\mu}$ and $p_{3\mu}$ and $g_{\mu\nu}$. For example,

$$J_\mu = ap_{+\mu} + bp_{-\mu} + cp_{3\mu}. \tag{4.3}$$

By multiplying $p_{i\mu}$, one gets three coupled equations for three unknown coefficients a, b and c from the right hand side of Eq. (4.3). In the left hand side the numerators $(l \cdot p_i)$ can be expressed in terms of the propagators and a constant; thus the integrals become those with less denominators of the form

$$\begin{aligned} F &= \frac{1}{i\pi^2}\int d^4l \frac{1}{(+)(-)(3)}, & H &= \frac{1}{i\pi^2}\int d^4l \frac{1}{(+)(-)(0)}, \\ G &= \frac{1}{i\pi^2}\int d^4l \frac{1}{(0)(+)(3)}, & G' &= \frac{1}{i\pi^2}\int d^4l \frac{1}{(0)(-)(3)}, \end{aligned} \tag{4.4}$$

and J itself. By solving the coupled equations one has

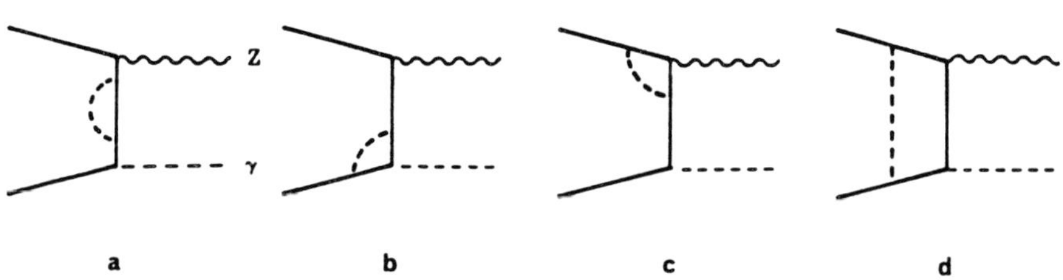

Fig. 4.1. Diagrams giving the QED one-loop correction to the diagam in Fig. 3.1a; the line of outgoing neutrino is omitted.

$$
\begin{aligned}
2stu \times a &= u(t-u)F - u(s+t)G + u^2G' + su(H+\kappa J),\\
2stu \times b &= (s(t+u) - t(t-u))F + (s-t)^2G - u(s+t)G' - s(s-t)(H+\kappa J),\\
2stu \times c &= -s(t+u)F - s(s-t)G + suG' + s^2(H+\kappa J).
\end{aligned} \tag{4.5}
$$

In this simple method the treatments of the integrals of higher rank tensors are rather tedious.

Quick method. In this method[23,24], the interference of the corrected amplitude with the lowest order amplitude is taken before the l-integral. The scalar products including l_μ in the numerator can be reduced to simpler integrals and treated as in the case of the covariant method. Or it can be treated as a whole. In this treatment it is easy to find at earlier stage whether the final result contains the mass-singularity or not. This method was used in our calculation of the correction in $SU(2) \times U(1)$ theory to $e^+e^- \to \gamma\gamma$[25]. The box diagram with a virtual Z-boson instead of the virtual photon in Fig. 4.1d was shown to make a correction free from the electron mass-singularity; it is needless to say that each integral contains the electron mass-singularity and that it disappears only when all the integrals are summed. Anyway, the merit of this method is that it is unnecessary to evaluate the integrals not appearing in the corrected cross section. The demerit of this method is that for different other purposes, e.g., to calculate the square of one-loop correction one must recalculate it since the explicit expression of the integral is not obtained.

Method á la Brown- Feynman. For the integral J_μ we start from the same expression as in Eq. (4.3). The difference from the covariant method is to use the vector variables for higher rank tensors[26]. For example, it is

$$
J_{\mu\nu} = a_\mu p_{+\nu} + b_\mu p_{-\nu} + c_\mu p_{3\nu} + d g_{\mu\nu} \tag{4.6}
$$

with unknown vectors a_μ, b_μ and c_μ and a scalar d. Furthermore we read off the solution for a, b and c in Eq. (4.5) as an equation for the independent integrals F, G, G' and H.

$$
J_\mu = F\xi_\mu + G\eta_\mu + G'\zeta_\mu + (H+\kappa J)\varepsilon_\mu \tag{4.7}
$$

with $\xi_\mu, \eta_\mu, \zeta_\mu$ and ε_μ being the linear combinations of $p_{i\mu}$. Then we have for the second rank tensor

$$
J_{\mu\nu} = F_\mu\xi_\nu + G_\mu\eta_\nu + G'_\mu\zeta_\nu + (H_\mu + \kappa J_\mu)\varepsilon_\nu + ds_{\mu\nu} \tag{4.8}
$$

In this result there appear instead of the integrals in Eq. (4.4) those with l_μ in the numerator;but their coefficients are the same. The last term is proportional to a second rank symmetric tensor $s_{\mu\nu}$. We have used this last method in the calculation of box contribution in Fig. 4.1d[16].

The sum of virtual corrections to NCR in Z is given by

$$
\frac{d^2\sigma^{VC}}{dxdy} = \frac{d^2\sigma^Z}{dxdy}\left(\frac{u}{t} + \frac{t}{u} + 2\frac{sQ^2}{tu}\right)^{-1} [C^{VC}(t) + C^{VC}(u)], \tag{4.9}
$$

$$
\begin{aligned}
C^{VC}(t) = \frac{\alpha}{\pi}\Bigg\{&\left(\frac{u}{t} + \frac{sQ^2}{tu}\right)\left[\left(\ln\frac{s}{m^2} - 1\right)\ln\frac{\lambda^2}{s} + \frac{1}{2}\ln^2\frac{s}{m^2} + \frac{1}{2}\ln\frac{s}{m^2} + \frac{2}{3}\pi^2 - 2\right]\\
&+ \frac{s^2 + (s+u)^2}{2tu}\left[\ln^2\left(\frac{s}{-t}\right) + \ln^2\left(\frac{-t+Q^2}{Q^2}\right) - \ln^2\left(\frac{-t}{Q^2}\right)\right.\\
&\qquad \left. + 2Sp\left(\frac{Q^2}{-t+Q^2}\right) + 2Sp\left(1 - \frac{Q^2}{s}\right) - \frac{\pi^2}{3}\right]\\
&+ \frac{Q^2}{s+u}\left(\frac{s}{2(s+u)} + \frac{1}{2} - \frac{3}{2}\frac{u}{t}\right)\ln\left(\frac{-t}{Q^2}\right)\\
&+ \frac{Q^2}{s-Q^2}\left(\frac{Q^2}{s-Q^2} + 3\frac{s}{t}\right)\ln\frac{s}{Q^2} + \left(1 + \frac{3}{2}\frac{u}{t}\right)\ln\left(\frac{-t}{s}\right)\\
&- \frac{s}{s-Q^2} + \frac{s}{2(s+u)} - \frac{s+u}{2t}\Bigg\},
\end{aligned} \tag{4.10}
$$

where $C^{VC}(u)$ is obtained by interchanging the role of t by u in $C^{VC}(t)$, λ is a fictitious photon mass to regulate the infrared divergence and the Spence function $Sp(z)$ is defined as

$$Sp(z) = -\int_0^z dt \frac{1}{t} \ln(1-t) \tag{4.11}$$

The virtual correction itself has no physical meaning because it is infrared divergent. One must add to it the corrections due to real photon emission process. For the soft photon which has small energy, it is well-known that the matrix element is given by the lowest order one multiplied by a classical current factor. This factor is composed of the terms due to the photon emission by external electron lines only. The photon emission from the internal line does not lead to the infrared divergence. Thus the correction due to the soft photon emission is given by a well-known factor.

$$\frac{d^2\sigma^{SOFT}}{dxdy} = R \times \frac{d^2\sigma^Z}{dxdy}, \tag{4.12}$$

$$\begin{aligned} R &= \frac{-e^2}{(2\pi)^3}\int_0^{\Delta_c} \frac{d^3k}{2\omega}\left(\frac{p_+}{p_+\cdot k} - \frac{p_-}{p_-\cdot k}\right)^2 \\ &= \frac{\alpha}{\pi}\left[2(\ln\frac{s}{m^2} - 1)\ln\frac{2\Delta_c}{\lambda} - \frac{1}{2}(\ln\frac{s}{m^2})^2 + \ln\frac{s}{m^2} - \frac{\pi^2}{3}\right], \end{aligned} \tag{4.13}$$

where the integral is performed symmetrically up to $\omega = \Delta_c$. This infrared divergence cancels that in the virtual correction.

In the double logarithmic approximation the sum of virtual correction and the correction due to the soft photon emission takes the following form:

$$\begin{aligned} C_{DLA}^{VC}(t) + C_{DLA}^{SOFT}(t) = \frac{\alpha}{\pi}\Bigg\{&\left(\frac{u}{t} + \frac{sQ^2}{tu}\right)\ln\frac{s}{m^2}\ln\frac{4\Delta_c^2}{s} \\ &+ \frac{s^2+(s+u)^2}{2tu}\left[\ln^2\left(\frac{s}{-t}\right) + \ln^2\left(\frac{-t+Q^2}{Q^2}\right) - \ln^2\left(\frac{-t}{Q^2}\right)\right]\Bigg\}, \end{aligned} \tag{4.14}$$

The first line in Eq. (4.14) is small when Δ_c is taken as large as $\sqrt{s}/2$ to mimic realistic case. The second line is quite small for $s \sim Q^2 \gg -t$ and also small for $s > -t \gg Q^2$ because $\ln^2(s/-t)$ is small. Therefore the sum behaves like a single logarithm instead of the double logarithm in appearance.

The calculations up to here, i.e., the virtual correction and the soft photon emission factor constitute the usal radiative correction though not directly applicable to experiments. Now let me turn to the lowest order cross section. The cross section for W-channel in PIA can be also obtained by the Fierz transformation.

$$\mathcal{L} = e^2\frac{g_W^2}{M_W^2}\bar{\psi}_{\nu_e}\gamma_\mu(1-\gamma_5)\psi_{\nu_e}\cdot\bar{\psi}_e\gamma^\mu(1-\gamma_5)\psi_e. \tag{4.15}$$

Since the Z-e interaction can be written as a sum of left and right helicity parts, the W-channel in PIA can be obtained from the left helicity part of the Z-channel by changing the coupling constant and the propagator or its analogue into the corresponding ones. Because the electron mass in the numerator can be neglected, the electron helicity does not flip even at the one-loop level. Therefore we can replace

$$\frac{d^2\sigma^Z}{dxdy} \quad by \quad \frac{d^2\sigma^{PIA}}{dxdy} \tag{4.16}$$

in the above results. This replacement makes the energy range of applicability of the result wider up to $\sqrt{s} \sim 110GeV$. In the following we perform the numerical calculations in PIA.

In realistic setup of experiments, there are beam pipes for incoming electron and positron. Even if energetic photon besides the detected photon is lost into beam pipes, we cannot calibrate the signal cross

cross section because the signal is a single photon. Therefore we must include the correction due to the hard photon emitted within θ_c along the beam direction and call it the collinear photon. Since the energy of photon is large, the exact square of matrix element must be used. To get the numerical values we used symbolic manupulation program and Monte Carlo multiple integration program. The result can be checked in the following way: if the limit on θ_c is removed, the sum of hard and soft photon emission cross sections should not show any dependence on Δ_c for suitable choices of the magnitude of Δ_c. Besides the collinear photon, one must consider another effect. The photon emitted along the detected photon cannot be discriminated from one another because of the finite angular resolution of detectors. We include the correction due to parallel photon emitted along the detected photon within θ_p. Depending on the variables used for analyses, it become necessary to include other contributions besides those considered above.

We show in Fig. 4.2 the corrected distribution $d\sigma/dx$ at $W = \sqrt{s} = 104GeV$ integrated over $10° \leq \theta \leq 170°$ together with the lowest order one. The inout parameters are taken as in Eq. (3.5) and the angle cuts are set to

$$\theta_c = 5°, \qquad \theta_p = 1°. \tag{4.17}$$

They are taken throughout the following calculations. The sum of virtual correction and soft photon contribution is negative and large in the absolute value. On the other hand, the real photon emission gives a large positive corrction, the most of which stems from the collinear singularity; the parallel photon makes only negligible contribution. The initinal e^+e^- system loses energy by emitting photons and hits the Z-boson peak when the invariant mass of $\nu\bar{\nu}$ system is equal to M_Z, i.e.,

$$x_1 + x_2(1 - x_1\frac{1-\cos\theta_{\gamma\gamma}}{2}) = 1 - \frac{M_Z^2}{s}, \tag{4.18}$$

where x_1 is the energy fraction of the detected photon and, x_2 is that of the missing photon and $\theta_{\gamma\gamma}$ is the angle between the momenta of two photons. Since the second term in Eq. (4.17) is positive definite, the Z-boson peak is always hit if

$$x_1 = x \leq 1 - \frac{M_Z^2}{s}. \tag{4.19}$$

This gives rise to a large radiative tail and even to the shift of the peak position. As a result the one-loop correction seems not sufficient for the analyses of the events in the tail region. It should be however noted

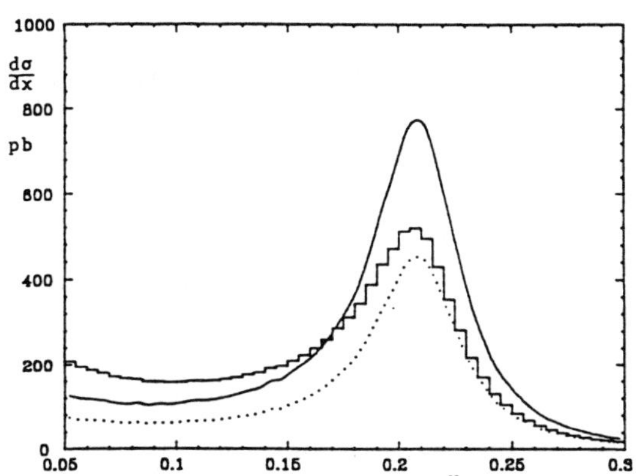

Fig. 4.2. The photon energy distribution of the neutrino counting reaction at $W = 104GeV$; (i) the Born cross section in PIA (solid line), (ii) the cross section with the virtual corrections and the soft photon emission contribution up to $0.5GeV$ (dotted line) and (iii) the cross section with the full correction (histogram).

that, although each contribution of virtual and real corrections is large, the area covered by the resonance increases only less than 20% and that the two-loop correction is expected to be by far smaller than the possible contribution of one more neutrino pair.

§ 5. Comparison of the results with the background

The background to NCR. In principle any reactions with the three or more final states with a photon can become the background to NCR. Among them the important processes are the annihilation into $\gamma\gamma\gamma$ and the radiative Bhabha scattering. Although the reactions with photons produced by two photon processes, e.g. $e^+e^- \to e^+e^-\pi^0(\gamma\gamma)$, have large cross section, they can be eliminated by taking a suitable kinematical cut on the energy of detected photon, say, $1GeV$[6].

The reaction $e^+e^- \to \gamma\gamma\gamma$ can be a background to NCR, when two photons are lost. Its possibility can be estimated from the diagram in Fig. 5.1a. However there is a potentially large contribution from the diagram in Fig. 5.1b. Rigorously speaking the latter contribution is not estimated yet. For the fermion loop, the V-V-V-V amplitude has been known from the calculation of the photon splitting [27] and the A-V-V-V amplitude is recently obtained to calculate the width $\Gamma(Z \to ggg)$[28]. We can expect from the result $\Gamma(Z \to ggg)$ ~several KeV that $\Gamma(Z \to \gamma\gamma\gamma)$ is of the order of eV and negligibly small. It should be noted that for $Z \to ggg$ and $\gamma\gamma\gamma$ there is no calculation of $W^{\pm}$-boson loop contribution[6].

It is well-known that the radiative Bhabha scattering makes a large background to NCR[11]. This is due to the very strong t- or t'-sigularity in the forward direction which is a result of the masslessness of virtual photon[29]. There the QED contibution dominates the cross section of radiative Bhabha scattering. In the following we take into account only the radiative Bhabha scattering as the background to NCR. For the relative contributions of $e^+e^- \to \gamma\gamma\gamma$ and $e^+e^-\gamma$, see refs. [11] and [29].

Comparison of NCR with the background. Now we compare the signal cross section of NCR with that of radiative Bhabha scattering in the energy region near the Z-boson peak[13]; the corrected cross section is also obtained in [30]. First we show the angular distribution $d\sigma/dy$ in Fig. 5.2. The latter cross section strongly depends on the vetoing angle θ_V which is assumed to be $1°, 3°$ and $5°$. In the case of higher energy, say, at $\sqrt{s} = 104GeV$ the background can be kinematically suppressed by making larger cut on the photon energy, which on the other hand does not change the signal itself because the Z-boson peak is by far higher, i.e., at $k = 10.95GeV$ ($x = 0.21$). This can be seen from the x-distribution in Fig. 4.2. This method of cut on photon energy does not work in lower energy because the cut suppresses the signal as well as the background because the energy of the initial system is almost on the peak.

The energy distribution $d\sigma/dx$ is shown in Fig. 5.3. The background is also strongly dependent on the vetoing angle. The limit of the range of background contribution can be roughly estimated from the configuration in Fig. 5.4a.

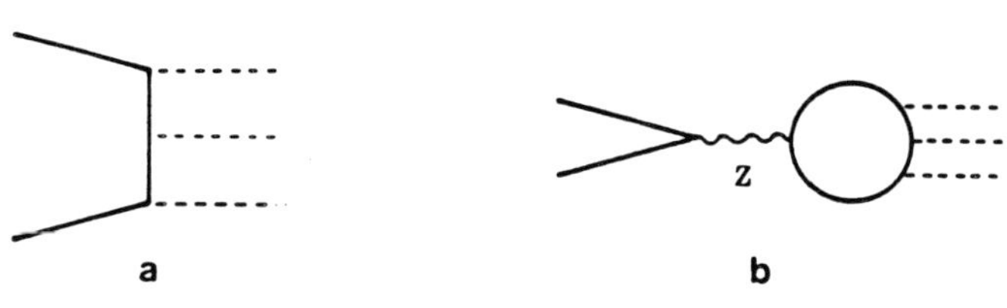

a b

Fig. 5.1. Typical Feynman diagrams of $e^+e^- \to \gamma\gamma\gamma$; the permutation of final photon is tacitly understood.

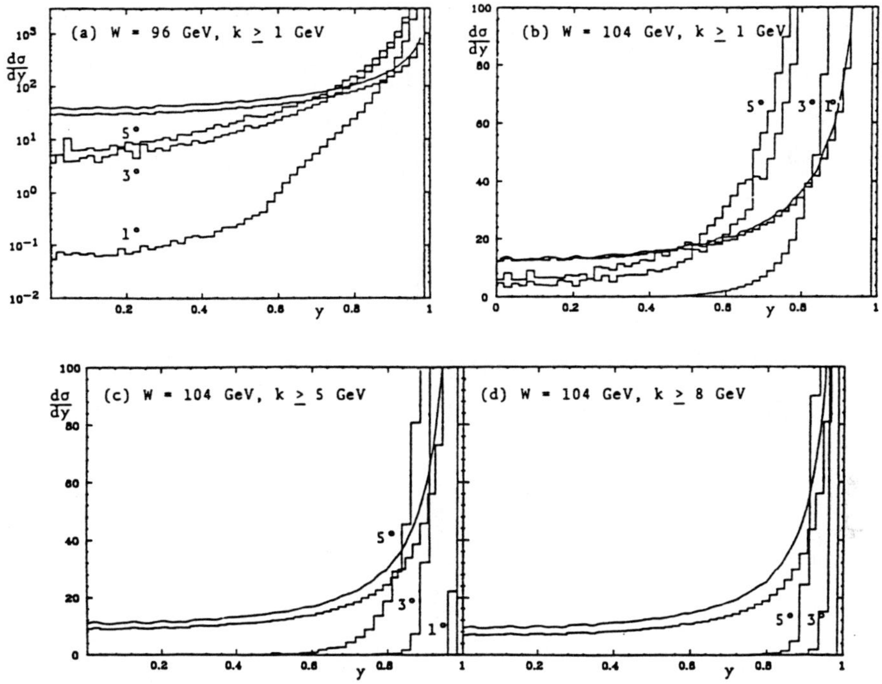

Fig. 5.2. The photon angular distribution of the neutrino counting reaction and the radiative Bhabha scattering in *pb*; the solid line is the Born cross section. The cross sections of radiative Bhabha scatterings are designated by the vetoing angle for $e^{\pm}$.

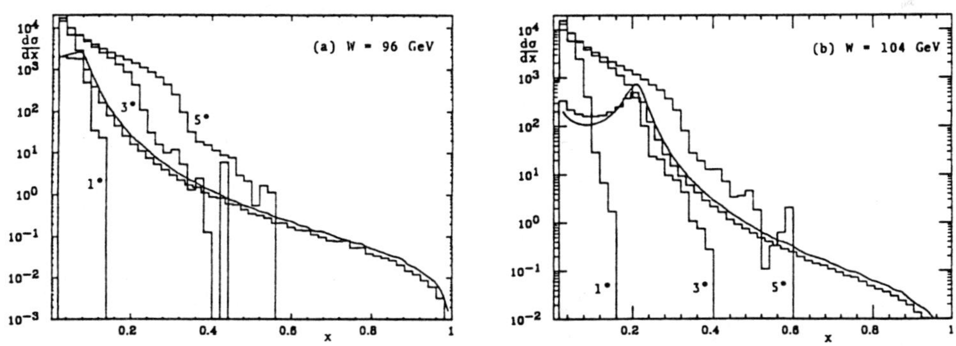

Fig. 5.3. The photon energy distribution of the neutrino counting reaction and the radiative Bhabha scattering; the notations are the same as in in Fig. 5.2.

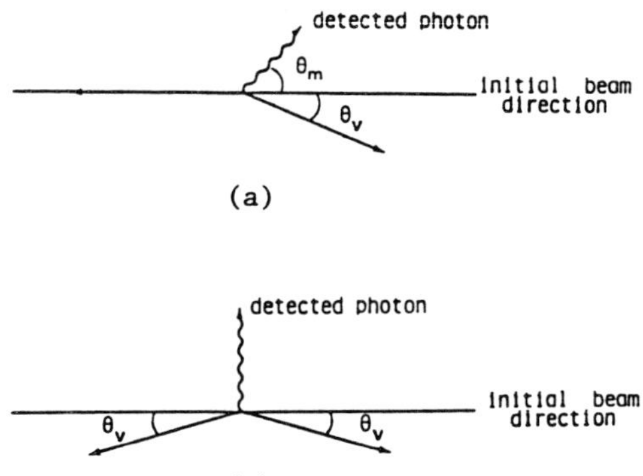

Fig. 5.4. Specific configurations of $e^{\pm}$ leading to the limiting cases of the backgronud cross section.

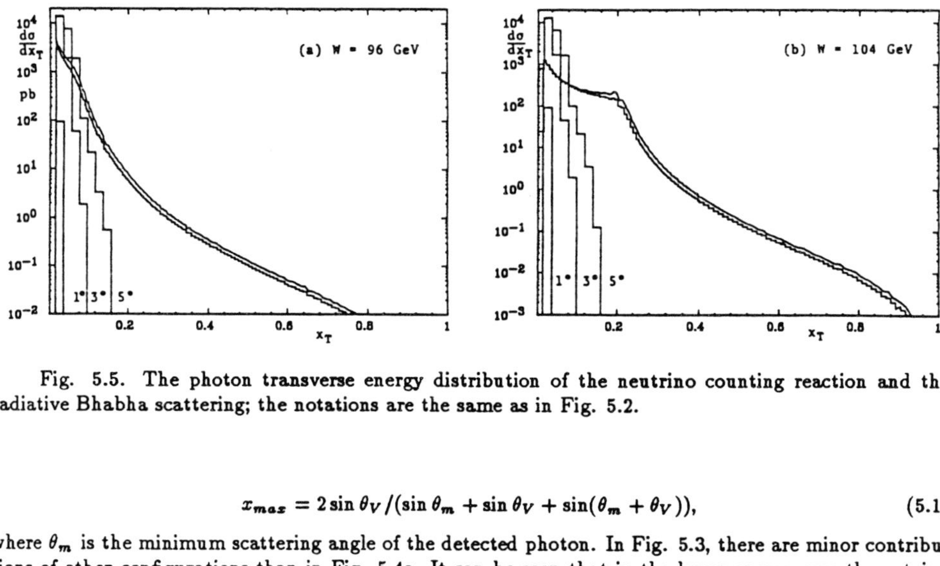

Fig. 5.5. The photon transverse energy distribution of the neutrino counting reaction and the radiative Bhabha scattering; the notations are the same as in Fig. 5.2.

$$x_{max} = 2\sin\theta_V/(\sin\theta_m + \sin\theta_V + \sin(\theta_m + \theta_V)), \tag{5.1}$$

where θ_m is the minimum scattering angle of the detected photon. In Fig. 5.3, there are minor contributions of other configurations than in Fig. 5.4a. It can be seen that in the lower energy case the vetoing angle must be kept as small as possible to get a sizable signal. Lastly we give the transverse photon energy distribution of

$$x_T = x\sin\theta. \tag{5.2}$$

In this case the vetoing angle is works effectively to suppress the background. Its limits are obtained from the configuration in Fig. 5.4b.

$$x_{T,max} = 2\sin\theta_V/(1 + \sin\theta_V). \tag{5.3}$$

Finally we summarize the magnitudes of effective signal cross sections in Table 5.1. Undoubtedly the best variable is the transverse photon energy. Nevertheless other variables may be used complementary to it.

As a result of these prescriptions, the renormalization is the same as in the usual treatment of QED. We do not need any extra factor for external lines for any fields. In the framework of this scheme we have calculated the one-loop correction to various electron-positron annihilation processes in the 't Hooft-Feynman gauge by treating the ultraviolet divergence in the dimensional regularization and by regularizing the infrared divergence by a fictitious photon mass λ.

In the one-loop correction there appears a large correction from the counterterm Y.

$$Y = 1 - \frac{\alpha}{4\pi}\{\frac{7}{2}(C_{UV} - \ln M_W^2) + \frac{1}{3} - \frac{2}{3}\sum_i Q_i^2(C_{UV} - \ln m_i^2)\} \tag{6.14}$$

where Q_i is the electric charge of fermion i and the color factor should be taken into account for the quark contribution. The finite part of δY is about 3.5% around the Z-boson mass. In the reaction $e^+e^- \to \mu^+\mu^-$, this large correction can be seen in the following way[4]. We have constructed the blobs of initial electron and final muon; they are composed of the vertex correction, half of the photon self-energy, half of the Z-boson self-energy and half of the γ-Z transition amplitude. In Table 6.1, I show the contributions of the QED and the weak interaction to the blob separately. Around the Z-boson pole, the photon channel can be neglected in comparison with the Z-boson channel and, furtheremore, the vector coupling of Z-boson is quite small compared with the axial-vector coupling. Thus the weak correction stems mainly from the axial-vector blob of Z-boson. Its magnitude can be read off from Table 6.1.

$$\frac{-465 + 519}{1769} = 3.05\% \tag{6.15}$$

W(GeV)		70	93	120	150
$F^{e\gamma}$	Re	765	809	839	873
(-3028)		(765)	(802)	(838)	(870)
	Im	-221	-223	-229	-234
		(-220)	(-222)	(-227)	(-232)
$G^{e\gamma}$	Re	1	-195	1	0
(0)		(0)	(0)	(0)	(0)
	Im	6	0	-13	-9
		(0)	(0)	(0)	(0)
F^{eZ}	Re	17	17	18	20
(-60.75)		(17)	(18)	(19)	(19)
	Im	5	8	11	14
		(-5)	(-5)	(-5)	(-5)
G^{eZ}	Re	-439	-465	-489	-508
(1769)		(-493)	(-519)	(-543)	(-564)
	Im	141	143	144	144
		(143)	(147)	(150)	(153)

Table 6.1. The numerical values of the blob in unit of 10^{-4} in the $SU(2) \times U(1)$ standard model; F and G are the vector and axial-vector blobs, respectively. The numbers in parentheses are those in QED. The numbers in the first column are those of the corresponding bare coupling constants for $M_Z = 93GeV$ and $M_W = 81GeV$. They are obtained by subtracting the infrared divergent terms proportional to $\ln(m_e/\lambda)$.

		W = 96 GeV	W = 104 GeV
x	x ≥ 0.15	3.05	34.9
	x ≥ 0.4	0.11	0.19
	x ≥ 0.6	0.02	0.03
x_T	x_T ≥ 0.04	49.8	46.2
	x_T ≥ 0.10	4.0	22.3
	x_T ≥ 0.16	0.7	9.9

Table 5.1. Effective signal cross sections in *pb*.

§ 6. Toward the$SU(2) \times U(1)$ correction to NCR

As I stated earlier the NCR $e^+e^- \to \nu\bar{\nu}\gamma$ can be also used for the study of weakly inertracting particles with electron number and/or the gauge coupling γ-W-W at the energy region off the Z-boson peak. In order to do so we must know the $SU(2) \times U(1)$ correction to it because these studies require high precision to find new physics. Before going into the correction itself, let me introduce you our renormalization scheme[22]. We take the $SU(2) \times U(1)$ standard model with three generations and one Higgs doublet as the basis of the calculation of one-loop full correction. Among the particles contained in this framework, the top quark and the Higgs particle are not found yet; thus these masses are free parameters at present.

An on-shell renormalization scheme. We adopt the on-shell scheme formulated by Kyoto group[31]. The divergent diagrams at loop level are the vertex, the fermion self-energy, the gauge boson self-energy and the Higgs particle self-energy ones. These divergences are cancelled with the divergences in the following counterterms.

$$\begin{aligned} e_0 &= Ye, \\ M_{W_0}^2 &= M_W^2 + \delta M_W^2, \\ M_{Z_0}^2 &= M_Z^2 + \delta M_Z^2, \\ m_{f_0}^2 &= m_f^2 + \delta m_f^2, \\ m_{\phi_0}^2 &= m_\phi^2 + \delta m_\phi^2, \end{aligned} \tag{6.1}$$

i.e.,the electric charge e, the W-boson mass M_W, the Z-boson mass M_Z, the fermion mass m_f and the mass of Higgs particle m_ϕ as is usual in the on-shell scheme. Here the quantities with and without 0 mean the unrenormalized and renormalized ones respectively. Besides them, one must introduce the wave function renormalization constants to obtain the finite Green's functions. A feature of the Kyoto scheme is to introduce these constants after the mixing of neutral gauge bosons, which is remarkably different from other on-shell schemes[32].

$$\begin{aligned} W_0^\pm &= Z_W^{1/2} W^\pm, \\ \begin{pmatrix} Z_0 \\ A_0 \end{pmatrix} &= \begin{pmatrix} Z_{ZZ}^{1/2} & Z_{ZA}^{1/2} \\ Z_{AZ}^{1/2} & Z_{AA}^{1/2} \end{pmatrix} \begin{pmatrix} Z \\ A \end{pmatrix}, \\ \phi_0 = Z_\phi^{1/2}\phi, &\qquad f_{L,R}^0 = Z_{L,R}^{1/2} f_{L,R}, \end{aligned} \tag{6.2}$$

where the letter means the corresponding field operator.

In order to see more details, let me consider the renormalized self-energy amplitudes of the neutral gauge boson sector. For the photon field, it is given by

$$A_R^A(q^2) = A^A(q^2) + A_C^A(q^2), \tag{6.3}$$

i.e., the sum of the amplitude at one-loop level and the counterterm. The latter is given by

$$A_C^A(q^2) = -2q^2 Z_{AA}^{1/2}. \tag{6.4}$$

Since the one-loop amplitude $A^A(q^2)$ is proportional to q^2 due to the remaining $U(1)$ gauge invariance, the property

$$A_R^A(q^2 = 0) = 0 \tag{6.5}$$

automatically follows. The renormalization constant $Z_{AA}^{1/2}$ is determined by the residue condition

$$A^{A\prime}(0) - 2Z_{AA}^{1/2} = 0. \tag{6.6}$$

Thus we have

$$\frac{1}{q^2} + \frac{1}{q^2} A_R^A(q^2) \frac{1}{q^2} \bigg|_{q^2 \to 0} \to \frac{1}{q^2} \tag{6.7}$$

and do not need any renormalization for the external photon field just like QED.

For the Z-boson field we have similarly

$$\begin{aligned} A_R^Z(q^2) &= A^Z(q^2) + A_C^Z(q^2), \\ A_C^Z(q^2) &= \delta M_Z^2 + 2Z_{ZZ}^{1/2}(M_Z^2 - q^2), \end{aligned} \tag{6.8}$$

and set the following on-shell renormalization conditions

$$\begin{aligned} A^Z(M^2) + \delta M_Z^2 &= 0, \\ A^{Z\prime}(M_Z^2) - 2Z_{ZZ}^{1/2} &= 0. \end{aligned} \tag{6.9}$$

Thus the renormalized amplitude takes the single pole behaviour at the Z-boson pole.

For the Z-A transition we have

$$\begin{aligned} A_R^{ZA}(q^2) &= A^{ZA}(q^2) + A_C^{ZA}(q^2), \\ A_C^{ZA}(q^2) &= 2Z_{ZA}^{1/2}(M_Z^2 - q^2) - Z_{AZ}^{1/2} q^2, \end{aligned} \tag{6.10}$$

and set the renormalization conditions

$$\begin{aligned} A^{ZA}(M_Z^2) - Z_{ZA}^{1/2} M_Z^2 &= 0, \\ A^{ZA}(0) + Z_{AZ}^{1/2} M_Z^2 &= 0. \end{aligned} \tag{6.11}$$

Therefore the renormalized Z-A transition amplitude

$$\frac{1}{q^2 - M_Z^2}[A^{ZA}(q^2) + A_C^{ZA}(q^2)]\frac{1}{q^2} \tag{6.12}$$

does not change the single pole properties of both the photon and Z-boson fields at their on-shell and require no wave function renormalization for the external fields. The charge renormalization constant Y is determined also in the same way as QED.

$$\bar{u}(m_e)\Gamma^\mu(k)u(m_e) \,|_{k^\mu = 0} = 0. \tag{6.13}$$

In the corrected cross section, a factor 2 due to the interference and another factor 2 of the electron and muon channels should be multiplied. Thus the weak correction around the Z-boson peak reaches to about 13% as expected from the magnitude of δY.

It should be noted that, although the large counterterm appears in our scheme from the vertex correction, the corresponding term with the same order of magnitude appears in other corrections depending on schemes; of course, the physical amplitudes are independent of schemes[32]. Another comment is that this counterterm is just the term taken into account in ref. [30] as the part of $SU(2) \times U(1)$ correction to NCR. However the amplitudes are not calculated at all.

Toward the $SU(2) \times U(1)$ correction. The one-loop corrections to NCR in $SU(2) \times U(1)$ are described by very many diagrams. This is due to the tripple and quartic gauge couplings of gauge bosons, especially those of charged W-boson. We first divide them into two classes, i.e., s-channel and t-channel diagrams by noticing the virtual gauge bosons. And at the moment we consider only the s-channel diagrams since, in the energy region up to LEP II, they are expected to make dominant contribution over the t-channel diagrams. Starting from the five point diagrams in Fig. 6.1a we consider a set of diagrams which are gauge-invariant with respect to the emitted real photon. Because this photon is attached to virtual lines, the photon momenta replaced in the polarization vector can be written as combinations of propagators and the rest. Thus the requirement of gauge-invariance give rise to so many diagrams, which is in a sharp contrast with the case that real photon is emitted from an external line as in the Born diagram in Fig. 3.1. The set contains the families of diagrams in which virtual W-boson line is in the final state like Fig. 6.1b and in the initial state like Fig. 6.1c. In the case of virtual W-boson line bridged between virtual massive gauge bosons its counter particle, i.e., the charged χ must be also taken into account. Furthermore, in the former case virtual photon can run in the s-channel besides Z-boson. It can be shown that seventeen diagrams of the former type and twelve diagrams of the latter type together with the five diagrams in Fig. 6.1a satisfy the requirement of gauge-invariance. This set already contains the diagrams derived from the self-energy correction and vertex correction.

In the evaluation of correction we must calculate the five point diagrams in Fig. 6.1a. Its treatment is known in literatures[33,34]. Because there is no rank five antisymmetric tensor or simply there is no five independent vectors in our four dimensional world, there is a relation among five propagators. It is the Schouten identity.

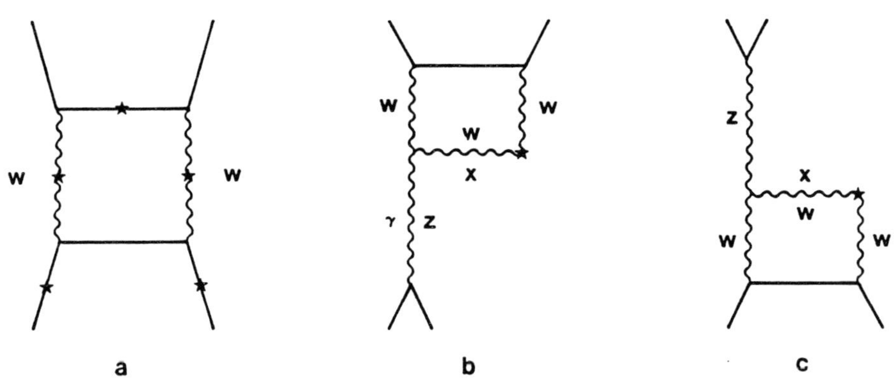

Fig. 6.1. Some Feynman diagrams of the $SU(2) \times U(1)$ correction to NCR; $\star$ means the attachment of emitted photon.

$$\epsilon^{p_1p_2p_3p_4}l_\mu = \epsilon^{\mu p_2p_3p_4}(l\cdot p_1) + \epsilon^{p_1\mu p_3p_4}(l\cdot p_2) + \epsilon^{p_1p_2\mu p_4}(l\cdot p_3) + \epsilon^{p_1p_2p_3\mu}(l\cdot p_4). \tag{6.16}$$

In ref. [33] it is proposed that this identity should be squared to get a relation among propagators. However you shall find that the relation obtained contains terms more than hundred terms because the sqaure of antisymmetric tensor ϵ leads to twenty four terms. A clever method is invented in ref. [34]: it is, instead of sqauring it, to simply multiply l_μ and $\epsilon_{p_1p_2p_3p_4}$ to Eq. (6.16). Then the five point function becomes a sum of five independent four point functions derived from it.

Another difficulty is related with the phase space integral, because the variables are contained in complicated functions of four point functions given by double logarithms and the Spence functions. Therefore this calculation becomes much more complicated than in the case of the calculation of the Born cross section in EC. In the worst case there may appear the case that the linear combination of the four point functions is zero and the divider also becomes zero at the same kinematical point.

§ 7. Summary

I reviewd some topics related with the neutrino counting reaction $e^+e^- \rightarrow \nu\bar{\nu}\gamma$. There are a few treatments of the lowest order cross section. I have compared them and showed that in the energy region higher than $110GeV$ it is better to use the exact calculation. The resonance nature of Z-boson makes the reaction really feasible but on the other hand requires the inclusion of radiative correction. It is shown that the presently known QED correction to the Z-boson channel can be also extended to the W-boson channel and the interference in the point interaction approximation of it. Finally this result is compared with the background, i.e., the radiative Bhabha scattering under realistic conditions.

Since the experiments at SLC and Lep have started, the main interest on the neutrino reaction is now changing from the original purpose of counting the number of almost massless neutrino species to the study of weakly interacting particles with electron number and/or the gauge coupling of γ-W-W. The former result shall give rise to a new constraint on cosmology and the latter in future may give clue to new physics. For the latter purpose it is inevitable to include the $SU(2) \times U(1)$ correction to it. The past experience on the corrections to the fermion pair production tells us that it is largely determined by the counterterm to the vertex correction[22,30]. However this case is different from them in the sense that it is a gauge boson pair production associated with a photon emission; as a result there appear very many one-loop diagrams. Therefore this conjecture may be wrong. Anyway in order to check it we must perform the calculation[35]. We are making efforts to it. I hope we can get the answer in very near future.

I sincerely thank the members of the TRISTAN e^+e^- project at KEK for encouragements and discussions.

REFERENCES

[1] M. Igarashi, Proc. of KEK Top. Conf. on e^+e^- Collision Physics, May, 1989, KEK, Japan.
[2] J. Yang et al., Astrophys. J. 281 (1984) 493.
[3] J. L. Hewett, Talk at this School (1989).
[4] Particle Data Group, Phys. Lett. 204B (1988) 1.
[5] R. Lich and D. R. Winn, Phys. Rev. D14 (1976) 1283.
[6] Proc. of the Workshop on Z Physics at LEP, edited by G. Altarelli, R. Kleiss and C. Versegnassi, CERN Report 89-08.
[7] K. Eggart, Talk at XIV Int'l Symp. on Lepton and Photon Interactions, Stanford, USA, August, 1989.
[8] L. DiLella, Talk at XIV Int'l Symp. on Lepton and Photon Interactions, Stanford, USA, August, 1989.
[9] E. Ma and J. Okada, Phys. Rev. Lett. 41 (1978) 287 and 1759(E).

[10] C. Hearty et al., Phys. Rev. Lett. 58 (1987) 1711.
[11] B. Barbiellini, B. Richter and J. L. Siegrist, Phys. Lett. 106B (1981) 414.
[12] L. Bento, J. Romao and A. Barroso, Phys. Rev. D33 (1986) 1488 and Phys. Lett. 194B (1987) 440.
[13] M. Igarashi, N. Nakazawa and K. Tobimatsu, preprint TKU-HEP 89/01 and KUDP 89/01 (1989) to be published in Prog. Theor. Phys. 82 (1989).
[14] G. V. Borisov, V. N. Larin and F. F. Tikhonin, Z. Physik C41 (1988) 287.
[15] D. Schildknecht, Talk at this School (1989) and M. Kuroda, F. H. Renard and D. Schildknecht, Phys. Lett. 183B (1987) 366.
[16] M. Igarashi and N. Nakazawa, Nucl. Phys. B288 (1987) 301 and in Essays in Honour of the 60th Birthday of Professor Y. Yamaguchi (World Scientific, Singapore, 1986).
[17] F. A. Berends, G. J. H. Burgers and W. L. van Neerven, Phys. Lett. 177B (1986) 191.
[18] A. Barroso, F. Boudjema, J. Cole and N. Dombey, Z. Physik C28 (1985) 149.
[19] A. I. Akhiezer and V. B. Berestetskii, Quantum Electrodynamics (Interscience, New York, 1965).
[20] F. A. Berends, R. Kleiss and S. Jadach, Nucl. Phys. B202 (1982) 63.
[21] M. Bohm and W. Hollik, Nucl. Phys. B204 (1982) 45.
[22] M. Igarashi, N. Nakazawa, T. Shimada and Y. Shimizu, Nucl. Phys. B263 (1986) 347.
[23] I. B. Kriplovich, Sov. J. Nucl. Phys. 17 (1973) 298.
[24] R. W. Brown. V. K. Cung, K. O. Mikaelian and E. A. Pascos, Phys. Lett. 43B (1973) 403.
[25] J. Fujimoto, M. Igarashi and Y. Shimizu, Prog. Theor. Phys. 77 (1987) 118.
[26] L. M. Brown and R. P. Feynman, Phys. Rev. 85 (1952) 231.
[27] V. Constantini et al., Nuovo Cim. A2 (1971) 753.
[28] E. W. N. Glover and J. J. van der Bij, Nucl. Phys. B313 (1989) 235.
[29] K. Tobimatsu and Y. Shimizu, Prog. Theor. Phys. 74 (1985) 567 and 76 (1986) 334 (E).
[30] F. A. Berends, G. Burgers, C. Mana, M. Martinez and W. L. Neerven, Nucl. Phys. B301 (1988) 583.
[31] K. Aoki, Z. Hioki, R. Kawabe, M. Konuma and T. Muta, Prog. Theor. Phys. Supplement No. 73 (1982).
[32] Radiative Corrections in $SU(2) \times U(1)$, ed. B.W. Lynn and J. F. Wheater (World Scientific, Singapore, 1984).
[33] G. 't Hooft and M. Veltman, Nucl. Phys. B153 (1979) 365.
[34] W. L. Neerven and J. A. M. Vermaseren, Phys. Lett. 137B (1984)241.
[35] Some parts are calculated in K. Kolodziej, J. Polak and M. Zralek, Acta Phys. Polonica B20 (1988) 591.

Standard Model
and Beyond

ISBN 1-56072-027-1

THE INFLUENCE OF A NEUTRINO MASS, RADIATIVE CORRECTIONS AND A COMPOSITE Z BOSON ON THE NEUTRINO COUNTING REACTION

H. Czyż, K. Kołodziej, J. Polak and M. Zrałek
University of Silesia, ul. Uniwersytecka 4, PL-40 007 Katowice, Poland

1. INTRODUCTION

A measurement of the reaction $e^+e^- \rightarrow \nu\bar{\nu}\gamma$ above the Z pole was proposed as one of the best methods of counting the light neutrino species [1,2,3].Its main advantage over other methods is that the cross section of $e^+e^- \rightarrow \nu\bar{\nu}\gamma$ would increase by 33% if there existed the light neutrino of the fourth family. In order to draw information about the number of the neutrino flavours from the neutrino counting experiments planned at LEP and SLC not only should the standard model (SM) cross section including radiative corrections be calculeted, but also the bacgkround coming mainly from the radiative Bhabha scattering $e^+e^- \rightarrow e^+e^-\gamma$ should be estimated and variety of possible non-standard effects should be taken into account.

The SM cross section of the process $e^+e^- \rightarrow \nu\bar{\nu}\gamma$ was found in Ref.[2] and the radiative corrections to it were calculeted in Refs.[4,5,6]. The background was discussed in Refs.[3,7]. Also, the radiative corrections to the process $e^+e^- \rightarrow e^+e^-\gamma$ which is the main source of the background were estimated [8] and they were found to be moderate. The SM predictions may be changed by some non-standard effects such as a big neutrino mass, discussed in Ref.[9], production of neutral supersymmetric particles, considered in Ref.[10] or the Z boson compositeness, investigated in Refs.[11,12].

In the present paper we consider the influence of a big neutrino mass, radiative corrections and a composite Z boson on the process $e^+e^- \to \nu\bar{\nu}\gamma$. We compare the effects caused by these three different factors and discuss the possibilities of distinguishing between them in the data analysis of the LEP/SLC neutrino counting experiments.

2. A BIG NEUTRINO MASS

In this section we discuss the influence of a neutrino mass on the neutrino counting reaction. In order to do this we generalize the SM formula of Ref.[2] for the cross section of the reaction

$$e^-(p_1) + e^+(p_2) \longrightarrow \nu(q_1) + \bar{\nu}(q_2) + \gamma(k) \tag{2.1}$$

in the Born approximation in such a way that it is valid not only for light neutrinos but also very heavy ones with the mass of either Dirac or Majorana origin.The Feynmann diagrams contributing to the cross section in the lowest order are depicted in Fig.1. Since the cross section for e^+e^- CM energies in the vicinity of the Z resonance is almost completely dominated by s-channel diagrams, it is convenient to split it into two parts.

$$\frac{d^2\sigma_0}{dxdy} = \frac{d^2\sigma_0^Z}{dxdy} + \frac{d^2\sigma_0^{T-Z}}{dxdy} \tag{2.2}$$

where the first term on the r.h.s. represents the contribution of the Z exchange and the second one corresponds to the contribution of W exchange and Z-W interference. The part related to the Z exchange reads (see Ref.[9])

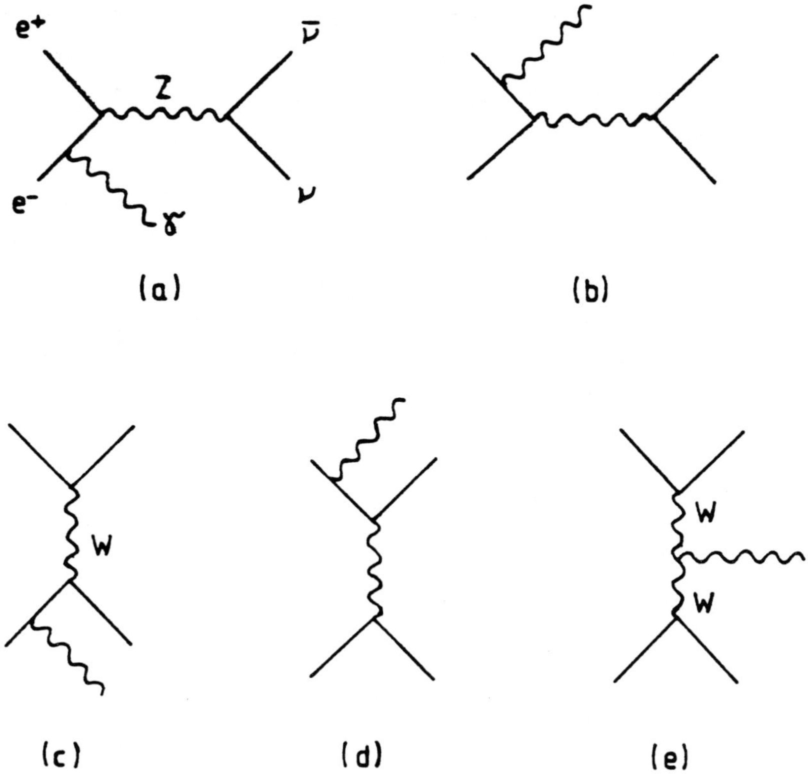

Fig.1. The SM Feynman diagrams of the process $e^+e^- \longrightarrow \nu\bar{\nu}\gamma$ in the lowest order.

$$\frac{d^2\sigma_0^Z}{dxdy} = \frac{\alpha G^2}{6\pi^2} M_Z^4(g_v^2+g_a^2)|Z(Q^2)|^2 \frac{Q^2}{x(1-y^2)} \left[\left(1-\frac{x}{2}\right)^2 + \frac{x^2y^2}{4}\right] \sum_i \beta_i \, , \tag{2.3}$$

where G is the Fermi coupling constant, α is the fine structure constant, $x = 2k/\sqrt{s}$ is the photon energy in units of the beam energy, $y = \cos\theta$ is the cosine of the photon angle with respect to the incident beam direction, M_Z is the Z mass, $g_v = -1/2 + 2\sin^2\theta_W$ and $g_a = -1/2$ are the vector and axial vector SM couplings, respectively, of the Z to the electron, $Z(Q^2)$ is the Z boson propagator

$$Z(Q^2) = (Q^2 - M_Z^2 + iM_Z\Gamma_Z)^{-1} , \tag{2.4}$$

where $Q^2 = (q_1 + q_1)^2 = s(1 - x)$ with s being the e^+e^- CM energy squared, Γ_Z is the total Z width, β_i(i = 1,2,...,N_ν- the number of neutrino flavours) are factors different for Dirac and Majorana neutrinos with

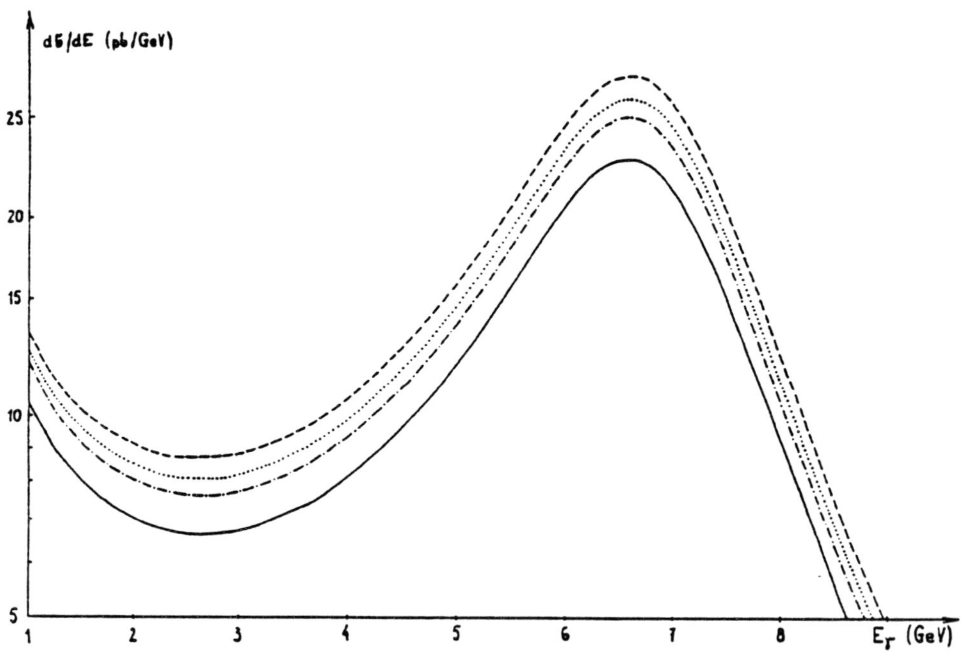

Fig.2. The photon energy distributions in the case of massless and massive neutrinos as referred to in the text.

mass m_i ($m_i^2 \le Q^2/4$) which are given by

$$\beta_i = \left(1 - \frac{4m_i^2}{Q^2} \right)^{\frac{1}{2}} \left(1 - \frac{m_i^2}{Q^2} \right) , \tag{2.5a}$$

for the Dirac, and

$$\beta_i = \left(1 - \frac{4m_i^2}{Q^2} \right)^{\frac{3}{2}} , \tag{2.5b}$$

for the Majorana neutrinos.It can be seen that for the neutrino mass $m_i^2 \le Q^2/4$ the factor $\beta_i \longrightarrow 1$ in both cases and the sum over the neutrino flavours in formula (2.3) becomes equal to N_ν. In the formula (2.3) the electron mass has been neglected. We do not present here the formula for $\frac{d^2\sigma_0^{T-Z}}{dxdy}$ which can be found in Ref. [5].

In order to illustrate the influence of the big neutrino mass on the neutrino counting reaction we have plotted the photon energy distribution (2.2), integrated over $y \le 0.94$ ($20^0 \le \theta \le 160^0$), at $\sqrt{s} = 100$ GeV, in Fig. 2. The solid and dashed lines represent the distribution for, respectively, 3 and 4 massless neutrino flavours. The dotted (dashed-dotted) line corresponds to the case where the 4-th neutrino is the Dirac (Majorana) one with a mass 30 GeV. The effect of a large neutrino mass is the decrease in the cross section, the position of the peak corresponding to the Z missing mass recoiling against the photon, however, remains almost unaltered. A clear difference in the value of the cross section for Dirac and Majorana neutrino can be seen. Unfortunately, the shape of the distribution in the vicinity of the resonance is almost the same which makes it impossible to distinguish between the final state with the heavy Dirac and a bit lighter Majorana neutrino of the 4-th family. For the neutrino mass of a dozen or so GeV the mass effect is less than 5%. Thus, the neutrino with the mass in this range would be probably detected as the massless one. On the other hand, if the mass was only few GeV smaller than $M_Z/2$ the neutrino would not be detected since the signal would be the same as in the case of 3 massless families.

3. RADIATIVE CORRECTIONS

In the present section we examine to which extent radiative corrections to process (2.1) may change the lowest order cross section and whether they may lead to wrong determination of N_ν or simulate effects of extended models, if not taken into account properly.

The radiative corrections can be devided into three groups

1. virtual electroweak correction
2. soft photon correction
3. hard photon correction.

The latter has been included since the reaction with two hard photons in the final state, in the case where one of the photons escapes detection, may imitate the reaction (2.1).

Since for e^+e^- CM energies a dozen or so GeV above the Z resonace cross section (2.2) is almost completely dominated by the part connected with the Z exchange given by eq. (2.3), it is enough to consider the radiative corrections to the s-channel diagrams of Fig.1 a,b only. The diagrams taken into account in evaluation of the virtual electroweak corrections are shown in Fig.3. The blobs on vertices and propagators represent the sums of all one-loop SM corrections, respectively, to vertices and propagators. The electroweak virtual corrections considered allow to perform a consistent renormalization of the eeZ three point Green function within the on-mass-shell renormalization scheme of the Glashow-Weinberg-Salam model [13] without imposing any additional renormalization condition on the axial-vector part of the eeZ coupling as it was done in Refs. [4,5]. It means that all ultra-violet divergences can be removed by renormalization constants determined from on-mass-shell conditions imposed on propagators and from the condition that a one loop correction to the $ee\gamma$ three point Green function vanishes for the momentum transfer equal to zero. Calculation of the electroweak correction has been done with the use of REDUCE [14] following the well known methods. Reduction to scalar integrals has been done according to

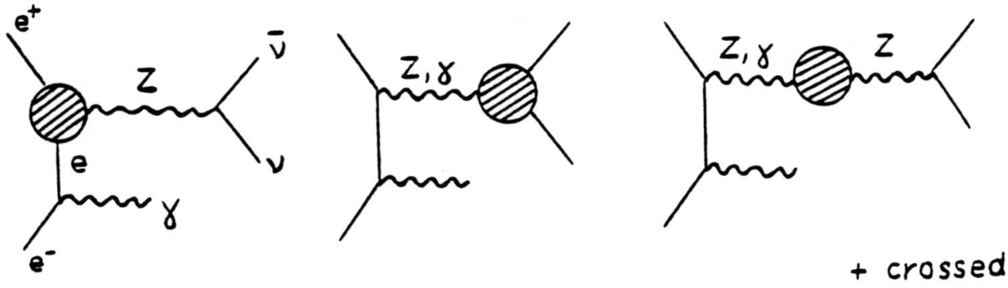

Fig.3. Feynman diagrams taken into account in evaluation of the virtual electroweak corrections.

the method of Passarino and Veltman [15]. The scalar integrals have been calculated following Ref.[16]. We do not present analytical results here since they are too lenghty. Although we do not take into account all the weak corrections to the process (2.1) we are able to estimate the range of them which is of a few per cent. For more detailed discussion see Ref.[6].

In order to obtain the infrared finite correction the soft photon correction has to be included. The diagrams corresponding to the emission of additional real photon are shown in Fig.4. The sum of the virtual electroweak and soft photon corrections depends on the cut energy of soft photons, E_m

The hard photon correction, which strongly depends on experimental cuts, is treated by the Monte Carlo methods. There are three phase space regions in which the final state of the reaction.

$$e^-(p_1) + e^+(p_2) \longrightarrow \nu(q_1) + \bar{\nu}(q_2) + \gamma(k_1) + \gamma(k_2) \qquad (3.1)$$

may imitate the final state of reaction (2.1). They are the following (see Refs.[5,6])

1. One photon fulfils detection conditions i. e. $E_1 \geq E_D$, $|\cos\theta_1| \leq \cos\theta_D$ (For the definition of angels see Fig.5) with E_D and θ_D being,

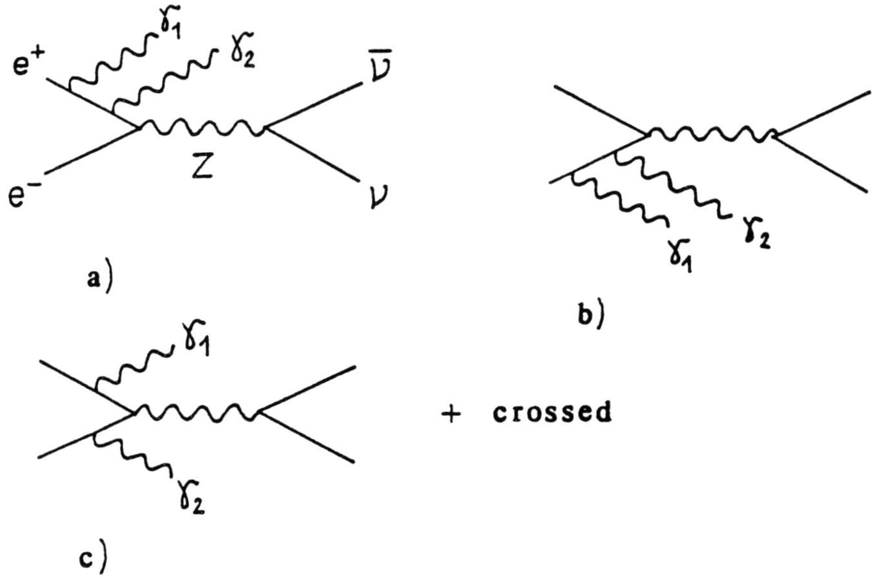

Fig.4. Feynman diagrams corresponding to the emission of an additional real photon.

respectively, the minimum photon energy to be detected and the photon tagging angle. The other photon is collinear with the initial beam, i.e. $|\cos\theta_2| \le \cos\theta_V$ with θ_V being the photon veto angle, and it is assumed to be hard, which means $E_2 \ge E_m$ where E_m is the cut energy of soft photons.

2. One photon fulfils the detection conditions as in 1 while the other, although satisfying $|\cos\theta_2| \le \cos\theta_V$ and being hard, has the energy below the detection threshold i.e. $E_m \le E_2 \le E_V$, where E_V is a veto energy of photons.
3. Both photons are in the detection range, but they are collinear within an angle of θ_S and because of that they are detected as one photon with an energy $E_1 + E_2$.

In region 1 the non zero electron mass m_e has to be kept because the terms $\sim m_e^2/(p_i k_j)^2$ $(i, j = 1, 2)$ become of the same order of magnitude as the terms $\sim 1/(p_i k_j)$ where $\vec{k}_j$ is parellel to $\vec{p}_i$. The non zero electron mass correction has been found according to the prescription given in

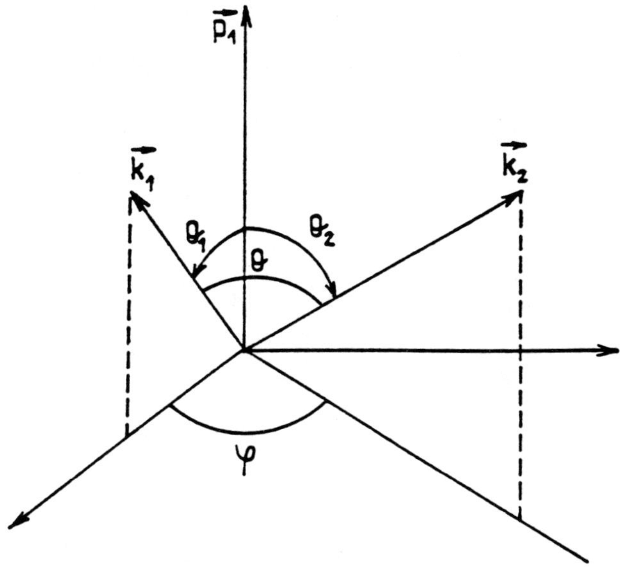

Fig.5. The centre of mass system frame used in this paper.

Ref.[17]. Due to this collinear singularity and the resonant character of the virtual Z boson the cross section of reaction (3.1) is very peaked. In order to obtain the correction to the photon energy distribution the integration over four variables has to be done. To do that we apply the standard Monte Carlo integration procedure RIWIAD [18]. However, in order to achieve good accuracy, we do not generate random numbers according to the uniform distribution, but according to the distribution

$$\frac{d\sigma_{app}}{dx_1} \sim Z(Q^2)^2 dQ^2 \frac{dy_1}{1-\beta^2 y^2} \frac{dy_2}{1-\beta^2 y^2} d\phi \ , \tag{3.2}$$

where $x_1 = 2E_1/\sqrt{s}$, Q^2 is the Z momentum transfer squared, $|Z(Q^2)|^2$ is given by eq. (2.4) and $\beta = \sqrt{1-4m_e^2/s}$ .The approximated distribution (3.2) is peaked in a similar way as the exact distribution corresponding to reaction (3.1), but it can be integrated analytically and the resulting expressions for the random variables satisfying (3.2) can be easily solved, which allows to save computational time. The factor on the r.h.s. of (3.2) has been omitted because the distribution has to be normalized to unity.

In order to obtain numerical results the following specific but realistic experimental set-up is assumed

1. veto angle of $\theta_V = 5^\circ$,
2. tagging angle of $\theta_D = 20^\circ$,
3. minimum photon energy to be detected of $E_D = 1$ GeV,
4. two photons separation angle of $\theta_S = 1^\circ$.

Although we use the renormalization scheme with α, gauge boson masses, the Higgs boson mass and fermion masses as initial parameters, we eliminate M_W, which is not supposed to be known with high precision even at LEP II, by means od the relation (see Ref.[19])

$$M_W^2 \left(1 - \frac{M_W^2}{M_Z^2} \right) = \frac{A}{1 - \Delta r}, \tag{3.3}$$

where $A = \pi\alpha/\sqrt{2}G_\mu = (37.281\,\text{GeV})^2$ with G_μ being the Fermi coupling constant calculated from the muon decay and Δr is a higher order correction mainly due to γ, W and Z self energy diagrams.There still remains the uncertainty of the top quark and Higgs boson masses which are involved in the radiative corrections through the Z boson self energy, the $Z\gamma$ mixing and the Δr of eq. (3.3), but fortunately it is rather moderate. The input values of the relevant parameters are $\alpha = 1/137.036$, $M_Z = 93$ GeV, $M_H = 100$ GeV and $m_t = 45$ GeV. For these values of M_H and m_t the $\Delta r = 0.0713 \pm 0.0013$ [20].

The numerical results for the radiative corrections are shown in Fig.6 where we have depicted the photon energy distribution at $\sqrt{s} = 100$ GeV, integrated over $|\cos\theta| \le 0.94$ for 3 massless families of the neutrinos. The dashed line corresponds to the Born approximation, the dotted line to the distribution including both the virtual electroweak

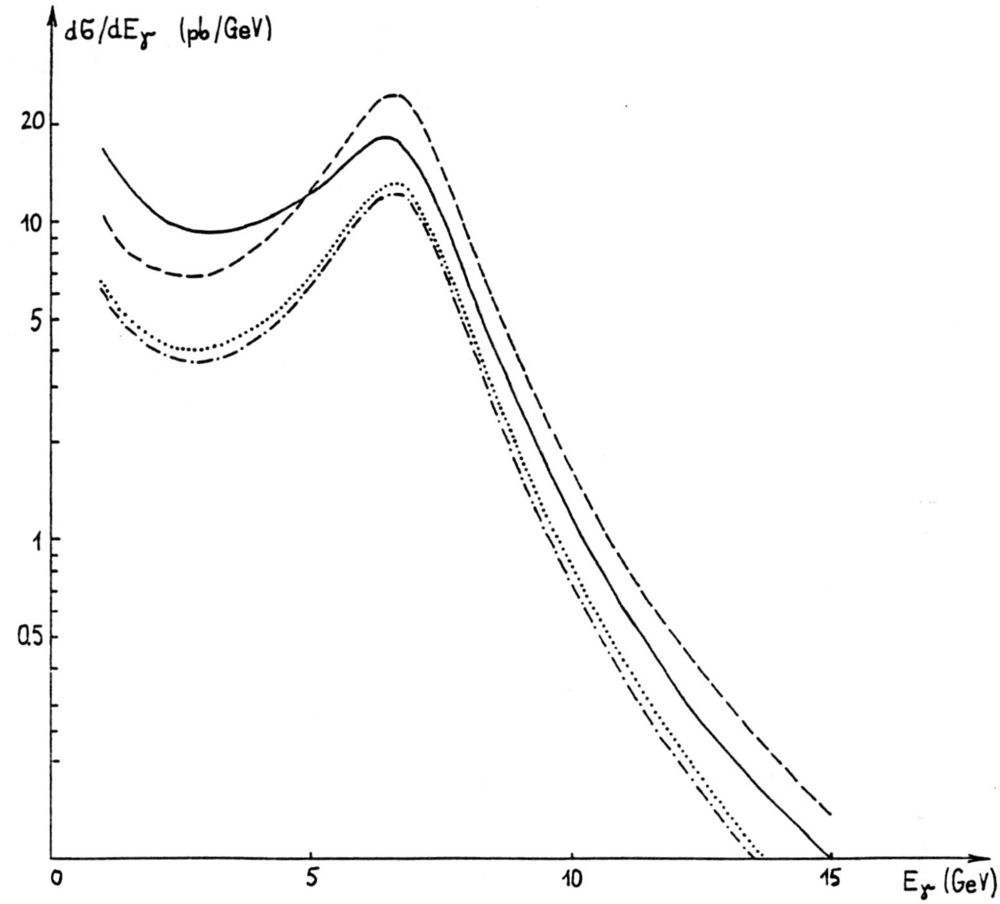

Fig.6. The photon energy distributions as reffered to in the text.

Although we use the renormalization scheme with α, gauge boson and soft photon corrections (E_m = 0.2 GeV) and the solid line to the distribution which, apart from the virtual and soft photon corrections, includes the hard photon correction. The virtual electroweak and soft photon corrections diminish the Born cross section in the photon energy region 4 GeV $\le E_\gamma \le$ 9 GeV, which is equal to 71 pb, by ~ 50% . The hard photon correction interferes with the latter which results in the total radiative correction of 18% . This corresponds to over half of the signal of the 4-th massless neutrino.

4. A COMPOSITE Z BOSON

In spite of the radiative corrections calculated in the frame of the standard electroweak model and a neutrino mass which occurs in many extensions of the SM, there is another nonstandard effect which may change the standard model expectations. The effect is related to the Z boson compositeness. Although the existing experimental evidence provides quite a good confirmation of the standard model it cannot be excluded that the incoming era of great accelerators will reveal some new physics. In particular, it may turn out that the Z and W are not really the gauge bosons of the SU(2) $\otimes$ U(1) model, but simply composite particles of a phenomenological theory. If the Z is composite, then the three gauge boson coupling $ZZ\gamma$ may differ from that resulting from the SM. The general form of such a coupling has been investigated in a series of papers [21,22,23]. In Ref.[11] the proces $e^+e^- \to \nu\bar{\nu}\gamma$ has been used for probing the $ZZ\gamma$ vertex with a nonvanishing electric dipole transition form factor only.

In this section we shall try to answer the questions:

(i) what is the influence of the general $ZZ\gamma$ coupling on counting of neutrinos with the use of the reaction $e^+e^- \to \bar{\nu}\nu\gamma$,

(ii) what is a chance to find a possible Z substructure in this reaction.

While studying the anomalous $ZZ\gamma$ coupling we shall take into account not only the CP conserving terms, but also the CP violating ones, leaving it to the taste of the reader whether he (or she) sets the latter ones to zero or not.

We start from presenting our parametrization of the $ZZ\gamma$ vertex, which is a little different from that previously used in the literature (see Ref.[23]). In the most general form, the Lorentz invariant $ZZ\gamma$ vertex involves 20 form factors. If the photon is on-mass-shell we are left only with 7 of them. If, moreover, one of the Z is on-mass-shell, the number of form factors is reduced to 4 (see Ref.[13]). Hence, in the case where only one of the Z (Z_α) is virtual, the vertex can be parametrized in the following way

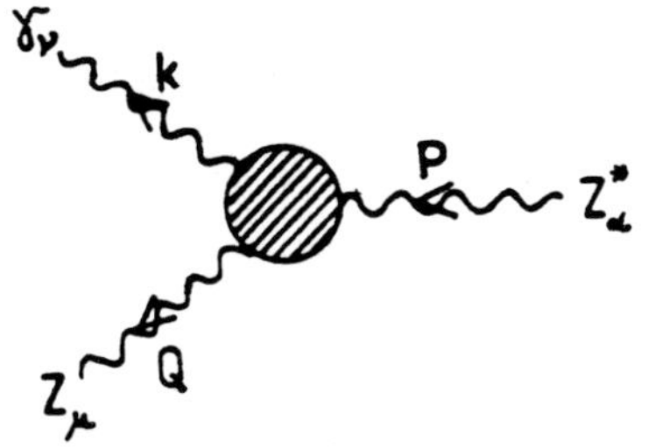

$$= e\Gamma^{\mu\nu\alpha}_{Z\gamma Z^*}(r,P),$$

where

$$\Gamma^{\mu\nu\alpha}_{Z\gamma Z^*}(r,P) = \frac{s - Q^2}{M_Z^2} \left\{ F_1 \left(g^{\mu\nu} r^\alpha - 2g^{\nu\alpha} P^\mu\right) \right.$$

$$+ \frac{F_2}{M_Z^2} P^\mu \left[P^\nu r^\alpha - (s - Q^2) g^{\nu\alpha} \right] \tag{4.1}$$

$$\left. + F_3 \varepsilon^{\mu\nu\alpha\beta} (r + P)_\beta + \frac{F_4}{M_Z^2} P^\mu \varepsilon^{\nu\alpha\beta\delta} r_\beta P_\delta \right\}$$

with $Q = p_1 + p_2 - k$, $r = k - Q$, $P = k + Q$, $s = (p_1 + p_2)^2$ and M_Z being the Z boson mass. The terms with F_3 and F_4 (F_1 and F_2) are CP-even (-odd). The factor $(s - Q^2)/M_Z^2$ on the right hand side of eq. (4.1) has been introduced to reproduce the behaviour of the standard model $ZZ\gamma$ coupling at the one loop level for $Q^2 \rightarrow M_Z^2$

In the frame of gauge theories, where unitarity is guaranted by gauge invariance, no constant coupling between neutral gauge bosons is not allowed. If the Z is composite then the form factors involved in

(4.1) are functions of energy which, in principle, can be calculated in the frame of any composite model (see e.g. [24]). In Ref. [25] we have found the unitarity bounds for $F_i(s)$ $(i = 1,\dots,4)$ which follow from the unitarity condition for partial wave amplitudes. They read

$$F_i \le \begin{cases} \dfrac{\varkappa}{\left(\varkappa^4 - 1\right)^{1/2}\left(\varkappa^2 - 1\right)} \times 82.6, & \text{for } i = 1,3 \\[2ex] \dfrac{2}{\left(\varkappa^2 - 1\right)^{5/2}} \times 82.6, & \text{for } i = 2,4 \end{cases} \qquad (4.2)$$

where $\varkappa = \sqrt{s}/M_Z^2$.

The presence of the $ZZ\gamma$ vertex at the tree level results in the existence of the additional Feynman diagram contributing to (2.1) (see Fig.7). Since we use the four form factors parametrization of the anomalous $ZZ\gamma$ coupling, one of the Z in Fig.7 should be on-mass-shell. The problem is which one? In paper [11], Barroso et al. ensure that the neutrino counting experiment will be done essentially at the

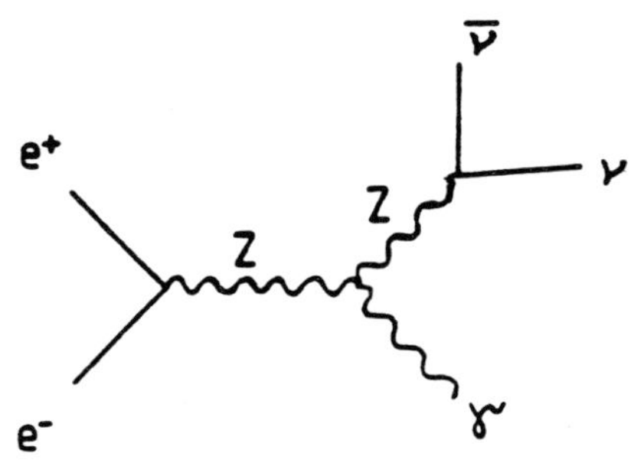

Fig.7. The anomalous Feynmam diagrams contributing to the $e^+e^- \to \nu\bar{\nu}\gamma$.

centre-of-mass energy equal to M_Z and they choose the Z which couples to e^+e^- as being on-mass-shell. That energy, however, does not seem to be convenient for carrying out the neutrino counting experiment for the two reasons. Firstly, the background coming from the radiative Bhabha scattering is much higher than the signal itself. Secondly, the peak connected with the Z resonance cannot be distinguished from the soft photon peak. Thus, it is more probable that the experiment will be done at higher energy. Up to now two regions in which the neutrino counting experiment could be done have been advocated in literature. One region is a few GeV ($\sqrt{s}$ = 96 - 98 GeV) and the second region is a dozen or so GeV ($\sqrt{s}$ = 104 - 106 GeV) above the Z threshold [26]. In the latter region the background can be almost completely eliminated. Since the cross section of reaction (2.1) has a very clear resonance for the momentum transfer $Q^2 \approx M_Z^2$ we shall assume that in the diagram of Fig.7 the Z boson decaying into the neutrinos is on-mass-shell. We think that our approximation is better for both regions of the neutrino counting experiment ($\Gamma_Z \approx 2.6$ GeV) than that of Ref.[11]. Considering the Z decaying into the neutrinos as being on-mass-shell we can write the formula for the Born cross section of reaction (2.1) in the presence of anomalous coupling (4.1).

$$\frac{d^2\sigma_0}{dxdy} = \frac{d^2\sigma_0^{SM}}{dxdy} + \frac{d^2\sigma_0^{A}}{dxdy} \qquad (4.3)$$

where the first term on the right hand side is given by (2.3) with the neutrino massess set to zero and the second one representing the anomalous contributions to (2.1) reads

$$\frac{d^2\sigma_A}{dxdy} = N_\nu \frac{\alpha G^2}{192\pi^2} \left(g_v^2 + g_a^2\right) M_Z^2 s^2 x^2 \left|Z(Q^2)\right|^2 \left|Z(s)\right|^2$$

$$\times \left\{ \frac{s^2 x^3}{M_Z^2} \left[\left(F_1^2 + F_3^2\right)\left(s(1+y^2) + 2Q^2(1-y^2)\right) \right.\right.$$

$$+ \frac{F_1 F_2}{M_Z^2} s^2 x (1+y^2) + \left(F_2^2 + F_4^2\right) \frac{s^3 x^2}{4M_Z^4}(1+y^2) \qquad (4.4)$$

$$\left.\left. - F_3 F_4 \frac{s^2 x}{M_Z^2}(1+y^2)\right] + 8M_Z\Gamma_Z\left(F_1 + \frac{sF_2}{2M_Z^2}\right) Q^2 sxy \right.$$

$$+ 16\left(s - M_Z^2\right) \frac{g_v g_a}{g_v^2 + g_a^2} Q^2 \left[\left(s + Q^2\right) F_3 - \frac{s^2 x F_4}{2M_Z^2}\right]\Bigg\} ,$$

where Z is given by (2.4), and the formfactors F_i $(i = 1,\ldots,4)$ are those of formula (4.1). The terms in the square brackets on the r.h.s. of (4.4) correspond to the square of the diagram in Fig.7 while the other terms on the r.h.s. of (4.4) represent the interference of the anomalous diagram with the diagrams of Fig 1a, b. We have neglected the interference of the diagram of Fig.7 with the W exchange diagrams of Fig. 1c, d, e, which is clearly justified in the energy region considered.

From formula (4.4) it is seen that the CP-odd form factors F_1 and F_2 introduce a forward-backward asymmetry into the photon distribution. We define the asymmetry by

$$A_{fb} = \frac{\int_0^{y_D} dy \int_{x_1}^{x_2} dx \frac{d^2\sigma_0^A}{dxdy} - \int_{-y_D}^{0} dy \int_{x_1}^{x_2} dx \frac{d^2\sigma_0^A}{dxdy}}{\int_{-y_D}^{y_D} dy \int_{x_1}^{x_2} dx \frac{d^2\sigma_0^A}{dxdy}} , \qquad (4.5)$$

where $\frac{d^2\sigma_0^A}{dxdy}$ is given by (4.4), $y_D = \cos\theta_D$ is the cosine of the photon tagging angle θ_D and the limits x_1 and x_2 will be discussed later. Asymmetry (4.5) can be a criterion of discrimination between CP-even and CP-odd terms in anomalous coupling (4.1) if it is big enough to be detected.

Formula (4.4) allows us to test the influence of particular form factors of the anomalous coupling (4.1) on the cross section of the neutrino counting reaction. To illustrate this influence we have plotted cross section (4.3) at $\sqrt{s}$ = 106 GeV, integrated over $|y| \leq 0.94$ $(20^\circ \leq \theta \leq 160^\circ)$, versus the photon energy, for 3 massless generations of neutrinos, in Figs.8 and 9. The solid lines in both figures correspond to the Born cross section. Dashed curves in Fig.8 show the distribution for F_3 = 9, 18, 27 and the other form factors set to zero, while in Fig.9 the distribution for F_4 = 60, 100, 160 and $F_1 = F_2 = F_3 = 0$. The

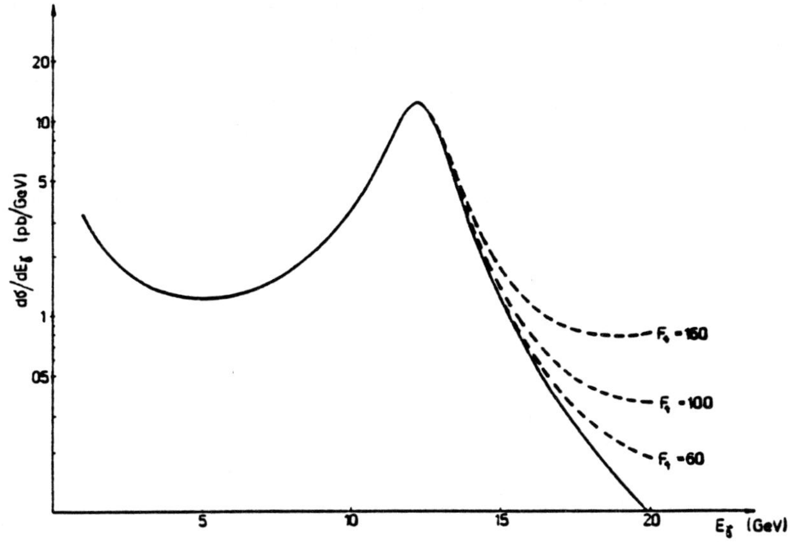

Fig.8. The photon energy distributions in the Born approximation at $\sqrt{s}$ = 106 GeV: solid line - $F_1 = F_2 = F_3 = F_4 = 0$, dashed lines - F_3 = 9, 18, 27 and $F_1 = F_2 = F_4 = 0$.

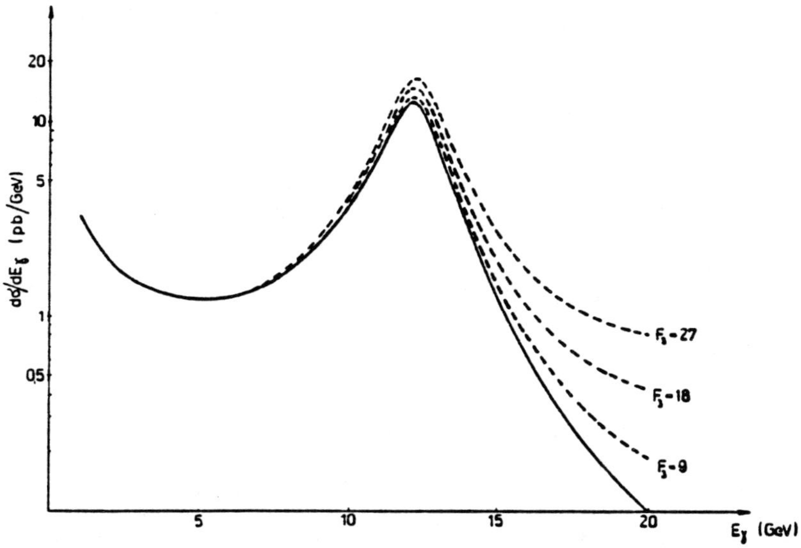

Fig.9. The photon energy distributions in the Born approximation at $\sqrt{s}$ = 106 GeV: solid line - $F_1 = F_2 = F_3 = F_4 = 0$, dashed lines - F_4 = 60, 100, 160 and $F_1 = F_2 = F_3 = 0$.

distributions for corresponding values of F_1 and F_2 are almost the same as the appropriate distributions for F_3 and F_4, respectively. The effect of the nonzero form factors is clearly visible for photon energies above the resonance peak.

The values of the form factors taken may seem to bee too large, but they are below the unitarity bounds following from formula (4.2) which, for $\sqrt{s}$ = 106 GeV are

$$F_i = \begin{cases} 380 & \text{for } i = 1,3, \\ 3400 & \text{for } i = 2,4 \end{cases} \tag{4.6}$$

Apart from fulfilling (4.6), they are not in contradiction with the CERN $p\bar{p}$ collider data (see Ref.[27]). Moreover, it should be stressed that the factor $(s - Q^2)/M_Z^2$ in (4.1) is of the range of 0.1 and the form factors F_2 and F_4 have been divided by M_Z^2 which is of the range of 10^4. Thus, the values of form factors chosen to ilustrate the influence of the anomalous $ZZ\gamma$ coupling on the neutrino counting reaction are not as unreal as they might seem to be at first glance.

In order to answer the question about the chance of finding a possible Z substructure in reaction (2.1) we define the quantity δ_A by

$$\delta_A = \int_{x_1}^{x_2} dx \int_{-y_D}^{y_D} dy \, \frac{d^2\sigma_0^A}{dxdy} \Bigg/ \int_{x_1}^{x_2} dx \int_{-y_D}^{y_D} dy \, \frac{d^2\sigma_0^{SM}}{dxdy} \,, \tag{4.7}$$

where $\frac{d^2\sigma_0^A}{dxdy}$ and $\frac{d^2\sigma_0^{SM}}{dxdy}$ are given by (4.4) and (2.2) (with the zero neutrino massess), respectively. The integration limits y_D, x_1, and x_2 are chosen in the following way

$$y_D = 0.94$$

$$x_1 = x_M + i\,\frac{2\text{GeV}}{\sqrt{s}} \,, \qquad i = 0,\, 1,\, 2,\, \ldots \tag{4.8}$$

$$x_2 = 1.0 \,,$$

where $x_M = 1 - M_Z^2/s$ is the value of x corresponding to the maximum of the SM cross section (2.2).

The quantity δ_A was computed for various values of the form factors at the CM energy $\sqrt{s}$ = 94,96, ...,106 GeV. The lower integration limit x_1 was shifted upwards until the total SM cross section was grater than 10 pb. The minimum value of δ_A to be detected is assumed to be of 0.05. In this way we can find the CM energy and the integration region $x_1 \le x \le x_2$ most sensitive for testing the Z substructure. The results are presented in Table 1. It is seen that regions corresponding to bigger values of x_1 are more suitable while searching for the Z substructure than the ones with lower values of x_1. On the other hand, the δ_A at lower CM energies seems to be a little bit more sensitive to the form factors than at higher CM energies. However, one should remember of a background, which is greater for energies closer to the M_Z.

$\sqrt{s}$	x_1	σ_{SM} (pb)	$F_{3min}(\delta_A(\%))$	$F_{4min}(\delta_A(\%))$
94	0.021	159	3 (5)	10 (8)
	0.042	50	2 (8)	5 (6)
	0.064	19	1 (5)	5 (17)
96	0.062	71	4 (7)	10 (6)
	0.082	28	3 (10)	10 (15)
	0.103	12	2 (11)	5 (9)
98	0.099	45	4 (6)	15 (11)
	0.120	19	3 (8)	10 (12)
100	0.135	33	4 (6)	15 (11)
	0.155	14	3 (8)	10 (11)
102	0.169	25	4 (7)	15 (11)
	0.188	11	3 (8)	10 (11)
104	0.200	20	4 (8)	10 (5)
106	0.230	17	3 (6)	10 (6)

Table 1. The minimum values of the form factors F_3 and F_4 that can be measured through the measurement of the δ_A under the conditions chosen in the text.

The assymmetry A_{fb} of formula (4.5) is also integrated in limits (4.8), but now $x_2 = x_M + 2\Gamma_Z/\sqrt{s}$ and we assume integrated cross section (4.4) grater than 10 pb. Since the systematic errors cancel in ratio (4.5) and one is left only with the statistical error, which can be reduced by lenghtening the time of an experimental run, the experiment should be sensitive to the values of the asymmetry of the range of 1 per cent. The results presented in Table 2 seem to be on the verge of the experimental sensitivity.

$\sqrt{s}$	x_1	F_1	σ(pb)	A_{fb}(%)	F_1	σ(pb)	A_{fb}(%)	F_2	σ(pb)	A_{fb}(%)
94	0.021	10	157	0.0	26	157	0.1	100	157	0.2
	0.042	10	47	0.1	26	47	0.3	100	47	0.5
	0.064	10	16	0.2	26	17	0.6	80	16	1.0
96	0.062	10	69	0.1	26	70	0.2	100	69	0.4
	0.082	10	26	0.2	26	26	0.4	100	26	0.8
	0.103	10	10	0.3	26	10	0.7	-	-	-
98	0.099	10	44	0.1	26	45	0.3	100	44	0.6
	0.120	10	17	0.2	26	18	0.5	90	17	1.0
100	0.135	10	32	0.2	26	33	0.4	100	31	1.0
	0.155	10	13	0.3	26	14	0.6	60	13	1.0
102	0.169	10	25	0.2	26	27	0.5	70	24	1.0
	0.188	10	10	0.3	26	12	0.8	80	10	1.7
104	0.200	10	20	0.3	26	24	0.7	50	20	1.0
	0.220	-	-	-	26	11	0.8	-	-	-
106	0.230	10	17	0.4	26	23	0.7	40	16	1.0
	0.249	-	-	-	26	11	0.8	-	-	-

Table 2. The minimum value of the form factors F_1 and F_2 that can be measured through the measurement of the A_{fb} under the conditions chosen in the text.

5. CONCLUSIONS

We have examined the influence of 3 different factors i.e. a neutrino mass, radiative corrections, and a composite Z boson on the neutrino process $e^+e^- \rightarrow \nu\bar{\nu}\gamma$.

It has been shown in Section 2 that an additional neutrino with mass of a dozen or so GeV would be detected as a massless one, but if the mass was bigger it would give a visible effect in the photon energy distribution diffrent in size for the Dirac and Majorana neutrinos.

However, since the shape of the distribution is almost the same in both cases, the neutrino counting reaction does not seem to be a good tool for making a distinction between the Dirac and Majorana neutrinos.

In Section 3 we have estimated the influence of the radiative corrections on the cross section. Apart from the photonic corrections considered in Refs. [4, 5] we have partly included the weak correction. In this way we have not only been able to perform a consistent renormalization within OS scheme of the GSW model, but also to estimate the range of weak correction to the process. The total effect of the radiative corrections in the vicinity of the Z resonance peak is of 18 % which is over half of the signal for the 4-th massless neutrinos.

The efect of the composite Z boson which may couple to the photon through the nonstandard $ZZ\gamma$ coupling has been discussed in Section 4. The anomalous cross section involves form factors used for parametrization of the vertex. It has been shown that the two CP violating form factors introduce the forward-backward asymmetry into the photon distribution.

Each of the 3 factors discussed here affects the photon energy distribution of the process $e^+e^- \rightarrow \nu\bar{\nu}\gamma$ in a different way. The big neutrino mass shifts the appropriate distribution in the massless case downwards without changing its shape. The radiative corrections diminish the resonance peak and increase the cross section for low energy photons. Finally, the anomalous $ZZ\gamma$ coupling increases the cross section for photons with energies above that corresponding to the resonance peak. Thus, even if the three factors are present together, it should be posssible to distinguish among them in the analysis of data coming from the LEP/SLC neutrino counting experiments.

REFERENCES

[1]. A. D. Dolgov, L. B. Okun, V. I. Zakharov, Nucl. Phys. B41 (1972) 197.

[2]. E. Ma, J. Okada, Phys. Rev. Lett. 41 (1978) 287, K. J. F. Gaemers, R. Gastmans, F. M. Renard, Phys. Rev. D19, (1979) 1605.

[3]. G. Barbiellini, B. Richter, J. L. Siegrist, Phys. Lett. 106 B, (1981) 414.
[4]. M. Igarashi, N. Nakazawa, Nucl. Phys. B288 (1987) 301.
[5]. F. A. Berends, G. J. H. Burgers, C. Mana, M. Martinez W. L. van Neerven, Nucl. Phys b 301 (1988) 583.
[6]. K. Kołodziej, J. Polak, M. Zrałek, Acta Physica Polonica B20 (1989) 591.
[7]. M. Caffo, R. Gatto, E. Remiddi, Phys. Lett. 173 B (1986) 91, Nucl. Phys. B286 (1987) 293;
C. Mana, M. Martinez, Nucl. Phys. B287 (1987) 601; Velthian, Nucl. Phys. B
[8]. D. Karlen, Nucl. Phys. B287 (1987) 601;
M. Martinez, R. Miquel, University of Barcelona preprint UAB - LFAE 87 - 01.
[9]. H. Czyż, K. Kołodziej, M. Zrałek, Phys. Lett. 207 B (1988) 511.
[10]. L. Bento, J. C. Romao, A. Barroso, Phys. Rev. D 33 (1986)1488;
P. L. Burke, SLAC p[reprint SLAC - PUB - 4284 (1987).
[11]. A. Barroso, P. Nogueria, J. C. Romao, Phys. Lett. B196 (1987)547.
[12]. H. Czyż, K. Kołodziej, M. Zrałek, "An anomalous ZZγ coupling and the neutrino counting reaction", work accepted for publication in Physica Scripta.
[13]. K. Aoki, Z. Hioki, R. Kawabe, M. Konuma, T. Muta, Prog. Theor. Phys. Suppl. 73(1981) 1.
[14]. A. C. Hearn, REDUCE 2 user's manual, UCP - 19, Univ. of Utah 1973
[15]. G. Passarino, M. Veltman, Nucl. Phys. B 160 (1980)151
[16]. G. 'tHooft, M. Veltman, Nucl. Phys. B 153 (1979) 365
[17]. F. A. Berends et al., Nucl. Phys. B 153 (1979) 365
[18]. B. Lautrup, RIWIAD, CERN - DD d 114.
[19]. A. Barroso et al., preprint CERN - EP/87-70 (1987)
[20]. F. Jegerlehner, Z. Phys. C32 (1986) 195, 425.
[21]. F. M. Renard, Nucl. Phys. B196 (1982) 93.
[22]. M. Kuroda et al., Nucl. Phys. B284 (1987) 271;
P. Mery,M. Perrottet and M. Renard, Z. Phys. C36 (1987) 249.
[23]. K. Hagiwara et al., Nucl. Phys. B282 (1987) 253.

Part II: Beyond the Standard Model

Standard Model
and Beyond

ISBN 1-56072-027-1

MODULI SPACE OF CALABI-YAU MANIFOLDS*

Philip Candelas and Xenia C. De La Ossa
Theory Group, Department of Physics, University of Texas, Austin, TX 78712 USA

ABSTRACT

We present an accessible account of the local geometry of the parameter space of Calabi–Yau manifolds. It is shown that the parameter space decomposes, at least locally, into a product with one factor the space of parameters of the complex structure and the other a complex extension of the parameter space of the Kähler class. It is also shown that each of these spaces is itself a Kähler manifold and is moreover a Kähler manifold of restricted type. There is a remarkable symmetry in the intrinsic structures of the two parameter spaces and the relevance of this to the conjectured existence of mirror manifolds is discussed. The two parameter spaces behave differently with respect to modular transformations and it is argued that the role of quantum corrections is to restore the symmetry between the two types of parameters so as to enforce modular invariance.

1. Introduction

The purpose of this article is to present an accessible account of aspects of the geometry of the moduli space of Calabi–Yau manifolds. The motivation for studying this topic arises in string theory. It is of interest to know what types of theories arise at low energies from the compactification of the heterotic string. If six of the ten dimensions of string theory are compactified on a Calabi–Yau manifold then there arises, at energies small compared to the Planck scale, an effective four–dimensional supergravity theory whose component fields correspond to the parameters that describe the possible deformations of the Calabi–Yau manifold. This is intuitively reasonable since the massless fields correspond to the parameters that take one vacuum state into a nearby equivalent one. The terms that

* Supported in part by the Robert A. Welch Foundation and N.S.F. Grants PHY-880637 and PHY-8605978.

make up the effective Lagrangian of the low energy theory turn out to have geometrical and topological significance. The parameter space admits an intrinsically defined metric and this metric appears in the kinetic terms of the effective field theory. Certain cubic quantities also have intrinsic meaning and these appear as the Yukawa couplings of the effective theory. The process by which these geometrical quantities appear in the low energy theory has been described in the literature[1-3] and will not be repeated here. We shall instead discuss the local geometry of the parameter space of Calabi–Yau manifolds. It seems that it is possible to compactify superstrings in ways that involve $c = 9$ conformal theories that may be more general than compactification for Calabi–Yau manifolds. At the time of writing it is not clear to what extent these compactifications are really more general since in many cases they turn out to correspond to compactifcation on Calabi–Yau manifolds after all. Our real justification however is that the techniques for treating the deformations of complex manifolds have been quite well developed and so allow us to go farther in this case. In any event it seems that much of the discussion carries over to the general case of conformal theories.

We shall find that the low energy effective theory enjoys a number of surprising properties:

(i) The parameter space factorizes, at least locally, into a product manifold $Z \times W$, with Z the manifold of complex structures and W a complexification of the space of parameters of the Kähler class.

(ii) There exists a natural gauge invariance on each of the two parameter spaces. A consequence of this is that each of these is itself a Kähler manifold and is moreover a Kähler manifold of restricted type. In other words although we expect and obtain $N = 1$ supersymmetry in spacetime nevertheless the effective theory enjoys $N = 2$ supersymmetry relations.

(iii) There is a remarkable symmetry in the structures of Z and W which lends credence to the existence of 'mirror manifolds'[4,5]. The precise conditions under which a mirror manifold $\mathcal{W}$ exists for a given Calabi–Yau manifold $\mathcal{M}$ have yet to be understood. Yet their existence would resolve a puzzle. The puzzle is that calculations based on Calabi–Yau manifolds are 'half–exact' in so far as Yukawa couplings, say, corresponding to the complex structure parameters are not corrected either in perturbation theory or by worldsheet instantons. On the other hand Yukawa couplings calculated for the Kähler class parameters are corrected by worldsheet instantons[6] . The existence of mirror manifolds opens up the possibility of calculating both types of couplings exactly. Given a Calabi–Yau manifold $\mathcal{M}$ and its mirror $\mathcal{W}$ one could calculate the couplings corresponding to the complex structure parameters for $\mathcal{M}$ and the remaining couplings as the couplings corresponding to the complex structure parameters of $\mathcal{W}$.

The present article is concerned with local properties of the moduli space. For a discussion of aspects of the global properties with particular emphasis on the existence of conifold points corresponding to the existence of certain singular Calabi–Yau manifolds that are common to more than one moduli space see[7]. A preliminary account of part of the present work was given in this reference.

2. The Metric on the Parameter Space

By a Calabi–Yau manifold we shall here mean a compact Kähler manifold of vanishing first Chern class and complex dimension three. A Kähler manifold is a complex manifold $\mathcal{M}$ whose Kähler form $J = \frac{1}{2}J_k{}^l g_{lm} dx^k \wedge dx^m$, obtained by lowering an index of the complex structure, is closed. In complex coordinates we have

$$J = i g_{\mu\bar{\nu}} dx^\mu \wedge dx^{\bar{\nu}} \ .$$

The pair $(\mathcal{M}, J)$ can be deformed by either deforming the complex structure of $\mathcal{M}$ or by deforming the components $g_{\mu\bar{\nu}}$ of the metric, *i.e.* deforming J. As a simple example of the deformation of the complex structure of $\mathcal{M}$ consider the manifold $\mathbb{P}_4(5)$ defined by the vanishing of a quintic polynomial in $\mathbb{P}_4$. Such a polynomial has 126 coefficients. There are 25 degrees of freedom associated with changes of coordinates $y^A \to M^A{}_B y^B$ of the five homogeneous coordinates of the embedding $\mathbb{P}_4$. Thus there are $126 - 25 = 101$ complex parameters in the defining quintic. In this case the naive count is correct. In more complicated examples there are a number of subtleties associated with the counting though in the following we are more concerned with the existence of these parameters and in the geometry of the parameter space rather than in the counting. The parameters corresponding to the Kähler class may be less familiar owing to the fact that in this the simplest example the number of harmonic $(1,1)$ forms, b_{11} is unity. The only $(1,1)$ form is the Kähler form induced from the embedding $\mathbb{P}_4$ and the variation of this single parameter just corresponds to the freedom to adjust the overall scale of the manifold. More generally, however, $b_{11} > 1$. A simple example is provided by the manifold[1]

$$\begin{matrix}\mathbb{P}_1\\ \mathbb{P}_4\end{matrix}\begin{pmatrix}1 & 1\\ 1 & 4\end{pmatrix}_{-168} \ .$$

Here $b_{11} = 2$ and we may think of the two (1,1)–forms J_1 and J_4 say as the forms induced by the Kähler–forms of the $\mathbb{P}_1$ and $\mathbb{P}_4$. In this case we may take

$$J = aJ_1 + bJ_4 \ .$$

The Kähler–form $J = i g_{\mu\bar{\nu}} dx^\mu \wedge dx^{\bar{\nu}}$ must correspond to a Kähler metric $g_{\mu\bar{\nu}}$ that is positive definite (such a form is said to be positive) but is otherwise arbitrary. In this case we have a two parameter family of Kähler classes and we may think of these two parameters as corresponding to the radii of the $\mathbb{P}_1$ and the $\mathbb{P}_4$.

---

[1] The notation denotes the manifold defined in the product $\mathbb{P}_1 \times \mathbb{P}_4$ by two polynomials whose degrees in the variables of the two projective spaces are given by the columns of the matrix. The number appended to the matrix is the Euler number.

Yau's theorem states that given a Kähler manifold $\mathcal{M}$ with associated Kähler–form J_0 there is a unique Ricci–flat Kähler metric for $\mathcal{M}$ whose associated Kähler–form J is in the same cohomology class as the given one J_0. In view of this we may consider the parameter space of Calabi–Yau manifolds to be the parameter space of Ricci–flat Kähler metrics. Let g_{mn} and $g_{mn} + \delta g_{mn}$ be two Ricci–flat metrics for $\mathcal{M}$

$$R_{mn}(g_{..}) = 0 \ , \quad R_{mn}(g_{..} + \delta g_{..}) = 0 \ ,$$

then with the coordinate condition $\nabla^n \delta g_{mn} = 0$ we have that δg_{mn} satisfies the Lichnerowicz equation

$$\nabla^k \nabla_k \delta g_{mn} + 2R_m{}^p{}_n{}^q \delta g_{pq} = 0 \ . \tag{2.1}$$

Owing to the special properties of Kähler manifolds the zero modes of this equation of mixed type, $\delta g_{\mu\bar{\nu}}$, and those of pure type, $\delta g_{\mu\nu}, \delta g_{\bar{\mu}\bar{\nu}}$, separately satisfy the equation. With a variation of the metric of mixed type we can associate the real (1,1)–form

$$i\delta g_{\mu\bar{\nu}}\, dx^\mu \wedge dx^{\bar{\nu}}$$

which is harmonic if and only if the variation satisfies Eq. (2.1). With a variation of the metric of pure type we may associate the complex (2, 1)-form

$$\Omega_{\kappa\lambda}{}^{\bar{\nu}} \delta g_{\bar{\mu}\bar{\nu}}\, dx^\kappa \wedge dx^\lambda \wedge dx^{\bar{\mu}}$$

which also is harmonic if and only if the variation of the metric satisfies (2.1). In other words the zero modes of the Lichnerowicz equation are in one–one correspondence with the elements of $H^{(1,1)}(\mathcal{M})$ and $H^{(2,1)}(\mathcal{M})$. The modes of mixed type are clearly variations of the Kähler class and give rise to b_{11} real parameters. The variations of pure type, which give rise to b_{21} complex parameters, correspond to variations of the complex structure. To see this note that $g_{mn} + \delta g_{mn}$ is a Kähler metric on a manifold close to the original one. There must therefore exist a coordinate system in which the pure parts of the metric vanish. Under a change of coordinates $x^m \rightarrow x^m + f^m(x)$ the metric variation transforms according to the familiar rule

$$\delta g_{mn} \rightarrow \delta g_{mn} - \frac{\partial f^r}{\partial x^m} g_{rn} - \frac{\partial f^r}{\partial x^n} g_{mr} \ . \tag{2.2}$$

If $f^\mu(x)$ is holomorphic then $\delta g_{\mu\nu}$ is invariant. Thus the pure part of the variation can be removed by a transformation of coordinates but it cannot be removed by a holomorphic coordinate transformation. In other words the pure part of the metric variation correspond precisely to changes of complex structure.

Along with $g_{\mu\bar{\nu}}$ and its variations, in the σ-model corresponding to compactification of superstrings, there occurs a two-form B on the internal space which is related to the metric by supersymmetry. The action functional suggests a complexification of the Kähler cone through inclusion of B. One is however always free to set $B = 0$ in order to recover the moduli space of Ricci-flat Calabi-Yau manifolds proper.

In virtue of the B field equation its variations are required to be harmonic. We take $b_{20} = 0$, which is certainly the case if the Euler number of $\mathcal{M}$ is nonzero, so that δB is of type $(1,1)$. We shall include the B field into the line element on the moduli space but will neglect : (1) derivative terms (these are systematically introduced by σ-model corrections) and (2) a possible term $(g^{mn}\delta g_{mn})^2$. This is done with a certain prescience since we anticipate that the natural combination is $B + iJ$. Up to these terms, the most general metric that can be written entirely in terms of the background quantities is

$$ds^2 = \frac{1}{4V}\int_{\mathcal{M}} g^{km}g^{ln}(\delta g_{kl}\delta g_{mn} + \delta B_{kl}\delta B_{mn})\, g^{\frac{1}{2}} d^6x\ , \tag{2.3}$$

The first physicist to introduce a metric on the space of metrics seems to have been deWitt[8] in connection with the quantization of the gravitational field. Our choice is essentially his metric. One might consider other choices such as (i) the metric that is the field theory limit of the Zamolodchikov metric[9] on the space of conformal field theories or (ii) the Weil–Petersson metric which can be written entirely in terms of cohomology classes. It is gratifying that these three choices coincide with (2.3). The demonstration of this for the Zamolodchikov metric is given in[10] while that for the Weil–Petersson metric is subsumed in the following.

It is natural to decompose the variations into type

$$ds^2 = \frac{1}{2V}\int_{\mathcal{M}} g^{\kappa\bar{\mu}}g^{\lambda\bar{\nu}}\left(\delta g_{\kappa\lambda}\delta g_{\bar{\mu}\bar{\nu}} + (\delta g_{\kappa\bar{\nu}}\delta g_{\lambda\bar{\mu}} + \delta B_{\kappa\bar{\nu}}\delta B_{\bar{\mu}\lambda})\right) g^{\frac{1}{2}} d^6x\ . \tag{2.4}$$

A simple yet important observation is that the metric is block-diagonal, with separate blocks corresponding to variations of the complex structure and of the Kähler class. This fact was previously noted[11,12] for Z_N-orbifolds[2]. This feature of the moduli space and some others was anticipated on the basis of so–called macroscopic reasoning[13]. This reasoning however is not altogether reliable so we turn now to study the metric on the moduli space of Calabi–Yau manifolds in its own right and in its own terms. It transpires that the metric (2.3) enjoys a number of surprising properties among these being the high degree of symmetry between the roles played by the (2,1)-forms and the (1,1)-forms (this also was anticipated in Ref.[13]). Since the metric separates into a direct product at least locally, we shall study these two parts in turn.

[2] The apparent discrepancy with the results of Ref.[12] was caused by a mislabeling of moduli. Upon a transformation of parameters the direct product form and agreement with our results are indeed achieved.

3. The (2, 1)–Forms

We turn first to the deformations corresponding to (2, 1)-forms largely following Tian[14]. We set

$$\chi_{\alpha\,\kappa\lambda\bar{\mu}} = -\frac{1}{2}\Omega_{\kappa\lambda}{}^{\bar{\nu}}\frac{\partial g_{\bar{\mu}\bar{\nu}}}{\partial z^\alpha}\ , \qquad \chi_\alpha = \frac{1}{2}\chi_{\alpha\,\kappa\lambda\bar{\mu}}\, dx^\kappa \wedge dx^\lambda \wedge dx^{\bar{\mu}}\ , \tag{3.1}$$

where the z^α, $\alpha = 1, \ldots, b_{21}$ are the parameters for the complex structure and a factor of $-1/2$ in the first of these relations has been chosen for later convenience. Each χ_α defines a (2, 1) cohomology class and we have also the inverse relation

$$\delta g_{\bar{\mu}\bar{\nu}} = -\frac{1}{\|\Omega\|^2}\bar{\Omega}_{\bar{\mu}}{}^{\rho\sigma}\chi_{\alpha\,\rho\sigma\bar{\nu}}\,\delta z^\alpha\ ,$$

where $\|\Omega\|^2 \stackrel{\text{def}}{=} \frac{1}{3!}\Omega_{\mu\nu\rho}\bar{\Omega}^{\mu\nu\rho}$ and is a constant. A little algebra reveals that

$$\begin{aligned} 2G_{\alpha\bar{\beta}}\delta z^\alpha \delta z^{\bar{\beta}} &\stackrel{\text{def}}{=} \frac{1}{2V}\int_{\mathcal{M}} g^{\kappa\bar{\nu}} g^{\mu\bar{\lambda}}\delta g_{\kappa\mu}\delta g_{\bar{\lambda}\bar{\nu}}\, g^{\frac{1}{2}} d^6x \\ &= -\frac{2i}{V\|\Omega\|^2}\delta z^\alpha \delta z^{\bar{\beta}}\int_{\mathcal{M}}\chi_\alpha \wedge \bar{\chi}_{\bar{\beta}} \end{aligned}$$

and hence we find

$$G_{\alpha\bar{\beta}} = -\frac{\displaystyle\int_{\mathcal{M}}\chi_\alpha \wedge \bar{\chi}_{\bar{\beta}}}{\displaystyle\int_{\mathcal{M}}\Omega \wedge \bar{\Omega}}\ . \tag{3.2}$$

Now the following is a useful observation which, in its original form, is attributed to K. Kodaira (see Ref.[14])

$$\frac{\partial \Omega}{\partial z^\alpha} = k_\alpha \Omega + \chi_\alpha, \tag{3.3}$$

where the k_α may depend on z^α but not on the coordinates of $\mathcal{M}$. To see this consider a system $\zeta^\mu(x, z)$ of holomorphic coordinates varying with z such that $\zeta^\mu(x, z_0) = x^\mu$. Then the holomorphic three-form for $\mathcal{M}_z$ takes the form

$$\Omega = \frac{1}{3!}\, h(\zeta)\, \epsilon_{\mu\nu\rho}\, d\zeta^\mu d\zeta^\nu d\zeta^\rho$$

and

$$\frac{\partial \Omega}{\partial z^\alpha} = \frac{1}{3!}\left(\frac{\partial h}{\partial z^\alpha}\right)\epsilon_{\mu\nu\rho}\, d\zeta^\mu d\zeta^\nu d\zeta^\rho + \frac{1}{2!}h\,\epsilon_{\mu\nu\rho}\, d\zeta^\mu d\zeta^\nu \left(\frac{\partial(d\zeta^\rho)}{\partial z^\alpha}\right)\ .$$

The first term is a $(3,0)$-form while the second contains the derivative of $d\zeta^\rho$. Clearly $d\zeta^\rho$ is a one-form and is in fact of type $(1,0)$ with respect to the complex structure corresponding to the z^α. Under a change of complex structure $d\zeta^\rho$ becomes partly of type $(1,0)$ and partly of type $(0,1)$. Thus the second term is a linear combination of $(3,0)$ and $(2,1)$ parts. Since d is independent of z^α we have that $\frac{\partial\Omega}{\partial z^\alpha}$ is closed and hence

$$\frac{\partial \Omega}{\partial z^\alpha} \in H^{(3,0)} \oplus H^{(2,1)} .$$

$H^{(3,0)}$ has dimension one for a Calabi–Yau manifold so we have established (3.3) for some $(2,1)$–form χ_α. It remains to show that this $(2,1)$–form is the one previously defined in Eq. (3.1).

The manifolds $\mathcal{M}_z$ are, of course all diffeomorphic and hence all identical as real manifolds since they differ only in their complex structure. We can attach to each point of the original manifold $\mathcal{M}_z$ a set of coordinate labels x^μ, say, and since we can identify points on the $\mathcal{M}_z$ for different z's we can attach the same labels to the points that are identified. We can also choose holomorphic coordinates $\zeta^\mu(x,z)$ for each $\mathcal{M}_z$. These two sets of labels will not coincide except on the original manifold. We sketch this in Fig. 1. The displacement between two points that carry the same holomorphic coordinates is decomposed into a lapse and a tangential shift.[15]

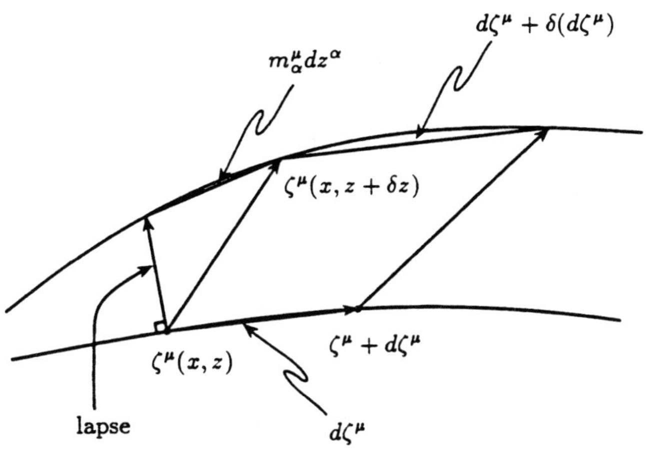

Fig.1 The displacement vector and the shift, $m^\mu_\alpha \delta z^\alpha$, for a deformation of the complex structure. The lapse vector joins two points that are identified if $\mathcal{M}_{z+\delta z}$ and $\mathcal{M}_z$ are identified as real manifolds.

We are interested in the $(0,1)$ part of the quantity $\frac{\partial}{\partial z^\alpha}(d\zeta^\mu)$. From the figure we see that $\delta(d\zeta^\mu) = \delta z^\alpha dm^\mu_\alpha$, hence we have

$$\frac{\partial}{\partial z^\alpha}(d\zeta^\rho)\bigg|_{(0,1)} = \bar\partial m^\rho_\alpha \ . \tag{3.4}$$

Also from the figure we see that

$$\zeta^\mu(x, z+\delta z) = x^\mu - m^\mu_\alpha \delta z^\alpha \ .$$

On $\mathcal{M}_{z+\delta z}$ there is a Ricci–flat metric $g_{mn} + \delta g_{mn}$ which is Kähler. The new metric in the new coordinates $\zeta^\mu(x, z+\delta z)$ has $g_{\bar\mu\bar\nu} = 0$ so in the original coordinates x^μ we have, by (2.2), that

$$\delta g_{\bar\mu\bar\nu} = -\Big(\frac{\partial m^\rho_\alpha}{\partial x^{\bar\mu}} g_{\rho\bar\nu} + \frac{\partial m^\rho_\alpha}{\partial x^{\bar\nu}} g_{\rho\bar\mu}\Big)\delta z^\alpha \ .$$

It can be shown that $\frac{\partial m^\rho_\alpha}{\partial x^{\bar\mu}} g_{\rho\bar\nu}$ is symmetric in $(\bar\mu, \bar\nu)$ so in fact

$$\frac{\partial g_{\bar\mu\bar\nu}}{\partial z^\alpha} = -2(\partial_{\bar\mu} m^\rho_\alpha) g_{\rho\bar\nu} \ . \tag{3.5}$$

From this relation and (3.4) we see that the $(2,1)$ part of $\frac{\partial\Omega}{\partial z^\alpha}$ is indeed χ_α.

Let us denote the space of the complex structure parameters by by Z. The manifolds $\mathcal{M}_z$ fit together to form a fiber bundle $\mathsf{M} = \{\mathcal{M}_z | z \in Z\}$. It is worth remarking that the derivative of the shift that appears in (3.5) is related to an important geometrical quantity, the extrinsic curvature, which describes the curvature of $\mathcal{M}_z$ as embedded in M. The nonvanishing components of the extrinsic curvature are $(\chi_{\alpha\,\bar\mu\bar\nu}\, , \chi_{\bar\alpha\,\mu\nu})$ and are given by

$$\chi_{\alpha\,\bar\mu\bar\nu} = (\partial_{\bar\mu} m^\rho_\alpha) g_{\rho\bar\nu} \ .$$

Thus

$$\chi_\alpha = \frac{1}{2}\Omega_{\kappa\lambda}{}^{\bar\nu}\chi_{\alpha\,\bar\mu\bar\nu}\, dx^\kappa \wedge dx^\lambda \wedge dx^{\bar\mu} \ .$$

Let us return now to the expression (3.2) for the metric. Given (3.3) it follows by straightforward differentiation that

$$G_{\alpha\bar\beta} = -\frac{\partial}{\partial z^\alpha}\frac{\partial}{\partial z^{\bar\beta}} \log\Big(i\int_{\mathcal{M}} \Omega\wedge\bar\Omega\Big) \tag{3.6}$$

so the space of complex structures is Kähler and

$$-\log\Big(i\int_{\mathcal{M}} \Omega\wedge\bar\Omega\Big)$$

is the Kähler potential.

We would like to describe this space in greater detail by making a particular choice of parameters. We shall describe the space of complex structures in terms of the periods of the holomorphic 3-form. To this end let (A^a, B_b) , $a, b = 0, \ldots, b_{21}$ be a canonical homology basis for $H_3(\mathcal{M}, \mathbb{Z})$ and let (α_a, β^b) be the dual cohomology basis so that

$$\int_{A^b} \alpha_a = \int_{\mathcal{M}} \alpha_a \wedge \beta^b = \delta_a{}^b , \qquad \int_{B_a} \beta^b = \int_{\mathcal{M}} \beta^b \wedge \alpha_a = -\delta_a{}^b .$$

Also,

$$z^a \stackrel{\text{def}}{=} \int_{A^a} \Omega , \qquad \mathcal{G}_a \stackrel{\text{def}}{=} \int_{B_a} \Omega \tag{3.7}$$

are called the *periods* of Ω. Bryant and Griffiths[16] show that, locally in the moduli space, the complex structure of $\mathcal{M}$ is entirely determined by the z^a, so that $\mathcal{G}_a = \mathcal{G}_a(z)$. It will be clear from the following that the z^a cannot all vanish, also it is clear that a rescaling $z^a \rightarrow \lambda z^a$ by a nonzero λ just corresponds to a rescaling of Ω. In other words we may regard the z^a as projective coordinates for the complex structure and Ω as being homogeneous of degree 1 in these coordinates.

$$z^a \in \mathbb{P}_{b_{21}} , \qquad \Omega(\lambda z) = \lambda \Omega(z) .$$

The holomorphic three-form may be expanded in terms of the basis

$$\Omega = z^a \alpha_a - \mathcal{G}_a(z) \beta^a . \tag{3.8}$$

Now, in virtue of Eq. (3.3) we have

$$\int_{\mathcal{M}} \left(\Omega \wedge \frac{\partial \Omega}{\partial z^a} \right) = 0 .$$

This yields the relation

$$2\mathcal{G}_a = \frac{\partial}{\partial z^a} (z^c \mathcal{G}_c) .$$

Thus $\mathcal{G}_a$ is the gradient of a function that is homogeneous of degree two

$$\mathcal{G}_a = \frac{\partial \mathcal{G}}{\partial z^a} , \quad \mathcal{G}(\lambda z) = \lambda^2 \mathcal{G}(z) . \tag{3.9}$$

It may also be shown that

$$\frac{\partial^2 \Omega}{\partial z^\alpha \partial z^\beta} \in H^{(3,0)} \oplus H^{(2,1)} \oplus H^{(1,2)}$$

and also that

$$\int_{\mathcal{M}} \left(\Omega \wedge \frac{\partial^2 \Omega}{\partial z^a \partial z^b} \right) = 0 , \qquad \int_{\mathcal{M}} \left(\frac{\partial \Omega}{\partial z^a} \wedge \frac{\partial \Omega}{\partial z^b} \right) = 0 .$$

These relations, however, are identically satisfied in virtue of Eq. (3.9) and give no further restrictions on $\mathcal{G}$.

Of considerable interest is the Yukawa coupling between (2, 1)–forms which is closely related to the inner product between Ω and its third derivatives. The Yukawa coupling is given[2] by

$$\kappa_{\alpha\beta\gamma} \stackrel{\text{def}}{=} -\int_{\mathcal{M}} \Omega \wedge \chi_\alpha^\mu \wedge \chi_\beta^\nu \wedge \chi_\gamma^\rho \Omega_{\mu\nu\rho} \tag{3.10}$$

where

$$\chi_\alpha^\mu = \frac{1}{2\|\Omega\|^2} \bar{\Omega}^{\mu\rho\sigma} \chi_{\alpha\,\rho\sigma\bar{\nu}} dx^{\bar{\nu}} = \chi_\alpha{}^\mu{}_{\bar{\nu}} dx^{\bar{\nu}} \; .$$

It follows by reasoning similar to that which leads to Eq. (3.3) that

$$\begin{aligned} \int_{\mathcal{M}} \Omega \wedge \chi_\alpha^\mu \wedge \chi_\beta^\nu \wedge \chi_\gamma^\rho \Omega_{\mu\nu\rho} &= \int_{\mathcal{M}} \left(\Omega \wedge \frac{\partial^3 \Omega}{\partial z^\alpha \partial z^\beta \partial z^\gamma}\right) \\ &= \frac{\partial^3 \mathcal{G}}{\partial z^\alpha \partial z^\beta \partial z^\gamma} \; . \end{aligned} \tag{3.11}$$

The Yukawa coupling, and hence $\mathcal{G}$, does not vanish in general. We have also the relation

$$i \int_{\mathcal{M}} \Omega \wedge \bar{\Omega} = i\left(\bar{z}^a \frac{\partial \mathcal{G}}{\partial z^a} - z^a \frac{\partial \bar{\mathcal{G}}}{\partial \bar{z}^a}\right) . \tag{3.12}$$

The Kähler potential for $G_{\alpha\bar{\beta}}$ is $-\log(i\int_{\mathcal{M}} \Omega \wedge \bar{\Omega})$ and is completely determined by $\mathcal{G}$. Thus the Kähler manifold is of a special type. In the following we shall refer to a Kähler manifold whose Kähler potential K is determined in terms of a holomorphic function $\mathcal{G}$, of homogeneity degree two, by a relation of the form (3.12), *i.e.* of the form

$$e^{-K} = -i\left(z^a \frac{\partial \bar{\mathcal{G}}}{\partial \bar{z}^a} - \bar{z}^a \frac{\partial \mathcal{G}}{\partial z^a}\right) \tag{3.13}$$

as a **Kähler manifold with holomorphic prepotential**. These manifolds are familiar from the literature of $N = 2$ supergravity[17] where they have come to be termed Kähler manifolds of restricted type. This usage conflicts with that of the mathematicians who use this term with a completely different meaning. Although cumbersome the terminology suggested here describes the manifolds in question as being precisely what they are: manifolds for which there is a potential for the Kähler potential. Eq. (3.13) is often written in terms of affine coordinates $\tau^\alpha = z^\alpha / z^0$, say, in the form

$$i \int_{\mathcal{M}} \Omega \wedge \bar{\Omega} = i\left[2(\mathcal{G} - \bar{\mathcal{G}}) - (\tau^\alpha - \bar{\tau}^\alpha)\left(\frac{\partial \mathcal{G}}{\partial \tau^\alpha} + \frac{\partial \bar{\mathcal{G}}}{\partial \bar{\tau}^\alpha}\right)\right] . \tag{3.14}$$

The prepotential $\mathcal{G}$ is itself largely determined by the Yukawa couplings. More precisely, the Yukawa couplings determine the third derivatives of $\mathcal{G}$. These, of course, determine $\mathcal{G}$ up to addition of a term $\frac{1}{2} S_{ab} z^a z^b$ with S_{ab} a symmetric constant matrix.

The vector space $H^{(3,0)} \oplus H^{(2,1)}$ is a linear subspace of $H^3(\mathcal{M})$ that varies holomorphically with z^a. A basis for $H^{(3,0)} \oplus H^{(2,1)}$ is

$$\frac{\partial \Omega}{\partial z^a} = \alpha_a - \frac{\partial^2 \mathcal{G}}{\partial z^a \partial z^b} \beta^b \ .$$

With respect to this basis the period matrix

$$\varpi = \left(\int_{A^b} \frac{\partial \Omega}{\partial z^a}, \int_{B_b} \frac{\partial \Omega}{\partial z^a} \right)$$

takes the simple form

$$\varpi = (\mathbf{1}\, ,\, \mathbb{G}) \quad , \quad \mathbb{G}_{ab} = \frac{\partial^2 \mathcal{G}}{\partial z^a \partial z^b} \ .$$

Now

$$i \int_{\mathcal{M}} \left(\frac{\partial \Omega}{\partial z^a} \wedge \frac{\partial \bar{\Omega}}{\partial z^{\bar{b}}} \right) = -2 \Im m\, \mathbb{G}_{ab}$$

we learn that $\Im m\, \mathbb{G}$ has signature $(b_{21}, 1)$ (b_{21} plus signs and one minus) a condition similar to lying in the Siegal upper half plane. The action of the symplectic group[3] on $\mathbb{G}$ is also analogous to its action on the period matrices of Riemann surfaces. Let (α, β) and $(\tilde{\alpha}, \tilde{\beta})$ be two symplectic bases with

$$\begin{pmatrix} \tilde{\alpha} \\ \tilde{\beta} \end{pmatrix} = \begin{pmatrix} A & B \\ C & D \end{pmatrix} \begin{pmatrix} \alpha \\ \beta \end{pmatrix} \quad , \quad \begin{pmatrix} A & B \\ C & D \end{pmatrix} \in \mathrm{Sp}(2b_{21} + 2; \mathbb{Z})$$

then

$$\begin{pmatrix} \widetilde{\partial \mathcal{G}} \\ \tilde{z} \end{pmatrix} = \begin{pmatrix} A & B \\ C & D \end{pmatrix} \begin{pmatrix} \partial \mathcal{G} \\ z \end{pmatrix} \tag{3.15}$$

and from this relation it is easy to show that

$$\widetilde{\mathbb{G}} = (A\mathbb{G} + B)(C\mathbb{G} + D)^{-1} \ . \tag{3.16}$$

It is tempting to conclude from this that $\mathrm{Sp}(2b_{21} + 2, \mathbb{Z})$ plays the role of a modular group. However it is not a priori clear either that the prepotential is such that (3.15) is satisfied for all symplectic matrices or that the symplectic transformations preserve the Yukawa couplings. This is an interesting topic about which we shall have a little more to say in §6.

---

[3] The term symplectic group is used with different meanings. Here by $\mathrm{Sp}(2n)$ we mean the group of $2n \times 2n$ matrices that preserves the matrix $\begin{pmatrix} 0 & 1 \\ -1 & 0 \end{pmatrix}$. The term is sometimes used in a different sense when it denotes the subgroup of the unitary group that preserves this matrix.

4. The $(1,1)$–Forms

In this section we shall show that the parameters for the Kähler class also comprise a Kähler manifold with holomorphic prepotential whose structure is analogous to the manifold of the complex structure parameters. We begin, as in the previous case, by recasting the metric in terms of cohomology classes.

The inner product corresponding to Eq. (2.4) on $H^{(1,1)}$ is

$$G(\varrho,\sigma) = \frac{1}{2V}\int_{\mathcal{M}} \varrho_{\mu\bar{\nu}}\sigma_{\rho\bar{\sigma}}g^{\mu\bar{\sigma}}g^{\rho\bar{\nu}}\, g^{\frac{1}{2}}d^6x = \frac{1}{2V}\int_{\mathcal{M}} \varrho\wedge *\sigma\ ,$$

for real $(1,1)$-forms, ϱ and σ. Note that, owing to the positivity of $g^{\mu\bar{\nu}}$, $G(\ ,\ )$ is positive. Apart from a factor of $1/V$ this is the metric adopted by Strominger[18] who observed that $G(\varrho,\sigma)$ may be rewritten entirely in terms of the cubic form

$$\kappa(\varrho,\sigma,\tau) \stackrel{\text{def}}{=} \int \varrho\wedge\sigma\wedge\tau$$

in virtue of the identities

$$V = \frac{1}{3!}\kappa(J,J,J)$$

$$*\sigma = -J\wedge\sigma + \frac{3}{2}\frac{\kappa(\sigma,J,J)}{\kappa(J,J,J)}J\wedge J\ .$$

This gives

$$G(\varrho,\sigma) = -3\left\{\frac{\kappa(\varrho,\sigma,J)}{\kappa(J,J,J)} - \frac{3}{2}\frac{\kappa(\varrho,J,J)\kappa(\sigma,J,J)}{\kappa^2(J,J,J)}\right\}\ . \tag{4.1}$$

Let e_A , $A = 1,\ldots,b_{11}$, be a basis for $H^2(\mathcal{M},\mathbb{Z})$. We write B and J in terms of this basis

$$B + iJ = w^A e_A\ ,\quad w^A = u^A + iv^A\ .$$

Since B and J are real they are parametrized by the real and imaginary parts of w^A. It is now straightforward to see that

$$G_{A\bar{B}} \stackrel{\text{def}}{=} \frac{1}{2}G(e_A,e_B) = -\frac{\partial}{\partial w^A}\frac{\partial}{\partial w^{\bar{B}}}\log\kappa(J,J,J)\ . \tag{4.2}$$

We have just shown that W, the space of the w^A is also a Kähler manifold. The Kähler potential is the logarithm of the volume of the Calabi–Yau manifold. Following Ferrara and Strominger[19], we note that if we set $\kappa_{ABC} = \kappa(e_A,e_B,e_C)$ and

$$f(w) \stackrel{\text{def}}{=} \frac{1}{3!}\kappa_{ABC}w^Aw^Bw^C\ ,$$

then

$$\begin{aligned}\kappa(J,J,J) &= \kappa_{ABC}\, v^A v^B v^C\ , \\ &= -\frac{3i}{4}\Big[2(f(w)-f(\bar{w})) - (w^A-\bar{w}^A)(f_{,A}(w)+f_{,A}(\bar{w}))\Big]\end{aligned} \tag{4.3}$$

and we find a striking resemblance with Eq. (3.14).

We can usefully develop this analogy by defining

$$\mathcal{F}(w) = -\frac{1}{3!}\frac{\kappa_{ABC}w^A w^B w^C}{w^0} \tag{4.4}$$

where we have introduced an extra coordinate w^0 in order that $\mathcal{F}$, which is the analog of the prepotential $\mathcal{G}$, should also be homogeneous of degree 2. w^0 can be identified with $1/\phi$ the inverse of the dilaton. For the moment however it is merely an extra coordinate which we will set to unity after differentiation. With this understanding we find

$$\exp(-K) = \frac{4}{3}\int_{\mathcal{M}} J^3 = -i\Big(w^j\frac{\partial\bar{\mathcal{F}}}{\partial w^{\bar{j}}} - w^{\bar{j}}\frac{\partial\mathcal{F}}{\partial w^j}\Big)$$

where we have shifted the Kähler potential by an inconsequential constant and j runs also over the value 0. Note also that if we construct from the basis forms e_A a set of forms analogous to the χ_α^μ

$$e_{A\mu} \stackrel{\text{def}}{=} e_{A\mu\bar{\nu}}dx^{\bar{\nu}} \in H_{\bar{\partial}}^{01}(\mathcal{M},\mathcal{T}^*)$$

then we have

$$\begin{aligned}-\kappa_{ABC} &= \int_{\mathcal{M}} \Omega\wedge e_{A\mu}\wedge e_{B\nu}\wedge e_{C\rho}\frac{\bar{\Omega}^{\mu\nu\rho}}{\|\Omega\|^2} \\ &= \frac{\partial^3\mathcal{F}}{\partial w^A\partial w^B\partial w^C}\end{aligned} \tag{4.5}$$

in precise correspondence to Eqs. (3.10) and (3.11).

Clearly, there are restrictions on the allowed values of v^A, one obvious condition being that

$$V = \frac{1}{3!}\kappa_{ABC}\, v^A v^B v^C > 0\ . \tag{4.6}$$

Another restriction arises from the metric (4.1) which must be positive definite. We rewrite this equality in the form

$$G(\varrho,\sigma) = -3\left\{\frac{\kappa(\varrho,\sigma,J)}{\kappa(J,J,J)} - \frac{\kappa(\varrho,J,J)\kappa(\sigma,J,J)}{\kappa^2(J,J,J)}\right\} + \frac{3}{2}\frac{\kappa(\varrho,J,J)\kappa(\sigma,J,J)}{\kappa^2(J,J,J)}\ . \tag{4.7}$$

Now any $(1,1)$–form can be decomposed into a part "orthogonal" to J and a part "parallel" to J

$$\varrho = \varrho^\perp + \varrho^\parallel = (\varrho - \frac{1}{6}\varrho_{mn}J^{mn}J) + \frac{1}{6}\varrho_{mn}J^{mn}J \tag{4.8}$$

with $\varrho^{\parallel}$ covariantly constant if ϱ is harmonic. (To a mathematician this is the decomposition $H^2 = H^2_{\perp} \oplus H^0$, where $H^2_{\perp}$ is the group of primitive two-forms.)

Eq. (4.7) is the decomposition of the metric into $G(\varrho^{\perp},\sigma^{\perp})$ and $G(\varrho^{\parallel},\sigma^{\parallel})$ both of which must be positive definite. $G(\varrho^{\parallel},\sigma^{\parallel})$ clearly is. Consider however

$$G(\varrho^{\perp},\sigma^{\perp}) = -3\frac{\kappa(\varrho^{\perp},\sigma^{\perp},J)}{\kappa(J,J,J)} \ . \tag{4.9}$$

On $H^2_{\perp}$ the matrix

$$\kappa_{AB}(v) \stackrel{\text{def}}{=} \kappa_{ABC}v^A$$

is negative definite, while from (4.9) we see that it is positive definite on $H^2_{\parallel}$. Thus it has signature $(1, b_{11}-1)$ (one plus and the rest minus). The signature can only change if one of the eigenvalues passes through zero. Hence a boundary of the allowed region is given by

$$\det \kappa_{AB}(v) = 0 \ .$$

The fundamental restriction on the parameters is that the Kähler form J be positive. This implies positive volume and positive G_{AB}. In Table 1 we summarize the analogies we have found thus far between the two parameter spaces.

The only entry of the Table that has not been established is the signature of $\Im m\frac{\partial^2\mathcal{F}}{\partial w^i\partial w^j}$. We shall now show that the signature of this matrix is $(b_{11},1)$, as required by the symmetry between $\mathcal{F}$ and $\mathcal{G}$. This follows in virtue of properties we have already established. We know that the matrix $\kappa = (\kappa_{AB}(v))$ has signature $(1, b_{11}-1)$ for allowed values of v. Now, by Eq. (4.4), we have

$$\Im m\left(\frac{\partial^2\mathcal{F}}{\partial w^i\partial w^j}\right) = \begin{pmatrix} -u^T\kappa u + 2V & \kappa u \\ u^T\kappa & -\kappa \end{pmatrix} .$$

A priori the signature of this matrix could depend on u. However it clearly has signature $(b_{11},1)$ for $u=0$. The signature of the matrix can change only if the determinant becomes either zero or infinite but

$$\det\begin{pmatrix} -u^T\kappa u + 2V & \kappa u \\ u^T\kappa & -\kappa \end{pmatrix} = 2V\det(-\kappa)$$

independent of u.

	Complexified Kähler Cone	Complex Structure
Number of projective coords.	$b_{11}+1$	$b_{21}+1$
Cohomology group	$H^1(\mathcal{M},\mathcal{T}^*)$	$H^1(\mathcal{M},\mathcal{T})$
Prepotentials	$\mathcal{F}(w)$	$\mathcal{G}(z)$
$\exp(-K)$	$\frac{4}{3}\int J^3 = i\left(w^{\bar{j}}\frac{\partial\mathcal{F}}{\partial w^j} - w^j\frac{\partial\bar{\mathcal{F}}}{\partial w^{\bar{j}}}\right)$	$i\int\Omega\bar{\Omega} = i\left(z^{\bar{a}}\frac{\partial\mathcal{G}}{\partial z^a} - z^a\frac{\partial\bar{\mathcal{G}}}{\partial z^{\bar{a}}}\right)$
Period matrices	$\Im m\left(\frac{\partial^2\mathcal{F}(w)}{\partial w^i\partial w^j}\right)$	$\Im m\left(\frac{\partial^2\mathcal{G}(z)}{\partial z^a\partial z^b}\right)$
Signature of the period matrix	$(b_{11},1)$	$(b_{21},1)$

Table 1: A table of closely analogous quantities describing the spaces of complex structures and Kähler–forms.

5. Formal Developments

It is clear from Table 1 that there is a close correspondence between the geometries of the parameter space of the complex Kähler class and that of the complex structure. We shall develop this analogy further here. The Hodge diamond for a Calabi–Yau manifold is

$$\begin{array}{ccccccc} & & & 1 & & & \\ & & 0 & & 0 & & \\ & 0 & & b_{11} & & 0 & \\ 1 & & b_{21} & & b_{21} & & 1 \\ & 0 & & b_{11} & & 0 & \\ & & 0 & & 0 & & \\ & & & 1 & & & \end{array} \tag{5.1}$$

and the symmetry that suggests itself is that of interchanging the central row with the central column. We shall introduce a quantity $\mho$ analogous to Ω which on variation of the parameters w^j moves among the cohomology groups of the central column

$$\mho \in H^{00} \oplus H^{11} \oplus H^{22} \oplus H^{33} \,.$$

First extend the previous basis e_A of H^2 to a basis

$$e_j \stackrel{\text{def}}{=} (e_0, e_A) \in H^0 \oplus H^2$$

and let e^j be the dual basis

$$e^j \stackrel{\text{def}}{=} (e^0, e^A) \in H^4 \oplus H^6$$

such that

$$\int_{\mathcal{M}} e_j e^k = \delta_j{}^k \,.$$

Since we are in integral cohomology $e_0 = 1$ and $e^0 = V^{-1} g^{\frac{1}{2}} d^6 x$. Let λ and μ be two anticommuting numbers and define an anticommuting basis analogous to the symplectic basis for H^3

$$\theta_j = \lambda e_j \quad , \quad \theta^j = \mu e^j$$

so that

$$\oint \theta_j \theta^k = - \oint \theta^k \theta_j = \delta_j{}^k$$

where the new integration symbol includes now also an integration over λ and μ. We now set

$$\mho = w^j \theta_j - \frac{\partial \mathcal{F}}{\partial w^j} \theta^j \,. \tag{5.2}$$

and our expression for the Kähler potential becomes

$$e^{-K} = i \oint \mho \bar{\mho} \ .$$

It is elementary to verify the relations of Table 2.

	Complex Structure	Complexified Kähler Cone
Ω , $\mho$	$\Omega = z^a \omega_a - \dfrac{\partial \mathcal{G}}{\partial z^a} \omega^a$	$\mho = w^j \theta_j - \dfrac{\partial \mathcal{F}}{\partial w^j} \theta^j$
Variational property	$\int \Omega \dfrac{\partial \Omega}{\partial z^a} = 0$	$\oint \mho \dfrac{\partial \mho}{\partial w^j} = 0$
$\exp(-K)$	$i \int \Omega \bar{\Omega} = i \left(z^{\bar{a}} \dfrac{\partial \mathcal{G}}{\partial z^a} - z^a \dfrac{\partial \bar{\mathcal{G}}}{\partial z^{\bar{a}}} \right)$	$i \oint \mho \bar{\mho} = i \left(w^{\bar{j}} \dfrac{\partial \mathcal{F}}{\partial w^j} - w^j \dfrac{\partial \bar{\mathcal{F}}}{\partial w^{\bar{j}}} \right)$
Yukawa couplings	$\int \Omega \dfrac{\partial^3 \Omega}{\partial z^a \partial z^b \partial z^c} = \dfrac{\partial^3 \mathcal{G}}{\partial z^a \partial z^b \partial z^c}$	$\oint \mho \dfrac{\partial^3 \mho}{\partial w^i \partial w^j \partial w^k} = \dfrac{\partial^3 \mathcal{F}}{\partial w^i \partial w^j \partial w^k}$
Period matrices	$\dfrac{\partial^2 \mathcal{G}}{\partial z^a \partial z^b}$	$\dfrac{\partial^2 \mathcal{F}}{\partial w^i \partial w^j}$
Signature of imaginary part	$(b_{21}, 1)$	$(b_{11}, 1)$

Table 2: Analogous relations for the spaces of parameters of the complex structure parameters and of the complex Kähler cone.

We can write an explicit expression for $\mho$ in virtue of Eq. (5.2) and the identities

$$e_A e_B = \kappa_{ABC} e^C \quad , \quad e_A e_B e_C = \kappa_{ABC} e^0 \ .$$

We find

$$\phi\mho = \lambda(1+\phi b) + \mu\left(\frac{1}{2}\phi^2 b^2 - \frac{1}{6}\phi^3 b^3\right)$$

where

$$b \stackrel{\text{def}}{=} B + iJ = w^A e_A$$

and $\phi = (w^0)^{-1}$ is the dilaton field which is a constant (*i.e.* a zero form) on $\mathcal{M}$ in virtue of its field equation. Two points need to be dealt with: (i) we need to explain the common mathematical structure that underlies the results that are summarized in Table 2 and (ii) we need to be able to express our results in a covariant formalism so as not to have to make explicit reference to the special coordinates that we have been using thus far. It turns out that these two issues are somewhat related. The key observation, made first by Tian[14], and further explained by Strominger[20], is that the three–form Ω is not a scalar over the parameter space but rather takes values in a line bundle L. What is being said is that for each Calabi–Yau manifold $\mathcal{M}_z$ the three–form Ω is undefined up to multiplication by a complex number $f(z)$.

$$\Omega \to f\Omega \tag{5.3}$$

It is convenient to change notation a little and to write

$$\langle\alpha,\beta\rangle = -i\int_{\mathcal{M}} \alpha\wedge\beta\ .$$

The inner product $\langle\ ,\ \rangle$ is then antisymmetric and linear in both its arguments. Now recall that

$$\langle\Omega,\bar{\Omega}\rangle = -e^{-K}$$

so under the replacement (5.3) the Kähler potential changes by a so called Kähler transformation

$$K \to K - \log|f|^2\ .$$

This, of course, leaves the metric invariant. To a physicist (5.3) is a generalized gauge invariance[4] and the metric we have adopted on the parameter space is the natural gauge invariant metric. This explains, perhaps, why the deWitt, Zamolodchikov and Weil–Petersson metrics all reduce to this one.

Recall also Eq. (3.3)

$$\frac{\partial\Omega}{\partial z^\alpha} = k_\alpha\Omega + \chi_\alpha \tag{5.4}$$

which expresses the fact that $\partial_\alpha\Omega \in H^{30}\oplus H^{21}$. Note that k_α plays the role of a gauge potential in this relation since k_α must change under the replacement according to the rule

$$k_\alpha \to k_\alpha + \frac{1}{f}\frac{\partial f}{\partial z^\alpha}$$

[4] To a mathematician it is of course just a special case of a gauge invariance.

(and also $\chi_\alpha \to f\chi_\alpha$) for the relation to be covariant. Actually k_α is simply related to the Kähler potential as we can see by taking the inner product of this equation with $\bar{\Omega}$. We find

$$\frac{\partial}{\partial z^\alpha}\langle \Omega, \bar{\Omega}\rangle = k_\alpha \langle \Omega, \bar{\Omega}\rangle$$

and hence

$$k_\alpha = -\frac{\partial K}{\partial z^\alpha} .$$

Following Strominger[20] we construct a gauge covariant derivative

$$\mathcal{D}_\alpha \Omega = \partial_\alpha \Omega + K_{,\alpha}\Omega$$

then under a gauge transformation

$$\mathcal{D}_\alpha \Omega \to f\mathcal{D}_\alpha \Omega .$$

Of importance in our considerations is the Yukawa coupling. A covariant expression follows from (3.10)

$$\kappa_{\alpha\beta\gamma} = -i\langle \Omega\, , \mathcal{D}_\alpha \mathcal{D}_\beta \mathcal{D}_\gamma \Omega\rangle .$$

Since it is only the term with three *partial* derivatives in $\mathcal{D}_\alpha\mathcal{D}_\beta\mathcal{D}_\gamma\Omega$ that has a $(0,3)$ part we see that $\kappa_{\alpha\beta\gamma}$ is symmetric in its indices, alternatively we know this in virtue of the fact that $\kappa_{\alpha\beta\gamma}$ is symmetric when written in special coordinates. For a general coordinate system however it is no longer the case that $\kappa_{\alpha\beta\gamma}$ is the third derivative a holomorphic prepotential. It can be shown[20][21] that there exists locally a function S such that

$$\kappa_{\alpha\beta\gamma} = \mathcal{D}_\alpha \mathcal{D}_\beta \mathcal{D}_\gamma S$$

however S is not in general holomorphic.

Note that $\kappa_{\alpha\beta\gamma}$ takes values in $\mathsf{L}\otimes\mathsf{L}$. We shall refer to this by saying that $\kappa_{\alpha\beta\gamma}$ has charge $(2,0)$. More generally a quantity $\Psi^{(a,b)}$ which takes values in $\mathsf{L}^a \otimes \bar{\mathsf{L}}^b$ will be said to have charge (a,b). Under a gauge transformation such a quantity transforms according to the rule

$$\Psi^{(a,b)} \to f^a \bar{f}^b \Psi^{(a,b)} .$$

We extend the definition of covariant derivative to such quantities so as to respect Leibnitz rule

$$\mathcal{D}_\alpha \Psi^{(a,b)} = (\partial_\alpha + aK_{,\alpha})\Psi^{(a,b)}$$
$$\mathcal{D}_{\bar{\beta}} \Psi^{(a,b)} = (\partial_{\bar{\beta}} + bK_{,\bar{\beta}})\Psi^{(a,b)} .$$

When acting on charged tensors such as $\kappa_{\alpha\beta\gamma}$ the partial derivatives are to be replaced by ordinary covariant derivatives $\nabla_\alpha\, , \nabla_{\bar{\beta}}$ that contain a Christoffel symbol for each tensor index in the usual way.

Let us return again to (5.4) which we write in the form

$$\mathcal{D}_\alpha \Omega = \chi_\alpha .$$

Observe that the $\mathcal{D}_\alpha\Omega$ which a priori could have resided in $H^{30}\oplus H^{21}$ reside in fact in H^{21} and moreover comprise a basis for H^{21}. We wish now to derive some further identities which greatly simplify the computation of the Riemann tensor of the moduli space. We shall show that the covariant derivatives of χ_β reside precisely in H^{12} and that the covariant derivatives of $\chi_{\bar\gamma}$ reside precisely in H^{03}. In the course of this we shall require a simple identity. In virtue of the relations above we have

$$[\mathcal{D}_\alpha, \mathcal{D}_{\bar\beta}]\Omega = -G_{\alpha\bar\beta}\Omega \ .$$

This is in fact just the statement that

$$c_1(L) = J$$

where $c_1(L)$ is the first Chern class of L and J is the Kähler form on the parameter space. Tian's original observation was, in fact, that $c_1(L)$ is the natural choice of Kähler form on the parameter space.

Now being a $(2,1)$–form χ_β has the structure $\chi_\beta = \frac{1}{2}\chi_{\beta\,\mu\nu\bar\rho}dx^\mu dx^\nu dx^{\bar\rho}$. The coordinates vary holomorphically with the parameters so we a priori expect $\mathcal{D}_\alpha\chi_\beta \in H^{21}\oplus H^{12}$. Since χ_γ and $\chi_{\bar\gamma}$ form a basis for $H^{21}\oplus H^{12}$ we write

$$\mathcal{D}_\alpha\chi_\beta = A_{\alpha\beta}{}^{\gamma}\chi_\gamma + B_{\alpha\beta}{}^{\bar\gamma}\chi_{\bar\gamma} \tag{5.5}$$

First we shall show that $A_{\alpha\beta}{}^{\gamma}$ vanishes. To this end we take the inner product of this equation with $\chi_{\bar\delta}$ and recall that

$$\langle\chi_\alpha, \chi_{\bar\beta}\rangle = G_{\alpha\bar\beta}e^{-K} \ .$$

We find

$$\begin{aligned} A_{\alpha\beta\bar\delta}e^{-K} &= \langle\mathcal{D}_\alpha\chi_\beta\,,\,\chi_{\bar\delta}\rangle \\ &= \mathcal{D}_\alpha\langle\chi_\beta, \chi_{\bar\delta}\rangle - \langle\chi_\beta\,,\,\mathcal{D}_\alpha\mathcal{D}_{\bar\delta}\bar\Omega\rangle \\ &= -\langle\chi_\beta\,,\,[\mathcal{D}_\alpha, \mathcal{D}_{\bar\delta}]\bar\Omega\rangle \\ &= 0 \ . \end{aligned}$$

It is perhaps not surprising that $B_{\alpha\beta}{}^{\bar\gamma}$ is related to the Yukawa coupling. To find an explicit expression for $B_{\alpha\beta}{}^{\bar\gamma}$ we take the inner product of (5.5) with χ_γ.

$$\begin{aligned} B_{\alpha\beta\gamma}e^{-K} &= \langle\chi_\gamma\,,\,\mathcal{D}_\alpha\chi_\beta\rangle \\ &= \mathcal{D}_\gamma\langle\Omega\,,\,\mathcal{D}_\alpha\chi_\beta\rangle - \langle\Omega\,,\,\mathcal{D}_\gamma\mathcal{D}_\alpha\mathcal{D}_\beta\Omega\rangle \\ &= -i\kappa_{\alpha\beta\gamma} \ . \end{aligned}$$

Thus

$$\mathcal{D}_\alpha\chi_\beta = -ie^K\kappa_{\alpha\beta}{}^{\bar\gamma}\chi_{\bar\gamma}$$

and we see that the second covariant derivatives of Ω span H^{12}. Proceeding in the same fashion it is easy to see that

$$\mathcal{D}_\alpha\chi_{\bar\gamma} = G_{\alpha\bar\gamma}\bar\Omega$$

and it is elementary that

$$\mathcal{D}_\alpha \bar{\Omega} = 0 \ .$$

We summarize these results in the form of a table:

Derivatives of the Basis	Spans
Ω	H^{30}
$\mathcal{D}_\alpha \Omega = \chi_\alpha$	H^{21}
$\mathcal{D}_\alpha \chi_\beta = -ie^K \kappa_{\alpha\beta}{}^{\bar{\gamma}} \chi_{\bar{\gamma}}$	H^{12}
$\mathcal{D}_\alpha \chi_{\bar{\gamma}} = G_{\alpha\bar{\gamma}} \bar{\Omega}$	H^{03}
$\mathcal{D}_\alpha \bar{\Omega} = 0$	

Table 3: A table of derivatives of the basis $(\Omega, \chi_\alpha, \chi_{\bar{\beta}}, \bar{\Omega})$.

With these results in hand it is now straightforward to proceed to a computation of the curvature tensor. Since χ_γ has charge $(1,0)$ we have the identity

$$[\mathcal{D}_\alpha, \mathcal{D}_{\bar{\beta}}]\chi_\gamma = -G_{\alpha\bar{\beta}}\chi_\gamma + R_{\alpha\bar{\beta}\gamma}{}^{\delta}\chi_\delta$$

with the first term on the right hand side corresponding to the gauge field strength. Given the identities of the table we find that

$$R_{\alpha\bar{\beta}\gamma\bar{\delta}} = G_{\alpha\bar{\beta}}G_{\gamma\bar{\delta}} + G_{\alpha\bar{\delta}}G_{\gamma\bar{\beta}} - e^{2K}\kappa_{\alpha\gamma}{}^{\bar{\lambda}}\kappa_{\bar{\beta}\bar{\delta}\bar{\lambda}} \ . \tag{5.6}$$

An entirely analogous expression can be obtained for the Riemann tensor for the complex Kähler cone. The inner product is now given by

$$\langle \alpha, \beta \rangle = \oint \alpha \wedge \beta \ .$$

The analogue of the decomposition of H^3 into Hodge type

$$H^3 = H^{30} \oplus H^{21} \oplus H^{12} \oplus H^{03}$$

is the decomposition of the space

$$\widetilde{H}^3 \stackrel{\text{def}}{=} H^{00} \oplus H^{11} \oplus H^{22} \oplus H^{33}$$

into the subpaces spanned by $\mho$ and $\bar{\mho}$ and their derivatives

$$\widetilde{H}^3 = \text{span}(\mho) \oplus \text{span}(\mathcal{D}_A\mho) \oplus \text{span}(\mathcal{D}_{\bar{B}}\bar{\mho}) \oplus \text{span}(\bar{\mho})$$

and the metric on this space is the natural gauge invariant metric corresponding to the gauge invariance $\mho \to f(w)\mho$. The mathematical theory appropriate to the structures we have found is that of Variation of Hodge Structure[22]. For the case of the complex structure parameters the Hodge structure is that induced from the complex structure of the Calabi–Yau manifold. A Hodge structure and its dependence on parameters can however be understood independently of an underlying complex structure. What we have found is that there is an alternative Hodge structure consequent on regarding $\mho$ as a section of a line bundle over the complex Kähler cone[5]. The fact that Ω is a $(3,0)$–form is, as is easily seen, equivalent to the properties

$$\left\langle \Omega, \frac{\partial \Omega}{\partial z^a} \right\rangle = 0 , \quad \langle \Omega, \bar{\Omega} \rangle < 0 .$$

$\mho$ is therefore the analogue of a $(3,0)$–form in virtue of the fact that it satisfies the analogous relations. If mirror manifolds indeed exist, as we suspect, then the reason for the existence of the alternative Hodge structure is that it is the Hodge structure corresponding to the complex structure of the mirror manifold.

[5] With benefit of hindsight we see that there is a more transparent derivation of the fact that the signature of the imaginary part of $\Im m \frac{\partial^2 \mathcal{F}}{\partial w^i \partial w^j}$ is $(b_{11}, 1)$, which precisely parallels the corresponding demonstration for the complex structure parameters, by noting that $\frac{\partial \mho}{\partial w^j}$ and $\frac{\partial \bar{\mho}}{\partial w^{\bar{j}}}$ form a basis for $\widetilde{H}^3$ and

$$\left\langle \frac{\partial \mho}{\partial w^j}, \frac{\partial \bar{\mho}}{\partial w^{\bar{k}}} \right\rangle = 2\Im m \frac{\partial^2 \mathcal{F}}{\partial w^j \partial w^k} .$$

6. The Modular Group

We gather together here some preliminary observations regarding the role of the symplectic group and its subgroups as groups of modular transformations. This is relevant to several interesting topics. Firstly it seems that this is the proper context in which to understand the 'duality transformations' and '$R \to 1/R$' symmetries that were first found in the context of orbifolds[23-26]. Support for this is provided by the interesting article of Shevitz[27] who shows, in the context of a particular orbifold, that the duality transformations of the orbifold are particular cases of the symplectic transformations to be discussed below. A second point is that modular transformations seem to be inextricably intertwined with the issue of quantum corrections. The reason being that although the two parameter spaces W and Z exhibit a compelling correspondence we have, at this point, to recall that the correspondence is not perfect. For example the Yukawa couplings for the Kähler class parameters, κ_{ABC}, have a topological significance[18] and are a set of constants. This is not true of the corresponding quantities $\kappa_{\alpha\beta\gamma}$ for the complex structure parameters. Presumably related to this is the fact that the κ_{ABC} are corrected by quantum effects, such as worldsheet instantons, whereas the $\kappa_{\alpha\beta\gamma}$ are not.

We are interested in the question of when two distinct points of W or Z can correspond to the same compactification and hence to the same physical theory. We know from our previous discussion that the Kähler potentials of W and Z are invariant under symplectic transformations. A simple but essential point, however, is that the three dimensional case is more complicated than the cases of dimension one and dimension two due to the existence of the Yukawa couplings

$$\kappa_{\alpha\beta\gamma} = \langle \Omega \,,\, \partial_\alpha \partial_\beta \partial_\gamma \Omega \rangle \;, \quad \kappa_{ABC} = \langle \mho \,,\, \partial_A \partial_B \partial_C \mho \rangle \;.$$

Since the couplings have three indices we cannot expect them to be preserved by a group action unless the couplings, or more precisely the prepotential from which they derive, have special properties.

As a preliminary, note that the submatrices of a symplectic matrix $\begin{pmatrix} A & B \\ C & D \end{pmatrix}$ satisfy the relations

$$\begin{aligned} &AB^T \;,\; CD^T \;,\; B^T D \;,\; A^T C \;\; \text{symmetric} \\ &AD^T - BC^T = D^T A - B^T C = 1 \;. \end{aligned} \tag{6.1}$$

We wish to study the transformation properties of

$$Y(\phi) \stackrel{\text{def}}{=} \frac{\partial^3 \mathcal{G}}{\partial z^a \partial z^b \partial z^c} \phi^a \phi^b \phi^c$$

where the ϕ^a transform like dz^a. It transpires, quite remarkably, that if the prepotential is such that the transformation rule

$$\begin{pmatrix} \widetilde{\partial \mathcal{G}} \\ \tilde{z} \end{pmatrix} = \begin{pmatrix} A & B \\ C & D \end{pmatrix} \begin{pmatrix} \partial \mathcal{G} \\ z \end{pmatrix} \tag{6.2}$$

is satisfied then $Y(\phi)$ is invariant. To see this one can write $Y(\phi)$ in a mixed matrix/tensor notation in the form

$$Y(\phi) = \phi^a \phi^T \frac{\partial \mathbb{G}}{\partial z^a} \phi \ .$$

Under a symplectic transformation

$$\begin{aligned} \mathbb{G} \to \tilde{\mathbb{G}} &= (A\mathbb{G} + B)(C\mathbb{G} + D)^{-1} \\ \phi \to \tilde{\phi} &= (C\mathbb{G} + D)\phi \end{aligned}$$

while the combination $\phi^a \frac{\partial}{\partial z^a}$ remains invariant. A little algebra[6], which makes use of the identities (6.1), reveals that $Y(\phi)$ is invariant. The prepotential itself is not invariant under the full group

$$\mathcal{G} = \frac{1}{2} z^T \mathbb{G} z \to \frac{1}{2} z^T (\mathbb{G} C^T + D^T)(A\mathbb{G} + B) z \tag{6.3}$$

but it is invariant under a subgroup. Consider a symplectic matrix that has $C = 0$. The identities tell us that in this case $D = (A^T)^{-1}$. Also we have that $D^T B = A^{-1} B$ is symmetric so $B = AS$ for some symmetric matrix S. Thus a symplectic matrix which has $C = 0$ is of the form

$$\begin{pmatrix} A & AS \\ 0 & (A^T)^{-1} \end{pmatrix} = \begin{pmatrix} A & 0 \\ 0 & (A^T)^{-1} \end{pmatrix} \begin{pmatrix} \mathbf{1} & S \\ 0 & \mathbf{1} \end{pmatrix} \ . \tag{6.4}$$

Under such a transformation we see that $\mathcal{G}$ is invariant up to the addition of a quadratic term

$$\mathcal{G}(Dz) = \mathcal{G}(z) + \frac{1}{2} z^T S z \ . \tag{6.5}$$

The extra term is of no consequence since, being quadratic, it clearly leaves the Yukawa couplings unchanged. Moreover, S being real, it also has no effect on the Kähler potential as may be seen from (3.13). In fact we may as well take $\mathcal{G}(Dz) = \mathcal{G}(z)$. For if (6.2) holds for a group of matrices D then $S = S(D)$ and the composition rule is

$$S(MN) = N^T S(M) N + S(N) \ .$$

---

[6] Under the transformation

$$Y(\phi) \to \widetilde{Y}(\tilde{\phi}) = \phi^a \phi^T \left\{ (\mathbb{G} C^T + D^T) A - (\mathbb{G} C^T + D^T)(A\mathbb{G} + B)(C\mathbb{G} + D)^{-1} C \right\} \frac{\partial \mathbb{G}}{\partial z^a} \phi \ .$$

The first term inside the parentheses, $(\mathbb{G} C^T + D^T) A$, can be replaced by $\mathbf{1} + (\mathbb{G} A^T + B^T) C$ in virtue of the identities so we have

$$\widetilde{Y} - Y = \phi^a \phi^T \left\{ (\mathbb{G} A^T + B^T)(C\mathbb{G} + D) - (\mathbb{G} C^T + D^T)(A\mathbb{G} + B) \right\} (C\mathbb{G} + D)^{-1} C \frac{\partial \mathbb{G}}{\partial z^a} \phi$$

and the quantity inside the parentheses vanishes by the identities.

Using the fact that $S(\mathbf{1}) = 0$ it is easily seen that the general form for $S(D)$ is

$$S(D) = c(D^T D - \mathbf{1})$$

with c a number independent of D. It follows that

$$\mathcal{G}(Dz) - cz^T D^T Dz = \mathcal{G}(z) - cz^T z$$

in other words $\mathcal{G}_{new}(z) = \mathcal{G}(z) - cz^T z$ is invariant. The burden of these remarks is that $\mathcal{G}$ is invariant under the $GL(b_{21}+1, \mathbb{Z})$ subgroup of the symplectic group consisting of matrices of the form (6.4). We note in passing that $\mathcal{G} \to -\mathcal{G}$ under the transformation $\begin{pmatrix} 0 & 1 \\ -1 & 0 \end{pmatrix}$.

There are two cases:

(i) There exist maps $z \to \tilde{z}(z)$ compatible with the prepotential in such a way that the transformation rule (6.2) obtains for all symplectic matrices. The modular group is then $\mathrm{Sp}(2b_{21}+2, \mathbb{Z})$. The prepotential is not invariant under the full group but $\mathcal{G}(Dz) = \mathcal{G}(z) + \frac{1}{2} z^T S(D) z$ for each $D \in GL(b_{21}+1, \mathbb{Z})$, such a prepotential could be said to be effectively invariant under the $GL(b_{21}+1, \mathbb{Z})$ subgroup. We believe this to be the general case for the complex structure parameters. At least we believe that the mirror manifold of $\mathbb{P}_4(5)$ furnishes an example[28].

(ii) Maps $z \to \tilde{z}$ exist as above but not for all symplectic matrices. This is the situation for the Kähler class parameters. Indeed it is clear that a prepotential of the form

$$\mathcal{F}(w) = -\frac{1}{3!} \frac{\kappa_{ABC} w^A w^B w^C}{w^0}$$

with constant coefficients κ_{ABC} cannot satisfy $\mathcal{F}(Dw) = \mathcal{F}(w)$ for all D's so the modular group cannot be the full symplectic group and is typically a small subgroup.

It is tempting to suggest that it is the quantum corrections that restore complete symmetry between W and Z. In a sense still to be made precise one might suppose that the fully corrected prepotential $\mathcal{F}$ is equal to the prepotential $\mathcal{G}$ for the mirror manifold

$$\mathcal{G}_{\mathcal{W}} = \mathcal{F}_{\mathcal{M}} + \Delta\mathcal{F}_{\mathcal{M}} ,$$

where $\Delta\mathcal{F}_{\mathcal{M}}$ is the correction which to leading order is due to instantons. If this is so then since $\mathcal{G}_{\mathcal{W}}$ is invariant under a larger modular group than $\mathcal{F}_{\mathcal{M}}$ it is tempting further to conclude that the role of the quantum corrections is to restore the modular invariance of $\mathcal{F}_{\mathcal{M}}$. As functions of the parameters the Yukawa couplings, and hence the prepotentials are much more complicated for the complex structure parameters than for the Kähler class parameters. It has been said that this is due to the fact that the Yukawa couplings for the Kähler class prameters are the 'bare' quantities, which are then corrected by quantum effects, while the corresponding quantities for the complex structure parameters already contain all quantum effects. The conjecture we have made here is that the Yukawa couplings and prepotential for the complex structure parameters are complicated because they are automorphic functions.

Note Added: After the work reported here was substantially complete we received the second of Refs.[5], in which the authors come to essentially the same conclusions as those expressed in §6 above, and Ref.[29] in which the authors calculate the corresponding couplings for a pair of mirror manifolds, in a large radius large complex structure limit, and show that the results agree.

Acknowledgements

It is a pleasure to acknowledge extensive discussions with B. Greene, P. S. Green, T. Hübsch and A. Strominger.

References

1. P. Candelas, G. Horowitz, A. Strominger and E. Witten, Nucl. Phys. **B258** (1985) 46.
2. A. Strominger and E. Witten, Commun. Math. Phys. **101**,(1985)341.
3. M. B. Green, J. H. Schwarz and E. Witten, *Superstring theory* (Cambridge University Press 1987).
4. P. Candelas, M. Lynker and R. Schimmrigk, *Calabi–Yau Manifolds in Weighted* $\mathbb{P}_4$, University of Texas report UTTG–37–89.
5. B. R. Greene and M. R. Plesser, *(2,2) and (2,0) Superconformal Orbifolds*, Harvard University report HUTP–89/B241; *Duality in Calabi–Yau Moduli Space*, Harvard University report HUTP–89/A043.
6. M. Dine, N. Seiberg, X. G. Wen and E. Witten, Nucl. Phys. **B278** (1986) 769; Nucl. Phys. **B289** (1987) 319.
7. P. Candelas, P. Green and T. Hübsch, *Rolling Among Calabi–Yau Vacua*, University of Texas report UTTG–10–89.
8. B. S. deWitt, Phys. Rev.**160**(1967)1113.

9. A. B. Zamolodchikov, JEPT Lett.**43**(1986)730, Sov. J. Nucl. Phys.**46**(1987)1090.

10. P. Candelas, T. Hübsch and R. Schimmrigk, *Relation Between the Weil–Peterson and Zamolodchikov Metrics*, University of Texas report UTTG–17–89.

11. S. Ferrara and M. Porratti Phys. Lett. **216B**, (1989)289.

12. M. Cvetič, J. Molera and B. Ovrut: University of Pennsylvania report UPR-0376T.

13. N. Seiberg, Nucl. Phys. **B303**(1988)286.

14. G.Tian in *Mathematical Aspects of String Theory*, p.629, S.-T.Yau (ed.) (World Scientific, Singapore, 1987).

15. For a somewhat more detailed account of aspects of the theory of deformation of complex structures from this point of view see: P. Candelas, Nucl. Phys. **B298** (1988)458. For a thorough account of the lapse and shift and their use in the description of embedded hypersurfaces see: C. W. Misner, K. S. Thorne and J. A. Wheeler, *Gravitation* (W. H. Freeman and Co., San Francisco 1970). The standard text for deformations of complex structure is K. Kodaira, *Complex Manifolds and Deformations of Complex Structure* (Springer Verlag 1985).

16. R. Bryant and P. Griffiths, *Progress in Mathematics* **36**pp.77–102, (Birkhäuser, Boston, 1983).

17. S. J. Gates, Jr., Nucl. Phys. **B238**(1984)349.
J. E. Bagger and E. Witten, Nucl. Phys. **B222**(1988)286.

18. A. Strominger, Phys. Rev. Lett. **55**, (1985)25947.

19. S. Ferrara and A. Strominger, *$N = 2$ Spacetime Supersymmetry and Calabi–Yau Moduli Space*, CERN–TH5291/89, UCLA/89/TEP/6, to appear in the Proceedings of the "Strings '89" meeting in College Station, Texas.

20. A. Strominger, *Special Geometry*, University of California at Santa Barbara report, UCSBTH–89–61.

21. L. J. Dixon, V. Kaplunovsky and J. Louis, *On Effective Field Theories Describing (2,2) Vacua of the Heterotic String*, SLAC–PUB–4959(1989).

22. P. Griffiths, *Topics in Transcendental Algebraic Geometry* (Princeton University Press, Princeton NJ, 1984).
C. Peters and J. Steenbrink, *Classification of Algebraic and Analytic Manifolds*, Katata Symposium Proceedings 1982, K. Ueno, J. Coates and S. Helgason eds.. *Infinitessimal Variation of Hodge Structure and the Generic Torelli Problem for Projective Hypersurfaces*, Progress in Mathematics **39** pp399–464, (Birkhäuser, Boston).

23. V. P. Nair, A. Shapere, A. Strominger and F. Wilczek, Nucl. Phys. **B287** (1987) 402

24. A. Shapere and F. Wilczek, Nucl. Phys. **B320** (1989) 669.

25. A. Giveon, E. Rabinovici and G. Veneziano, Nucl. Phys. **B322** (1989) 167.

26. M. Dine, P. Huet, and N. Seiberg, *Large and Small Radius in String Theory*, Institute for Advanced Study preprint IASSNS-HEP-88/54

27. D. Shevitz, *The Global Structure of a $c = 9$ Type (2,2) Moduli Space*, University of California at Santa Barbara report UCSBTH-89-60.

28. P. Candelas, X. de la Ossa, P. S. Green and L. Parkes, paper in preparation.

29. P. A. Aspinwall, C. A. Lütken and G. G. Ross, *Construction and Couplings of Mirror Manifolds*, University of Oxford report.

Standard Model
and Beyond

ISBN 1-56072-027-1

ALL THE STRING'S VACUUA

Tristan Hübsch*
*Theory Group, Department of Physics, University of Texas,
Austin, TX 78712 USA*

ABSTRACT

Non-linear sigma models offer a framework in which all ground states in the string theory can be treated on the similar (if not the same) footing. Explorers of each such genre have devised more or less specific techniques of analysis (and even jargon) quite inapplicable and incomprehensible in general. It thus appears to be an imperative :

Models of all String Vacua, Unite !

The Programme

*Supported by the Robert A. Welch Foundation and the NSF Grant PHY 8605978. On leave of absence from the Institute „Ruđer Bošković", Bijenička 54, 41000 Zagreb, Croatia, Yugoslavia.

0. CREDO

Whether (super)strings are going to become the standard foundation of all physics or not, they undoubtedly feature a beautiful and amazingly rich structure the analysis of which is both challenging and rewarding in its own right. Occasionally, (super)strings exhibit properties that are remarkably similar to experimentally observed properties of our real world. This leads us to suspect that Nature itself might have recognized this splendour and actually employed it; or maybe it is just that (super)strings may offer a very acurate, albeit not exact, model of our real world. In any case, unlike for anything else of similar scope in theoretical physics, nobody has been able so far to disprove the ability of (super)strings to yield phenomenologically consistent effective models [1,2]. We shall give them the benefit of doubt and try to explore those families of such models that seem to be closest to the real world.

Before delving into an analysis of (super)string models, it should be noted that "(super)string *theory*" is really a misnomer. We know just about nothing as regards the first principles underlying this 'theory'. We do, however, have an embarrasingly large—and still growing—number of ground states around which perturbation theory can be done consistently. All the chances are that there are many more ground states, some of which we may never learn to describe acurately enough to satisfy an experimentalist. Nevertheless, we can study the physically observable properties of (super)strings in each of the known ground states to an acuracy that varies from case to case; this is mainly what is being done in the literature.

How does this unknown '(super)string theory' pick one out of this infinitude of ground states, or, does it do so at all ? Honestly, nobody knows. In my mind, this is the strongest motivation for trying to uncover the first principles of what we now see as an incredibly numerous collection of (super)string *models*. While there seems to be no significant progress in this direction at the moment, I do believe that increasing the *quantity* of information about individual models and the possible ways to stitch them together is bound to yield new *quality* in our understanding of (super)strings.

Recall, to describe the pattern of symmetry breaking in Grand Unified Theories, for example, one introduces a set of suitable Higgs fields, writes a (phenomenological, á la Landau–Ginzburg) potential for it and then minimizes the temperature-dependent free energy of the system. The result of minimization is determined by the shape and parameters of the Higgs potential, which are more or less chosen by hand—hoping that the underlying fundamental theory determines them dynamically.

Now, in the case of '(super)string theory', none of the choices can be blamed on a more fundamental theory since (super)strings should be *the fundamental theory.*

Maybe this is asking too much, but that is what we do expect of '(super)string theory'. However, not only have we no knowledge of the 'Higgs potential'—we don't even know what the relevant parameter space really is. My main aim here is to present some recent progress in assembling the jigsaw puzzle of this parameter space and also some of the processes that appear to enable the (super)strings to move around in it.

To get ourselves oriented, we begin with a lightning review of results obtained in the so-called 'point field limit' (another misnomer; see later) in Section 1. To take into account the spatial extendedness of (super)strings, we consider a suitable non-linear σ-model in Section 2 and discuss the relation to the 'point field limit'. In Section 3, we discuss some of the methods and techniques used to construct and analyze families of ground states for (super)strings. As a curiousity, we include a brief discussion of the ubiquitous phenomenon that the families of ground states have common members and connect thereby. In Section 4, we created σ-models in the image of our pointillistic models. For a number of such constructions, we analyze in Section 5 the parameters and the spaces that they span and identify the connecting phenomena to correspond to (somewhat tumultuous) voyages in the parameter space. In Section 6, we review the arguments indicating that the distance traversed in such voyages to new worlds is finite. Finally, Section 7 presents a discussion of prospects and tasks that have spawned and offers some (unwarranted?) remarks and conclusions.

1. LESSONS OF POINTILLISTIC COMPACTIFICATION

With a point, well taken, into the point-blank range.

Following the historical evolution of the topic [1,2], we consider the $E_8 \times E_8$ heterotic string [3]. As defined initially, it propagates in 9+1-dimensional Minkowski spacetime and a modification á la Kałuża–Klein is necessary. In doing so, one replaces the 9+1-dimensional Minkowski spacetime with another which is (at least locally) $M_{3+1} \times \mathcal{M}$. M_{3+1} denotes the 3+1-dimensional Minkowski spacetime so that the choice of such vacuua boils down to choosing among candidates for $\mathcal{M}$.

1.1 Choosing the 'internal' space

For one thing, since no experimental apparatus seems detect more then three spacelike directions, the physically observable size of $\mathcal{M}$ better be confined. Except for some strange spaces of dubious phenomenological relevance [4], it follows that $\mathcal{M}$ should be compact. The fields of the original 9+1-dimensional model therefore expand into Fourier series and the mass of the various Fourier-modes is quantized in units of the inverse size. It transpires that this mass scale, M_C, must be within a few orders of magnitude below the Planck mass $M_{Pl} \sim 10^{19} GeV$ [5].

One requires the effective model in 3+1-dimensions to have also a local N=1 supersymmetry surviving from the local N=1 supersymmetry in the 9+1-dimensions. The reason for this requirement is that supersymmetry provides the only generally accepted mechanism to protect the weak mass-scale ($\sim 10^2 GeV$) vs. M_C from radiative corrections, once such a hierarchy is somehow achieved. (In the composite models which are capable of generating and protecting mass hierarchies [6], local N=1 supersymmetry is essential for other reasons.)

The fist such compactification attempt [7] started from the massless modes of the $E_8 \times E_8$ heterotic string

$$g_{MN},\ \psi_M{}^A,\ B_{MN},\ \chi_A,\ \phi\ , \qquad A_M{}^\theta,\ \lambda^{A\theta}\ . \tag{1.1}$$

The $_{M,N}$ indices take values $0,\ldots,9$ and label 9+1-vectors, while ^{A} ($_A$) labels right-handed (left-handed) spinors in 9+1-dimensional spacetime; θ labels the 496 generators of the $E_8 \times E_8$.

According to the compactification ansatz, the indices pertaining to the 9+1-dimensional spacetime have to be adapted to describe $M_{3+1} \times \mathcal{M}$:

$$\{V^M,\ M=0,\ldots,9\} \ \rightarrow\ \{V^{\underline{\mu}} \oplus V^m,\ \underline{\mu}=0,1,2,3,\ m=4,\ldots,9\}\ , \tag{1.2}$$

$$\{\psi^A,\ A=1,\ldots,16\} \ \rightarrow\ \{\psi^{\alpha a} \oplus \psi^{\dot\alpha}{}_a,\ \alpha,\dot\alpha=1,2\ ,\ a=1,\ldots,4\}\ . \tag{1.3}$$

Next we need to determine the background values of the quantities (1.1). To preserve Lorentz invariance of the 3+1-dimensional spacetime, only quantities that are Lorentz-scalars may acquire a non-vanishing background, including $\langle g_{\underline{\mu\nu}}\rangle = \mathrm{diag}(+,-,-,-)$, the Minkowski metric. It remains to determine the admissible background values of g_{mn}, B_{mn} and ϕ.

To this end, Candelas et al. [7] (see also Ref. [8]) require the supersymmetry transformations, of the general form

$$\delta_\epsilon\,\mathrm{fermion} = F(\mathrm{bosons},\mathrm{fermions}^2)\,\epsilon\ , \tag{1.4}$$

to preserve the vanishing of the background value of fermions. Hence

$$F(\mathrm{bosons},\mathrm{fermions}^2)\epsilon = 0 \tag{1.5}$$

for the background values of the bosonic and fermion-bilinear fields involved and for precisely one of the four components of ϵ. (Owing to the spinor decomposition law (1.3), the parameter of the N=1 local supersymmetry in 9+1-dimensions, ϵ, gives rise to four N=1 local supersymmetry parameters in 3+1-dimensions.)

Solving the integrability equations for (1.5), Candelas et al. conclude that the M_{3+1} factor in spacetime must be Minkowskian if it is to be maximally symmetric.

Further, they assume that $\mathcal{M}$ is a manifold (i.e., everywhere smooth) and that the quantity

$$H_{pqr} \stackrel{\text{def}}{=} d_{[p}B_{qr]} - \tfrac{1}{30}\mathrm{tr}(A_{[p}F_{qr]} - \tfrac{1}{2}A_{[p}A_qA_{r]}) \;+\; \mathrm{tr}(\Gamma_{[p}R_{qr]} - \tfrac{1}{2}\Gamma_{[p}\Gamma_q\Gamma_{r]}) \;, \qquad (1.6)$$

which is necessary for anomaly cancellation and manifestly obeys

$$d_{[p}H_{rqs]} = \mathrm{tr}(R_{[pq}R_{rs]}) \;-\; \tfrac{1}{30}\mathrm{tr}(F_{[pq}F_{rs]}) \;, \qquad (1.7)$$

actually has a vanishing background value. Here A_m is the $E_8 \times E_8$ Yang–Mills connection 1-form restricted to $\mathcal{M}$ and F_{mn} is its field strength. Γ_m is the Cristoffel connection on $\mathcal{M}$ and R_{mn} is its field strength, i.e., the Riemann curvature 2-form.

Subject to these assumptions, $\mathcal{M}$ is derived[♮1] to be a complex manifold that admits a Ricci-flat Kähler metric. Equivalently, $\mathcal{M}$ admits a Kähler metric for which the Cristoffel connection 1-form is $SU(3)$ holonomy valued. Such manifolds were only recently proven to exist [11,12] and are typically called Calabi-Yau manifolds.

When discussing compactifications on Calabi-Yau manifolds á la Candelas et al. [7], one should bear in mind that only the massless states of the $E_8 \times E_8$ heterotic string (1.1) are being considered. In certain aspects, this looks like approximating the string with its center of momentum but does not really involve a continuous limiting process; rather, it is a simple restriction[♮2]. The results in this approach necessarily have to be corrected for the effect of the ignored states. A number of general results exists regarding these corrections [13,14] but explicit, case by case computations of many of the physically observable quantities is still necessary.

The requirement of smoothness was soon shown to be unnecessary : consistent heterotic string models were constructed on toroidal orbifolds [15] which are flat except at an isolated number of cyclic quotient singular points. In point of fact, these models are *exact* solutions to the string equations of motion, suggesting that one should also consider singular and not just smooth candidates for $\mathcal{M}$. Unfortunately, we do not know the complete class of singularities that are innocuous for strings and hence do not know just how singular may the 'internal' $\mathcal{M}$ become. We shall consider candidates that are smooth except maybe at a finite number of isolated and rather mildly singular points.

Many other stringy constructions have been suggested meanwhile, such as those involving : covariant lattices [16], free fermions [17], $c<3$ N=2 superconformal models [18,19] ... All these are exactly soluble : all string tree-level but also some string loop results can (in principle) be obtained exactly. Hence their technical superiority to any

---

[♮1]This also follows from the classification of Berger [9,10].

[♮2]A "0-mode restriction" would therefore describe the analysis better then "point field limit" but we shall abide by the latter locution since it has became universally accepted.

generic Calabi-Yau manifold compactification where even string tree results are typically obtainable, at best, in some perturbative scheme. In many instances, they however seem to be related to Calabi-Yau manifolds as their special (possibly singular) forms [20,21,22]. We therefore focus on Calabi-Yau manifolds, aiming at generality, but bear in mind such relations and use them whenever possible.

1.2 Massless particles and harmonic forms

We start from the massless modes of the $E_8 \times E_8$ heterotic string (1.1).

1. Assuming the compactification ansatz (1.2) and (1.3), we decompose the tensors in (1.1) accordingly. In addition, since $\mathcal{M}$ is a complex manifold, we can choose a basis of complex coordinates

$$v^m,\ m=4,\ldots,9 \ \rightarrow\ v^\mu \oplus v_\mu,\ \mu=1,2,3\ , \tag{1.8}$$

$$\psi^a \oplus \psi_a,\ a=1,\ldots,4 \ \rightarrow\ \psi^\mu \oplus \psi^4 \oplus \psi_\mu \oplus \psi_4\ . \tag{1.9}$$

As usual, contravariant μ indices (superscripts) denote $\mathcal{T}_\mathcal{M}$-valuedness while covariant μ indices (subscripts) denote $\mathcal{T}^*_\mathcal{M}$-valuedness, where $\mathcal{T}_\mathcal{M}$ is the (holomorphic) tangent bundle and $\mathcal{T}^*_\mathcal{M}$ the (holomorphic) cotangent bundle on $\mathcal{M}$.

2. Recall that the standard ansatz requires H_{pqr}, as defined in Eq. (1.6), to vanish. The integrability condition for this is $d_{[p}H_{qrs]}=0$, so that Eq. (1.7) yields

$$30(R^{mn}{}_{[pq}R_{rs]mn}) \ =\ \sum_{\theta=1}^{496}(F^\theta_{[pq}F^\theta_{rs]}) \tag{1.10}$$

for the background values of $\boldsymbol{R}$ and $\boldsymbol{F}$. The holonomy of $\mathcal{M}$ being $SU(3)$, $\langle\boldsymbol{R}\rangle$ is $SU(3)$ valued. Choosing $\langle\boldsymbol{F}\rangle$ to vanish except for a $SU(3)\subset E_8\times E_8$ and requiring $\langle\boldsymbol{F}\rangle=\langle\boldsymbol{R}\rangle$ solves Eq. (1.10) straightforwardly.

More precisely, the ansatz identifies $\mathcal{T}_\mathcal{M}$ with the **3**-fibre of an $SU(3)$ subbundle of the original $E_8\times E_8$ Yang–Mills principal bundle (see also Ref. [23]).

We now consider fluctuations of the fields (1.1).

3. The component-tensors obtained thus far are functions on $M_{3+1}\times\mathcal{M}$ and it is only sensible to Fourier-expand them on the 'internal' space $\mathcal{M}$. Consider, e.g., the Dirac equation for some spinor that is massless in the 9+1-dimensional sense. According to (1.2) and (1.3),

$$i\not{\nabla}_{9+1}\Psi(x,y)=0\ \rightarrow\ i\sum_k(\not{\nabla}_{3+1}+\not{\nabla}_6)\psi_k(x)Y_k(y)=0\ , \tag{1.11}$$

where $\not{\nabla}_{3+1}$ and $\not{\nabla}_6$ are suitably chosen Dirac operators on the M_{3+1} and $\mathcal{M}$ factors (see Ref. [1], Vol. II, p.366). Since $\not{\nabla}_6 Y_k(y)=m_kY_k(y)$, Eq. (1.11) yields

$$i(\not{\nabla}_{3+1}+m_k)\psi_k(x)=0\ . \tag{1.12}$$

By supersymmetry, the analogous is true for the bosonic fields.

Massless fields in the 3+1-dimensional sense precisely correspond to 0-modes $Y_k^0(y)$ on $\mathcal{M}$ ($m_k = 0$). Since $\mathcal{M}$ is Kähler, we also have

$$\triangle_{\mathcal{M}} Y_k^0(y) = 0 \quad \Longleftrightarrow \quad \overline{\partial}^{\dagger} Y_k^0(y) = \overline{\partial} Y_k^0(y) = 0 \ , \tag{1.13}$$

where $\triangle_{\mathcal{M}}$ is the Laplacian in $\mathcal{M}$, $\overline{\partial}$ is the Dolbeault operator and $\overline{\partial}^{\dagger}$ its adjoint. The 0-modes $Y_k^0(y)$ are harmonic forms on $\mathcal{M}$, i.e., elements of Dolbeault cohomology groups.

Imposing $\langle \boldsymbol{F} \rangle = \langle \boldsymbol{R} \rangle$, the gauge symmetries of the model have been reduced to those that commute with $\langle \boldsymbol{F} \rangle$ and they form an $E_6 \times E_8$ Lie algebra. The latter factor, E_8, turns out to interact with the rest of the massless fields only through gravity and we can safely ignore it for our present purposes.

As an example, we here consider the decomposition of the E_8 connection according to steps 1 and 2.

$$\begin{array}{lllllllll}
\mathbf{248} & \rightarrow & (\mathbf{78},\mathbf{1}) & \oplus & (\mathbf{27},\mathbf{3}) & \oplus & (\mathbf{27^*},\mathbf{3^*}) & \oplus & (\mathbf{1},\mathbf{8}) \\
A_M{}^{\theta} & \rightarrow & A_{\underline{\mu}}{}^{I\overline{I}} & \oplus & A_{\underline{\mu}}{}^{I\sigma} & \oplus & A_{\underline{\mu}}{}^{\overline{I}\overline{\sigma}} & \oplus & A_{\underline{\mu}}{}^{\sigma\overline{\sigma}} \ \oplus \\
 & & A_{\mu}{}^{I\overline{I}} & \oplus & A_{\mu}{}^{I\sigma} & \oplus & A_{\mu}{}^{\overline{I}\overline{\sigma}} & \oplus & A_{\mu}{}^{\sigma\overline{\sigma}} \ \oplus \\
 & & A_{\overline{\mu}}{}^{I\overline{I}} & \oplus & A_{\overline{\mu}}{}^{I\sigma} & \oplus & A_{\overline{\mu}}{}^{\overline{I}\overline{\sigma}} & \oplus & A_{\overline{\mu}}{}^{\sigma\overline{\sigma}} \ .
\end{array} \tag{1.14}$$

Since the $\mathbf{3}$-fibre of $SU(3) \subset E_8$ has been identified with $\mathcal{T}_{\mathcal{M}}$, an σ superscript denotes $\mathcal{T}_{\mathcal{M}}$-valuedness, while an $\overline{\sigma}$ superscript is equivalent to an σ subscript and denotes $\mathcal{T}_{\mathcal{M}}^*$-valuedness.

Quite manifestly, the quantities in the second row are CPT-conjugates of those in the third row and we need not consider them separately. For the third row, e.g., we now take step 3.

$$A_{\overline{\nu}}{}^{I\overline{I}}(x,y)\, dz^{\overline{\nu}} = \sum_k A^{k\,I\overline{I}}(x)\ Y_k(y) \tag{1.15}$$

is expanded into (0,1)-forms on $\mathcal{M}$. Since there are no harmonic (0,1)-forms, these are all massive.

$$A_{\overline{\nu}}{}^{I\mu}(x,y)\, dz^{\overline{\nu}} = \sum_{\alpha} \Phi^{\alpha\,I}(x)\ \varphi_{\alpha}{}^{\mu}(y) \tag{1.16}$$

is expanded into $\mathcal{T}_{\mathcal{M}}$-valued (0,1)-forms : massless $\mathbf{27}$-plets of E_6 correspond to elements (labeled by α) of the $H^1(\mathcal{M},\mathcal{T}_{\mathcal{M}})$ cohomology group which is nonempty for a general Calabi-Yau manifold $\mathcal{M}$.

$$A_{\overline{\nu}}{}^{\overline{I}\overline{\mu}}(x,y)\, dz^{\overline{\nu}} = \sum_{A} \Phi^{A\,\overline{I}}(x)\ e_{A\overline{\mu}}(y) \tag{1.17}$$

is expanded into $\mathcal{T}^*_{\mathcal{M}}$-valued (0,1)-forms : massless **27***-plets of E_6 correspond to elements (labeled by A) of the $H^1(\mathcal{M},\mathcal{T}^*_{\mathcal{M}}) = H^{(1,1)}(\mathcal{M})$ cohomology group which is always nonempty since a Calabi-Yau manifold $\mathcal{M}$ always has at least the Kähler (1,1)-form.

$$A_{\bar\nu}{}^{\mu\bar\rho}(x,y)dz^{\bar\nu} = \sum_k \Phi^k(x)\phi_k{}^\mu{}_\rho(y) \tag{1.18}$$

is expanded into $\mathrm{End}\mathcal{T}_{\mathcal{M}}$-valued (0,1)-forms : these massless E_6-neutral matter fields correspond to elements of the $H^1(\mathcal{M},\mathrm{End}\mathcal{T}_{\mathcal{M}})$ cohomology group, for which only the lower bound

$$\dim H^1(\mathcal{M},\mathrm{End}\mathcal{T}_{\mathcal{M}}) \;\geq\; \dim H^1(\mathcal{M},\mathcal{T}_{\mathcal{M}}) \tag{1.19}$$

exists in general [10]. ($\mathrm{End}\mathcal{T}_{\mathcal{M}}$ is the bundle of traceless mappings $\mathcal{T}_{\mathcal{M}} \to \mathcal{T}_{\mathcal{M}}$: $\delta_{\alpha\bar\alpha}A_{\bar\mu}{}^{\alpha\bar\alpha} = 0$.)

The same is to be done with all the fields (1.1). By the N=1 supersymmetry preserved in compactification, the resulting spectrum assembles into N=1 superfields as summarized in Table 1.

Superfield	spin	E_6	cohomology	origin
Supergravity	$(2,3/2)$	**1**	$H^0(\mathcal{M})$	$g_{MN},\psi_M{}^A$
Yang–Mills	$(1,1/2)$	**78**	$H^0(\mathcal{M})$	$A_M{}^\theta,\lambda^{A\theta}$
S	$(1/2,0)$	**1**	$H^0(\mathcal{M})$	$g_{MN},\psi_M{}^A,B_{MN}$
T	$(1/2,0)$	**1**	$H^0(\mathcal{M})$	B_{MN},χ_A,ϕ
Ψ	$(1/2,0)$	**1**	$H^1(\mathcal{M},\mathcal{T}_{\mathcal{M}})$	$g_{MN},\psi_M{}^A$
Ψ_*	$(1/2,0)$	**1**	$H^1(\mathcal{M},\mathcal{T}^*_{\mathcal{M}})$	B_{MN},χ_A
$\Phi^{\alpha I}$	$(1/2,0)$	**27**	$H^1(\mathcal{M},\mathcal{T}_{\mathcal{M}})$	$A_M{}^\theta,\lambda^{A\theta}$
$\Phi^{A\bar I}$	$(1/2,0)$	**27***	$H^1(\mathcal{M},\mathcal{T}^*_{\mathcal{M}})$	$A_M{}^\theta,\lambda^{A\theta}$
Φ^k	$(1/2,0)$	**1**	$H^1(\mathcal{M},\mathrm{End}\mathcal{T}_{\mathcal{M}})$	$A_M{}^\theta,\lambda^{A\theta}$

Table 1: The spectrum of massless fields in Calabi-Yau compactification.

Calabi-Yau manifolds have a covariantly constant (3,0)-form Ω, because of which

$$H^q(\mathcal{M},\mathcal{T}_{\mathcal{M}}) \approx H^{(2,q)}(\mathcal{M}) \;, \quad q=0,1,2,3 \tag{1.20}$$

and since $H^q(\mathcal{M},\mathcal{T}^*_{\mathcal{M}}) = H^{(1,q)}(\mathcal{M})$, one has the essential part of the spectrum

$$b_{2,1}\,\mathbf{27} \oplus b_{1,1}\,\mathbf{27^*}\;, \qquad b_{p,q} \stackrel{\mathrm{def}}{=} \dim H^{(p,q)}(\mathcal{M})\;. \tag{1.21}$$

A **27**-plet superfield accommodates the 15 fermions of a Standard Model generation and contains in addition 12 exotic fermions and also their 27 superpartners. **27***-plets contain mirror-fermions. It is a convention to say that the model has $\frac{1}{2}|\chi_E| =$

$|b_{1,1} - b_{2,1}|$ generations of **27**s and $\min(b_{1,1}, b_{2,1})$ **27**–**27*** mirror pairs. χ_E is the Euler characteristic.

For a large class of Calabi-Yau manifolds [10,24,25,26] these numbers have been computed (see also Ref. [27]) and computer programs exist to assist the computation for each particular case. The computation of $\dim H^1(\mathcal{M}, \mathrm{End}\mathcal{T}_{\mathcal{M}})$ has been an unsolved problem until recently. It was known for a number of simple cases [1,28,29]. The technique of exact and spectral sequences was recently employed in its full force [30] to compute $H^q(\mathcal{M}, \mathrm{End}\mathcal{T}_{\mathcal{M}})$ for several physically interesting cases and implying that the same is possible for each Calabi-Yau manifold of Ref. [10,24,25].

It should be noted that while $b_{1,1}$ and $b_{2,1}$ are stable under smooth deformations, $\dim H^q(\mathcal{M}, \mathrm{End}\mathcal{T}_{\mathcal{M}})$ is semi-continuous, i.e., may (and in fact often does) increase for special choices of the complex structure[♯3].

1.3 Effective couplings

As derived in Eqs. (1.14)–(1.18), matter fields stem from $A_M{}^\theta$. Their couplings are found in an analogous decomposition of the interaction terms in the action of the original 9+1-dimensional model [31].

$$d_{IJK}\,\Phi^{\alpha\,I}(x)\Phi^{\beta\,J}(x)\Phi^{\gamma\,K}(x)\,\kappa_{\alpha\beta\gamma}\,, \qquad \text{"}\mathbf{27}^3\text{"}\,, \tag{1.22}$$

$$d_{\bar I\bar J\bar K}\,\Phi^{A\,\bar I}(x)\Phi^{B\,\bar J}(x)\Phi^{C\,\bar K}(x)\,\kappa_{ABC}\,, \qquad \text{"}\mathbf{27}^{*3}\text{"}\,, \tag{1.23}$$

$$\delta_{I\bar J}\,\Phi^{\alpha\,I}(x)\Phi^{B\,\bar J}(x)\Phi^k(x)\,\kappa_{\alpha Bk}\,, \qquad \text{"}\mathbf{27}\cdot\mathbf{27}^*\cdot\mathbf{1}\text{"}\,, \tag{1.24}$$

$$\Phi^i(x)\Phi^j(x)\Phi^k(x)\,\kappa_{ijk}\,, \qquad \text{"}\mathbf{1}^3\text{"} \tag{1.25}$$

are the renormalizable couplings. Here d_{IJK} is the symmetric coupling tensor of E_6 and

$$\kappa_{\alpha\beta\gamma} = \int_{\mathcal{M}} \Omega \wedge \varphi^\mu_\alpha \wedge \varphi^\nu_\beta \wedge \varphi^\rho_\gamma\, \Omega_{\mu\nu\rho}\,, \tag{1.26}$$

$$\kappa_{ABC} = \int_{\mathcal{M}} \Omega \wedge e_{A\,\mu} \wedge e_{B\,\nu} \wedge e_{C\,\rho}\, \frac{i\bar\Omega^{\mu\nu\rho}}{\|\Omega\|^2}\,, \tag{1.27}$$

$$\kappa_{\alpha Bk} = \int_{\mathcal{M}} \Omega \wedge \varphi^\mu_\alpha \wedge e_{B\,\nu} \wedge \phi_k{}^\nu{}_\mu\,, \tag{1.28}$$

$$\kappa_{ijk} = \int_{\mathcal{M}} \Omega \wedge \phi_i{}^\mu{}_\nu \wedge \phi_j{}^\nu{}_\rho \wedge \phi_k{}^\rho{}_\mu\,. \tag{1.29}$$

For the computation of the first two of these, there exist methods which are not universally valid, but can often be applied (see Ref. [32,27] for $\kappa_{\alpha\beta\gamma}$ and Ref. [33] for κ_{ABC}). The results of Ref. [30] indicate that a suitable generalization is possible whenever the $\mathcal{M}$ is embedded in a product of homogeneous spaces. Clearly, quite some more work needs to be done here.

[♯3] Details of this phenomenon will be reported elsewhere.

As if this were not enough trouble, the effective Lagrangian for any of the massless matter fields takes the form

$$\mathcal{L}_{\text{eff.}} = N_{ab}\int d^4\theta\ \Phi^a(\Phi^b)^\dagger\ +\ \kappa_{abc}\Big(\int d^2\theta\ \Phi^a\Phi^b\Phi^c\ +\ h.c.\ \Big)\ , \tag{1.30}$$

where E_6 indices are suppressed and a, b, c are some generic labels for the various massless superfields. N_{ab} is not much harder to compute [34], but it gets renormalized through the inclusion of non-zero modes and quantum corrections. Changing the Φ^a basis to the physical one, where δ_{ab} replaces N_{ab}, the coupling coefficients become "$(N)^{-3/2}\kappa$". Fortunately, symmetries often simplify the necessary change of basis and also ratios of couplings tend to be less sensitive to N_{ab} so that some reliable predictions can be derived without the knowledge of N_{ab}.

There are many more issues of phenomenological issue that we simply have no spacetime to discuss here and I refer the interested reader to Ref. [1,2,7,10,13,14,27,31, 32,33,35], to mention a few. See also the contributions of B. Greene and C.A Lütken in these Proceedings.

2. TOWARDS σ-MODEL COMPACTIFICATION

Instead of following the historical path, where one takes the $E_8 \times E_8$ heterotic string in 9+1-dimensional spacetime, restricts to massless modes, compactifies on some (possibly singular) Calabi-Yau space, computes physically relevant quantities and then keeps correcting them for the host of fields and modes that have been ignored in the process—we might as well (try to) compactify the strings.

2.1 σ-Modelling Strings

Strings propagating through some target space can also be considered as a σ-model which maps the world sheet (swept out by the string) into the target space. This point of view was useful, e.g., for developing a first quantized description of strings in 25+1-dimensional spacetime [36], providing part of the motivation. Also, we should expect to improve on the computability of the physical predictions of our compactification models, moreover by using standard field theoretic methods. Finally, a number of rather intriguing results regarding Calabi-Yau spaces, compactifications and relevant parameter spaces have been derived [21,22], [37] — [55], unfortunately at very different levels of generality and in very different approaches. The σ-model appears to be capable of bridging this jazz [56].

Also, more and more relations are being found among compactifications that have been constructed in totally different ways. For example, orbifolds are related to certain corresponding Calabi-Yau manifolds through the process of blowing up [20]. Even a huge number of topologically distinct Calabi-Yau manifolds have their relevant moduli spaces connected into a fine web [41] where the "interfaces" are provided by moduli spaces of mildly singularized Calabi-Yau manifolds. This process appears to head towards uncovering more and more of a universal Calabi-Yau moduli space whereby different Calabi-Yau manifolds may be thought of as merely different phases of a "master theory". Again, σ-models allow the 'stringification' of this process.

Heterotic strings feature a local (1,0)-supersymmetry on the world sheet. Such a unidexterous supersymmetric action for the original heterotic string [3] in 9+1-dimensional spacetime is given by

$$S_{Het.} = S^{9+1}_{SpT} + S_{YM} , \tag{2.1}$$

where the string action, S^{9+1}_{SpT}, describes 9+1-dimensional spacetime degrees of freedom,

$$S^{9+1}_{SpT} = \int_\Sigma d^2\sigma\, d\varsigma^- E^{-1} \left[i\tfrac{1}{2}\eta_{MN} (\nabla_+ \boldsymbol{X}^M)(\nabla_{--}\boldsymbol{X}^N) \right] , \tag{2.2}$$

with $M = 0,\ldots,9$.

Here $\nabla_{\hat{A}} = E_{\hat{A}}{}^{\hat{M}} D_{\hat{M}} + \omega_{\hat{A}} M - ig\Gamma^i{}_{\hat{A}} t_i$ is the covariant superderivative, with $E_{\hat{A}}{}^{\hat{B}}$, $\omega_{\hat{A}}$ and $\Gamma^i{}_{\hat{A}}$ the zweibein, spin- and gauge-connection superfields, respectively, and M and t_i the world sheet Lorentz and the Yang-Mills generators, respectively. $\hat{A}, \hat{M} = (++, --, -)$ are super world sheet indices and E is the standard superdeterminant of the zweibein superfield. Also,

$$\boldsymbol{X}^M(\varsigma^+, \sigma, \tau) = X^M(\sigma,\tau) + \varsigma^+ \psi_+{}^M(\sigma - \tau) . \tag{2.3}$$

The remaining S_{YM} introduces the Yang–Mills degrees of the theory. The simplest is the fermionic representation

$$S^{(32)}_{YM} = -\tfrac{1}{2} \int_\Sigma d^2\sigma\, d\varsigma^- E^{-1}\, \boldsymbol{\eta}_-{}^i \Big(\delta_{ij} \nabla_+ \;+\; (\nabla_+ \boldsymbol{X}^M) A_{M\,[ij]}(\boldsymbol{X}) \Big) \boldsymbol{\eta}_-{}^j , \tag{2.4}$$

with $i = 1,\ldots,32$ and where only the (right-handed) lowest component field of each $\boldsymbol{\eta}_-{}^i(\varsigma^+, \sigma, \tau)$ propagates. The 496 10-vectors $A_{M\,[ij]}(\boldsymbol{X})$ gauge an $SO(32)$ Yang–Mills symmetry. To make all 496 gauge vectors of $E_8 \times E_8$ appear as manifestly in an action, a slightly more complicated construction can be used [57] and we give it for completeness

$$\begin{aligned} S^{(R)}_{YM} &= -\frac{i}{2} \int_\Sigma d^2\sigma\, d\varsigma^- E^{-1} \mathbf{Tr} \Big\{ U_+ U_{--} \;+\; \Lambda_{--}{}^{++} U_+ U_{++} \;+\; \frac{i}{3} \Lambda_{--}{}^{++} U_+ U_+ U_+ \\ &\quad + \int_0^1 dy \Big(e^{-i\tilde{\boldsymbol{\Phi}}_R} \frac{d}{dy} e^{i\tilde{\boldsymbol{\Phi}}_R} \Big) \big[\nabla_+ U_{--} \;-\; \nabla_{--} U_+ \big] \Big\} , \end{aligned} \tag{2.5}$$

where

$$U_{\hat{A}} \stackrel{\text{def}}{=} \Big(e^{-i\tilde{\boldsymbol{\Phi}}_R} \nabla_{\hat{A}} e^{i\tilde{\boldsymbol{\Phi}}_R} \Big) . \tag{2.6}$$

y is the usual bosonic extension of the world sheet and $\tilde{\Phi}$ the corresponding extension of the superfield, needed for the Wess-Zumino-Novikov-Witten term. Actually, another term has to be added to this action to ensure proper cancellation of the Siegel super anomaly in a covariant manner [58], but we shall not discuss this here.

Here $\Phi(\varsigma^+, \sigma, \tau)$ is an $E_8 \times E_8$ algebra-valued superfield only the lowest component of which propagates and is right-moving, i.e., σ^{++}-independent on shell. We may expand it in generators of $E_8 \times E_8$,

$$\Phi \stackrel{\text{def}}{=} \Phi^i T_i , \qquad i = 1, \dots, 496 , \tag{2.7}$$

obtaining an explicit representation of all 496 degrees in S_{YM}. The generators T_i are normalized

$$\text{Tr}(T_i T_j) = 2k_G \delta_{ij} , \tag{2.8}$$

our case being $G = E_8 \times E_8$ and $k_G = 1$.

The (super)conformal symmetry of the combined system (2.2), (2.4) is crucial for consistent quantization and its stringy interpretation [1,2]. It is signalled by the vanishing of the trace of the conserved energy-momentum tensor (see subsection 3.2.3 of Ref. [1], Vol. II). Being a coefficient of the characteristic polynomial of the energy-momentum matrix, the trace is coordinate independent. Now, quantum corrections generally violate the classical (super)conformal invariance, which can be seen by computing the one-loop contribution to the trace of the conserved energy-momentum tensor. In the above system, however, the various contributions cancel precisely and the system is (super)conformally invariant as a quantum field theory.

For future reference, Table 2 lists the anomaly contributions of some basic quantities that we shall need. One should note the special property of 1+1-dimensional field theories : all fields can be decomposed in left- and right-movers in a Lorentz-invariant way. The contributions to the conformal anomaly are therefore computed separately among left- and right-movers. With the use of Table 2 we now obtain for $S^{9+1}_{SpT} + S^{(32)}_{YM}$ as given in (2.2) and (2.4) :

$$\begin{aligned} \nu_L &= -26 + 11 + 10 \cdot 1 + 10 \cdot \tfrac{1}{2} = 0 , \\ \nu_R &= -26 + 0 + 10 \cdot 1 + 32 \cdot \tfrac{1}{2} = 0 . \end{aligned} \tag{2.9}$$

Apart from (super)reparametrization, contribution are added from the left- and right-moving part of the ten string coordinates X^M, their ten (1,0)-superpartners $\psi_+{}^M$ and the 32 right-handed spinors from $S^{(32)}_{YM}$.

2.2 Compactifying σ-Models

The idea of compactification is to reinterpret six of the string superfields X^M as 'internal' degrees of freedom. Still being σ-model fields, they map into the 'part' of

Source of anomaly	ν_R	ν_L
World sheet reparametrization invariance	-26	-26
World sheet (1,0)-fermionic reparametrization	0	$+11$
Right- (left-)moving (abelian) scalar	$1\,(0)$	$0\,(1)$
Right- (left-)moving (abelian) spinor	$\frac{1}{2}\,(0)$	$0\,(\frac{1}{2})$
G-algebra valued right-moving scalar[a]	$d_G(1+c_2/2k)^{-1}$	0

[a] d_G and $c_2(G)$ are the dimension and 2nd Casimir of the adjoint rep. of G and k is defined in Eq. (2.8).

Table 2: Contributions to the trace anomaly.

spacetime that is now being considered as 'internal'. As in the pointillistic description, the 9+1-dimensional spacetime is considered to be $M_{3+1} \times \mathcal{M}$ except that, instead of this 9+1-dimensional spacetime, we now need to deal with the (infinite dimensional) space of σ-model maps

$$\boldsymbol{X} : \Sigma \;\rightarrow\; M_{3+1} \times \mathcal{M} \tag{2.10}$$

of the world sheet into $M_{3+1} \times \mathcal{M}$. Such compactifications are termed 'geometrical' since the geometrical interpretation is obvious.

Ultimately, we wish to construct a quantum field theory. In the first-quantized picture, we therefore define the partition function and path-integrate over the space of sigma model maps (2.10). *A posteriori*, we couldn't have cared less whether the 'internal' sector had a geometric description or not. Indeed, switching on this factor of negligence, we may choose just about anything for this sector, as long as the contribution to the trace-anomaly, together with the (possibly modified) S_{YM}, adds up to $(\nu_L, \nu_R) = (9, 22)$. It is however quite obvious that *some* geometric interpretation must exist[14].

To describe compactification on a Calabi-Yau manifold, one considers a non-linear σ-model action

$$S^{(2,0)}_{CY} = \tfrac{i}{4}\int_\Sigma d^2\sigma\, d\varsigma^- E^{-1}\, g_{\mu\bar\nu}(\boldsymbol{Y},\overline{\boldsymbol{Y}})\,\Big[(\nabla_+\boldsymbol{Y}^\mu)\,(\nabla_{--}\boldsymbol{Y}^{\bar\nu}) + \text{ h.c. }\Big] \tag{2.11}$$

with $\boldsymbol{Y}^{\bar\nu} \stackrel{\text{def}}{=} (\boldsymbol{Y}^\nu)^\dagger$, as a part of the heterotic string action

$$S_{4D} = S^{9+1}_{SpT} + S^{(2,0)}_{CY} + S^{(32)}_{YM}\,. \tag{2.12}$$

[14]Our knowledge of possible geometries is at the moment confined to the cases where most rudimentary, nearly smooth spaces are the target of σ-model mappings (2.10).

The three (1,0) superfields Y^μ are complex. By virtue of this alone [59] they *automatically* provide a representation of world sheet (2,0)-supersymmetry. The bosonic components of these superfields are identified with coordinates of a Calabi-Yau manifold, $\mathcal{M}$, on which the metric $g_{\mu\bar{\nu}}$ has been chosen. More precisely, the bosonic components of $Y^\mu(\sigma,\varsigma)$ are the standard σ-model maps from the world sheet into the Calabi-Yau manifold. The three complex left-handed spinor components are valued in $\mathcal{T}_{\mathcal{M}}$, the holomorphic tangent space of $\mathcal{M}$, i.e., map the world sheet into $\mathcal{T}_{\mathcal{M}}$.

The action (2.12) however describes strings in 3+1-dimensional spacetime with $E_8 \times E_8$ Yang-Mills local symmetry, which is known to be anomalous. To cancel the anomaly, the curl of B_{mn} has to be modified as in Eq. (1.6). This, however, induces Eq. (1.10) and one way to solve it is to identify $\mathcal{T}_{\mathcal{M}}$ with an $SU(3)$ vector subbundle of the $E_8 \times E_8$ principal bundle. In plain English, the bosonic components of Y^μ are to acquire new superpartners (since these must map into $\mathcal{T}_{\mathcal{M}}$), stolen from S_{YM} .

Thus, the (2,0)-supersymmetry is being promoted into a left–right symmetric (2,2)-supersymmetry by stealing six real fermions from S_{YM}. Except, there are no fermions in S_{YM} suitable for $E_8 \times E_8$ (2.5). There are two ways out. In both of them, add six fermions to $S^{(2,0)}_{CY}$ to turn it into the required $S^{(2,2)}_{CY}$ and then

1. make S_{YM} into a coset construction of proper anomaly contribution, or
2. leave S_{YM} as is (for $E_8 \times E_8$) and constrain an $SU(3)$-generating subset to equal the fermionic bilinear generators of an $SU(3)$ built from the newly acquired superpartners in $S^{(2,2)}_{YM}$.

In any case, the σ-model action is of the form

$$S_{4D} = S^{3+1}_{SpT} \; + \; S^{(2,2)}_{CY} \; + \; S^{(13,0)}_{YM} \; , \tag{2.13}$$

where the '(13,0)' superscript on S_{YM} is its (ν_R, ν_L) anomaly content. Work along these lines is in progress and will be reported elsewhere.

2.3 Specifying a compactification

So far, we have not been specific about the choice of the complex manifold $\mathcal{M}$ to which S_{4D} refers. Indeed, any definition of a partition function using S_{4D} would have to specify the range of integration for the bosonic components Z^μ of $\boldsymbol{Z}^\mu$, the (2,2)-superfields in $S^{(2,2)}_{YM}$. Corresponding to coordinates of a topologically non-trivial manifold, these variables have a well-defined meaning for the purpose we are intending only if the manifold $\mathcal{M}$ is covered with a suitable number of $\mathbb{C}^3$-like open neighbourhoods $\mathcal{U}_i$ in each of which then a separate collection $\{Z_i{}^\mu\}$ corresponds to local coordinates.

The particular pattern in which the open neighbourhoods are patched together and also the identifications in the overlap regions impose certain patching conditions on the collections of superfields $Z_i{}^\mu$. To avoid double integration over the overlap regions, we may shrink these into infinitesimally thin "layers" interfacing the various neighbourhoods. Thereafter, the patching conditions on the collections of $Z_i{}^\mu$'s become boundary conditions on these "layers". To be specific, one may triangularize $\mathcal{M}$, break up the path-integral in so many corresponding summands and then impose appropriate boundary conditions on each of these[15], reflecting the pattern of gluing the simplices together into $\mathcal{M}$.

It should be obvious that, in the case of any Calabi-Yau manifold, this process is only a "gedanken-construction" : while possible in principle, such constructions have actually never been carried out. This however reveals that all the global information on our internal manifold $\mathcal{M}$ lies in specifying the range of the path-integration rather then in the action itself. We shall soon find a remedy to this and show how to lift global information on $\mathcal{M}$ into the action.

On the face of it, the action (2.13) harbours two unfortunate properties : There is no such thing as "the most general Calabi-Yau manifold" which clearly seems to prevent us from constructing a unified action for Calabi-Yau compactification that would encompass all particular examples as regions in its "universal" moduli space. The constructive approach of Ref. [41], whereby moduli spaces of as many Calabi-Yau manifolds as possible are pieced together, may in fact accumulate enough information for surmounting this obstruction and may eventually lead to a unified "master theory", but it seems premature to rely on this here.

The second problem is that, even for particular Calabi-Yau manifolds, explicit metrics to be plugged into $S_{CY}^{(2,2)}$ are not known. Recall however that physical observables in such models depend only on some fairly specific information (such as the choice of the complex structure and the Kähler class) rather then the actual metric. This opens a by-pass to this problem : techniques have been developed [31,60,27] to give explicit constructions of a large class of Calabi-Yau manifolds where the physically relevant information can be extracted without reference to the unknown metric. In these constructions, the Calabi-Yau manifold appears as the common solution to a number of constraints in a product of complex projective spaces, enabling us to use all the bundled hardware of algebraic geometry, including simple geometric intuition, along with

[15]Note that these are boundary conditions for the superfields Z^μ in the boundary of the simplex that they parametrize, not boundary conditions for Z^m's as functions of world-sheet coordinates. A trivial example is provided by a complex 3-torus, where one collection of Z^μ's is enough and are subject to the periodicity conditions $Z^\mu \cong Z^\mu + 2\pi$, $\mu = 1,2,3$.

constrained dynamics for the analysis of which techniques are being devised ever since Lagrange. For one thing, the number of homogeneous coordinates, Z, on the complex projective spaces *exceeds* those necessary to describe a Calabi-Yau manifold. This, in fact, may be turned into an advantage : they allow more symmetries to be realized linearly in the world sheet action and permit more leeway in general.

3. A HITCHHIKER'S GUIDE THROUGH GROUND STATES

When eating an elephant, take a bite at a time.

We here review some of the constructions of Kähler Ricci-flat, so-called Calabi-Yau spaces[#6] Any such construction turns out to be (possibly a combination of)

1. embedding in a bigger, preferably well understood space,
2. complex 3-D surgery and/or deformation.

We will give examples of both these processes but wish to make it clear that no universal algorithm exists and that undoubtedly many other constructions will be invented.

3.1 Configurations and CICYs

An important class of embeddings starts out with a product of complex projective spaces $X = \mathbb{P}_1^{n_1} \times \cdots \times \mathbb{P}_m^{n_m}$. In this space, one considers a set of K polynomial constraints $\{P^a = 0\}_{a=1...K}$. The zero-set of every such polynomial defines a hypersurface

$$P_a(x) = 0 \ , \qquad \forall x \in \Xi_a \subset X \ . \tag{3.1}$$

The subspace of X of *simultaneous* solutions to the whole system of polynomial constraints corresponds to the complete intersection of the individual hypersurfaces

$$\{P_a(x) = 0\}_{a=1,\dots,K} \ , \qquad \forall x \in \bigcap_{a=1}^{K} \Xi_a \subset X \ . \tag{3.2}$$

We represent this by a *configuration matrix,* defined to be an integer-valued matrix of the general form :

$$[X \| M] = \begin{bmatrix} n_1 \\ \vdots \\ n_m \end{bmatrix}\!\!\!\left.\begin{matrix} \| \\ \| \\ \| \end{matrix}\right. \!\! \begin{matrix} q_1^1 & \cdots & q_K^1 \\ \vdots & \ddots & \vdots \\ q_1^m & \cdots & q_K^m \end{matrix}\Bigg] , \qquad q_a^r \geq 0, \tag{3.3}$$

[#6]Reinterpreting the Kählerness and Ricci-flatness property for singular spaces is a technical which matter we shall not delve into here.

where n_r is the dimension of $\mathbb{P}_r^{n_r}$, the rth factor-space in $\boldsymbol{X}$, and q_a^r is the degree of homogeneity of the ath polynomial with respect to the homogeneous coordinates of $\mathbb{P}_r^{n_r}$. $\boldsymbol{M}$ simply denotes the matrix of the degrees q_a^r.

The subspace $\bigcap_{a=1}^{K} \Xi_a$ of $\boldsymbol{X}$ is itself a compact complex (possibly singular) variety $\mathcal{V}$ of dimension $\dim \mathcal{V} = \dim \boldsymbol{X} - K$. Typically, there are many possible polynomials with given degrees of homogeneity, defining a whole family of varieties. It is in this sense that we say that a particular variety *belongs* to a configuration matrix, or call it a *member* of the configuration matrix and write $\mathcal{V} \in [\boldsymbol{X}\|\boldsymbol{M}]$. For a generic choice of (i.e., almost all) parameters in the defining equations, $\mathcal{V}$ is a manifold [24]; that is, a generic member of a configuration matrix is a manifold.

For smooth varieties (manifolds) $\mathcal{V} \in [\boldsymbol{X}\|\boldsymbol{E}]$, the Euler characteristic, χ_E, is computed as a coefficient in the formal expansion :

$$c_{\text{top}}(\mathcal{V}) \cdot \prod_{a=1}^{K} \Big(\sum_{r=1}^{m} q_a^r \eta^r\Big) = \ldots + \chi_E(\mathcal{V}) \cdot \prod_{r=1}^{m} (\eta^r)^{n_r} \tag{3.4}$$

where η^r generates $H^2(\mathbb{P}_r^{n_r}, \mathbb{Z})$, $r = 1, \ldots, m$ and is the cohomology class of the curvature two-form on $\mathcal{V}$. The top Chern class, $c_{\text{top}}(\mathcal{V})$, is obtained from the total Chern characteristic polynomial :

$$c(\mathcal{V}) = \frac{\prod_{r=1}^{m} (1+\eta^r)^{n_r+1}}{\prod_{a=1}^{K} (1 + \sum_{r=1}^{m} q_a^r \eta^r)} = 1 + \sum_{i=1}^{\dim \mathcal{V}} c_i(\mathcal{V}), \qquad c_{\text{top}}(\mathcal{V}) \equiv c_{\dim \mathcal{V}}(\mathcal{V}), \tag{3.5}$$

where $c_i(\mathcal{V})$, the ith Chern class, is the sum of terms of total degree i in η^r's. In the extreme case $\dim \mathcal{V} = 0$, i.e., when $\mathcal{V}$ is a discrete set of points, $c(\mathcal{V}) = 1$ and $\chi_E(\mathcal{V})$ reduces to the coefficient of $\prod_{r=1}^{m} (\eta^r)^{n_r}$ in $\prod_{a=1}^{K} (\sum_{r=1}^{m} q_a^r \eta^r)$.

It is straightforward from (3.5) that

$$c_1(\mathcal{V}) = \sum_{r=1}^{m} \Big(n_r + 1 - \sum_{a=1}^{K} q_a^r\Big)\eta^r , \tag{3.6}$$

which is the cohomology class of the Ricci two-form on $\mathcal{V}$. The condition for a manifold $\mathcal{V} \in [\boldsymbol{X}\|\boldsymbol{M}]$ to have vanishing first Chern class (which has an appropriate generalization in the singular case) is :

$$\sum_{a=1}^{K} q_a^r = (n_r + 1), \qquad \forall r. \tag{3.7}$$

Configuration matrices $[\boldsymbol{X}\|\boldsymbol{M}]$ satisfying Eq. (3.7) represent (complete intersection) Calabi-Yau manifolds (CICYs). The number of configuration matrices that satisfy Eq. (3.7) is infinite but they represent a finite collection of topologically distinct CICY [24] and a thorough computer search [25] finds this to be nearly 8000.

We shall also consider configurations for which

$$\sum_{a=1}^{K} q_a^r \leq (n_r + 1), \qquad \forall r \tag{3.8}$$

is satisfied instead of (3.7), with the inequality being strict at least for one r. Such varieties of complex dimension three were termed almost Fano complete intersections (AFCIs) [41].

In simple cases, the $b_{1,1}$ Hodge number is easily computed by the Lefschetz Hyperplane theorem; see Ref. [24] for application and discussion of applicability. Similarly, b_{21} can be obtained by the simple but not universally valid polynomial deformation method (PDM); in the general case we compute b_{pq} by universally valid techniques discussed in Ref. [27].

3.2 The Web of CICYs

Candelas et al. Ref. [25] found a peculiar process in the zoo of CICYs. They considered a generic CICY embedded in $\mathbb{P}^3 \times \mathbb{P}^3 \times \mathbb{P}^1$

$$\check{\mathcal{M}} \in \begin{bmatrix} 2 \\ 2 \\ 1 \end{bmatrix}\!\!\begin{Vmatrix} 3 & 0 \\ 0 & 3 \\ 1 & 1 \end{bmatrix} \quad : \quad \begin{array}{rcl} S(x)z_1 + P(x)z_2 & = & 0 \, , \\ Q(y)z_1 + R(y)z_2 & = & 0 \, , \end{array} \tag{3.9}$$

Here x, y and z are homogeneous coordinates on $\mathbb{P}^3 \times \mathbb{P}^3 \times \mathbb{P}^1$ and S, P, Q and R are some generic cubic polynomials. Smooth members of this family have $\chi_E = 0$ and $b_{1,1} = b_{2,1} = 19$ [61,27] and so does $\check{\mathcal{M}}$.

Over every point (x, y) of $\mathbb{P}^3 \times \mathbb{P}^3$, the above two equations are a system of two homogeneous equations in z_1 and z_2 which cannot both vanish as they live on $\mathbb{P}^1$. It must be therefore that

$$\Delta(x,y) \stackrel{\text{def}}{=} P(x)Q(y) - R(y)S(x) \; = 0 \, . \tag{3.10}$$

This, however, is a singular bi-cubic hypersurface $\mathcal{M}^\sharp \in \mathbb{P}^3 \times \mathbb{P}^3$. At precisely 81 isolated points both Δ and $\mathrm{d}\Delta$ vanish but the matrix of second derivatives is positive definite. Such singularities are called nodes [62] and are the simplest that can occur [63]. Spaces that are smooth apart from a set of isolated nodes were termed 'conifolds' [46].

Since the singular bi-cubic manifestly belongs to the family of bi-cubics in $\mathbb{P}^3 \times \mathbb{P}^3$, we may deform

$$\Delta(x,y) + \varepsilon(x,y) = 0 \, , \qquad : \quad {\mathcal{M}^\flat}_\varepsilon \in \begin{bmatrix} 2 \\ 2 \end{bmatrix}\!\!\begin{Vmatrix} 3 \\ 3 \end{bmatrix} \, , \tag{3.11}$$

to obtain a family of smooth CICYs of $\chi_E = -162$, $b_{1,1} = 2$ and $b_{2,1} = 83$.

It was then noted [41] that this process is in fact ubiquitous and connects all CICYs to each other.

Consider first the simple generalization of the process that takes us from (3.9) to (3.10) and then to (3.11), or rather the respective families of CICYs as represented by their configurations

$$\begin{bmatrix} X \\ 1 \end{bmatrix}\begin{Vmatrix} a & b & M \\ 1 & 1 & 0 \end{Vmatrix} \leftrightarrow [X \| a + b\ M] \, . \tag{3.12}$$

Here we employ a block notation with X and M as in Eq. (3.3) and a and b are column vectors. In view of Eq. (3.7), if a, b and M are chosen so that the l.h.s. represents a 3-dimensional Calabi-Yau manifold, so does the r.h.s. The relationship denoted by "$\leftrightarrow$" is that the two families of (possibly singular) varieties have common members, the conifolds. This becomes clear on writing out the equations that involve the $\mathbb{P}^1$ and then eliminating its coordinates by means of a determinant.

There is a generalization of this relation [25,41]

$$\left[\begin{matrix} X \\ n \end{matrix}\ \middle\|\ \begin{matrix} a_1 & a_2 & \dots & a_{n+1} & M \\ 1 & 1 & \dots & 1 & 0 \end{matrix}\right] \leftrightarrow [X \| \sum_{k=1}^{n+1} a_k \ , \ M] \, . \tag{3.13}$$

The matrix on the left has a $\mathbb{P}^n$ row which consists entirely of 1's. These equations are $n+1$ linear equations in the $n+1$ homogeneous coordinates of the $\mathbb{P}^n$. This leads to a determinantal equation, of degree $\sum_{k=1}^{n+1} a_k$ which, taken together with the other polynomials, describes a conifold which may be smoothed to give a manifold belonging to the configuration on the right. We refer to the this as *determinantal contraction*, while passing from the r.h.s. configuration to the l.h.s. one will be called a *determinantal splitting*. In the following we shall also use the terms contraction and splitting for the replacement of one moduli space by another.

It is theorem [41], quite simple prove, that all configurations representing CICYs are connected to one another by iteration of the relation 3.13. For any CICY, represented by some configuration matrix $[X\|M]$ as in Rel. (3.3), perform determinantal splittings iteratively, introducing only $\mathbb{P}^1$'s into the embedding space, in a way that finally leaves each constraint at most linear in at most one $\mathbb{P}_r^{n_r}$ factor in X with $n_r \neq 1$. Perform next determinantal contractions of all $\mathbb{P}_r^{n_r}$'s, $n_r \neq 1$. This leaves a CICY which is embedded in a product of only $\mathbb{P}^1$'s. Finish with all possible determinantal contractions of $\mathbb{P}^1$'s :

$\left[\begin{matrix} 1 \\ 1 \\ 1 \\ 1 \end{matrix}\middle\|\begin{matrix} 2 \\ 2 \\ 2 \\ 2 \end{matrix}\right]$ is the result of the above algorithm, regardless of the initial CICY. Thus, every CICY can be deformed continuously (allowing only for occasional occurrence of nodes) into this one. Hence all CICYs may be deformed continuously into each other. In other words, all the families of CICYs have certain conifolds as common members through which this <u>web</u> of CICYs is connected.

3.3 Double Covers of $\mathbb{P}^3$ and Generalizations

Many other Calabi-Yau manifolds are connected in a way similar to the CICYs. Consider the octic hypersurface in a weighted complex projective space [64][17]

$$z_0^2 + \sum_{k=1}^{4} z_k^8 = 0 \tag{3.14}$$

where with (z_0, z_k) the quasi-homogeneous coordinates subject to

$$(z_0, z_k) \cong (\lambda^4 z_0, \lambda z_k)\ , \quad \lambda \neq 0\ , \qquad (z_0, z_k) \in \mathbb{P}^4_{(4,1,1,1,1)} \tag{3.15}$$

The space of solutions to Eq. (3.14) is smooth and has zero first Chern class as can be seen from the formula (3.5), suitably generalized for weighted complex projective spaces

$$c(\mathcal{V}) \;=\; \frac{\prod_{r=1}^{m}\prod_{\mu_r=0}^{n_r}(1 + w_{\mu_r}\eta^r)}{\prod_{a=1}^{K}(1 + \sum_{r=1}^{m} q_a^r \eta^r)}\ . \tag{3.16}$$

The defining equation (refdoublecover) obviously has deformations and they form the family of octics in $\mathbb{P}^4_{(4,1,1,1,1)}$.

The same space can also be viewed in another way. For each $z_k \in \mathbb{P}^3 \subset \mathbb{P}^4_{(4,1,1,1,1)}$, Eq. (3.14) yields two values for z_0 except when z_k lies in the octic surface $\sum_{k=1}^4 z_k^8 = 0$ for which there is the unique value $z_0 = 0$. The manifold thus defined may be deformed in ways that keep the branching locus smooth and of degree 8, generating the family of double covers of $\mathbb{P}^3$ branched over a smooth octic surface which we denote by $[\![3]\!](8)$. Let $\mathcal{M}_1^\flat$ be a member of this latter family. (It may well be that the family of such double covers of $\mathbb{P}^3$ is identical to the family of octics in $\mathbb{P}^4_{(4,1,1,1,1)}$ but is irrelevant here.)

The Euler characteristic of $\mathcal{M}_1^\flat$ is

$$\chi_E\big([\![3]\!](8)\big) = 2\cdot\chi_E(\mathbb{P}^3) - \chi_E\big([3\|8]\big) = 2\cdot 4 - 304 = -296,$$

where $\chi_E\big([3\|8]\big) = 304$ is found using standard techniques (see, e.g., Ref. [25,10]). Also, $b_{1,1}(\mathcal{M}_1^\flat) = 1$ [65], so that $b_{2,1}(\mathcal{M}_1^\flat) = 149$.

Consider now a seemingly unrelated CICY

$$\check{\mathcal{M}}_1 \in \begin{bmatrix} 3 \\ 1 \end{bmatrix}\!\!\begin{bmatrix} 4 \\ 2 \end{bmatrix}, \quad Q_{ij}(x)\,y^i y^j = 0, \quad i,j = 0,1, \tag{3.17}$$

[17] Here we follow the analysis of A. Strominger and E. Witten [31].

where the $Q_{ij}(x) = Q_{ji}(x)$ are quartic in x, the homogeneous coordinates of $\mathbb{P}^3$. We have $\chi_E(\check{\mathcal{M}}_1) = -168$, $b_{1,1}(\check{\mathcal{M}}_1) = 2$ and $b_{2,1}(\check{\mathcal{M}}_1) = 86$. Clearly, this manifold is different from $[3\|(8)$.

Let $\mathcal{C}(x) = Q_{ij}(x)Q_{kl}(x)\,\epsilon^{ik}\epsilon^{jl}$ denote the discriminantal octic polynomial. For x in $\mathbb{P}^3$, the space of solutions to (3.17) consists of

1. two points in $\mathbb{P}^1$ when $\mathcal{C}(x) \neq 0$;
2. one point in $\mathbb{P}^1$ when $\mathcal{C}(x) = 0$, but at least one of $Q_{ij}(x)$ does not vanish;
3. an entire copy of $\mathbb{P}^1$ when $Q_{ij}(x) = 0$.

$\check{\mathcal{M}}_1$ is therefore a double cover of $\mathbb{P}^3$ (condition 1.), branched over the octic surface $B \subset \mathbb{P}^3$ defined by $\mathcal{C}(x) = 0$ (condition 2.), with the exception of singular points of this octic surface (condition 3.), at each of which $\check{\mathcal{M}}_1$ is smooth and contains a $\mathbb{P}^1$.

Projecting along $\mathbb{P}^1$ (or, heuristically, letting its radius shrink to zero), we obtain $\mathcal{M}^\sharp{}_1$. This space is identical to $\check{\mathcal{M}}_1$ except it has nodes where $\check{\mathcal{M}}_1$ has the copies of $\mathbb{P}^1$ located by condition 3. There both $\mathcal{C}(x)$ and $d\mathcal{C}(x) = 2[dQ_{ij}(x)]Q_{kl}(x)\,\epsilon^{ik}\epsilon^{jl}$ vanish, which is bound to happen since three generic quartics, Q_{ij}, intersect at $4 \cdot 4 \cdot 4 = 64$ points in $\mathbb{P}^3$.

We next deform $\mathcal{M}^\sharp{}_1$ through deforming the equation which defines the octic branching surface :

$$\mathcal{C}(x) + \varepsilon(x) = 0, \tag{3.18}$$

where $\varepsilon(x)$ is a generic octic perturbation to $\mathcal{C}(x) = 0$. Obviously, this brings us into $[3](8)$.

The family of double coverings of $\mathbb{P}^3$ with an octic branching set (overlapping and possibly identical to the family of octics in the $\mathbb{P}^4_{(4,1,1,1,1)}$ weighted projective space) and the family of CICY hypersurfaces in $\mathbb{P}^3 \times \mathbb{P}^1$ have the conifold $\mathcal{M}^\sharp{}_1$ as a common member.

This simple *discriminantal contraction* may be generalized by considering double covers of more general three-dimensional manifolds [41].

$$\check{\mathcal{M}} \in \begin{bmatrix} \boldsymbol{X} & \| & q & \ell_2 & \cdots & \ell_n & \boldsymbol{M} \\ n & \| & 2 & 1 & \cdots & 1 & 0 \end{bmatrix} \leftrightarrow [\boldsymbol{X}\|\boldsymbol{M}](2q + 2\Sigma\ell_i) \ni \mathcal{M}^\flat \tag{3.19}$$

means that a family of CICYs (l.h.s) and the family of double covers of AFCIs belonging to $[\boldsymbol{X}\|\boldsymbol{M}]$, branched over (smooth or possibly nodal) degree-$2d$ surfaces $B^\flat$ have at least one common member. This is a double cover of an AFCI that belongs to $[\boldsymbol{X}\|\boldsymbol{M}]$, branched over a nodal $B^\sharp$ of degree $2q$ or, alternatively, a CICY that belongs to the configuration on the l.h.s., in which a number of $\mathbb{P}^1$s have been squashed to a node. While $B^\sharp$ is defined by the discriminantal equation

$$Q_{i_0 j_0} Q_{i_1 j_1} L^2_{i_2} L^2_{j_2} \cdots L^n_{i_n} L^n_{j_n} \varepsilon^{i_0 \cdots i_n} \varepsilon^{j_0 \cdots j_n} = 0 \tag{3.20}$$

and has nodes, $B^\flat$ is defined by a generically perturbed equation and is smooth.

Complete intersections in products of weighted projective spaces seem to have become fairly popular in the recent physics literature [26,47,48,49,54]. In fact, a thorough computer search lists about 6000 examples only in weighted $\mathbb{P}^4_{(a,b,c,d,e)}$, which moreover exhibit a fascinating $b_{1,1} \leftrightarrow b_{2,1}$ symmetry [26][♯8]. A suitable generalization of the process described here may connect the families of all these Calabi-Yau manifolds in much the same manner. (It does seem, though, that singularizations worse than nodes need to be allowed.)

3.4 Fibred Products Joining the Web

We have so far constructed Calabi-Yau manifolds mostly by embedding. Instead of that, Chad Schoen discusses Ref. [61] fibre products of elliptic surfaces along a $\mathbb{P}^1$. Each elliptic surface is realized as a cubic hypersurface in $\mathbb{P}^2$. The fibre products of two such elliptic surfaces are then related to the conifold $\mathcal{M}^\sharp{}_S \subset \mathbb{P}^2_x \times \mathbb{P}^2_y$, defined by

$$\mathcal{M}^\sharp{}_S \ : \quad A_1(x) B_2(y) - A_2(x) B_1(y) = 0, \tag{3.21}$$

where $A_i(x)$ and $B_i(y)$ are some cubic polynomials in the homogeneous coordinates $x \in \mathbb{P}^2_x$ and $y \in \mathbb{P}^2_y$. Following our analysis above, we view $\mathcal{M}^\sharp{}_S$ as a nodal deformation of a generic (hence smooth) bi-cubic hypersurface $\mathcal{M}^\flat{}_S \in \left[\begin{smallmatrix} 2 \\ 2 \end{smallmatrix} \middle\| \begin{smallmatrix} 3 \\ 3 \end{smallmatrix}\right]^{b_{11}=2}_{\chi_E=-162}$.

The generic member of Schoen's family of fibred products is constructed from smooth elliptic surfaces (tori). It is a smooth manifold, $\check{\mathcal{M}}_{S(0)}$, and is obtained from $\mathcal{M}^\sharp{}_S$ by a small resolution of the 81 nodes (where $A_{ij}(x_{(i)}) = 0$), a form of which is by the above analysis embeddable as

$$\check{\mathcal{M}}_{S(0)} \in \left[\begin{matrix} 2 \\ 2 \\ 1 \end{matrix} \middle\| \begin{matrix} 3 & 0 \\ 0 & 3 \\ 1 & 1 \end{matrix}\right]^{b_{11}=19}_{\chi_E=0} , \tag{3.22}$$

with defining equations

$$A_i(x)\, z^i = 0 \ , \quad B_i(y)\, z^i = 0 \ , \tag{3.23}$$

with $z \in \mathbb{P}^1$ so that $\check{\mathcal{M}}_{S(0)}$ is in fact a member of the CICYs discussed in the beginning of this section.

[♯8]See also Brian Greene's contribution to this proceedings.

Schoen however considers the various possible nodal forms of $\check{\mathcal{M}}_{S(0)}$ and classifies them in terms of an existing classification of singular elliptic surfaces. The nodes of nearly all of these conifolds can be repaired by small resolution so as to lead to other Kähler manifolds (see Lemma 3.1 of Ref. [61]). A list of Calabi-Yau manifolds, with many values $0 < \chi_E \leq +168$, is thus obtained as a list of small resolutions of various nodal deformations of $\check{\mathcal{M}}_{S(0)}$ of $\chi_E = 0$, which is a common member of the family of CICYs and Schoen's family of fibred products[#9].

Additional examples of similar kind are provided by F.Hirzebruch and J.Werner in Ref. [66] some of which also reach Calabi-Yau manifolds with positive Euler characteristics.

4. ALGEBRAIC σ-GEOMETRY

Threading strings through algebraic geometry

We now concentrate on the $S_{CY}^{(2,2)}$ part of the world sheet action that describes the Calabi-Yau sector and possesses a *rigid* (2,2) supersymmetry. A (1,0) subsector of this is actually local as required by the dynamics of the heterotic string but we shall not attempt to make it manifest from now on.

4.1 CICY σ-Model Actions

Most of the algebraic constructions of Calabi-Yau manifolds begin by taking a larger space X and then embedding the Calabi-Yau manifold into X. In the case of so-called complete intersection Calabi-Yau (CICY) manifolds [10,60], X is taken to be some product $\mathbb{P}_1^{n_1} \times \cdots \times \mathbb{P}_m^{n_m}$ of complex projective spaces. The embedding is then defined by a system of homogeneous algebraic equations to which our Calabi-Yau manifold is the space of simultaneous solutions.

These constructions can straightforwardly be translated into σ-model language [49,67]. To begin with, one takes a sum of actions, one for every $\mathbb{P}^n$ factor in X. There are several candidate Lagrangians, known as $\mathbb{C}\mathrm{P}^n$-models, all of them equivalent classically.

For each $\mathbb{P}_r^{n_r}$ factor in X, we introduce a set of (2,2) superfields $\mathcal{Z}_{(r)}^{\mu_r}$, with $\mu_r = 0, \ldots, n_r$. The constant modes of the bosonic components $Z_{(r)}^{\mu_r}$ are identified as the homogeneous coordinates of $\mathbb{P}_r^{n_r}$. We can choose the

$$S_{FS}^{(r)} = \int_\Sigma d^2\sigma\, d^2\varsigma\, d^2\bar{\varsigma}\; \mathbf{K}_{FS}^{(r)}(\mathcal{Z},\bar{\mathcal{Z}})\ , \qquad \mathbf{K}_{FS}^{(r)}(\mathcal{Z},\bar{\mathcal{Z}}) = \log\Big(\sum_{\mu_r=0}^{n_r} \big|\mathcal{Z}_{(r)}^{\mu_r}\big|^2\Big) \tag{4.1}$$

[#9] I am indebted to Sheldon Katz for explaining this to me.

action for each $\mathbb{P}_r^{n_r}$ model and the

$$S_{kin.} = \sum_{r=1}^{m} \int_\Sigma d^2\sigma\, d^2\varsigma\, d^2\bar{\varsigma}\; w^r \mathbf{K}_{FS}^{(r)}(Z,\bar{Z}) \tag{4.2}$$

sum for $\mathcal{W}$. The coupling constants w^r are non-vanishing in general and their imaginary parts may be thought of as the radii of the $\mathbb{P}_r^{n_r}$'s with respect to the Fubini-Study metrics induced by Eq. (4.2).

An important property of this action is the fact that the Kähler metric, written here for a generic $\mathbb{P}^n$ factor in $\mathcal{W}$,

$$\mathbf{G}_{\mu\bar{\nu}} \stackrel{\text{def}}{=} \mathbf{K}_{,\mu\bar{\nu}} = \frac{\delta^{\mu\bar{\nu}}}{(Z.\bar{Z})} - \frac{Z^\mu Z^{\bar{\nu}}}{(Z.\bar{Z})^2}\,, \qquad Z.\bar{Z} \stackrel{\text{def}}{=} \sum_{\mu=0}^{n} |Z^\mu|^2 \tag{4.3}$$

occurs explicitly in the action upon integration over $\varsigma, \bar{\varsigma}$. With this metric, the kinetic terms (4.1) are invariant under local complex scale invariance

$$Z^{\mu_r} \mapsto \Lambda_r\big(Z^{\mu_r}\big)\,, \tag{4.4}$$

since the Z^{μ_r} correspond to the homogeneous coordinates on $\mathbb{P}_r^{n_r}$ and they are not all independent. One possibility is to choose a gauge, such as fixing one of $Z^{\mu_r} = 1$ for each r, which corresponds to choosing a coordinate patch (and then summing properly over all possible such patches). Assuming this has been done, the quadratic form

$$i\mathbf{K}^{(r)}_{,\mu\bar{\nu}}\, dz^\mu \wedge d^{\bar{\nu}} \tag{4.5}$$

is positive and in fact the standard Kähler (1,1)-form on $\mathbb{P}_r^{n_r}$. It generates the (1,1)-cohomology of $\mathbb{P}_r^{N_r}$.

Except for a few cases which can be modified to avoid the caveat (see the first article in Ref. [33]), these forms are always inherited by the Calabi-Yau manifold and correspond to massless states (the $\Phi^{A\bar{I}}$ and the Ψ_* in Table 1) in the effective model. In favourable cases moreover, the *entire* (1,1)-cohomology of $\mathcal{M}$ is spanned by such forms whereby all the $\mathbf{27}^*$'s of E_6 and all the Ψ_* moduli fields are represented by these terms in the action. With explicit terms in the action, we can set up a standard field theoretic computation of n-point Green's functions to obtain corrections to point field limit results for observables such as Yukawa couplings ($n = 3$). As these depend mainly on cohomological data on the Calabi-Yau manifold rather then an explicit choice of the metric, the action (4.8) appears therefore just tailored for this task.

Instead of $S_{FS}^{(r)}$, we may choose

$$S_{lin}^{(r)} = \int_\Sigma d^2\sigma\, d^2\varsigma\, d^2\bar{\varsigma}\, \Big(\sum_{\mu_r=0}^{n_r} \big|Z^{\mu_r}\big|^2 e^{\boldsymbol{A}_{(r)}} - c\boldsymbol{A}_{(r)} \Big) \tag{4.6}$$

for each $\mathbb{P}_r^{n_r}$ factor in $\mathcal{W}$, and the

$$S_{lin.} = \sum_{r=1}^{m} w^r S_{lin.}^{(r)} \tag{4.7}$$

for the entire $\mathcal{W}$.

In the realm of quantum theory, the two actions, (4.1) and (4.6) have somewhat different properties and the choice will clearly be determined by the particular computation for which one intends to use it. For example, all the superfields in $S_{lin.}$ are unconstrained, so that the measure in the path integral for the partition function is just the straightforward one; the $U(1)$ gauge superfield $\boldsymbol{A}$ has no kinetic term on its own and acts as a Lagrange multiplier. Such an action is known to be well suited for e.g. the $1/N$-expansion computations.

We shall here use the Lagrangian (4.2) involving the Fubini-Study metric—for definiteness and proximity to the algebro-geometric analysis, but bear in mind that it may be preferable to use the (4.6) for some explicit computations.

In order to ensure that we are describing the Calabi-Yau submanifold rather then the embedding manifold $\boldsymbol{X}$, a constraint term is added to enforce the polynomial constraints $P_a(Z) = 0$, $a = 1, \ldots, K$, $K = \dim \mathcal{W} - 3$. Since Lagrange's work, it is most natural to employ a set of Lagrange multiplier superfields Λ^a for this task. Thus, in the rigid limit of the world sheet supersymmetry of the heterotic string (see also Ref. [46,49,67]),

$$S_{CICY}^{(2,2)} = S_{kin.} + S_{con.} \tag{4.8}$$

$$S_{con.} = \int_\Sigma d^2\sigma\, d^2\varsigma\, d^2\bar{\varsigma}\, \Big[-i \int_\Sigma d^2\sigma\, d^2\varsigma\, \sum_{a=1}^{K} \boldsymbol{\Lambda}^a\, P_a(\boldsymbol{Z}; t) \;+\; \text{h.c.} \Big] . \tag{4.9}$$

These polynomials are homogeneous and of degrees prescribed by Eq. (3.7) :

$$\sum_{a=1}^{K} \deg_r(P_a) = n_r + 1\ , \tag{3.7$'$}$$

where $\deg_r$ denotes the degree of homogeneity with respect to the homogeneous coordinates of $\mathbb{P}_r^{n_r}$. We really need not appeal to algebraic geometry to derive this. Eq. (3.5), (3.6) and so (3.7) encapsulate a straightforward but tedious computation which is actually physically motivated also. Eq. (3.7′) follows from a straightforward computation of the β-functions which we soon discuss.

The partition function is of course

$$\mathcal{Z}_{CICY} \stackrel{\text{def}}{=} \int D[\boldsymbol{Z}]\, D[\overline{\boldsymbol{Z}}]\, D[\boldsymbol{\Lambda}]\, D[\overline{\boldsymbol{\Lambda}}]\; e^{-S_{CICY}^{(2,2)}}\ , \tag{4.10}$$

from which a first-quantized description of the Calabi-Yau sector of the compactified string model should be no sweat to formulate (see below). We have, for definiteness, assumed a Euclidean formulation of the world sheet actions, whence the minus in the exponents in (4.10).

4.2 Quantum behaviour

The quantum field theory described by the action (4.8) undergos renormalization, changing the form of the chosen function $\sum_{r=1}^{m} w^r \mathbf{K}^{(r)}(Z,\bar{Z})$ ($S_{con.}$ does not get renormalized because of supersymmetry). To measure this change, one computes the corresponding β-functions. For a generic Calabi-Yau σ-model

$$S_{CY}^{(2,2)} = \int_{\Sigma} d^2\sigma\, d^2\varsigma\, d^2\bar{\varsigma}\; \mathbf{K}_{CY}(Z,\bar{Z}) \;, \tag{4.11}$$

this was done in Ref. [68]. As also shown in general so-called background field expansion [69], up to four loops, the β-function is proportional to the Ricci tensor computed from $\mathbf{K}$. For $S_{CICY}^{(2,2)}$ in (4.8), it is again the vanishing of the Ricci tensor, except that it now has to be computed from the Fubini-Study metric on the embedding space $\mathcal{W}$ which is then properly restricted to the Calabi-Yau submanifold.

This straightforward computation is however quite involved and lengthy. One first computes the Riemann tensor, which occurs in the 4-fermion term of the component-field expansion of $S_{CICY}^{(2,2)}$. We note here that the Ricci tensor $R_{\mu\bar{\nu}}$ on $\mathbb{P}_r^{n_r}$, computed from the Fubini-Study metric $\mathbf{K}^{(r)}_{,\mu\bar{\nu}}$ is

$$R^{(r)}_{\mu\bar{\nu}} = -(n_r+1)\mathbf{K}^{(r)}_{,\mu\bar{\nu}} \;, \tag{4.12}$$

for each $\mathbb{P}_r^{n_r}$ factor in $\mathcal{W}$, and so will be the restrictions to the submanifold $\mathcal{M}$. The Riemann tensor (and so also the Ricci tensor) however, also picks up corrections from derivatives of the constraint polynomials once the Lagrange multiplier has been integrated out. These corrections turn out to yield a contribution to the Ricci tensor that is, up to total derivative terms, again proportional to the restriction of the Fubini-Study metric (see Ref. [70], p.368), so that

$$R^{(r)}_{\mu\bar{\nu}}\Big|_{\mathcal{M}} = \Big(\sum_{a=1}^{K} \deg_r(P_a) \;-\; (n_r+1)\Big)\,\mathbf{K}^{(r)}_{,\mu\bar{\nu}} \;, \qquad \forall r \;. \tag{4.13}$$

Since total derivative terms can always be absorbed in a redefinition of the metric through adding local counterterms, we obtain the condition (3.7) for the Ricci tensor on $\mathcal{M}$ and the β-functions to vanish.

Once a Ricci-flat metric is obtained, the vanishing of the β-function can be ensured through further modification of the metric, order by order in perturbation theory [13,71]. Understood in this way, Eq. (3.7′) becomes the condition for the σ-model action $S_{CICY}^{(2,2)}$ in Eq. (4.8) to describe the Calabi-Yau sector of a consistent string model.

Under this renormalization flow, the system (4.8) should be driven to the conformally invariant fixed point whereby the physical metric is reached and replaces the (restriction of the) sum of Fubini–Study metrics. The process envisioned here is similar to that analyzed in Ref. [48,72] but is much harder to deal with.

The difficulties stem (1) from the use of the highly non-trivial kinetic term with the Fubini-Study rather then the flat metric and (2) from the use of Lagrange multiplier fields. We now discuss these in turn.

The kinetic terms In Ref. [48,72], the authors argue that the $\int_\Sigma d^2\sigma\, d^2\varsigma\, d^2\bar\varsigma$ term in $S^{(2,2)}_{CICY}$, with either $S^{(r)}_{lin.}$ or $S^{(r)}_{FS}$, is irrelevant for the renormalization group flow. The argument is based on the observation that in

$$\int_\Sigma d^2\sigma\, d^2\varsigma\, d^2\bar\varsigma\, \mathcal{O}\ , \tag{4.14}$$

whatever we choose $\mathcal{O}$ to be, it must be that its canonical dimension vanishes. Quite generally, the anomalous dimension of $\mathcal{O}$ will then be positive (in units of mass) so that counterterms will enter suppressed by factors of the inverse renormalization scale and can therefore be neglected.

This, however, may not be true of the terms $S^{(r)}_{FS}$ (4.1) because of the use of the Fubini-Study metric (4.3). It renders (4.1) not only locally scale-invariant, which is also true of the action (4.6), but also homogeneous of homogeneity zero. Because of this rather curious property, the classical canonical dimension of Z^μ remains undetermined when using (4.1). In other words, $S^{(r)}_{FS}$ are not usual quantum field theory operators and they need not follow the crowd as regards renormalization. In fact, because of the complex scaling gauge invariance, they are ill defined without choosing a gauge (coordinate patch). Gauge fixing introduces ghosts in turn, which cancel out the contribution of the unphysical Z^{μ_r}s and it would be a little surprise if none of this affected the β-functions at all.

Now, similar surprises have been encountered before but this time we have some circumstantial evidence that $S^{(r)}_{FS}$ do affect the β-functions. Recall that the (1,1)-forms (4.5) span (part of) the (1,1)-cohomology of the CICY. Elements of this cohomology correspond to the massless Ψ_* moduli fields. These correspond to exactly marginal operators in the underlying 1+1-dimensional theory (see e.g. [73]).

It transpires therefore that the operators in (4.1) do get renormalized to become a set of marginal operators. If all of the (1,1)-cohomology of $\mathcal{M}$ is inherited from the embedding space X, we have a complete set of $b_{1,1}$ marginal operators corresponding to the Ψ_* moduli fields which allow to 'deform' the background, parametrized by the coefficients w^r in $S_{kin.}$ (4.2) or $S_{lin.}$ (4.7). This corresponds to a choice of the Kähler class on the compactifying manifold $\mathcal{M}$ and its possible 'deformations'. Thus, we describe the choice of the Kähler class by a particular choice w_0 and describe variations by some δw^A.

Also, focusing on the right-handed fermions in the Z^{μ_r} which were stolen from the Yang–Mills sector, we have marginal operators corresponding to the $\mathbf{27}^*$massless fields in the 3+1-dimensional effective model. These occur explicitly in the action $S^{(2,2)}_{CICY}$ and allow straightforward n-point function, i.e., Yukawa coupling computations.

The constraint terms We take $P_a(Z;t)$ to be the most general polynomial, i.e., a linear combination of all possible monomials of the proper degrees of homogeneity, where the coefficients in this polynomial are denoted by t^α. Choosing a particular value for each t, i.e., a particular polynomial corresponds to a specific choice of the complex structure on $\mathcal{M}$.

The choice of a slightly different assembly of coefficients, $t^\alpha + \delta t^\alpha$ will lead to a slightly different set of constraints and therefore to a σ-model with a slightly different choice of the complex structure for the Calabi-Yau target space. Since $P_a(Z;t)$ is linear in t^α, we may rewrite

$$P_a(Z;t_0+\delta t) \;=\; P_a(Z;t_0) \;+\; \delta t^\alpha M_{\alpha a}(Z)\;, \tag{4.15}$$

where $M_{\alpha a}$ is a basis of independent polynomials of the same degree as $P_a(Z;t)$.

As the superpotential does not get renormalized (by world-sheet supersymmetry), both the particular operator $P_a(Z;t_0)$ and the set of perturbation operators $\delta t^\alpha M_{\alpha a}(Z)$ are marginal. Thus, we describe the choice of the complex structure by a particular choice t_0 and describe variations by some δt^α.

When polynomial deformations are effective (for details see Ref. [27]), the monomials in the most general $\delta P_a(Z;t)$, when taken modulo certain reparametrizations, will correspond to harmonic (2,1)-forms, which in turn correspond to massless $\mathbf{27}$s in the 3+1-dimensional effective model. Again, these operators are manifestly present in $S^{(2,2)}_{CICY}$. Of course, it would be quite interesting to find out how are defects of the polynomial deformation method manifested, i.e., to identify the operators which remain hidden in this picture.

In Ref. [48], the authors write (constant) coupling constants instead of Lagrange multipliers in $S_{con.}$ (4.9). Intuitively, this means that only the ground state is constrained to live in $\mathcal{M}$, while perturbations live in the embedding space X, which makes it questionable if this treatment leads to a superconformal quantum theory. Granting that the renormalization flow will nevertheless 'find' the superconformal fixed point as the analysis of Ref. [48] indicates, this can be understood as an approximation of $S^{(2,2)}_{CICY}$, where the Lagrange multiplier superfields have been expanded

$$\Lambda^a \rightarrow g^a + \lambda^a \ , \tag{4.16}$$

where g^a are the coupling constants and 'λ^a' represent the ignored quantum fluctuations of the Lagrange multiplier superfields.

Turning this around, we may take the results of Ref. [48] as the tree-level results for the $S_{con.}$ part of the $S^{(2,2)}_{CICY}$ action. Corrections are to be obtained through a systematic inclusion of the λ^a fields. It is quite dubious if such a (weak coupling) expansion is sensible at all and I should rather advise a $1/N$ expansion treatment of $S^{(2,2)}_{CICY}$.

5. MODUS MODULI

We have seen so far that there are certain parameters on which the compactification on Calabi-Yau manifolds depends. In particular, the σ-model action $S^{(2,2)}_{CICY}$ makes it obvious that there are two sets of such parameters.

1. Parameters that were denoted by w^r and occur in $S_{kin.}$ (4.2). These correspond to the choice of the Kähler class on $\mathcal{M}$, i.e., the corresponding background kinetic term for $S^{(2,2)}_{CICY}$. Variations of this choice, parametrized by δw, introduce marginal perturbations (by the argument of the previous section) of this background action and correspond to massless particles in the effective 3+1-dimensional model, which in turn correspond to harmonic (1,1)-forms on $\mathcal{M}$[♯10].

2. Parameters that were denoted by t^α and occur in $S_{con.}$ (4.9). These correspond to the choice of the complex structure on $\mathcal{M}$, i.e., the corresponding background constraint term for $S^{(2,2)}_{CICY}$. Variations of this choice, parametrized by δt, introduce marginal perturbations (by the argument of the previous section) of this background action and correspond to massless particles in the effective 3+1-dimensional model, which in turn correspond to harmonic (2,1)-forms on $\mathcal{M}$[♯11].

Noting the symmetry between the above two paragraphs, we are reminded of the remarkable and still somewhat mysterious symmetry between the rôles of these to sectors of our models and are tempted to pry further into this Pandora's box (largely borrowing from Ref. [55]), stumbling again into the same symmetry.

[♯10] Lots of correspondences to keep track, but also to make use of!

[♯11] Lots of correspondences to keep track, but also to make use of!

5.1 Moduli space of a Calabi-Yau manifold

We recall once again that a Calabi-Yau compactification relies on specifying a Calabi-Yau threefold $\mathcal{M}$, together with a Ricci-flat Kähler metric, which is uniquely determined by its Kähler class, J. $\mathcal{M}$ admits a holomorphic three-form Ω which is determined up to a multiplicative constant. We are interested in assigning a metric to the space which is obtained by letting $\mathcal{M}$ vary within its deformation class while J ranges over the positive real (1,1)-cohomology of $\mathcal{M}$. This metric respects the continuity of the volume, the holomorphic 3-form and the Kähler form all of which are essential from both the mathematical and the physical point of view.

We begin by presenting physical motivation for our choice of the metric on the moduli space.

Effective Kinetic Terms In the 3+1-dimensional low-energy effective theory there are $b_{2,1}$ massless families $\mathbf{27}^{\alpha}$ of Standard Model particles accompanied by gauge-singlet "moduli superfields" Ψ and also $b_{1,1}$ massless families $\mathbf{27}^{*A}$ of mirror-particles likewise accompanied by gauge-singlet "moduli superfields" Ψ_*.The effective Lagrangian density is obtained as in Eq. (1.30) for which the triple couplings were given in Eqs. (1.26–1.29).

The point fields limit expressions for the kinetic terms are [34]

$$N_{\alpha\bar{\beta}} = \int_{\mathcal{M}} \varphi_\alpha \wedge \overline{\varphi_\beta} \,, \tag{5.1}$$

$$N_{AB} = -\frac{3}{2}\left\{\frac{\int_{\mathcal{M}} e_A \wedge e_B \wedge J}{\int_{\mathcal{M}} J^3} - \frac{3}{2}\frac{\int_{\mathcal{M}} e_A \wedge J^2 \int_{\mathcal{M}} e_B \wedge J^2}{\int_{\mathcal{M}} J^3 \int_{\mathcal{M}} J^3}\right\} . \tag{5.2}$$

we have here denoted the Kähler class on $\mathcal{M}$ by J. At this moment, the attentive reader must have deemed at least one of the above two formulae wrong—they are not even similar. While indeed not similar, they are in fact equal !!! To see that, one must train one's eye better.

The Natural Line Element We may take the moduli space to be the space of Ricci-flat metrics of a deformation class of Calabi-Yau manifolds. Let g_{mn} be a Ricci-flat metric for $\mathcal{M}$ and δg_{mn} its variation. With the coordinate condition $\nabla^n \delta g_{mn} = 0$, δg_{mn} satisfies

$$\left(\delta^p_m\, \delta^n_q \nabla^k \nabla_k + 2R_m{}^p{}_n{}^q\right)\delta g_{pq} = 0 \,. \tag{5.3}$$

The zero modes of this equation, $\delta g_{\mu\bar{\nu}}$ and $\delta g_{\mu\nu}, \delta g_{\bar{\mu}\bar{\nu}}$, separately satisfy the equation.

We define a (1,1)-form

$$\omega \stackrel{\text{def}}{=} i\delta g_{\mu\bar{\nu}}\, dx^\mu \wedge dx^{\bar{\nu}} \tag{5.4}$$

which is harmonic if and only if $\delta g_{\mu\bar\nu}$ satisfies (5.3). Also,

$$\varphi \stackrel{\text{def}}{=} \tfrac{1}{2}\Omega_{\kappa\lambda}{}^{\bar\nu}\delta g_{\bar\mu\bar\nu}\, dx^\kappa \wedge dx^\lambda \wedge dx^{\bar\mu} \tag{5.5}$$

is a harmonic (2,1)-form if and only if $\delta g_{\bar\mu\bar\nu}$ satisfies (5.3).

Along with $g_{\mu\bar\nu}$ and its variations, in compactification of superstrings there occurs a two-form B on $\mathcal{M}$ which is related to the metric by supersymmetry (recall Eq. (1.1)). This suggests a complexification of the Kähler cone through inclusion of B. By the field equations, δB must be harmonic and since $b_{20} = 0$ on a Calabi-Yau manifold, δB is of type $(1,1)$.

The most general[112] line element in the space of these variations is

$$\begin{aligned} ds^2 &= \frac{1}{V}\int_{\mathcal{M}} g^{km}g^{ln}(\delta g_{kl}\delta g_{mn} + \delta B_{kl}\delta B_{mn})\, g^{\frac{1}{2}} d^6x \ , \\ &= \frac{2}{V}\int_{\mathcal{M}} g^{\kappa\bar\mu}g^{\lambda\bar\nu}\left(\delta g_{\kappa\lambda}\delta g_{\bar\mu\bar\nu} + (\delta g_{\kappa\bar\nu}\delta g_{\lambda\bar\mu} + \delta B_{\kappa\bar\nu}\delta B_{\bar\mu\lambda})\right) g^{\frac{1}{2}} d^6x \ . \end{aligned} \tag{5.6}$$

It is, somewhat surprisingly, block-diagonal, with a separate block corresponding to (2,1)-forms, i.e., variations of the complex structure and to (1,1)-forms, i.e., variations of the Kähler class, respectively. We note here that this structure agrees with the result found well beyond the point-field limit, for the case of blown-up toroidal Z_N-orbifolds [43,40].

Knowing there are $b_{2,1}$ harmonic (2,1)-forms, we expand φ and invert Eq. (5.5)

$$\delta g_{\bar\mu\bar\nu} = \frac{1}{2\|\Omega\|^2}\overline{\Omega}_{\bar\mu}{}^{\rho\sigma}\varphi_{\alpha\,\rho\sigma\bar\nu}\,\delta t^\alpha \ , \tag{5.7}$$

where $\|\Omega\|^2 \stackrel{\text{def}}{=} \frac{1}{3!}\Omega_{\mu\nu\rho}\overline{\Omega}^{\mu\nu\rho}$ and is constant on $\mathcal{M}$. A little algebra reveals that

$$\frac{2}{V}\int_{\mathcal{M}} g^{\kappa\bar\nu}g^{\mu\bar\lambda}\delta g_{\kappa\mu}\delta g_{\bar\lambda\bar\nu}\, g^{\frac{1}{2}}d^6x = \frac{i}{4V\|\Omega\|^2}\delta t^\alpha \delta t^\beta \int_{\mathcal{M}} \varphi_\alpha \wedge \bar\varphi_{\bar\beta} \stackrel{\text{def}}{=} 2\mathrm{G}_{\alpha\bar\beta}\delta t^\alpha \delta t^\beta \tag{5.8}$$

with

$$\mathrm{G}_{\alpha\bar\beta} = \frac{\int_{\mathcal{M}} \varphi_\alpha \wedge \overline{\varphi}_{\bar\beta}}{\int_{\mathcal{M}} \Omega \wedge \overline{\Omega}} \ . \tag{5.9}$$

Note that the metric on the space of complex structures $G_{\alpha\bar\beta}$ coincides with the matrix $N_{\alpha\bar\beta}$ in Eq. (5.1) up to normalization.

The following is a useful observation which is attributed to K. Kodaira (see Ref. [37])

$$\frac{\partial\Omega}{\partial t^\alpha} = k_\alpha\Omega + \varphi_\alpha, \tag{5.10}$$

[112] Well, not quite ... See Ref. [55].

where the k_α may depend on t^α but not on the coordinates of $\mathcal{M}$. Given these relations it follows by straightforward differentiation that

$$\mathbf{G}_{\alpha\bar{\beta}} = -\frac{\partial}{\partial t^\alpha}\frac{\partial}{\partial t^{\bar{\beta}}}\log\left(i\int_{\mathcal{M}}\Omega\wedge\overline{\Omega}\right) \tag{5.11}$$

so the space of complex structures, $\mathcal{B}$, is Kähler and $\log\left(i\int_{\mathcal{M}}\Omega\wedge\overline{\Omega}\right)$ is the Kähler potential.

5.2 The $(1,1)$–forms

The inner product corresponding to Eq. (5.6) on $H^{(1,1)}$ is

$$\mathbf{G}_{A\overline{B}} = \frac{1}{V}\int_{\mathcal{M}} e_{A\,\mu\bar{\nu}}e_{B\,\rho\bar{\sigma}}g^{\mu\bar{\sigma}}g^{\rho\bar{\nu}}\,g^{\frac{1}{2}}d^6x \;= \frac{2}{V}\int_{\mathcal{M}} e_A\wedge *e\ ,$$

for real $(1,1)$-forms, e_A and e_B. Note that, owing to the positivity of $g^{\mu\bar{\nu}}$, $\mathbf{G}_{A\overline{B}}$ is positive. Apart from a factor of $1/V$ this is the metric adopted by Strominger [34,42] who observed that $\mathbf{G}_{A\overline{B}}$ may be rewritten entirely in terms of the triple product integral, to obtain Eq. (5.2).

Let the basis $\{e_A\}$ used in Eq. (5.2) be integral and expand

$$B + iJ = w^A e_A\ , \quad w^A = u^A + iv^A\ . \tag{5.13}$$

Since B and J are real they are parametrized by the real and imaginary parts of w^A. It is now straightforward to see that

$$\mathbf{G}_{A\bar{B}} \stackrel{\text{def}}{=} \mathbf{G}(e_A, e_B) = -\frac{\partial}{\partial w^A}\frac{\partial}{\partial w^{\bar{B}}}\log\kappa(J,J,J)\ . \tag{5.14}$$

We have just shown that $\mathcal{W}$, the space of the w^A is also a Kähler manifold. The Kähler potential is the logarithm of standard volume of the Calabi-Yau manifold.

Having rewritten Eqs. (5.1) and (5.2), the symmetry between Eq. (5.11) and (5.14) is now manifest.

Moreover, this can be carried even further Ref. [52,55]. It turns out that there exists a holomorphic function of the parameters of the complex structure, $\xi(t)$, and another holomorphic function of the parameters of the Kähler class, $\kappa(w)$, such that

$$i\int_{\mathcal{M}}\Omega\wedge\overline{\Omega} \;=\; i\Big[2(\xi(\tau)-\overline{\xi}(\overline{\tau})) - (\tau^\alpha - \overline{\tau}^\alpha)\big(\xi_{,\alpha}(\tau)+\xi_{,\alpha}(\overline{\tau})\big)\Big]\ . \tag{5.15}$$

$$\int_{\mathcal{M}} J^3 \;=\; -\tfrac{i}{8}\Big[2(\kappa(w)-\kappa(\overline{w})) - (w^A - \overline{w}^A)(\kappa_{,A}(w)+\kappa_{,A}(\overline{w}))\Big]\ . \tag{5.16}$$

	Complexified Kähler cone	Complex structure
Number of affine coordinates	$b_{11}-1$	b_{21}
Cohomology group	$H^1(\mathcal{M},\mathcal{T}^*_{\mathcal{M}})$	$H^1(\mathcal{M},\mathcal{T}_{\mathcal{M}})$
Analytic functions	$\kappa(w)$	$\xi(z)$
exponential of the Kähler potential[a]	$\int_{\mathcal{M}} J\wedge J\wedge J = 2(\kappa-\overline{\kappa})-(w^A-\overline{w}^A)(\kappa_{,A}+\overline{\kappa}_{,A})$	$i\int_{\mathcal{M}}\Omega\wedge\overline{\Omega} = 2(\xi-\overline{\xi})-(\tau^\alpha-\overline{\tau}^\alpha)(\xi_{,\alpha}+\overline{\xi}_{,\alpha})$
Period matrices	$\Im m\left(\frac{\partial^2\kappa(w)}{\partial w^A\partial w^B}\right)$	$\Im m\left(\frac{\partial^2\xi(z)}{\partial z^\alpha\partial z^\beta}\right)$
Signature of the period matrix	$(b_{11}-1,1)$	$(b_{21},1)$

[a] See Eq. (5.11), (5.15), (5.14) and (5.16) in text.

Table 3: Closely analogous quantities which describe the moduli space

5.3 Calabi-Yau σ-models and the Zamolodchikov Metric

Consider now the action functional on the ("Euclideanized") $1+1$-dimensional world-sheet Σ determining the propagation of strings in an internal space :

$$\begin{aligned} S_{CY} &= -\frac{1}{2\pi\alpha'}\int_\Sigma d^2\sigma\,\Big\{(G_{mn}+iB_{mn})(\overline{\partial}X^m(\sigma))(\partial X^n(\sigma)) \\ &\qquad + (G_{mn}-iB_{mn})(\partial X^m(\sigma))(\overline{\partial}X^n(\sigma))\Big\}\,. \end{aligned}$$

For Calabi-Yau manifolds, it is standard to choose a Kähler metric $G_{\mu\bar\nu}$ and set $B=0$. Alternatively, requiring world sheet (2,2)-supersymmetry sets $B=0$ and makes G_{mn} Kähler. The relevant σ-model describing a Calabi-Yau background action is then

$$S_0 = -\frac{1}{\pi\alpha'}\int_\Sigma d^2\sigma\, G_{\mu\bar\nu}(Z,\overline{Z})\Big[(\overline{\partial}Z^\mu(\sigma))(\partial Z^{\bar\nu}(\sigma)) + (\partial Z^\mu(\sigma))(\overline{\partial}Z^{\bar\nu}(\sigma))\Big]\,. \qquad (5.17)$$

Any choice of such a $G_{\mu\bar\nu}(Z,\overline{Z})$ simply specifies a point in the relevant moduli space. To determine the metric on the moduli space at that point, we shall need to consider deformations[♯13] of this choice and evaluate second order effects. We briefly describe this here; a more complete and careful analysis will be presented elsewhere.

[♯13] It is important to note and indeed is the essence of our analysis that the deformations of the σ-model action (5.17) describing Calabi-Yau vacuua of superstring theories may be parametrized through deformations of the target Calabi-Yau manifold, i.e., deformations of the complex structure and of the Kähler class.

To deform the action (5.17), add

$$S_1 = +\frac{1-i}{2\pi\alpha'}\int_\Sigma \Big[\mathcal{G}_{\mu\bar\nu}\Big((\bar\partial Z^\mu(\sigma))(\partial Z^{\bar\nu}(\sigma)) + i\,(\partial Z^\mu(\sigma))(\bar\partial Z^{\bar\nu}(\sigma))\Big)$$
$$-i\,\overline{\mathcal{G}}_{\mu\bar\nu}\Big((\bar\partial Z^\mu(\sigma))(\partial Z^{\bar\nu}(\sigma)) - i\,(\partial Z^\mu(\sigma))(\bar\partial Z^{\bar\nu}(\sigma))\Big)\Big]\ , \tag{5.18}$$

$$S_2 = -\frac{1}{\pi\alpha'}\int_\Sigma \Big[\mathcal{G}_{\mu\nu}(\bar\partial Z^\mu(\sigma))(\partial Z^\nu(\sigma)) + \mathcal{G}_{\bar\mu\bar\nu}(\partial Z^{\bar\mu}(\sigma))(\bar\partial Z^{\bar\nu}(\sigma))\Big] \tag{5.19}$$

to $\mathcal{A}_0$. $\mathcal{A}_0 + \mathcal{A}_1 + \mathcal{A}_2$ is the most general expression for an action that still has the geometrical meaning of embedding the world sheet into the target (Calabi-Yau) manifold.

The peculiar form of the term (5.18) is better understood if we recall that $J = iG_{\mu\bar\nu}\,dZ^\mu \wedge dZ^{\bar\nu}$ is the Kähler form defined by the Kähler metric $g_{\mu\bar\nu}$. The deformations of the Kähler form, maintaining its positivity, are parametrized by a real forward cone (the Kähler cone) of dimension $b_{1,1}$. Since $B_{\mu\bar\nu}$ is antihermitean, the combination $B+iJ = (B_{\mu\bar\nu} - G_{\mu\bar\nu})dZ^\mu \wedge dZ^{\bar\nu}$ offers a straightforward complexification of the Kähler cone (see also Ref. [42]). Variations of the background parametrized by this complexified Kähler cone take the form $\delta B_{\mu\bar\nu} \pm \delta G_{\mu\bar\nu}$, which is precisely accounted for by $\mathcal{G}_{\mu\bar\nu}$ and $\overline{\mathcal{G}}_{\mu\bar\nu}$, respectively. Massless modes of this form correspond to harmonic (1,1)-forms.

The terms (5.19) can be brought in the form of those in (5.17) only by a non-holomorphic change of variables $Z' = Z'(Z,\bar Z)$ and thus clearly correspond to deformations of the complex structure. Such deformations of a Calabi-Yau manifold are parametrized by (2,1)-forms which in turn correspond to massless modes in $\mathcal{G}_{\bar\mu\bar\nu}$. It transpires that the parameter space relevant to the above σ-model coincides with the combined space of complex structures and the Kähler cone (complexified through the inclusion of δB). In addition, it is precisely these geometrical data on the Calabi-Yau manifold that determine the essential physical observables, such as the Yukawa couplings, in the low-energy limit.

Given a conformally invariant theory, Zamolodchikov's metric [74] is the coefficient of the $|\sigma|^{-4}$ singularity in the irreducible two-point function $\langle\Phi_I(\sigma)\Phi_J(0)\rangle$ where the set $\{\Phi_I(\sigma)\}$ parametrizes the deformations of the background action density.

From the point field limit we know that the deformations induced by inclusion of S_1 and of S_2 admit a finite basis and we introduce external world sheet currents $W^A(\sigma)$, $W^{\bar A}(\sigma)$, $T^\alpha(\sigma)$ and $T^{\bar\alpha}(\sigma)$ to probe for these deformations :

$$\mathcal{G}_{\mu\bar\nu} = W^A(\sigma)\,\mathcal{G}_{A\mu\bar\nu}(Z,\bar Z)\ , \qquad \overline{\mathcal{G}}_{\mu\bar\nu} = W^{\bar A}(\sigma)\,\mathcal{G}_{\bar A\mu\bar\nu}(Z,\bar Z)\ ,$$
$$\mathcal{G}_{\mu\nu} = T^\alpha(\sigma)\,\mathcal{G}_{\alpha\mu\nu}(Z,\bar Z)\ , \qquad \mathcal{G}_{\bar\mu\bar\nu} = T^{\bar\alpha}(\sigma)\,\mathcal{G}_{\bar\alpha\bar\mu\bar\nu}(Z,\bar Z)\ .$$

These currents are Zamolodchikov's $J^I(\sigma)$s and the corresponding operators are the $\Phi_I(Z,\bar Z)$s.

Since we are evaluating the metric at the point in the moduli space specified by S_0, the expectation values are computed from

$$\langle \mathcal{O} \rangle \stackrel{\text{def}}{=} \frac{\int_{\mathcal{M}} g^{\frac{1}{2}} d^6 z \int \left[\prod_\sigma d^6 \tilde{Z}(\sigma)\right] \mathcal{O}\, e^{-S_0}}{\int_{\mathcal{M}} g^{\frac{1}{2}} d^6 z \int \left[\prod_\sigma d^6 \tilde{Z}(\sigma)\right] e^{-\mathcal{A}_0}} , \tag{5.20}$$

where we have indicated a separation

$$Z(\sigma) = z + \tilde{Z}(\sigma) ,$$

into the constant mode and a remainder in a Fourier-type expansion.

To one loop order, i.e., quadratic order in $\tilde{Z}(\sigma)$ and the fermionic fields then we may replace

$$G_{\mu\bar{\nu}}(Z, \bar{Z}) \rightarrow g_{\mu\bar{\nu}}(z, \bar{z})$$

in S_0 and have

$$\langle (\bar{\partial} Z^\mu(\sigma))(\partial Z^{\bar{\nu}}(0)) \rangle = \langle (\partial Z^\mu(\sigma))(\bar{\partial} Z^{\bar{\nu}}(0)) \rangle = -\frac{\pi\alpha'}{|\sigma|^2} g^{\mu\bar{\nu}}(z, \bar{z}) + \dots , \tag{5.21}$$

and all other correlation functions between derivatives of Z's and $\bar{Z}$'s vanish. We shall similarly approximate the functionals $\mathcal{G}$ with their limit that is independent of the non-constant modes $\tilde{Z}^\mu(\sigma)$ and $\tilde{Z}^{\bar{\nu}}(\sigma)$.

The standard definition of the irreducible two-point functions in terms of a second functional derivative of the logarithm of the partition function and the expectation values (5.21), it is immediate that the only non-vanishing entries of the Zamolodchikov metric in this limit are

$$\mathbf{G}_{A\bar{B}} = \frac{1}{V} \int_{\mathcal{M}} g^{\frac{1}{2}} d^6 x \left[\mathcal{G}_{A\,\mu\bar{\nu}}(z, \bar{z}) \mathcal{G}_{\bar{B}}{}^{\bar{\nu}\mu}(z, \bar{z})\right] , \tag{5.22}$$

$$\mathbf{G}_{\alpha\bar{\beta}} = \frac{1}{V} \int_{\mathcal{M}} g^{\frac{1}{2}} d^6 x \left[\mathcal{G}_{\alpha\,\mu\nu}(z, \bar{z}) \mathcal{G}_{\bar{\beta}}{}^{\nu\mu}(z, \bar{z})\right] . \tag{5.23}$$

The Weil–Petersson metric (5.11) and (5.14) obtained in the point field limit is thus the one-loop (lowest non-vanishing) contribution to the Zamolodchikov metric.

6. ANOTHER WORLD IN (LESS THEN) EIGHTY DAYS

The process that was reviewed in Section 3, which provided an interfacing relation between most of the families of Calabi-Yau manifolds and which enabled us to stitch their moduli spaces together into a fine web seems also to signal phase transitions for the effective 3+1-dimensional mode and therefore also for the heterotic string [41].

6.1 Chameleonic σ-models

Let us consider a σ-model action

$$S = \int_\Sigma d^2\sigma\, d^2\varsigma\, d^2\bar{\varsigma}\; w_X \mathbf{K}(\boldsymbol{X},\overline{\boldsymbol{X}}) + \int_\Sigma d^2\sigma\, d^2\varsigma\, d^2\bar{\varsigma}\; w_Y \mathbf{K}(\boldsymbol{Y},\overline{\boldsymbol{Y}}) \tag{6.1}$$

$$+ \Big[\int_\Sigma d^2\sigma\, d^2\varsigma\; \boldsymbol{\Lambda}^1\big(S(\boldsymbol{X})\boldsymbol{Y}^1 + P(\boldsymbol{X})\boldsymbol{Y}^2\big) + \boldsymbol{\Lambda}^2\big(Q(\boldsymbol{X})\boldsymbol{Y}^1 + R(\boldsymbol{X})\boldsymbol{Y}^2\big) + \text{h.c.} \Big] \tag{6.2}$$

for a compactification on a Calabi-Yau manifold defined by the embedding

$$\mathcal{M} \hookrightarrow \boldsymbol{X} \times \mathbb{P}^1 , \qquad \mathcal{M} \in \left[\begin{array}{c\|cc} \boldsymbol{X} & q_1 & q_2 \\ \mathbb{P}^1 & 1 & 1 \end{array}\right] , \tag{6.3}$$

where q_1 and q_2 are column-vectors of homogeneity degrees of the two constraints with respect to homogeneous coordinates on $\boldsymbol{X}$, x. In the action (6.1) (6.2), $\boldsymbol{X}^\mu$ and $\boldsymbol{Y}^{1,2}$ are homogeneous coordinate (2,2)-superfields on the $\boldsymbol{X}$ and $\mathbb{P}^1$ factors of the embedding space, respectively.

It is instructive to do the fermionic ς-integration and obtain the component field action (see Ref. [75]). We use

$$\boldsymbol{X}^\mu = X^\mu + \varsigma^+\xi^\mu_+ + \varsigma^-\xi^\mu_- + \varsigma^+\varsigma^-\mathsf{X}^\mu , \tag{6.4}$$

$$\boldsymbol{Y}^{1,2} = X^{1,2} + \varsigma^+\xi^{1,2}_+ + \varsigma^-\xi^{1,2}_- + \varsigma^+\varsigma^-\mathsf{Y}^{1,2} , \tag{6.5}$$

$$\boldsymbol{\Lambda}^{1,2} = \Lambda^{1,2} + \varsigma^+\lambda^{1,2}_+ + \varsigma^-\lambda^{1,2}_- + \varsigma^+\varsigma^-\mathsf{L}^{1,2} , \tag{6.6}$$

where X^μ and $\mathsf{Y}^{1,2}$ are the auxiliary components of the $\boldsymbol{X}^\mu$ and $\boldsymbol{Y}^{1,2}$ superfields. The equations of motion for these and all the components of the $\boldsymbol{\Lambda}$ superfields are algebraic and are in fact the constraints imposed when path-integration is performed over the corresponding component fields.

Having created this σ-model in the image of the Calabi-Yau manifold (3.9), let us see what happens when we let $w_Y \to 0$. In the limit, all the components of the two $\boldsymbol{Y}$ superfields become non-propagating and their equations of motion are algebraic. Thus $\boldsymbol{Y}$ may be integrated out and we can rewrite the above action (6.1) and (6.2) as

$$S(w_Y = 0) = \int_\Sigma d^2\sigma\, d^2\varsigma\, d^2\bar{\varsigma}\; w_X \mathbf{K}(\boldsymbol{X},\overline{\boldsymbol{X}}) + \Big[S'_{con.} + \text{h.c.} \Big] \tag{6.7}$$

$$S'_{con.} = \int_\Sigma d^2\sigma\, d^2\varsigma\; \Big((\boldsymbol{\Lambda}^1\boldsymbol{Y}^1)S(\boldsymbol{X}) + (\boldsymbol{\Lambda}^1\boldsymbol{Y}^2)P(\boldsymbol{X}) + (\boldsymbol{\Lambda}^2\boldsymbol{Y}^1)Q(\boldsymbol{X}) + (\boldsymbol{\Lambda}^2\boldsymbol{Y}^2)R(\boldsymbol{X})\Big) . \tag{6.8}$$

A straightforward analysis shows that upon elimination of the $\boldsymbol{\Lambda}$ and the $\boldsymbol{Y}$ superfields, the

$$P(\boldsymbol{X})Q(\boldsymbol{X}) \;-\; R(\boldsymbol{X})S(\boldsymbol{X}) \;=0 \tag{3.10$'$}$$

is *the only constrain imposed on the $\boldsymbol{X}$ superfields.*

Therefore, at $w_Y = 0$, we might as well write

$$\begin{aligned} S^\sharp &= \int_\Sigma d^2\sigma\, d^2\varsigma\, d^2\bar{\varsigma}\; w_X \mathbf{K}(\boldsymbol{X},\overline{\boldsymbol{X}}) \;+\; \big[S^\sharp_{con.} + \text{h.c.}\big] \\ S^\sharp_{con.} &= \int_\Sigma d^2\sigma\, d^2\varsigma\; \boldsymbol{\Lambda}\big[\,P(\boldsymbol{X})Q(\boldsymbol{X}) - R(\boldsymbol{X})S(\boldsymbol{X})\,\big] \end{aligned} \tag{6.9}$$

in place of S of Eqs. (6.1) and (6.2). This is, no surprise, the σ-model on the Calabi-Yau conifold denoted $\mathcal{M}^\sharp$ in subsection 3.2.

Following the process there, we can deform the action (6.9) by adding a small perturbation to the constraint polynomial to replace

$$S^\sharp_{con.} \to S^\flat_{con.} = \int_\Sigma d^2\sigma\, d^2\varsigma\; \boldsymbol{\Lambda}\big[\,P(\boldsymbol{X})Q(\boldsymbol{X}) - R(\boldsymbol{X})S(\boldsymbol{X}) \;+\; t^\alpha \mathcal{O}_\alpha(\boldsymbol{X})\,\big]\;, \tag{6.10}$$

where $\mathcal{O}_\alpha(\boldsymbol{X})$ is a complete set of polynomials of the appropriate degree. Needless to explain, this action pertains to $\mathcal{M}^\flat$.

The two σ-model actions, one in Eqs. (6.1), (6.2) and the other in Eq. (6.10), define consistent *families* of compactification. In both cases, the families are swept out by marginal perturbations corresponding to (2,1)- and (1,1)-forms (see subsections 4.3 and 4.3). For finite such perturbations (such as $w_Y \to 0$ or $t^a \to 0$, respectively), the interfacing σ-model action $S^\sharp$ (6.9) is reached.

Since the physical observables on one and the other 'side' of $S^\sharp$ are dramatically different, this connecting process seems indeed to signal some sort of phase transition. We however need to check if the interfacing action $S^\sharp$ itself corresponds to a consistent string ground state. To this end, we should check (1) if the (2,2)-supersymmetry is still consistently defined and (2) if the action defines (presumably through renormalization flow, as by now should be deemed usual) a conformally symmetric quantum field theory.

At the moment, we can unfortunately only speculate about the second feature, based on the observation that compactifications on orbifolds are perfectly consistent, moreover exact, solutions of string equations of motion. Were one to try to repeat the renormalization flow argument (usual for Calabi-Yau manifolds) for a compactification on a conifold a technical problem arises. Conifolds are singular so the metric and the curvature tensors blow up at a number of isolated points. Consequently, the Ricci tensor and so the β-function is not well defined at those points since its definition involves the inverse metric most of the usual tensor calculus goes to dust and ashes at the singular points. The same is, however, true of orbifolds which makes one believe that this is merely a technical difficulty and I shall assume it resolved.

The issue of (2,2)-supersymmetry can be treated in the existing formalism, with just a little care. Consider $S^\sharp$ in Eq. (6.9) and use

$$\Delta(\boldsymbol{X}) \stackrel{\rm def}{=} P(\boldsymbol{X})Q(\boldsymbol{X}) - R(\boldsymbol{X})S(\boldsymbol{X}) \ . \tag{6.11}$$

Upon the fermionic integration, we obtain the following constraints on the component fields of $\boldsymbol{X}$, obtained as the equations of motion for the component fields in $\boldsymbol{\Lambda}$ and the auxiliary component of $\boldsymbol{X}^\mu$:

$$\delta\Lambda^a \ : \ 0 = \mathsf{X}^\mu\Delta_{,\mu}(X) \ - \ \xi_+^\mu\xi_-^\nu\Delta_{,\mu\nu}(X) \ , \tag{6.12}$$

$$\delta\lambda_\pm^a \ : \ 0 = \xi_\pm^\nu\Delta_{,\mu}(X) \ , \tag{6.13}$$

$$\delta\mathsf{L}^a \ : \ 0 = \Delta(X) \ , \tag{6.14}$$

$$\delta\mathsf{X}^\mu \ : \ 0 = \Lambda\Delta_{,\mu} \ - \ \left(\mathbf{K}_{,\mu\bar\nu}\mathsf{X}^{\bar\nu} - \mathbf{K}_{,\mu\bar\nu\bar\sigma}\xi^{\bar\nu}\xi^{\bar\sigma}\right) \ . \tag{6.15}$$

Eq. (6.14) embeds the manifold into the embedding space. $\Delta_{,\mu}$ is the gradient of the defining polynomial and represents the normal to the constrained hypersurface and does not vanish on a manifold (because it is smooth). Eq. (6.13) then says that ξ^μ are restricted to be orthogonal to $\Delta_{,\mu}$, i.e., tangent to the constrained hypersurface, which is what they have to be. Using Eqs. (6.12) and (6.15), we obtain a $\xi_+\xi_-\bar\xi_+\bar\xi_-$ term that involves the Riemann tensor on the constrained hypersurface consisting of the restriction of the Riemann tensor on the embedding space and the extrinsic curvature correction terms.

Now, $S^\sharp$ refers not to a manifold but to a conifold which is singular at a number of isolated points, where $\Delta_{,\mu}$ does vanish. Since Eq. (6.13) is vacuous there, it would appear that the fermions still live in the tangent space of the embedding space, rather than that of the constrained hypersurface. On a second glance, however, Eqs. (6.12) and (6.15) have also changed. In particular, Eq. (6.12) now implies that

$$\Delta_{,\mu\nu}\xi_+^\mu\xi_-^\nu = 0 \ .$$

As long as the matrix of second derivatives $\Delta_{,\mu\nu}$ does not vanish where $\Delta_{,\mu}$ and Δ do, the above quadratic equation is non-empty and in fact restricts the fermions at the conically singular point to live on the conical tangent space, which is what they should.

Somewhat curiously, Eq. (6.15) says that the $\xi_+\xi_-\bar\xi_+\bar\xi_-$ term involves the restriction of the Riemann curvature of the embedding space at the singularity, rather then the (divergent) curvature tensor of the constrained hypersurface. Maybe this is how the string avoids being bothered by the singularity of the hypersurface.

Finally, let us remark that the hamiltonian analysis (as in Ref. [76], for example) can be adapted for the holomorphic setting and yields qualitatively a similar result.

6.2 The pandimensional chameleonic σ-model

The *constants* w and t represent explicit choices of the Kähler class and the complex structure, respectively. In subsection 5.3, we have already introduced world sheet currents $W(\sigma,\tau)$ and $T(\sigma,\tau)$, with the relation

$$\langle W^A\rangle \stackrel{\rm def}{=} w^A\ , \qquad \langle T^\alpha\rangle \stackrel{\rm def}{=} t^\alpha\ . \tag{6.16}$$

Consider now an action

$$\begin{aligned} S^{(2,2)}_{CICY} &= S_{kin.}(\boldsymbol{W},\overline{\boldsymbol{W}}) + S_{kin.}(\boldsymbol{T},\overline{\boldsymbol{T}}) \\ &\quad + S_{kin.}(\boldsymbol{X},\overline{\boldsymbol{X}};\boldsymbol{W}) + \Big[\, S_{con.}(\boldsymbol{X};\boldsymbol{T}) + \text{ h.c.}\Big]\ , \end{aligned} \tag{6.17}$$

where the first two terms are some kinetic terms for the $\boldsymbol{W}$ and $\boldsymbol{T}$ superfields and

$$S_{kin.}(\boldsymbol{X},\overline{\boldsymbol{X}};\boldsymbol{W}) = \int_\Sigma d^2\sigma\, d^2\varsigma\, d^2\overline{\varsigma}\ \boldsymbol{W}^A \mathbf{K}_A(\boldsymbol{X},\overline{\boldsymbol{X}})\ , \tag{6.18}$$

$$S_{con.}(\boldsymbol{X};\boldsymbol{T}) = \int_\Sigma d^2\sigma\, d^2\varsigma\, d^2\overline{\varsigma}\ \boldsymbol{T}^\alpha \mathcal{O}_\alpha(\boldsymbol{X})\ . \tag{6.19}$$

We have assumed that a complete set of operators can be found to represent the (1,1)-forms (the $\mathbf{K}$s) and the (2,1)-forms (the $\mathcal{O}$s). Quite often, this is straightforward for complete intersection Calabi-Yau manifolds [27,32,33,60] and all the chances are that this situation will improve [30].

At this point there is no Lagrange multiplier to enforce the constraint on the $\boldsymbol{X}$ superfields, that is on the embedding space and its tangent space, spanned by the bosonic and fermionic component fields of the $\boldsymbol{X}$ superfields. However, if $S_{kin.}(\boldsymbol{T},\overline{\boldsymbol{T}})$ is chosen to have a complex scale invariance (all of $\boldsymbol{T}$ cannot vanish simultaneously as that would take us back to the embedding space rather then the embedded manifold), we can make a holomorphic change of variables to identify the 'scaling' superfield. If, for example,

$$S_{kin.}(\boldsymbol{T},\overline{\boldsymbol{T}}) = \int_\Sigma d^2\sigma\, d^2\varsigma\, d^2\overline{\varsigma}\ \log\Big(\sum_{\alpha=1}^{b_{2,1}} |\boldsymbol{T}^\alpha|^2\Big)\ , \tag{6.20}$$

we can chose a gauge (coordinate patch) in which $\boldsymbol{T}^0 \neq 0$ and rescale all $\boldsymbol{T}$s throughout. There is obviously no kinetic term for $\boldsymbol{T}^0$,

$$S_{con.}(\boldsymbol{X};\boldsymbol{T}) = \int_\Sigma d^2\sigma\, d^2\varsigma\, d^2\overline{\varsigma}\ \boldsymbol{T}^0\big(\tilde{\boldsymbol{T}}^\alpha \mathcal{O}_\alpha(\boldsymbol{X})\big) \tag{6.21}$$

and T^0 acts as a Lagrange multiplier. This can obviously be generalized if more then one constraint need to be enforced (for complete intersections of hypersurfaces rather then single hypersurfaces) by choosing $S_{kin.}(\boldsymbol{T},\overline{\boldsymbol{T}})$ to have an appropriate number of independent complex scaling invariances.

The constants t^α are vacuum expectation values of the rescaled $\tilde{\boldsymbol{T}}$ superfields and correspond to affine coordinates on the space of complex structures of the embedded manifold.

Quite similarly, all of the $\boldsymbol{W}$ superfields cannot vanish simultaneously as that would render all $\boldsymbol{X}$ superfields non-propagating. As it stands though, $S_{kin.}(\boldsymbol{W},\overline{\boldsymbol{W}})$ must not have complex scale invariances such as $S_{kin.}(\boldsymbol{T},\overline{\boldsymbol{T}})$. Such a symmetry would induce a Lagrange multiplier field in front of the (sum of kinetic) term(s) for the $\boldsymbol{X}$s and thereby constrain it to vanish. As a result, $\boldsymbol{X}$s wouldn't be propagating 1+1-dimensional fields. This suggest that the $\boldsymbol{W}$s are homogeneous coordinates on the (complexified) space of Kähler classes but correspond to affine coordinates on a slightly bigger parameter space[♯14].

The constants w^A are vacuum expectation values of the $\boldsymbol{W}$ superfields and correspond to homogeneous coordinates on the space of Kähler classes of the embedded manifold.

Quite manifestly, the σ-model maps the world sheet into a fibre bundle the base space of which is the combined moduli space (complex structures + Kähler classes) and a point in it determines this structure on a Calabi-Yau manifold which in turn is the fibre at that point of the base space. In other words, the σ-model we've just built maps into the entire family of Calabi-Yau manifolds. Needless to say, at special regions in the base space (such as $w_Y = 0$ above) the compactification space $\mathcal{M}$ undergos some singularization and these regions are common to more then one such family of Calabi-Yau manifolds and extends throughout the web discussed in section 3.

Consequently, the above pandimensional chameleonic σ-model maps the world sheet into the entire web of families of Calabi-Yau manifolds.

6.3 A tumultuous voyage

Using the Weil-Petersson metric on the moduli space, it was shown in Ref. [45,55] that there exist paths of finite length between any two Calabi-Yau manifolds (with any particular choices of the complex structure and the Kähler class) in the connected web

[♯14] P. Candelas and X. de la Ossa have recently suggested that the inclusion of the dilaton superfield may account for the apparent lack of perfect symmetry between the rôles of (2,1)- and (1,1)-forms, polishing up the results of [52,55].

described in section 3, even though this includes several thousands of topologically distinct manifolds.

The relevant parameter space of σ-models, which provide a stringy description of the compactification, is closely related (if not actually identical) to the moduli space examined in Ref. [45,55]. Also, the physically relevant Zamolodchikov metric on this moduli space equals the Weil-Petersson metric to the lowest non-vanishing order [56].

It appears only sensible to assume that the finiteness result pertains also at the level of fully fledged string model, with the actual value of the distance, of course, being renormalized.

Typically, compactifications on the two 'sides' of an interfacing region feature a different number of massless fields (compare the $b_{p,q}$ numbers of the example given in section 3 and the spectrum of massless particles in Table 1). Preliminary Yukawa coupling computations[♮15] indicate that the 3+1-dimensional model undergos a strong coupling regime at the conifold interface regions. This is rather typical of phase transitions and it is not totally unexpected that some massless fields could form (massive) bound states while others might emerge massless from through decay of some composite fields.

The σ-model analysis presented here suggests that strings can propagate consistently on conifolds much the same as on smooth manifolds and orbifolds. There seems to be therefore no obstruction to set out on any of the gedanken-voyages of Ref. [41,45,46,55], starting in one compactification and ending up in another. In fact, our Universe may have actually undertaken such voyages in its youth (hopefully, before the first three minutes), hovering through several compactifications and settling finally in the best (?) of all[♮16].

It may therefore be necessary to include this possibility in our extrapolations from the now experimentally accessible $10^2\ GeV$ to the Planck scales $10^{19}\ GeV$ where strings (may) reign.

7. UNITED STRINGS

It may seem that I have ignored most of other compactifications [15,16,17,18, 19,20]. More precisely, I have ignored to describe those techniques as they are expertly treated elsewhere in this proceedings and also because of the lack of spacetime. Instead I offer my own conjecture[♮17] :

♮15 I thank Philip Candelas for discussions on this point.

♮16 Admittedly, as observed by Emil Martinec, quantum particles may still quantum-fluctuate into a compactification where they do not exist, but this is not certain.

♮17 Refining Doron Gepner's conjecture which appeared in [22].

The connected web of Calabi-Yau manifolds, with all singularizations innocuous for the string included, is the web of all world sheet $(2,2)$*-supersymmetric string compactifications.*

Admittedly, it is difficult to specify just which singularities are innocuous for strings and proceeding by examples is clearly helpful. As regards hypersurface singularities, which can be described by an algebraic equation

$$\Delta(X^{\mu}) = 0$$

in some local coordinates X^{μ}, the analysis in subsection 6.1 suggests that the matrix of second derivatives $\Delta_{,\mu\nu}$ ought to be non-vanishing, i.e., of rank at least 1.

The recent result [78] is that actually Calabi-Yau spaces with worse singularities are also at finite distance from smooth ones. The simple (2,2)-supersymmetric may therefore provide us only with a necessary condition and a Hamiltonian analysis appears necessary; for more and more singular cases, Dirac's method for constrained systems involves higher and higher derivatives of the constraint polynomial. Cyclic quotient singularities seem to be rather well understood and are indeed innocuous to a very high degree [15,47,48,54]. Combining these two ways (i.e., piling singularities atop each other), essentially all singularizations can be obtained (see M. Reid in Ref. [63]) and combining the two lines of analysis (I hope) all consistent compactifications can be achieved. This extends Reid's original conjecture [77], which included only nodal singularizations, and gives it a natural physically (somewhat) relevant arena.

Finally, I cannot resist noting that all the ground states that fit into the connected web (and also many more that quite probably share the same fate) are degenerate and feature local supersymmetry in 3+1-dimensional spacetime. All supergravity models feature the same degeneracy (of a less infinite degree, however) and quantum fluctuations inevitably lift this degeneracy once supersymmetry is broken. Again appears the supersymmetry breaking at the heart of the vacuum degeneracy problem; we can't keep sweeping it under the rug.

ACKNOWLEDGEMENTS

Most of the fun and work reported here has happened in collaboration with P. Candelas, P. Green and S.J. Gates Jr., but they had no opportunity this time to correct my mistakes and misconceptions; these are solely my fault just as are typos and grammatical errors. I have also benefited from often quite intense discussions with V. Kaplunovski and B. Warr.

My warmest thanks to the people who organized and ran all the minute details of this meeting and provided an excellent ambient to communicate and learn about the Theory of Everything (except maybe our World).

BIBLIOGRAPHY

[1] M.B. GREEN, J.H. SCHWARZ and E. WITTEN: *Superstring Theory*, (Cambridge University Press, Cambridge, 1987).

[2] M. KAKU: *Introduction to Superstrings* (Springer-Verlag, New York, 1988).

[3] D. GROSS, J. HARVEY, E. MARTINEC and R. ROHM: *Phys. Rev. Lett.* **54** (1985)502, *Nucl. Phys.* **B265** (1985)253.

[4] C. WETTERICH: *Nucl. Phys.* **B255** (1985)480, in the *Proceedings of the Jerusalem Winter School 1984*, p.204;
H. NICOLAI and C. WETTERICH *Phys. Lett.* **150B** (1985)347;
S. RANDJBAR-DAEMI and C. WETTERICH *Phys. Lett.* **159B** (1985)22.

[5] V. KAPLUNOVSKI: *Phys. Rev. Lett.* **55** (1985)1036.

[6] J.C. PATI: *University of Maryland report* UMDEPP-89-190 and references therein.

[7] P. CANDELAS, G. HOROWITZ, A. STROMINGER and E. WITTEN: *Nucl. Phys.* **B258** (1985)46.

[8] S.J. GATES, JR. and H. NISHINO: *Phys. Lett.* **173B** (1986)52.

[9] M. BERGER: *Soc. Math. Franc. Bull.* **83**(1953)79.

[10] T. HÜBSCH: *Commun. Math. Phys.* **108** (1987)291.

[11] E. CALABI: *Proc. Int. Congr. Math. Amsterdam* **2**(1954)206–207, in *Algebraic Geometry and Topology, A Symposioum in Honor of S. Lefschetz*, p.78–89 (Princeton University Press, Princeton, NJ, 1955).

[12] S.-T. YAU: *Proc. Natl. Acad. Sci. USA* **74**(1977)1798.

[13] M. DINE, N. SEIBERG, X.-G. WEN and E. WITTEN: *Nucl. Phys.* **B278** (1986)769, *ibid.* **B289** (1987)319.

[14] J. DISTLER and B.R. GREENE: *Nucl. Phys.* **B309** (1988)295.

[15] L. DIXON, J.A. HARVEY, C. VAFA and E. WITTEN: *Nucl. Phys.* **B261** (1985)678;
M. MUELLER and E. WITTEN: *Phys. Lett.* **182B** (1986)28;
K.S. NARAIN, M.H. SARMADI and C. VAFA: *Nucl. Phys.* **279** (1987)369.

[16] K.S. NARAIN: *Phys. Lett.* **169B** (1986)41;
D. GEPNER and E. WITTEN: *Nucl. Phys.* **B278** (1987)369;
M. MUELLER and E. WITTEN: *Phys. Lett.* **182B** (1986)28;
A.N. SCHELLEKENS and N.P. WARNER: *Nucl. Phys.* **B308** (1988)397;
W. LERCHE, D. LÜST and A.N. SCHELLEKENS: *Nucl. Phys.* **B288** (1987)1;
W. LERCHE and A.N. SCHELLEKENS: CERN-TH.4925 (1987), in the *Proceedings of the Summer School on Strings and Superstrings*, Poiana Brasov, Romania, Sep. 1–12, 1987.

[17] I. DI VECCHIA, V. KNIZHNIK, J. PATERSEN and P. ROSSI: *Nucl. Phys.* **B253** (1985)701;
I. ANTONIADIS, C. BACHAS, C. KOUNNAS and P. WINDEY: *Phys. Lett.* **171B** (1986)51;
H. KAWAI, D. LEWELLEN and S.-H. TYE: *Phys. Rev. Lett.* **57** (1986)1832.

[18] W. BOUCHER, D. FRIEDAN and A. KENT: *Phys. Lett.* **172B** (1986)316;
A.B. ZAMOLODCHIKOV and V.A. FATEEV: *Zh. Eksp. Theor. Fiz.* **90**(1986)1553;
S. NAM: *Phys. Lett.* **172B** (1986)323;
P. DI VECCHIA, J.L. PETERSEN and H.B. ZHENG: *Phys. Lett.* **174B** (1986)280;
Z. QIU: *Phys. Lett.* **188B** (1987)207, *Phys. Lett.* **198B** (1987)497.

[19] D. GEPNER: *Nucl. Phys.* **B296** (1988)732;
M. LYNKER and R. SCHIMMRIGK: *Phys. Lett.* **208B** (1988)216, *ibid.* **215B** (1988)681.

[20] S.-S. ROAN and S.-T. YAU: *Acta Math. Sin.* **3**(1987)256;
A. FONT, L.E. IBAÑEZ, H.P. NILLES and F. QUEVEDO: *Nucl. Phys.* **B278** (1986)577;
M. CVETIČ: in *Proceedings of the Eight Workshop on Grand Unification*, Syracuse, New York, April 1987; in *Superstrings, Unified Theories and Cosmology 1987* p.138 (World Scientific, Singapore, 1988), *Phys. Rev. Lett.* **59** (1987)1795, *ibid.* **59** (1987)2989;
C.A. LÜTKEN: *J. Phys.* **A21** (1988)1889.

[21] D. GEPNER: *Phys. Lett.* **199B** (1987)380.

[22] D. GEPNER: *Princeton University Report* Print-88-0085 (1987), (unpublished).

[23] B. MCINNES: *Class. Quant. Grav.* **5** (1988)561.

[24] P. GREEN and T. HÜBSCH: *Commun. Math. Phys.* **109** (1987)99.

[25] P. CANDELAS, A.M. DALE, C.A. LÜTKEN and R. SCHIMMRIGK: *Nucl. Phys.* **B298** (1988)493.

[26] P. CANDELAS, M. LYNKER and R. SCHIMMRIGK: *University of Texas report* UTTG-37-89.

[27] P. GREEN and T. HÜBSCH: *Commun. Math. Phys.* **113** (1987)505;
P. GREEN, T. HÜBSCH and A. LÜTKEN: *Class. Quant. Grav.* **6** (1989)105.

[28] P. CANDELAS and C.A. LÜTKEN: private communication;
T. HÜBSCH and C.A. LÜTKEN, unpublished.

[29] J. DISTLER, B. GREENE, K. KIRKLIN and P. MIRON:
Commun. Math. Phys. **112** (1989)117.

[30] M.G. EASTWOOD and T. HÜBSCH: *University of Texas report* UTTG-25-89.

[31] A. STROMINGER and E. WITTEN: *Commun. Math. Phys.* **101** (1985)341.

[32] P. CANDELAS: *Nucl. Phys.* **B298** (1988)458.

[33] J. DISTLER, B. GREENE, K. KIRKLIN and P. MIRON: *Phys. Lett.* **195B** (1987)41;
P. GREEN and T. HÜBSCH: *Class. Quant. Grav.* **6** (1989)311.

[34] A. STROMINGER: *Phys. Rev. Lett.* **55** (1985)2547;
E. CREMMER, C. KOUNNAS, A. VAN PROEYEN, J.–P. DERENDINGER,
S. FERRARA, B. DE WIT and L. GIRARDELLO *Nucl. Phys.* **B250** (1985)385.

[35] Y. HOSOTANI: *Phys. Lett.* **129B** (1983)193;
T. HÜBSCH: *J. Phys.* **A21** (1988)3051;
G.G. ROSS: *Phys. Lett.* **221B** (1988)315.

[36] A.M. POLYAKOV: *Phys. Lett.* **103B** (1981)207, *ibid.* **103B** (1981)211.

[37] G. TIAN: in *"Mathematical Aspects of String Theory"*, p.629, S.-T. Yau (ed.) (World Scientific, Singapore, 1987).

[38] A. SEN: *Nucl. Phys.* **284B** (1987)423.

[39] N. SEIBERG: *Nucl. Phys.* **B303** (1988)286.

[40] M. CVETIČ, J. LOUIS and B. OVRUT: *Phys. Lett.* **206B** (1988)227.

[41] P.S. GREEN and T. HÜBSCH: *Phys. Rev. Lett.* **61** (1988)1163,
Commun. Math. Phys. **119** (1989)431.

[42] S. CECOTTI, S. FERRARA and L. GIRARDELLO: *Phys. Lett.* **213B** (1989)443.

[43] S. FERRARA and M. PORRATTI: *Phys. Lett.* **216B** (1989)289;
M. CVETIČ, J. MOLERA and B. OVRUT: *Phys. Rev.* **D40** (1989)1140.

[44] M. Cvetič, J. Louis and B. Ovrut: *Phys. Rev.* **D40** (1989)684.

[45] P. Candelas, P. Green and T. Hübsch: *Phys. Rev. Lett.* **62** (1989)1956.

[46] P. Candelas, P. Green and T. Hübsch: in "Strings '88", p.155, eds. S.J. Gates Jr., C.R. Preitschopf and W. Siegel (World Scientific, Singapore, 1989).

[47] E. Martinec: *Phys. Lett.* **217** (1989)431.

[48] B.R. Greene, C. Vafa and N.P. Warner: *Nucl. Phys.* **B324** (1989)371.

[49] J.I. Latorre and C.A. Lütken: : *Phys. Lett.* **222B** (1989)55.

[50] C. Vafa and N. Warner: *Phys. Lett.* **218B** (1989)51;
W. Lerche, C. Vafa and N. Warner: *Nucl. Phys.* **B324** (1989)427.

[51] C. Vafa: *Harvard University Report* HUTP-89/A021.

[52] S. Ferrara and A. Strominger: to appear in the Proceedings of the "Strings '89" meeting in College Station, Texas.

[53] L. Dixon, V. Kaplunovski and J. Louis: *University of Texas report* UTTG-19-89.

[54] B.R. Greene: *Harvard University Report* HUTMP-87/B235.

[55] P. Candelas, T. Hübsch and R. Schimmrigk: *Nucl. Phys.* **B** , in press.

[56] P. Candelas, P.S. Green and T. Hübsch: *Nucl. Phys.* **B** , in press.

[57] S.J. Gates, Jr., R. Brooks and F. Muhammad: *Phys. Lett.* **194B** (1987)35;
S.J. Gates, Jr., and W. Siegel: *Phys. Lett.* **206B** (1988)631;
L. Mezincescu and R.I. Nepomechie: *Phys. Rev.* **D37** (1988)3067;
R.I. Nepomechie: in *Boulder Superstring Workshop*, p.229 (1987).

[58] C.M. Hull: *Phys. Lett.* **206B** (1988)234; *ibid.* **212B** (1988)437;
S. Bellucci, D.A. Depireux and S.J. Gates, Jr.: *University of Maryland report* UMDEPP-89-123;
D.A. Depireux, S.J. Gates, Jr. and B. Radak: *University of Maryland report* UMDEPP-90-036.

[59] R. Brooks, F. Muhammad and S.J. Gates, Jr.: *Nucl. Phys.* **B268** (1986)599.

[60] S.-T. Yau: in Proc. of *Symposium on Anomalies, Geometry, Topology*, eds. W.A. Bardeen and A.R. White (World Scientific, Singapore, 1985);
B.R. Greene, K.H. Kirklin, P.J. Miron and G.G. Ross
Nucl. Phys. **B278** (1986)667;

P.S. GREEN and T. HÜBSCH: *Commun. Math. Phys.* **115** (1988)231;
R. SCHIMMRIGK: *Phys. Lett.* **193B** (1987)175;
P.S. ASPINWALL, B.R. GREENE, K.H. KIRKLIN and P.J. MIRON
Nucl. Phys. **B294** (1987)193;
P. CANDELAS, C.A. LÜTKEN and R. SCHIMMRIGK: *Nucl. Phys.* **B306** (1988)113.

[61] C. SCHOEN: *Math.Z.* **197**(1988)177.

[62] M.F. ATIYAH: *Proc. Royal Soc.* **A 247**(1958)237.

[63] V.I. ARNOLD: in *Proceedings of the International Congress of Mathematicians*, p.19, Vancouver, 1974;
M. REID *Proc. Sump. Pure Math.* **46**(1987)345.

[64] J.K. KIM, I.G. KOH and Y. YOON: *Phys. Rev.* **D33** (1986)2893.

[65] H. CLEMENS: *Publ. Math. IHES* **58** (1983)19.

[66] F. HIRZEBRUCH and J. WERNER: *Max-Planck-Institute Report* SFB/MPI-85–58.

[67] S.J. GATES JR. and T. HÜBSCH: *Phys. Lett.* **226** (1989)100.

[68] D. GROSS and E. WITTEN: *Nucl. Phys.* **B277** (1986)1;
J. ATICK, A. DHAR and B. RATRA: *Phys. Lett.* **169B** (1986)54;
J. ATICK and A. DHAR: SLAC report SLAC-PUB-3964 ;
M.T. GRISARU, A. VAN DE VEN and D. ZANON: *Phys. Lett.* **173B** (1986)423, *Nucl. Phys.* **277** (1986)388, *ibid.* **277** (1986)409;
C.N. POPE, M.F. SOHNIUS and K.S. STELLE: Imperial College report Imperial/TP/85-86/16;
M.D. FREEMAN and C.N. POPE: *Phys. Lett.* **174B** (1986)48;
M.D. FREEDMAN, C.N. POPE, M.F. SOHNIUS and K.S. STELLE:
Phys. Lett. **178B** (1986)199.

[69] For a recent review, see S.MUKHI: in the Proc. of the TIFR Winter School, Panchgani 1986, eds. V.Singh and S.Wadia (World Scientific, Singapore, 1986); see also
L.ALVAREZ-GAUMÉ, D.Z.FREEDMAN and S.MUKHI:
Ann. Phys.(N.Y.) **134** (1981)85;
S.MUKHI: *Nucl. Phys.* **B264** (1986)640.

[70] T. EGUCHI, P.B. GILKEY and A.J. HANSON: *Phys. Rep.* **C66** (1980)213–393.

[71] D. NEMESCHANSKY and A. SEN: *Phys. Lett.* **178B** (1986)365;
P. CANDELAS, M.D. FREEMAN, C. POPE, M.F. SOHNIUS and K.S. STELLE:
Phys. Lett. **177B** (1986)341.

[72] E. Martinec: *University of Chicago report* April 1989, to appear in the V.G. Knizhnik memorial volume ed. L. Brink.

[73] L. Dixon: in *Superstrings, Unified Theories and Cosmology 1987*, p.67–127 eds. G.Furlan et. al. (World Scientific, Singapore, 1988).

[74] A.B. Zamolodchikov: *JEPT Lett.* **43**(1986), *Sov. J. Nucl. Phys.* **46**(1987)1090.

[75] S.J. Gates, Jr., M.T. Grisaru, M. Roček and W. Siegel: *Superspace* (The Benjamin/Cummings Pub. Co., Reading, Massachusetts, 1983).

[76] N.K. Pak and R. Percacci: *International School for Advanced Studies*, Trieste, Italy, Ref. SISSA 17 EP (1989).

[77] M. Reid: *Math. Ann.* **278**(1987)329.

[78] P. Green: in preparation.

Standard Model
and Beyond

ISBN 1-56072-027-1

STRING THEORY OF CALABI-YAU COMPACTIFICATIONS

C.A. Lütken[†]

NORDITA, Blegdamsvej 17, DK-2100 Copenhagen Ø, Denmark

ABSTRACT

The conformal field theory description of Calabi-Yau compactifications of the heterotic superstring from 10 to 4 dimensions is outlined. The basic ideas of ordinary (bosonic) conformal field theory are explained before descibing the exactly solvable N=2 superconformal minimal models which are needed in the tensor construction of certain particularly simple string vacua. Using a simple sigma-model construction of algebraic varieties and drawing on insight gained from the Landau-Ginzburg description of critical phenomena, it is explained how the critical behaviour of these 2-dimensional solvable quantum field theories with complex supersymmetry may be regarded as string compactification on a Calabi-Yau background. The virtue of this is to provide a tool for computing exact (tree level) results for strings in these highly non-trivial vacua, including all the Yukawa couplings needed in the construction of the low-energy effective field theory.

[†] Address after Oct.1, 1989 during leave of absence: Oxford University, Department of Theoretical Physics, 1 Keble Road, Oxford OX1 3NP, U.K.

1. INTRODUCTION

The idea that Nature may look a lot simpler if viewed from a higher dimensional vantage point is not a new one, but it fits perfectly into the framework of string theory where it is an essential ingredient.

Although it has by no means been proven, it is generally believed that string theory is a consistent way of quantizing and unifying gravity with all other forces. The most pressing problem in string theory is how to make contact with reality, by which we here mean particle phenomenology. Perhaps surprisingly, the major obstacle to doing this is our lack of understanding of the classical theory. Since superstrings naively only make sense in ten dimensions, the first step is to identify ground states of string which can serve as backgrounds on which to 'compactify' the theory from ten to the observed four dimensions. If we require that supersymmetry survives to low energy and that the compact space is smooth, then the background manifold must be of the socalled Calabi-Yau type.

In this scenario the properties of the low energy model (i.e. the observable part) are intimately connected to the topology and geometry of the compact manifold. In principle therefore, given the correct Calabi-Yau manifold we should be able to 'derive' the 20-parameter standard model. The question of why the string chooses one particular of the large number of available ground states, which at least classically are degenerate (the 'vacuum degeneracy' problem), cannot be addressed with todays primitive string technology, but that need not obstruct progress. After all, using 'macroscopic' (i.e. phenomenological) information to prune down the tree of possible 'microscopic' (i.e. more fundamental) theories is always how real physics is done. The situation is somewhat analogous to that of gauge-theories. There is no theoretical reason to choose the standard $U_1 \times SU_2 \times SU_3$ gauge group, it is experiment that compels us to do so. Similarly, we hope to one day be able to pin down the correct compact space. This requires that we develop powerful methods for extracting the parameters of the low-energy effective field theory from a string theory on a Calabi-Yau background.

Efforts of this type can be concentrated in two directions. The oldest and conceptually simplest is the spacetime approach (which was implicitly adopted above), where the string is regarded as moving through an ambient spacetime. In this 'extrinsic' picture the observable string modes are zero-modes on the compact manifold.

The zero-modes carry topological and geometrical information about the the manifold, which can be extracted using methods from the algebraic and differential geometry of Kähler manifolds. The great virtue of this approach is that some features of the low-energy theory are easily obtained in an intuitive manner which does not obscure their geometric origin. Thus, such hitherto unexplained phenomena as the duplication of generations and the pattern of Yukawa couplings emerge as simple consequences of the topology of the compact space.

Although semi-realistic three and four generation models have been built along these lines, the technical problems involved in extracting information from the Calabi-Yau manifold in sufficient detail are enormous. However, recently these obstacles have largely been overcome due to the discovery of a surprising connection between Calabi-Yau spaces and certain simple representations of the N=2 superconformal algebra, which describes the operators of the current algebra on the 2-dimensional world-sheet.

In this alternative 'intrinsic' picture of string physics, spacetime coordinates are radically reinterpreted as scalar fields on the string world-sheet. This leads directly to the idea that strings can be represented, order by order in the string loop expansion, by 2-dimensional non-linear sigma-models. These provide the explicit link between the intrinsic algebraic and the extrinsic geometric construction of classical string vacua, and we shall see that the two constructions are in fact equivalent. More precisely, we consider sigma-models

$$S = \int d^2z \sqrt{g} g^{\alpha\beta} \partial_\alpha X^\mu \partial_\beta X^\nu G_{\mu\nu} + (N = 2 \ \ SUSY) \tag{1.1}$$

which map a 2-dimensional surface $M = M(z, \bar{z})$ (the world-sheet) with metric $g_{\alpha\beta}(z, \bar{z})$ into a 10-dimensional target space $\mathcal{M} = \mathcal{M}(X^1, X^2, \ldots, X^{10})$ (spacetime) with metric $G_{\mu\nu}(X)$. At the moment, in the absence of *any* understanding of the basic principles of string theory, this is what passes for an action. As we see, it is clearly an effective description which only encodes the few degrees of freedom associated with perturbations around a fixed and arbitrarily chosen solution of the true string equations, which remain to be discovered.

The compactification assumption is that $\mathcal{M}^{10} = \mathcal{M}^4 \times K^6$, where $\mathcal{M}^4$ is usually taken to be flat 4-dimensional Minkowski spacetime and K^6 is a 6-dimensional Calabi-Yau manifold. Notice that the dimension of spacetime is explicitly put in by hand; it is the first piece of phenomenological input which is needed. This is sometimes claimed to be a major failing of a theory where spacetime should be part of the solution, not part of the initial data. Now that may or may not be the case, we certainly all hope that string theory will eventually perform this remarkable feat, but it can hardly be held against a theory that it doesn't immediately explain everything. If the same criteria were applied to *any* traditional field theory it would also fail, and in a much more serious way since its very formulation cuts off any chance of addressing the problem. In string theory the question can at least be asked, and quite possibly answered in the forseeable future. And even if we fail to do so, we are certainly no worse off than before, all else being equal. Now all else is not equal of course, and it is our first job to remedy this. That is what the 'tinker toy' approach to string theory presented here is all about. We do not pretend to try to derive all current and future observations from first principles (which are not understood in the case of string theory). It is an attempt to see if string theory can correlate known data using a novel picture of spacetime which is naturally suggested by the theory.

In the next section we start by giving the traditional, and by far the simplest and most intuitive picture of these string solutions as spacetime manifolds. Rather than elaborate on the mathematical details of algebraic geometry, which will be explained by other speakers, we only introduce the basic concepts before moving on to the intrinsic conformal field theory description of the Calabi-Yau solutions. An attempt is made to make the presentation selfcontained by also giving a short introduction to conformal field theory (CFT). The reader who is already familiar with CFT should therefore skip Section 3.

In Section 4 the discrete series of conformal field theories with 2 supersymmetries is introduced. As shown by Gepner, they can be used to build solutions of the super- and heterotic string. We give some simple examples of how to implement his ideas, after which we are compelled to conclude that these constructions are in some sense equivalent to the Calabi-Yau construction described in Section 2. In other words, we have overwhelming circumstantial evidence for an unexpected connection between infinite dimensional N=2 current algebras and the algebraic geometry of compact Kähler manifolds.

Although one might have expected that a proper understanding of this astonishing phenomenon would have to await an elucidation of the basic principles of string theory, it turns out that the sigma-model description (1.1) is sufficient, as explained in Section 5.

With this knowledge in hand, we can proceed to exploit the tensor construction to compute low energy parameters of the effective 4-dimensional field theory, which is a more or less traditional grand unified theory of supergravity Yang-Mills with a very large gauge group, typically containing $E_8 \times E_6$ at the compactification scale. In Section 6 it is explained how the computation of couplings can be done exactly since the minimal models are exactly solvable.

Finally, in Section 7 the leading candidate for a realistic model is presented, and a natural mechanism for breaking gauge symmetries is briefly described.

The reader is hereby warned that the words 'compactification', 'phase transition' and 'renormalization group flow' should be understood in the same sense as 'spontaneous symmetry breaking'. No dynamical process is implied by this customary abuse of language.

A roadmap of these talks is presented in Fig.1.

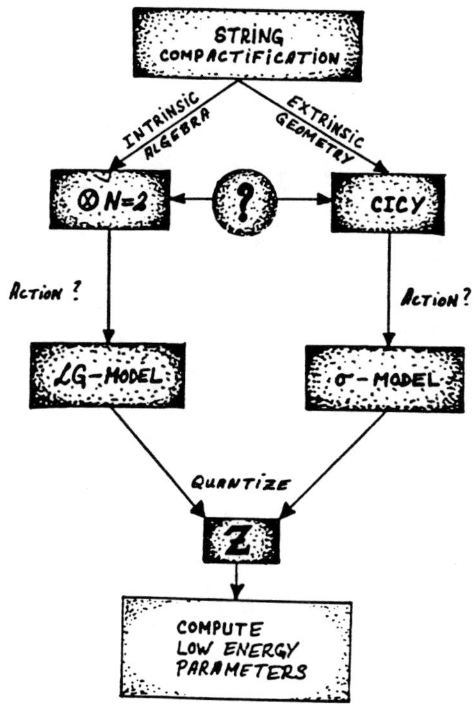

2. ALGEBRAIC TARGET SPACES

In 1984, when it was first realized that Calabi-Yau manifolds play an important rôle in string theory [15], none were known to the physics community. This was soon rectified by Yau [71], and his constructions were quickly generalized. Almost all known Calabi-Yau manifolds can be constructed as 'complete intersections' in complex projective spaces, and furthermore, all such manifolds are known [49][40][3][9]. Consequently, a large number ($\sim 10^4$) of Calabi-Yau manifolds are today available for model builders [41][28].

In real geometry we are used to considering compact manifolds as subsets of R^n. Calabi-Yau spaces are complex manifolds*, but the only compact complex submanifold of C^n is a point so C^n is not a viable ambient space from which to extract Calabi-Yau manifolds. The correct analogue of R^n turns out to be CP^n, complex projective space in n dimensions, whose points are lines through the origin of $C^{n+1}(Z_A)$. The redundant coordinates Z_A ($A = 1, 2, \ldots, n+1$) are called homogeneous coordinates of CP^n, while a complete set of n independent (inhomogeneous) coordinates on the patch $U_j(Z_j = const \neq 0)$ ($j = 1, 2, \ldots n+1$) can be chosen as:

$$\zeta_i^{(j)} = \frac{Z_i}{Z_j} \quad (i = 1, 2, \ldots, \hat{j}, \ldots, n+1). \tag{2.1}$$

An algebraic hypersurface of degree k in $CP^n(Z_1, Z_2, \ldots, Z_{n+1})$ is the zero-locus of a homogeneous polynomial p in the homogeneous coordinates; for example:

$$p = Z_1^k + Z_2^k + \ldots + Z_{n+1}^k = 0. \tag{2.2}$$

All the ('discrete') topological information about the manifold is contained in k. If $k = n+1$ the hypersurface $p = 0$ is a Calabi-Yau manifold. The coefficients of the various monomials that can appear in p determine the shape and size of the space and are called moduli. In (2.2) most of the moduli have been chosen to vanish so that the space has a very large number of discrete symmetries. These are of great interest in phenomenology.

---

* For details on the material in this section, see the excellent review by Candelas [7], or the second volume of ref.[42].

A simple example is provided by the constraint $Z_1^k + Z_2^k + Z_3^k = 0$ in $CP^2(Z_1, Z_2, Z_3)$, which is the surface $\mathcal{M}$ with Euler number $\chi = k(3-k)$ shown (heuristically) in Fig.2.*

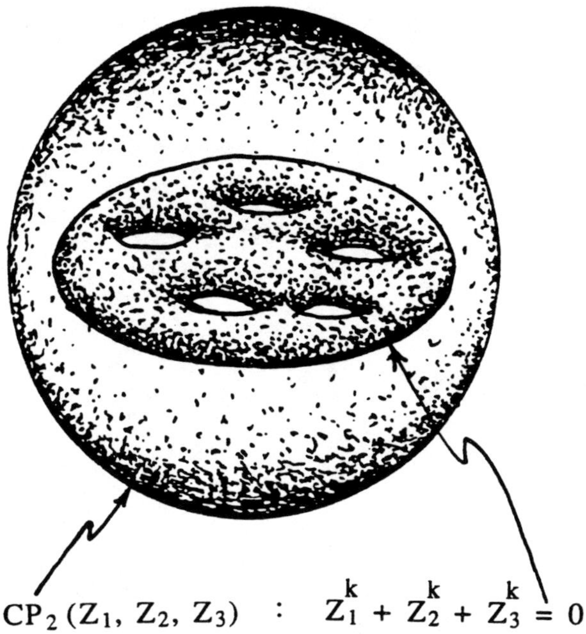

When $k = 3$ the Euler number vanshies so $\mathcal{M}$ is a torus, which is the only Calabi-Yau space in two dimensions.

A more interesting six-dimensional Calabi-Yau space is given by a quintic polynomial hypersurface in CP^4, say:

$$p = Z_1^5 + Z_2^5 + Z_3^5 + Z_4^5 + Z_5^5 = 0. \tag{2.3}$$

Since the purpose of this talk is just to explain the recent conceptual advances in string compactification (rather than equip the reader with all the tools of the trade), we shall only need to refer to this simple example. The complications associated with generalizing the following to complete intersections in products of (possibly weighted) projective spaces are only technical and will be ignored here.

* The observant reader will find an error in this picture.

We shall need to know some of the most important topological properties of these spaces, in particular the number of harmonic forms of various types [15]. This is because the phenomenology in which we are interested at our extremely low energies (compared to the string scale) only involves the zero-modes on the Calabi-Yau manifold, since excitations will represent particles with Planck (or compactification scale) masses. From the equation of motion for zero-mode perturbations of the metric it can be seen directly that they correspond to the harmonic forms, which are unique representatives of the cohomology classes. This is not unreasonable in view of the intuitive argument that zero-modes 'feel out' all of the manifold, i.e. its global properties or topology, so that they must somehow correspond to harmonic forms which according to de Rham's theorem carry the topological information about the manifold.

To see this notice first that locally closed and exact forms are the same, because on flat space they are the same. Their global relationship is described by de Rham's theorem:

$$\omega = \alpha + d\beta + d^{\dagger}\gamma, \tag{2.4}$$

which says that any form ω can be split into an exact piece $d\beta$, a coexact piece $d^{\dagger}\gamma$ and a harmonic piece α ($\Delta\alpha = 0$). Choosing ω to be closed ($d\omega = 0$) eqn.(2.4) implies that the coexact piece vanishes, and we see that globally the closed and exact form differ by a harmonic form, which must therefore contain the topological information about the manifold. The study of obstructions to extending local equivalences to global ones is called cohomology. An equivalence class of closed forms that differ only by exact forms is called a cohomology class or group (the group operation is addition of forms), and given a metric it is uniquely labeled by a harmonic form.

In order to count the number of topologically distinct modes (harmonic forms) on our algebraically constructed spaces, we exploit the fact that $(D-1,1)$ forms on a D-dimensional Calabi-Yau manifold can be represented by deformations of the defining polynomial. All such deformations represent closed forms, but some of them differ only by an exact form and should therefore only be counted once. They are the deformations that can be achieved by a change of coordinates. The quintic manifold (2.3) provides a good example. There are a total of 126 quintic monomials in five variables: $C_{(ABCDE)}Z^AZ^BZ^CZ^DZ^E$. Of these, 25 can be written as $C_A^{\ B}Z^A\partial_B p$, which can be interpreted as the lowest order difference (in a Taylor expansion) between the original polynomial and the polynomial obtained after an infinitesimal coordinate transformation. Hence there are a total of 101 topologically distinct (2,1) forms on the quintic hypersurface. In the heterotic string these are used to build 101 chiral matter superfields transforming as **27** under the GUT group E_6, as well as 101 gauge singlet fields. A very similar deformation argument tells us that there are an additional 224 gauge singlet fields which originate in the cohomology group of 1-forms which transform as mixed rank 2 tensors under the holonomy group SU_3 ($\in EndT$). We shall also need to know that a simple application of the Lefschetz hyperplane theorem [47] tells us that there is a unique harmonic (1,1) form. It gives rise to one chiral matter superfield transforming as a **27** under the GUT group E_6, and one gauge singlet mode. All told there are 326 gauge singlet modes which are guaranteed by the topology of the Calabi-Yau space. As we shall see, there may also be some 'accidentally massless' modes at particularly symmetric points in the parameter (moduli) space.

The set of polynomials which contains the objects $\partial_A p$ is sometimes called the 'ideal' or 'ring' of trivial deformations of p. They can be identified with certain marginal operators in the CFT description, to which we now turn.

3. SCALE-INVARIANT QUANTUM FIELD THEORY

In order to explain the intrinsic picture where the spacetime coordinates of the string are regarded as scalar fields on a Riemann surface, we must first familiarize ourselves with the beautiful theory of conformally invariant 2-dimensional fields *. Such field theories have unique properties. The requirement of conformal invariance (together with unitarity) is so strong in two dimensions that the spectrum of allowed theories can be parametrized by a single number c (the central charge of the Virasoro algebra) which can be computed algebraically. Even more remarkable is the fact that if c is sufficiently small, the allowed spectrum is discrete (minimal models) and in one-to-one correspondence with the simple Lie groups (ADE-classification) [38][16][51].

* For details see one of the many good reviews that now exist [17][60][18][39], and especially the remarkable paper that initiated the subject [5].

Conformal Transformations

We first consider the general definition of conformal transformations in any dimension. They are a special kind of coordinate transformation or reparametrization which can be written infinitesimally as:

$$x^\mu \to x'^\mu = x^\mu + \epsilon^\mu(x) + \ldots \tag{3.1}$$

We can define conformal transformations as coordinate transformations that change the local scale of the metric:

$$g'_{\mu\nu}(x') = \lambda(x) g_{\mu\nu}(x). \tag{3.2}$$

This may or may not change the *geometry* of the space coordinatized by x^μ. Mathematically, they are maps that preserve the *angles* between tangent vectors at every point of the space coordinatized by x^μ. Without proof we simply list the explicit (infinitesimal) form of the conformal transformations for $d > 2$:

$$\epsilon^\mu(x) = a^\mu + \omega^\mu_\nu x^\nu + \lambda x^\mu + b^\mu(x_\nu x^\nu) - 2x^\mu(b_\nu x^\nu) \tag{3.3}$$

where the first two terms are translations and rotations (i.e. Poincare transformations), λ parametrizes scale transformations (dilatations), and the rest are called 'special conformal transformations', which can also be viewed as a translation combined with an inversion:

$$\frac{x'^\mu}{x'^2} = \frac{x^\mu}{x^2} + b^\mu. \tag{3.4}$$

The conformal group in $d > 2$ is locally isomorphic to $O(d+1, 1)$, which consists of translations, rotations, dilatations and inversions, all told $\frac{1}{2}(d+1)(d+2)$ transformations. In $d = 2$ these transformations are called Möbius transformations, and they are the only ones that map the complex plane onto itself in a regular way. However, in $d = 2$ there are an infinite number of other conformal transformations. They can be thought of as a local generalization of the renormalization group transformations, which involves a global rescaling.

It is easy to see what these extra transformations are in $d = 2$. Introduce a new set of convenient complex coordinates z and $\bar{z}$ in terms of the usual Cartesian coordinates x and y of flat space:

$$z = x + iy; \qquad \bar{z} = x - iy \tag{3.5}$$

in terms of which the line element (which defines the flat metric) reads:

$$ds^2 = dzd\bar{z}. \tag{3.6}$$

The transformation $z \to z'$ is a rescaling:

$$dz'd\bar{z}' = \lambda(z,\bar{z})dzd\bar{z} \tag{3.7}$$

provided the 'crossterms' vanish:

$$\frac{\partial z'}{\partial \bar{z}} = 0; \quad \frac{\partial \bar{z}'}{\partial z} = 0. \tag{3.8}$$

This means that the new coordinates must be analytic (holomorphic) and anti-analytic (anti-holomorphic) functions of the old coordinates:

$$z' = f(z); \quad \bar{z}' = \bar{f}(\bar{z}), \tag{3.9}$$

and that the scaling *factorizes*:

$$\lambda(z,\bar{z}) = (\frac{\partial f}{\partial z})(\frac{\partial \bar{f}}{\partial \bar{z}}). \tag{3.10}$$

More precisely, a wellknown theorem from complex analysis states that:

f is conformal if f is analytic and $f'(z) \neq 0$.

Since there are infinitely many analytic functions on the plane, the $d = 2$ conformal group contains, in addition to the 6-parameter Möbius group, an infinite number of holomorphic transformations.

An arbitrary analytic transformation:

$$z \to z' = f(z) = z + \epsilon(z) + \ldots \tag{3.11}$$

is parametrized by the coefficients ϵ_n $(n \in Z)$ of its Laurent expansion:

$$\epsilon(z) = \sum_{n=-\infty}^{\infty} \epsilon_n z^{n+1}, \tag{3.12}$$

where

$$\epsilon_n = \oint \frac{dz}{2\pi i} z^{-n-2} \epsilon(z). \tag{3.13}$$

Algebraically we can think of the z^{n+1} as a set of basis-functions, and the differential operators $l_n = z^{n+1}\partial_z$ as the generators of infinitesimal conformal transformations of the complex plane:

$$l_n : \quad z \rightarrow z^{n+1}. \tag{3.14}$$

The algebraic properties of the operators l_n are collected in the commutation relation:

$$[l_n, l_m] = (n-m)l_{n+m}. \tag{3.15}$$

This is the classical (Virasoro) algebra of the conformal group. The Möbius transformations are recovered as the subgroup generated by $\{l_0, l_{\pm 1}\}$ (they span a group because $[l_0, l_{\pm 1}] = \mp l_{\pm 1}$ and $[l_1, l_{-1}] = 2l_0$). More precisely:

Translations: $l_{-1} = \partial_z = \partial; \ \bar{l}_{-1} = \partial_{\bar{z}} = \bar{\partial}$

Rotations: $l_0 - \bar{l}_0 = z\partial - \bar{z}\bar{\partial}$ (q.m. eigenvalue=spin)

Dilatations: $l_0 + \bar{l}_0 = z\partial + \bar{z}\bar{\partial}$ (q.m. eigenvalue=scaling dimension)

Special: $l_1 = z^2\partial; \ \bar{l}_1 = \bar{z}^2\bar{\partial}.$

When we quantize conformally invariant 2-dimensional field theories these operators will become proper quantum operators, i.e. operators acting on the Hilbert space of states of the theory, and the algebra (3.15) will pick up an extra ('anomalous', 'Schwinger') term, whose exact form will depend on which theory is being quantized. Hence the importance of the full (quantum) Virasoro algebra is that it carries detailed information about the field content of the the quantum field theory, not just general information about spacetime symmetries, as one would expect from the classical result above.

To summarize this section: demanding conformal symmetry of a 2-dimensional theory imposes an infinite number of constraints on the field theory. Together with the additional requirements for a consistent quantization this turns out to be so restrictive that many 2-dimensional conformal quantum field theories of interest in string theory, and all the ones of interest in statistical mechanics, can be classified. We will spend the rest of this section trying to understand this amazing result.

Symmetries and Currents

We start with a brief review of the basic principles of classical field theory. All information about the system is encoded in the Lagrangian $\mathcal{L}$, which is built from the fields $\phi(x)$ and their derivatives: $\mathcal{L} = \mathcal{L}(\phi(x), \partial_\mu\phi(x))$. The equations of motion derive from the condition that the action $S = \int d^dx \mathcal{L}(\phi(x), \partial_\mu\phi(x))$ be stationary ($\delta S = 0$). The form of the action is constrained by investigating how it transforms when the fields transform, and demanding that it be invariant under some transformations which are called symmetries. If the action changes by a local quantity under transformations $t : \phi \to \phi + \delta\phi$, i.e.

$$\delta S = \int d^dx \Delta_t(\phi(x), \partial_\mu\phi(x)), \tag{3.16}$$

then by the Noether theorem there exists a local current j_t^μ such that:

$$\partial_\mu j_t^\mu = \Delta_t. \tag{3.17}$$

If $\Delta_t = 0$ then t is called a symmetry, and the corresponding Noether current is conserved: $\partial_\mu j_t^\mu = 0$, as well as the charge $Q = \int d^dx j_t^0(x)$. The most important example of the Noether construction is when t is a spacetime transformation, which includes conformal transformations:

$$x^\mu \to x^\mu + \epsilon^\mu(x), \tag{3.18}$$

in which case the current is associated with the energy-momentum or stress tensor:

$$j_\mu = T_{\mu\nu}\epsilon^\nu. \tag{3.19}$$

Recall that the stress tensor itself is the current associated with some special space-time transformations, rotations and translations, so it is reasonable that it generates conformal transformations. It is therefore the object of most interest in conformal field theory. In 2 dimensions it has the extremely useful property of being an (anti-)analytic function. This is easily seen by using the complex coordinates introduced in (3.5) , in terms of which the derivatives are:

$$\partial = \partial_z = \frac{1}{2}(\partial_x - i\partial_y); \quad \bar{\partial} = \partial_{\bar{z}} = \frac{1}{2}(\partial_x + i\partial_y), \tag{3.20}$$

and the flat metric we are assuming takes the form:

$$\eta_{\alpha\beta}(z, \bar{z}) = \frac{1}{2}\begin{pmatrix} 0 & 1 \\ 1 & 0 \end{pmatrix}. \tag{3.21}$$

Conservation of the stress tensor becomes:

$$\begin{aligned} \partial T^{zz} + \bar{\partial} T^{\bar{z}z} &= 0 \\ \partial T^{z\bar{z}} + \bar{\partial} T^{\bar{z}\bar{z}} &= 0. \end{aligned} \tag{3.22}$$

Since (global) scale invariance requires $T_{\alpha\beta}$ to be traceless, we also have:

$$T^{z\bar{z}} = 0. \tag{3.23}$$

Therefore:

$$\begin{aligned} \partial T^{zz} &= 0 \Rightarrow T^{zz}(z,\bar{z}) = \bar{T}(\bar{z}) \\ \bar{\partial} T^{\bar{z}\bar{z}} &= 0 \Rightarrow T^{\bar{z}\bar{z}}(z,\bar{z}) = T(z). \end{aligned} \tag{3.24}$$

In other words, in 2 dimensions conformal invariance means that we can split the stress tensor into a holomorphic (analytic) and an anti-holomorphic part:

$$T_{\alpha\beta}(z,\bar{z}) = \begin{pmatrix} T(z) & 0 \\ 0 & \bar{T}(\bar{z}) \end{pmatrix}. \tag{3.25}$$

This 'holomorphic factorization', which is a consequence of imposing the infinite number of conformal symmetries available in 2 dimensions, means that we have all the power of complex geometry at our disposal when we now proceed to analyse 2-dimensional quantum field theories .

Quantization

We could now proceed immediately with the formal machinery of canonical quantization, but in order to keep things as familiar as possible we consider first the following reparametrization of C^2 †:

$$z = e^{\tau + i\sigma}; \quad \bar{z} = e^{\tau - i\sigma}. \tag{3.26}$$

If $\sigma, \tau \in R$ and $0 < \sigma \leq \pi$, then we can interpret this as a (conformal) map of the cylinder to C:

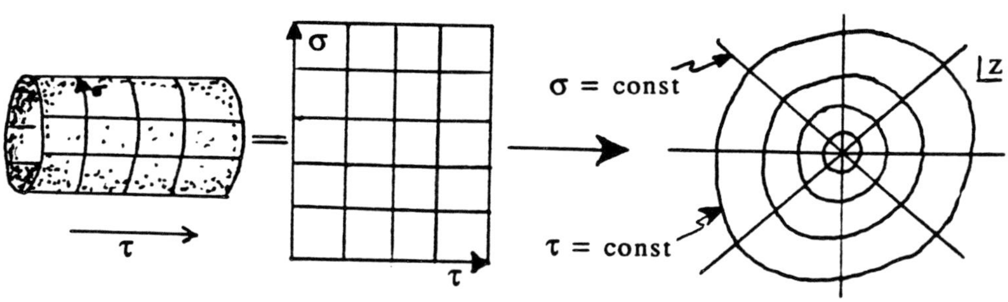

† The literature on conformal field theory is a bit confusing at first sight because all the machinery is developed for $C(z) \times C(\bar{z})$, i.e. for independent right and left movers in string language, with the understanding that physical quantities are obtained by restricting to a real section, i.e by setting $\bar{z} = z^*$.

The cylinder is the worldsheet of a closed string, and we can therefore interpret σ as the position on the string and τ as the time coordinate. Clearly, constant time slices are circles of fixed radius e^{τ} on C, while constant space slices are lines radiating from the origin at angle σ. Hence *time translations* $\tau \rightarrow \tau + \lambda$ become *dilatations* on the z-plane:

$$\lambda : \quad z \rightarrow e^{\lambda} z = z + \lambda z + \ldots \tag{3.27}$$

and space translations become rotations on the z-plane:

$$\theta : \quad z \rightarrow e^{i\theta} z. \tag{3.28}$$

According to the axioms of quantum mechanics, the basic problem of quantization is to identify the operators which when acting on the Hilbert space relate states in such a way that the symmetries of the theory are preserved. For example, we all know that the operator generating time translations is the Hamiltonian (H) of the system. In the Heisenberg picture it generates the time evolution of any operator ϕ according to the 'operator equation of motion':

$$\dot{\phi} = [H, \phi]_{ETC}, \tag{3.29}$$

or equivalently:

$$\delta_{\lambda}\phi = [H_{\lambda}, \phi]_{ETC}. \tag{3.30}$$

Since time translations can be viewed as dilatations in two (Euclidean) dimensions, we can interpret H as the charge of the scaling current ($\epsilon(z) = \lambda z$):

$$H_{\lambda} = \lambda H = \lambda \oint_{|z|=const} \frac{dz}{2\pi i} z T(z). \tag{3.31}$$

In view of this, the general statement we are after can only be that the charge

$$T_{\epsilon} = \oint_{|z|=const} \frac{dz}{2\pi i} \epsilon(z) T(z), \tag{3.32}$$

associated with the classical conformal transformation $z \rightarrow z + \epsilon(z)$, generates (operator-) field transformations in the quantum theory according to:

$$\delta_{\epsilon}\phi = [T_{\epsilon}, \phi]_{ETC}. \tag{3.33}$$

Notice that all charges are canonically defined on 'constant time slices', which with our parametrization are circles ($|z| = const$) around the origin. Since the line-integral of an analytic function is independent of choice of path (up to homotopy), these charges are conserved, as required by the canonical formalism.

Notice also that for simplicity we are only exhibiting the holomorphic properties of the fields $\phi = \phi(z, \bar{z})$. A completely analogous anti-holomorphic term is implicit in every equation, e.g. (3.33) should be read as:

$$\delta_{\epsilon,\bar{\epsilon}}\phi(w,\bar{w}) = [T_\epsilon, \phi(w,\bar{w})]_{ETC} + [\bar{T}_{\bar{\epsilon}}, \phi(w,\bar{w})]_{ETC} \tag{3.34}$$

for a general conformal transformation $(z \to z + \epsilon(z), \bar{z} \to \bar{z} + \bar{\epsilon}(\bar{z}))$.

To complete the specification of the quantum theory, we must as usual resolve the 'operator ordering ambiguity'. Since two classical fields $A(z)$ and $B(w)$ commute, while the operator fields $\hat{A}(z)$ and $\hat{B}(w)$ do not if the arguments can coincide, there is an ambiguity when we want to turn products of fields into 'composite' operators: do we promote $A(z)B(w)$ or $B(w)A(z)$ or some combination of these to an operator product? The problem arises above because we use equal time commutators (ETC), i.e. $|z| = |w|$, so that $z \to w$ when $arg(z) \to arg(w)$:

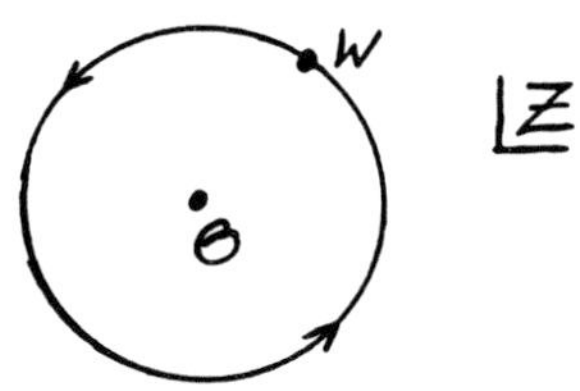

A standard way to resolve the operator ordering problems in quantum field theory is to use Schwinger's time splitting procedure. This means that we define the equal time commutator used above by evaluating the charge at slightly different times from the field operator $\phi(w, \bar{w})$:

$$[T_\epsilon, \phi(w,\bar{w})]_{ETC} \stackrel{\text{def}}{=} \lim_{|z| \to |w|} \{ \oint_{|z|>|w|} \frac{dz}{2\pi i} \epsilon(z) T(z) \phi(w,\bar{w}) - \oint_{|z|<|w|} \frac{dz}{2\pi i} \epsilon(z) T(z) \phi(w,\bar{w}) \}, \tag{3.35}$$

where the operator products are understood to be time (radial) ordered. We can deform the contours of integration:

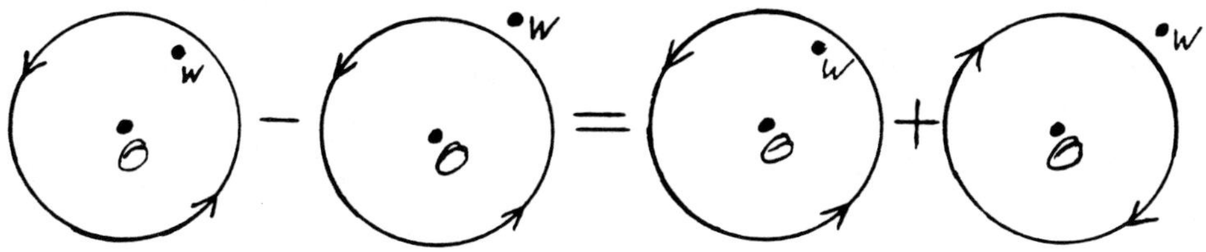

to obtain:

$$[T_\epsilon, \phi(w,\bar{w})]_{ETC} = \lim_{|z|\to|w|} \oint_{\odot} \frac{dz}{2\pi i} \epsilon(z) T(z) \phi(w,\bar{w}). \tag{3.36}$$

This integral is non-vanishing only if there is a singularity in the operator product:

$$\lim_{z\to w} T(z)\phi(w,\bar{w}) \tag{3.37}$$

whose residue can contribute to the integral according to Cauchy's theorem:

$$\oint_{\odot} \frac{dz}{2\pi i} \frac{f(z)}{z-w} = f(w). \tag{3.38}$$

We can now figure out the operator product (3.37) by comparing the quantum operator $\delta_\epsilon \hat{\Phi}(w,\bar{w})$ to the classical transformation $\delta_\epsilon \Phi(w,\bar{w})$ to which it must correspond. Consider first a *primary field* $\phi(w,\bar{w})$, which by definition transforms as a tensor of weight $(h,\bar{h})$ under general conformal transformations $(z \to f(z), \bar{z} \to \bar{f}(\bar{z}))$:

$$\phi(z,\bar{z}) \to (\frac{\partial f}{\partial z})^h (\frac{\partial \bar{f}}{\partial \bar{z}})^{\bar{h}} \phi(f(z), \bar{f}(\bar{z})). \tag{3.39}$$

With $f(z) = z + \epsilon(z) + \ldots$ we have:

$$\delta_\epsilon \phi(w,\bar{w}) = h(\partial_w \epsilon(w))\phi(w,\bar{w}) + \epsilon(z)\partial_w \phi(w,\bar{w}) + \ldots. \tag{3.40}$$

Comparing (3.40) and (3.36) we find:

$$T(z)\phi(w,\bar{w}) = \frac{h}{(z-w)^2}\phi(w,\bar{w}) + \frac{1}{z-w}\partial_w \phi(w,\bar{w}) + (\textit{regular terms}). \tag{3.41}$$

This expression is called an operator product expansion (OPE), and it contains all information about the quantum properties of the fields under conformal transformations. *En lieu* of explicit Lagrangians, 2-dimensional quantum field theories are defined either thru OPE's or, equivalently, by giving the 'operator algebra': the commutators of all the currents. You will therefore see many papers start with a list of commutators/OPE's, rather than giving a Lagrangian which you are used to.

The double pole singularity in the OPE of $\phi(w,\bar{w})$ with $T(z)$ is therefore as good a definition of a primary conformal field as any. A secondary conformal field has by definition a higher order singularity in its OPE with $T(z)$ or $\bar{T}(\bar{z})$. The most important secondary field is $T(z)$ itself. The simplest way to see this is to determine the functional form of the 2-point function (correlator) of T's by naive dimensional analysis (which works because T is a conserved current):

$$\langle T(z)T(0)\rangle = \frac{c/2}{z^4}. \tag{3.42}$$

The constant c will depend on which theory T is computed for, and has been normalized so that $c = 1$ for a free scalar field theory ($\mathcal{L} = \partial_\mu\phi\partial^\mu\phi$). It will actually turn out to parametrize different *universality classes* of 2-dimensional quantum field theories , which means that many apparently very different theories will turn out to have the same value of *the central charge c*. It measures the conformal anomaly of the theory, which means that if $c \neq 0$ then the classical scale-invariance of the theory has been broken by quantum effects. (On a curved background it would suffice to study the 1-point function, i.e. the VEV of T, since the trace anomaly c is the coefficient of the scalar curvature R ($\langle T\rangle \sim cR$). In string theory, where we want scaling to be an exact (unbroken) symmetry, c must always vanish (after ghosts etc. are accounted for).

The result (3.42) also follows directly from the OPE:

$$T(z)T(w) = \frac{c/2}{(z-w)^4} + \frac{2}{(z-w)^2}T(w) + \frac{1}{z-w}\partial_w T(z). \tag{3.43}$$

The last two terms are exactly what we would expect if $T(z)$ were a proper $(2,0)$ tensor (which it is classically, we are seeing the effects of quantum anomalies here). Unlike for $T(z)\phi(w,\bar{w})$, dimensional analysis now allows the additional term, and its existence can easily be verified in specific models, e.g. for a free scalar field. The anomaly means that the primary fields and stress are not true tensors, and may transform with anomalous scaling dimensions. In particular $h(\bar{h})$ may not be integers any longer in the full quantum theory.

We can, if we wish, continue to elucidate the quantum structure of 2-dimensional conformal transformations by studying their operator content. However, it is now more convenient to examine the spectrum of states in these theories. They must be organized into irreducible representations of the conformal group. As usual these are obtained by examining the algebra of infinitesimal generators near unity, which for conformal symmetries is the infinite dimensional Virasoro algebra.

Virasoro Algebra

We have seen that conformal transformations are generated in the quantum theory by the stress tensor $T(z)$ (we discuss only the holomorphic part here). Since $T(z)$ is an analytic function, it can be Laurent expanded:

$$T(z) = \sum_{n=-\infty}^{\infty} L_n z^{-n-2} \tag{3.44}$$

where the quantum operators generating conformal transformations

$$L_n = \oint \frac{dz}{2\pi i} z^{n+1} T(z) \tag{3.45}$$

are called Virasoro operators. They satisfy the Virasoro algebra:

$$[L_m, L_n] = (m-n)L_{m+n} + \frac{c}{12}(m-1)m(m+1)\delta_{m,-n}. \tag{3.46}$$

Comparing with the classical algebra of generators of conformal transformations (3.15) we see that the quantum algebra has picked up a socalled 'central extension' or 'Schwinger term'. As already mentioned the value of the central charge c depends on which specific theory is being considered. Therefore, unlike in the classical case where only information about spacetime symmetries is encoded in the conformal algebra, we see that the representations of the quantum algebra (3.46) will also carry more detailed information about the field content of the theory.

Notice that the subgroup generated by $\{L_0, L_{\pm 1}\}$ is not anomalous, so the Möbius transformations remain exact symmetries of the vacuum (i.e. unbroken by quantum effects).

The 'in' vacuum state is defined as usual (think of the L_n's as Fourier coefficients a_k) by:

$$L_n|0\rangle = 0, \quad n \geq -1. \tag{3.47}$$

Since T is hermitean the adjoint of L_n is $L_n^\dagger = L_{-n}$, and the 'out' vacuum is annihilated by:

$$\langle 0|L_n^\dagger = 0, \quad n \geq -1. \tag{3.48}$$

In analogy to atomic physics, we now define a highest weight state (hws) as a state created from the vacuum by a primary operator $\phi(0)$ of weight h:

$$|h\rangle = \phi(0)|0\rangle. \tag{3.49}$$

It has the required properties of being an eigenstate of the 'Hamiltonian' L_0:

$$L_0|h\rangle = h|h\rangle \tag{3.50}$$

and annihilated by half of the L_n's:

$$L_n|h\rangle = 0 \quad (n > 0). \tag{3.51}$$

By 'stepping down' with the 'ladder' operators L_{-n} ($L_0, L_{\pm n}$ are analogous to the SO_3 operators $J_z, J_\pm$ used in atomic physics) we obtain all states 'descended' from the primary state $|h\rangle$:

$$L_{-n_1}L_{-n_2}\ldots L_{-n_k}|h\rangle \;\; (n_i > 0, i = 1, 2, \ldots, k). \tag{3.52}$$

Since

$$L_0 L_{-n}|h\rangle = (h + n)L_{-n}|h\rangle \tag{3.53}$$

these are all eigenstates of the 'Hamiltonian'.

It can easily be seen that all states in a positive Hilbert space are linear combinations of primary and descendant states. We call the set of one primary state together with all its secondary states an *irreducible representation* (or Verma module):

$$\begin{array}{ccc}
\textit{Level} & L_0 - \textit{eigenvalue} & \textit{States} \\
 & & \\
0 & h & |h\rangle \\
1 & h+1 & L_{-1}|h\rangle \\
2 & h+2 & L_{-2}|h\rangle, L_{-1}^2|h\rangle \\
3 & h+3 & L_{-3}|h\rangle, L_{-1}L_{-2}|h\rangle, L_{-1}^3|h\rangle \\
\cdot & \cdot & \cdot \\
\cdot & \cdot & \cdot \\
\cdot & \cdot & \cdot \\
n & h+n & P(n) \; \textit{states}
\end{array} \tag{3.54}$$

where the degeneracy $P(n)$ is determined by the generating function:

$$\frac{1}{\prod_{n=1}^{\infty}(1-q^n)} = \sum_{n=0}^{\infty} P(n)q^n. \tag{3.55}$$

The properties of the secondary fields are completely determined by the primary field to which they belong. It can also be seen that the spectrum of primary fields completely determine the field theory. It is because the infinite dimensional Virasoro algebra in this way drastically reduces and organizes the number of essentially independent fields that conformal symmetry is such a powerful tool in 2 dimensions. In fact, since the 'minimal models' to be discussed next have a *finite* number of primary fields, these theories can be classified and exhaustively studied. Here is a graphical representation of the spectrum of such a model organized into Verma modules (irreps):

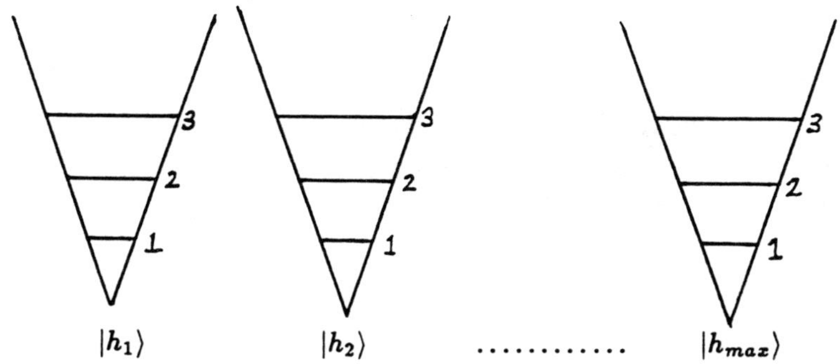

Minimal Models

The main result above was that a conformally invariant quantum field theory in 2 dimensions is uniquely specified by the values of the conformal dimensions $\{h_i\}$ of the primary fields, which depend on the value of the central charge c: $h_i = h_i(c)$. We shall now see that not all $h_i(c)$ are possible. In fact, in a unitary theory (positive definite Hilbert space) with $c < 1$ only a *discrete* set of c-values are allowed, to which are associated only a finite number of *rational* scaling dimensions.

The point is that the construction of irreps above does not automatically produce independent positive norm states. The descendants may be linearly dependent (null states), in which case we throw them out of the representation, or they may even have negative norm, in which case we throw out the representation.

In order to investigate for which values of the c and h the Virasoro algebra has unitary representations, we compute the inner products of all the $P(n)$ states at each level n. They are naturally organized into a $P(n) \times P(n)$ matrix $M_n(c,h)$, whose determinant (the socalled 'Kač determinant') can be written:

$$detM_n(c,h) = \prod_{pq \leq n} (h - h_{p,q}(c))^{P(n-pq)}, \tag{3.56}$$

where p and q are positive integers. The eigenvalues $h_{p,q}(c)$ are :

$$h_{p,q}(m) = \frac{[(m+1)p - mq]^2 - 1}{4m(m+1)}, \tag{3.57}$$

where m is a convenient reparametrization of c:

$$c(m) = 1 - \frac{6}{m(m+1)}, \tag{3.58}$$

which in general may be a complex number. If $M_n(c,h)$ has a zero eigenvalue the eigenstate must vanish for the Hilbert space to be positive definite. If $M_n(c,h)$ has negative eigenvalues then the representation is not unitary. A detailed analysis of the vanishing set of $detM_n(c,h)$ (as a function of c and h), shows that unitary representations are possible only if

$$c \geq 1, \; h > 0 \quad or \quad c = c(m) < 1 \quad with \;\; m = 3, 4, \ldots. \tag{3.59}$$

Explicit (lattice) models with $c < 1$ have been constructed for all m. As promised there are only a finite number ($\frac{1}{2}m(m+1)$) of primary fields with anomalous scaling dimensions given by (3.57) with

$$\begin{array}{c} 1 \leq p \leq m-1 \\ 1 \leq q \leq p. \end{array} \tag{3.60}$$

To summarize, constructing a unitary and conformally invariant quantum field theory is so hard that for $c < 1$ only a discrete choice is available. These theories model statistical mechanical systems near criticality. The simplest of them, which has $m = 3$ and $c = \frac{1}{2}$, is in fact the well-known Ising model, to which we now turn in order to illustrate the physical meaning of some of the concepts introduced above.

Ising Model

If we want to model an Ising system which is a lattice of spins pointing either up or down:

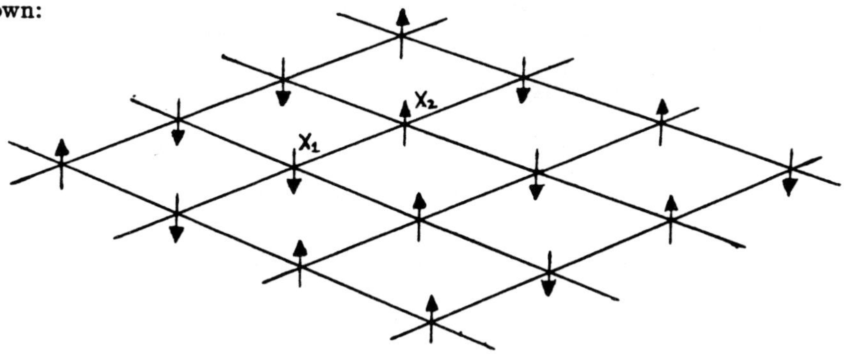

we expect to describe it in the continuum limit by scalar field whose values describe the average local magnetization (which is a number) of the system.

In general the primary fields (ϕ) of a model depend on both the scaling weights $h_{p,q}$ and $\bar{h}_{\bar{p},\bar{q}}$:

$$\phi = \phi_{p,q}^{\bar{p},\bar{q}}(z, \bar{z}). \tag{3.61}$$

For the minimal model ($m = 3, c = \frac{1}{2}$) it can be shown (by considering modular invariance, i.e. global conformal transformations) that only fields with $p = \bar{p}$ and $q = \bar{q}$ appear in the theory. Because $L_0 - \bar{L}_0$ measures the spin of a field, this means that this model only has spinless fields. Furthermore, from (3.60) we see that there in this case are only two non-trivial conformal families labeled by the primary fields $\phi_{2,2}$ and $\phi_{2,1}$ ($\phi_{1,1}$, which has scaling dimension $h_{1,1} = 0$, is the identity operator and has no physical signifigance.) They have scaling dimensions $h_{p,q} = \bar{h}_{\bar{p},\bar{q}} = \frac{1}{16}$ and $\frac{1}{2}$, respectively.

Since the Ising model is characterized by two local lattice quantities, the local magnetization and the local energy density, we see that the simplest of the minimal models has the right type and number of fields to mimic the Ising model near criticality.

In order to identify the lattice model described by this field theory, as well as the physical meaning of the scaling operators, it is sufficient for us to study the 2-point functions (correlators) of the scaling operators.

The functional form of the 2-point function

$$A^{(2)}(z_1, z_2, \bar{z}_1, \bar{z}_2) = \langle \phi_1(z_1, \bar{z}_1)\phi_2(z_2, \bar{z}_2)\rangle \tag{3.62}$$

(which until we take the real section of $C(z) \times C(\bar{z})$ depends on four points) is in fact completely fixed by conformal invariance (in any number of dimensions, i.e. by the finite dimensional Möbius group alone). By translational invariance (3.62) can depend only on the differences between the points where the fields are evaluated

$$A^{(2)} = F(z_1 - z_2, \bar{z}_1 - \bar{z}_2), \tag{3.63}$$

and by rotational invariance it can depend only on the value, not the direction, of these vectors:

$$A^{(2)} = F(|z_1 - z_2|, |\bar{z}_1 - \bar{z}_2|). \tag{3.64}$$

Furthermore, since the fields are primary they transform as 'quasi-tensors' and the correlator transforms as:

$$\begin{aligned} &\langle \phi_1(z_1, \bar{z}_1)\phi_2(z_2, \bar{z}_2)\rangle \to \\ &f'(z_1)^{h_1} f'(z_2)^{h_2} f'(\bar{z}_1)^{\bar{h}_1} f'(\bar{z}_2)^{\bar{h}_2} \langle \phi_1(f(z_1), f(\bar{z}_1))\phi_2(f(z_2), f(\bar{z}_2))\rangle. \end{aligned} \tag{3.65}$$

Hence, with $x = |z_1 - z_2|$ and $\bar{x} = |\bar{z}_1 - \bar{z}_2|$, from dilatation invariance we must have:

$$F(\lambda x, \bar{\lambda}\bar{x}) = \lambda^{-(h_1+h_2)}\bar{\lambda}^{-(\bar{h}_1+\bar{h}_2)} F(x, \bar{x}). \tag{3.66}$$

In other words the correlator is a quasi-homogeneous function of $(x, \bar{x})$ with scaling weights $(-(h_1 + h_2), -(\bar{h}_1 + \bar{h}_2))$ (Widom scaling). The simplest choice is:

$$A^{(2)} \sim |z_1 - z_2|^{-(h_1+h_2)} |\bar{z}_1 - \bar{z}_2|^{-(\bar{h}_1+\bar{h}_2)}. \tag{3.67}$$

The physical correlation function is obtained by taking the real section $\bar{z}_i = z_i^*$ ($i = 1, 2$), whence

$$A^{(2)} \sim |z_1 - z_2|^{-(\Delta_1+\Delta_2)}, \tag{3.68}$$

where the total scaling dimensions are defined as $\Delta_i = h_i + \bar{h}_i$. Finally it can be seen that invariance under the special conformal transformations requires that two spinless fields correlate iff they have the same scaling dimension $\Delta = \Delta_1 = \Delta_2$.

So all told Möbius invariance implies that:

$$\langle \phi_{p,q}(z_1)\phi_{p,q}(z_2)\rangle \sim x^{-2\Delta_{p,q}}, \tag{3.69}$$

where

$$2\Delta_{2,2} = 4h_{2,2} = \frac{1}{4}; \quad 2\Delta_{2,1} = 4h_{2,1} = 2. \tag{3.70}$$

We wish to compare this with the critical behaviour of the Ising model with nearest neighbor (n.n.) interactions:

$$S = -\beta \sum_{bonds} \sigma(x_1)\sigma(x_2) \tag{3.71}$$

where x_1 and x_2 are sites on the lattice and $\sigma(x) = \pm 1$ is the value of the Ising spin at site x. The order parameter is the value of the local magnetization at x, $\langle\sigma(x)\rangle$, which vanishes in the disordered (high temperature) phase, fluctuates wildly at the critical point, and freezes into a non-vanishing value in the ordered (low temperature) phase. The local energy density at position x_1 is given by:

$$\epsilon(x_1) = \sum_{x_2=n.n.} \sigma(x_1)\sigma(x_2). \tag{3.72}$$

The two basic critical exponents η and ν are defined by the behaviour of the corrlators of σ and ϵ as a function of $x = |x_1 - x_2|$ near criticality:

$$\begin{aligned} \langle\sigma(x_1)\sigma(x_2)\rangle &\sim x^{-\eta} \\ \langle\epsilon(x_1)\epsilon(x_2)\rangle &\sim x^{-2(2-\frac{1}{\nu})}. \end{aligned} \tag{3.73}$$

Exponents which describe the behaviour of correlation functions, like η and ν, are called microscopic critical exponents. They are related to the exponents which describe the behaviour of thermodynamic quantities near criticality (macroscopic exponents) thru the *scaling laws* which are necessarily true if the thermodynamic functions also are quasi-homogeneous functions of the reduced variables:

$$\begin{aligned} \alpha + 2\beta + \gamma = 2 \quad &(Rushbrooke) \\ \gamma = \beta(\delta - 1) \quad &(Widom) \\ \gamma = (2-\eta)\nu \quad &(Fisher) \\ 2\nu = 2 - \alpha \quad &(Josephson). \end{aligned} \tag{3.74}$$

In the case of the Ising model we know from Onsager's exact solution (or numerical simulations) that

$$\eta = \frac{1}{4}; \quad \nu = 1. \tag{3.75}$$

Hence by comparing (3.73) and (3.69), we conclude that $\sigma = \phi_{2,2}$ and $\epsilon = \phi_{2,1}$.

With this identification in hand, one can completely 'solve' the model, in the sense that any correlation function (and therefore any observable) can be computed in the minimal conformal field theories.

For the other minimal models we can also carry out the necessary identification of the physical meaning of the field theoretic scaling operators by comparing their properties under discrete symmetries (like Z_2) with the symmetries of the local lattice operators. In this way we can compute all the critical exponents from the exact scaling dimensions (3.57). The most familiar (i.e. Ising type) exponents are given by (+(−) refer to $m = 2k + 2$ ($m = 2k + 1$) with $k = 2, 3, \ldots$):

$$\begin{aligned} \alpha_{\pm} &= \frac{2k \pm 2}{3k \pm 1} \\ \beta_{\pm} &= \frac{k \mp 1}{12k}, \end{aligned} \tag{3.76}$$

together with the scaling laws (3.74). Notice that the discrete choice of allowed c-values ($c < 1$) accounts for the observed universality of critical phenomena, while the result (3.76) accounts for the observed rationality of critical exponents.

This should be ample motivation for students of statistical mechanics to study this beautiful subject. In the next section we turn to a totally orthogonal application: building unified models of all the fundamental interactions in Nature.

4. SOLVABLE STRING VACUA

We now adopt the intrinsic picture and regard the spacetime coordinates of the string as scalar fields on a Riemann surface. The physics of the string must of course be invariant uner reparametrizations, and in particular conformal transformations, of the world sheet coordinates. In other words we represent the string by a conformally invariant 2-dimensional scalar field theory. As already explained such field theories have unique properties. If we demand that our unitary scale-invariant field theories also carry N=2 supersymmetry, there is still a discrete series of minimal models [1][6][24][25]. The result analogous to (3.58) applies when $c < 3$:

$$c(k) = \frac{3k}{k+2} \quad (k = 1, 2, \ldots). \tag{4.1}$$

This is the case relevant for superstring theory, but unfortunately anomaly cancellation requires $c = 12$. Three units of the central charge are responsible for the 4-dimensional part which we assume to be flat, so our task is to construct $c = 9$ representations of the N=2 algebra which describes the physics of the remaining degrees of freedom. Gepner [32]-[37] has suggested an elegant way of generating such representations, by simply tensoring r minimal models at levels k_i together so that the total central charge adds to 9:

$$c = \sum_{i=1}^{r} \frac{3k_i}{k_i + 2} = 9. \tag{4.2}$$

Rather than going through all the details of Gepner's formal argument, which can be found in a readable form in his recent review article [37], we shall here just define the N=2 models and concentrate on how to carry out the tensor construction in practice.

In the previous section we learned that a conformally invariant quantum theory is completely specified up to a c-number by the OPE's of the basic currents associated with the symmetries, or equivalently by the commutator algebra of the moments of these currents. The two supersymmetries in the N=2 models are carried by superpartners $G^{\pm}$ of the bosonic current T (the stress tensor). It turns out that the resulting algebra will only close if there is also a U_1 current J present, and all told we have the following superconformal extension of the Virasoro algebra:

$$\begin{aligned}
[L_m, L_n] &= (m-n)L_{m+n} + \frac{c}{12}(m-1)m(m+1)\delta_{m,-n} \\
[L_m, J_n] &= -nJ_{m+n} \\
[L_m, G_r^{\pm}] &= (\frac{m}{2} - r)G_{m+r}^{\pm} \\
[J_m, J_n] &= -nJ_{m+n} \\
[J_m, G_r^{\pm}] &= \pm G_{m+r}^{\pm} \\
\{G_r^{+}, G_s^{-}\} &= 2L_{r+s} + (r-s)J_{r+s} + \frac{c}{3}(r^2 - \frac{1}{4})\delta_{r,-s}.
\end{aligned} \tag{4.3}$$

The indices n, m are integers, while r, s are half-integers.

The minimal N=2 model at level k has primary fields which may be labeled by the 'isospin' (l) of the underlying SU_2 Kac-Moody algebra, the 'magnetic' quantum number q which determines the charge of the U_1-symmetry which is automatically enforced by the two superconformal symmetries on the world sheet, and a number s which labels the *sector* to which the field belongs. Odd $s(= \pm 1)$ denote primary

Ramond (RA) fields, $s=0$ labels primary Neveu-Schwarz (NS) fields, and fields labeled by $s=\pm 2$ are obtained by acting on primary fields with the lowest moments $G^{\pm}_{-\frac{1}{2}}$ of the two superconformal partners $G^{\pm}$ of the stress tensor. This may result in a null state (e.g. $(l,q,s)=(0,0,\pm 2)$), but if it does not the resulting mode may contribute to the massless spectrum of the tensored model and must therefore be considered when searching for such states. This exhausts the set of 'atomic' fields which can contribute to the massless matter spectrum of the 'composite' model, which must all have conformal dimension $H_L=\frac{1}{2}$ in the (spacetime) supersymmetric sector, and $H_R=\frac{1}{2}$ or 1 in the gauge sector depending on whether the mode represents a vector or scalar particle.

The conformal dimension (analogous to (3.57)in the bosonic case) and U_1-charge (Q) of these minimal primary fields are given by:

$$\begin{aligned} h &= \frac{l(l+2)-\tilde{q}^2}{4(k+2)}+\frac{s^2}{8} \\ Q &= \frac{\tilde{q}}{k+2}-\frac{s}{2} \end{aligned} \tag{4.4}$$

where $\tilde{q}=q+s$ in the RA-sector, and $\tilde{q}=q$ in the NS-sector.

The symmetries of the superstring are given by a product of two Virasoro algebras, or the corresponding $SU_2^L\times SU_2^R$ current algebra, so each primary field (Φ) of the solvable (minimal) model at level k is labeled by two sets of quantum numbers (l,q,s) and $(\bar{l},\bar{q},\bar{s})$ for the left and right movers, respectively, which we denote by:

$$\begin{pmatrix} l & q & s \\ \bar{l} & \bar{q} & \bar{s} \end{pmatrix}_k \equiv \Phi^{l;\bar{l}}_{q,s;\bar{q},\bar{s}}(z,\bar{z}). \tag{4.5}$$

Because of the $Z_k\times\bar{Z}_k$ symmetry of the parafermionic theory to which the N=2 superconformal theory at level k is intimately related, the primary fields have some important symmetry properties which identifies some operators with distinct labels:

$$\begin{aligned} \begin{pmatrix} l & q & s \\ \bar{l} & \bar{q} & \bar{s} \end{pmatrix}_k &\simeq \begin{pmatrix} l & q\pm 2(k+2) & s\pm 4 \\ \bar{l} & \bar{q} & \bar{s} \end{pmatrix}_k \simeq \begin{pmatrix} l & q & s \\ \bar{l} & \bar{q}\pm 2(k+2) & \bar{s}\pm 4 \end{pmatrix}_k \\ &\simeq \begin{pmatrix} k-l & q\pm(k+2) & s\pm 2 \\ k-\bar{l} & \bar{q}\pm(k+2) & \bar{s}\pm 2 \end{pmatrix}_k . \end{aligned} \tag{4.6}$$

Using these symmetries we can always bring the SU_2 quantum numbers into the 'standard range':

$$\begin{aligned} 0 \leq & l \leq k \\ |q| &\leq l, \end{aligned} \tag{4.7}$$

but not necessarily simultaneously in both the left and right sector.

Gepner showed that it is possible to build a modular invariant partition function from these minimal representations of the extended superconformal algebra on the world sheet without destroying the N=1 spacetime supersymmetry which is highly desirable for phenomenological reasons. This is achieved by keeping only those composite states in the tensor product which have odd integer U_1 charge. The resulting superstring theory can easily be mapped onto a heterotic one.

Spacetime supersymmetry and modular invariance of the partition function are extremely restrictive requirements. It is therefore perhaps surprising that a consistent heterotic string can be built from *any* product of solvable models ($c < 3$) as long as they together carry the central charge $c = 9$ required for anomaly cancellation in 4 spacetime dimensions. It is even more surprising that there is sometimes more than one way to do this. While it is always possible to tie together the left and right partition function in a modular invariant way by using the diagonal affine invariant ($A_{l,\bar{l}} \sim \delta_{l,\bar{l}}$) which exists at every level k, at level $k = 10, 16$ and 28 there is also an exceptional invariant available (whose properties are related to E_6, E_7 and E_8, respectively). The massless spectrum of the resulting 228 models is known [56][58]. In addition, a very large number of other models is available by 'twisting' the simple A or E type model with some discrete group action. In this way one can construct the models based on the only other affine invariants, the D invariants which exist at levels $k = 4j$ and $k = 4j - 2$ ($j = 1, 2, \ldots$), as well as other twisted models formed by using the discrete symmetries of the theta function invariants.

We wish to construct a modular invariant partition function using only states that appear in the tensor product of r minimal models. In this we are aided by the fact that under modular transformations the transformation of the partition function for a N=2 theory factorises into a l piece, which is just an affine transformation, a q piece which is a level $(k+2)$ theta function system, and an s piece which is a level 2 theta function. The complete set of invariants for the theta functions at level m is known and consists of the left-right symmetric solution and quotients of it by any subgroup of the full Z_m symmetry. The set of affine invariants is also known, as discussed above.

Thus it is straightforward to write a modular invariant partition function for the tensor theory. This is then constrained to include only N=1 supersymmetric representations in the D uncompactified dimensions to yield a N=1 spacetime supersymmetric theory. It was found possible to do this, in a manner consistent with modular invariance , by identifying the supersymmetry charge with $Q = exp(i\phi)$, where $J = \partial_z \phi$ is the U_1 current algebra part of the N=2 superconformal algebra, and by requiring that this charge be an odd integer.

The final step in the construction is to turn the superstring compactification into a heterotic one. The right sector (now describing the bosonic string) uses precisely the same tensor product of minimal N=2 superconformal blocks as the left sector (this allows for the use of the same modular invariants), contributing the same amount to the conformal anomaly. The remaining conformal anomaly is made up by $c = 2$ from the transverse bosons and $c = 13$ from the free bosons of an $E_8 \times SO_{10}$ Lie group. The observation that allows for immediate use of the modular invariant solutions found for the superstring is that the representation of the modular group for $E_8 \times SO_{10}$ theta functions is precisely the same as the representation for SO_2 if one interchanges the vector representation, v, with the singlet, o, and changes the signs of the spinor $(s, \bar{s})$ partition functions. (Here, for the right sector, the notation denotes the SO_{10} representation multiplied by the singlet of E_8.)

The resultant partition function has the form (without the contribution from the transverse bosons)

$$W = \sum_{\substack{V, \bar{V}, l, \bar{l} \\ V - \bar{V} \in \Lambda \\ V\beta_0, \bar{V}\beta_0 \in \frac{1}{2} + Z \\ V\beta_i, \bar{V}\beta_i \in Z}} \pm A_{l,\bar{l}} Z^{l}_{V} Z^{\bar{l}\,*}_{\mu+\bar{V}} \tag{4.8}$$

where $A_{l,\bar{l}}$ represents a product of the multiplicities of the $\mathbf{A}_1^{(1)}$ invariants

$$A_{l,\bar{l}} = \prod_{i=1}^{r} A^{(k_i)}_{l_i, \bar{l}_i}. \tag{4.9}$$

The quantity V $(\bar{V})$ is a vector $(a; b; c)$ whose first index (first 5 indices), a, labels the SO_2 spacetime (SO_{10} internal) degrees of freedom. The remaining $2r$ indices b, c of V $(\bar{V})$ are the q, s $(\bar{q}, \bar{s})$ indices of the r minimal N=2 superconformal theories. The condition $V - \bar{V} \in \Lambda$ in (4.8) restricts the sum of states to include right-left symmetric states and those connected by either the supersymmetry generator $\beta_0 = (\bar{s}; 1; 1)$ or the vectors $\beta_i = (v; 0; 2 \textit{ on } i, 0 \textit{ elsewhere})$ which span the RA and NS sectors separately for a given l and q. (We use the notation that $o, s, \bar{s}, v$ refer to the weight vectors of the

SO_2 spacetime (left sector) or SO_{10} gauge (right sector) degrees of freedom.) Thus Λ is the lattice spanned by these vectors over the integers:

$$\Lambda = \sum_{i=0}^{r} \sum_{n_i \in Z} n_i \beta_i \tag{4.10}$$

The notation introduced above in (4.5), in terms of which the non-gauge part of Λ is generated by:

$$\begin{aligned} \beta_0 &= \begin{pmatrix} 0 & 0 & 0 \\ 0 & 1 & 1 \end{pmatrix}_{k_1} \begin{pmatrix} 0 & 0 & 0 \\ 0 & 1 & 1 \end{pmatrix}_{k_2} \cdots\cdots \begin{pmatrix} 0 & 0 & 0 \\ 0 & 1 & 1 \end{pmatrix}_{k_r} \\ \beta_i &= \begin{pmatrix} 0 & 0 & 0 \\ 0 & 0 & 0 \end{pmatrix}_{k_1} \cdots \begin{pmatrix} 0 & 0 & 0 \\ 0 & 0 & 2 \end{pmatrix}_{k_i} \cdots \begin{pmatrix} 0 & 0 & 0 \\ 0 & 0 & 0 \end{pmatrix}_{k_r} \end{aligned} \tag{4.11}$$

$(i = 1, 2 \ldots, r)$ is perhaps more transparent.

The condition $V\beta_0, \bar{V}\beta_0 \in \frac{1}{2} + Z$ in (4.8) projects onto integer supercharge states, and the condition $V\beta_i, \bar{V}\beta_i \in Z$ projects all the subsectors of the fields in the theory onto either the RA or the NS sector. This is required in order to conserve the total N=1 superconformal gauge condition. Here the product of two vectors n, m is given by $nm = \sum_{i=1}^{r} \frac{n_i m_i}{2p_i}$, where the i'th entry labels a level p_i theta function. Finally, in (4.8) the appearance of the vector $\mu = (v; 0; 0)$ implements the heterotic construction.

It is now straightforward to determine the massless spectrum of the theory. It consists of the states included in the sum in (4.8) which, in addition, satisfy the masslessness conditions $H^R_{TOT} = 2H^L_{TOT} = 1$ for the right and left sector, respectively, where the contribution to H_{TOT} from each of the subtheories is given by (4.4).

Consider first the left sector component of the chiral supermultiplets of the N=1 spacetime theory (4 flat dimensions). Using charge conjugation we may always choose these to be left-handed in which case the spacetime index a of the left sector is o or s. For the scalar component the contribution to H from spacetime vanishes so that $H_{TOT} = H^L_{INT} = \frac{1}{2}$. Using the definition for H and Q from (4.4) together with (4.2) immediately shows that the only possible massless states have $Q_{TOT} = 1$ and $l_i = q_i, s_i = 0$ for their primary field components. (Note that the states $l_i = q_i = 0, s_i = 2$ are null states.) It is now easy to scan through the possible tensor products of such states to determine which give $H = \frac{1}{2}$. Applying the supersymmetry generator β_0 one may always construct the massless spinor partners. (Using (4.4) and (4.2) one easily verifies that these indeed have $H^L_{INT} = \frac{3}{8}$ and integer supercharge as required.)

Using this set of left sector states we now construct the right sector states by operating on them with the elements of Λ. The gauge non-singlet fields transforming as the **27** of E_6 are immediately given for if $V = (o; b; c)$ is an L-state with $H^L_{TOT} = \frac{1}{2}, Q^L_{TOT} = 1$, then $\tilde{V} = (v; \tilde{b}; \tilde{c})$ (where the tilde denotes that they may not be in the standard range) will be an R-state with $H^R_{TOT} = 1, Q^R_{TOT} = 1$. (Note that the gauge contribution to the charge is obtained by replacing v by o.) The state $V \otimes \tilde{V}$ is a massless scalar transforming as the **10** of SO_{10}, and operating with β_0 on the R-state gives first the **16** and then the **1** of SO_{10} which together make up the 27-dimensional representation of E_6.

The states transforming as the $\overline{\mathbf{27}}$ are found by searching for R-states with $H^R_{TOT} = 1, Q^R_{TOT} = -1$, starting with the $\tilde{V}$ states and adding elements of Λ which keep the vector gauge component (i.e. $a = v$). This gives the component of the $\overline{\mathbf{27}}$ representation of E_6 transforming as the $\overline{\mathbf{10}}$, under the SO_{10} subgroup, and the remaining $\overline{\mathbf{16}}$ and **1** are found by applying β_0 as before.

Finally the chiral superfields transforming as singlets under E_6 are found by starting with the left-sector states $\tilde{V}$ and operating with Λ-vectors as before, but this time requiring that the gauge component of the state be singlet (i.e. $a = o$). The contribution to H_{TOT} and Q_{TOT} from the non-gauge indices are then $H^R_{INT} = 1, Q^R_{INT} = 0, \pm 2$.

The simplest 6-dimensional example is provided by taking five copies of the representation at level 3 ($n = 3$). The massless chiral superfields determine the particle spectrum of the effective 4-dimensional theory. In this particular case (but not in general!) counting the number of massless chiral fields is almost trivial and can be done by hand, because all the matter fields have left-right symmetric quantum numbers. We can without loss of generality always choose to start with the scalar component of the left moving mode, which as explained above are built only from primary fields with left quantum numbers $(l, l, 0)$ $(l = 0, 1, 2, 3 = k)$. There are only five combinations that can contribute to massless matter fields (i.e. that have $H_L = \frac{1}{2}, Q_L = 1$), so there is a total of 101 chiral matter fields transforming as a **10** under SO_{10} (**27** under the full E_6), which are permutations of the five left-right symmetric modes:

$$\phi_1 = \begin{pmatrix} 0 & 0 & 0 \\ 0 & 0 & 0 \end{pmatrix}_3 \begin{pmatrix} 0 & 0 & 0 \\ 0 & 0 & 0 \end{pmatrix}_3 \begin{pmatrix} 0 & 0 & 0 \\ 0 & 0 & 0 \end{pmatrix}_3 \begin{pmatrix} 2 & 2 & 0 \\ 2 & 2 & 0 \end{pmatrix}_3 \begin{pmatrix} 3 & 3 & 0 \\ 3 & 3 & 0 \end{pmatrix}_3 \quad (20)$$

$$\phi_2 = \begin{pmatrix} 0 & 0 & 0 \\ 0 & 0 & 0 \end{pmatrix}_3 \begin{pmatrix} 0 & 0 & 0 \\ 0 & 0 & 0 \end{pmatrix}_3 \begin{pmatrix} 1 & 1 & 0 \\ 1 & 1 & 0 \end{pmatrix}_3 \begin{pmatrix} 1 & 1 & 0 \\ 1 & 1 & 0 \end{pmatrix}_3 \begin{pmatrix} 3 & 3 & 0 \\ 3 & 3 & 0 \end{pmatrix}_3 \quad (30)$$

$$\phi_3 = \begin{pmatrix} 0 & 0 & 0 \\ 0 & 0 & 0 \end{pmatrix}_3 \begin{pmatrix} 0 & 0 & 0 \\ 0 & 0 & 0 \end{pmatrix}_3 \begin{pmatrix} 1 & 1 & 0 \\ 1 & 1 & 0 \end{pmatrix}_3 \begin{pmatrix} 2 & 2 & 0 \\ 2 & 2 & 0 \end{pmatrix}_3 \begin{pmatrix} 2 & 2 & 0 \\ 2 & 2 & 0 \end{pmatrix}_3 \quad (30)$$

$$\phi_4 = \begin{pmatrix} 0 & 0 & 0 \\ 0 & 0 & 0 \end{pmatrix}_3 \begin{pmatrix} 1 & 1 & 0 \\ 1 & 1 & 0 \end{pmatrix}_3 \begin{pmatrix} 1 & 1 & 0 \\ 1 & 1 & 0 \end{pmatrix}_3 \begin{pmatrix} 1 & 1 & 0 \\ 1 & 1 & 0 \end{pmatrix}_3 \begin{pmatrix} 2 & 2 & 0 \\ 2 & 2 & 0 \end{pmatrix}_3 \quad (20)$$

$$\phi_5 = \begin{pmatrix} 1 & 1 & 0 \\ 1 & 1 & 0 \end{pmatrix}_3 \begin{pmatrix} 1 & 1 & 0 \\ 1 & 1 & 0 \end{pmatrix}_3 \begin{pmatrix} 1 & 1 & 0 \\ 1 & 1 & 0 \end{pmatrix}_3 \begin{pmatrix} 1 & 1 & 0 \\ 1 & 1 & 0 \end{pmatrix}_3 \begin{pmatrix} 1 & 1 & 0 \\ 1 & 1 & 0 \end{pmatrix}_3 \quad (1). \tag{4.12}$$

There is only one anti-family mode ($\overline{\mathbf{27}}$ of E_6) satisfying the above conditions. It is obtained by adding the Λ-lattice vector $-2\beta_0 + \beta_1 + \beta_2 + \beta_3 + \beta_4 + \beta_5$ to ϕ_5, which gives:

$$\begin{pmatrix} 1 & 1 & 0 \\ 1 & -1 & 0 \end{pmatrix}_3 \begin{pmatrix} 1 & 1 & 0 \\ 1 & -1 & 0 \end{pmatrix}_3 \begin{pmatrix} 1 & 1 & 0 \\ 1 & -1 & 0 \end{pmatrix}_3 \begin{pmatrix} 1 & 1 & 0 \\ 1 & -1 & 0 \end{pmatrix}_3 \begin{pmatrix} 1 & 1 & 0 \\ 1 & -1 & 0 \end{pmatrix}_3 . \tag{4.13}$$

Some of the 330 gauge singlet modes are much harder to find, because in addition to the 310 obvious ones which are obtained by changing one $\bar{s}$ quantum number in each of the above family (anti-family) modes by $2(-2)$, there are 20 left-right asymmetric modes which are permutations of

$$\begin{pmatrix} 0 & 0 & 0 \\ 0 & -2 & -2 \end{pmatrix}_3 \begin{pmatrix} 1 & 1 & 0 \\ 1 & -1 & 0 \end{pmatrix}_3 \begin{pmatrix} 1 & 1 & 0 \\ 1 & -1 & 0 \end{pmatrix}_3 \begin{pmatrix} 1 & 1 & 0 \\ 1 & -1 & 0 \end{pmatrix}_3 \begin{pmatrix} 2 & 2 & 0 \\ 2 & 0 & 0 \end{pmatrix}_3 . \tag{4.14}$$

This mode is obtained by adding the Λ-lattice vector $-2\beta_0 + \beta_2 + \beta_3 + \beta_4 + \beta_5$ to ϕ_4.

Notice now that the massless spectrum of the 3^5 model is almost precisely the signature of a compactification on a quintic hypersurface in CP^4. An apparent discrepancy of 4 in the number of gauge singlets can be accounted for by noticing that the tensor construction also gives rise to 4 extra gauge bosons, since each minimal piece contributes a U_1 operator (and one is needed in the total N=2 representation). Presumably the extra scalars provide the longitudinal component needed to give the extra bosons a mass as we move away from the highly symmetric vacuum obtained by the tensor construction.

By generalizing the simple combinatorics evident in this example ($\sum_{i=1}^{r} l_i = r$), it is easy to show that all the superconformal field theories with integer central charge obtained by homogeneously tensoring minimal models have the same spectrum as that of the 'Fermat surfaces' in CP^n which we discussed in Section 2 (a hypersurface of degree $n+1$ in CP^n always has 1 harmonic (1,1) form, and the deformation counting of harmonic $(n-2,1)$ modes is exactly the same as for the quintic).

Thus we are compelled to conclude that the algebraic construction of string vacua, which is essentially an excercise in constructing representations of infinite dimensional super-Kac-Moody algebras, has somehow recovered the information which we obtained in the extrinsic approach to string theory. In the 'low-brow' (or 'low-energy') language of ordinary spacetime field theory, the result is again that strings must compactify on Calabi-Yau manifolds.

It is possible to see why this is the case by expressing both the geometric Calabi-Yau compactification and the algebraic tensor construction of N=2 superconformal representations in the Lagrangian formalism. Once the geometric information is encoded in a non-linear sigma-model action and the operator content of the N=2 representations is encoded in a Landau-Ginzburg action, the two pictures can be related through the partition functions which encode the full quantum content of the theories.

5. NON-LINEAR SIGMA-MODELS

Let us now write down sigma models with target spaces of the algebraic type discussed in Section 2 [54]. In the generic form of such a sigma-model, recorded in (1.1), the target space metric $G_{\mu\nu}$ appears explicitly. Since no metric is known for any nontrivial algebraic target space it would seem impossible to proceed. However, we can get around this obstacle by encoding the metric information in the action in a less direct way.

The only indirectly available metric on these manifolds is the one induced by the canonical metric on the ambient space $CP^n(\zeta\bar{\zeta} = \sum_{i=1}^{n} \zeta_i \zeta^i)$:

$$G_{i\bar{j}}^{FS} = \partial_i \partial_{\bar{j}} K^{FS} = \frac{1}{1+\zeta\bar{\zeta}}(\delta_{i\bar{j}} - \frac{\zeta_i \zeta_{\bar{j}}}{1+\zeta\bar{\zeta}}), \tag{5.1}$$

which is called the Fubini-Study metric. Because CP^n is a Kähler manifold the metric is just two derivatives on a single scalar function K^{FS} (the Kähler potential):

$$K^{FS} = ln(1 + \zeta\bar{\zeta}). \tag{5.2}$$

We can not give an equally explicit expression for the induced FS-metric on an algebraic hypersurface $p = 0$. But in physics we have an easy way of imposing a constraint on a system, by simply adding a term λp (and its complex conjugate) to the FS-action [66]. The Lagrange multiplier λ is an independent unphysical (non-propagating, 'auxiliary') field. It can in principle be eliminated by using its equation of motion, in which case the induced FS-metric would appear explicitly in the new action. Hence at string tree level, because the world-sheet metric is conformally flat, the bosonic piece of the action is just:

$$S = \int d^2z\{-2G^{FS}_{i\bar{j}}(\partial\zeta^i\bar{\partial}\zeta^{\bar{j}} + \bar{\partial}\zeta^i\partial\zeta^{\bar{j}}) + \lambda p(\zeta) + \bar{\lambda}\bar{p}(\bar{\zeta})\} \tag{5.3}$$

where $\partial = \frac{\partial}{\partial z}$ and $\bar{\partial} = \frac{\partial}{\partial \bar{z}}$.

It is now quite surprising that extending this to an N=2 supersymmetric action is very easy. This follows from the remarkable fact that N=2 supersymmetry on the world-sheet is *equivalent* to Kähler geometry on the target space [74][4]. Since all Calabi-Yau manifolds are Kähler , it is now automatic that the sigma-model admits a unique N=2 extension, which is precisely what we want. Even more remarkable is the fact that the N=2 invariant action is just the 'super Kähler potential'. More precisely, extend the world-sheet $M(z,\bar{z})$ to a superspace $\tilde{M}(z,\bar{z},\theta,\bar{\theta})$, where $(\theta,\bar{\theta})$ are complex Weyl spinors associated with the two world-sheet supersymmetries. We should then regard the scalar fields $\zeta^i(z,\bar{z})$ as the lowest components of N=2 chiral scalar superfields (we follow the conventions of [68]):

$$X^i = \zeta^i + \sqrt{2}\theta\psi^i + \theta\bar{\theta}F^i, \tag{5.4}$$

where ψ^i and F^i are Weyl spinors and auxiliary fields, respectively. If $K(\zeta^i,\zeta^{\bar{\imath}})$ is the Kähler potential on $\mathcal{M}(\zeta^i,\zeta^{\bar{\imath}})$ then

$$S = \int d^2z d^4\theta K(X^i, X^{\bar{\imath}}) \tag{5.5}$$

is the N=2 invariant action, whose bosonic piece appears in (1.1). We could hardly have asked for anything simpler!

The best example is the CP^n-model itself. According to (5.2) and (5.5) the N=2 supersymmetric sigma-model is

$$S^{FS} = \int d^2 z d^4 \theta ln(1 + X\bar{X}), \tag{5.6}$$

which is indeed the action originally derived by Di Vecchia and Ferrara [23][2] and Witten [69] in 1977. This D-term is the N=2 invariant extension of the kinetic piece of the bosonic action (5.3). The constraint λp should be generalized to a 'potential' term (F-term).

To see how to do this consider the superspace constraint

$$\Pi = 1 + \sum_{i=1}^{n} (X^i)^k = 0. \tag{5.7}$$

Since a product of chiral superfields is itself a chiral superfield, this is really three constraints, one of which is $p = 0$. In order to enforce the other two in the action, we need two more Lagrange multiplier fields. Clearly λ must be extended to the chiral scalar superfield

$$\Lambda = f + \theta\chi + \theta\theta\lambda, \tag{5.8}$$

and the N=2 invariant action for the constrained CP^n-model can be written:

$$S^{CCP} = \int d^2 z d^4 \theta ln(1 + X\bar{X}) + \int d^2 z d^2 \theta \Lambda \Pi + \int d^2 z d^2 \bar{\theta} \bar{\Lambda} \bar{\Pi}. \tag{5.9}$$

To uncover the geometrical meaning of this superspace magic, we integrate out the Grassmann variables $(\theta, \bar{\theta})$ and eliminate the auxiliary fields F and f which have algebraic equations of motion. We can then write the action in the more familiar but also much more complicated form ($S^{CCP} = \int d^2 z \mathcal{L}(z, \bar{z})$):

$$\begin{aligned} \mathcal{L} = & - 2G^{FS}_{i\bar{j}}(\partial\zeta^i \bar{\partial}\zeta^{\bar{j}} + \bar{\partial}\zeta^i \partial\zeta^{\bar{j}}) + \lambda p + \bar{\lambda}\bar{p} \\ & - iG^{FS}_{i\bar{j}} \psi^i \overleftrightarrow{\not{\mathcal{D}}} \psi^{\bar{j}} - \chi p_i \psi^i - \bar{\chi} \bar{p}_{\bar{i}} \psi^{\bar{i}} \\ & + \frac{1}{4} R^{FS}_{i\bar{k}j\bar{l}} (\psi^i \psi^j)(\psi^{\bar{k}} \psi^{\bar{l}}) + \frac{1}{4} (K^{FS}_{ij} \psi^i \psi^j)(K^{FS}_{\bar{k}\bar{l}} \psi^{\bar{k}} \psi^{\bar{l}}). \end{aligned} \tag{5.10}$$

Here $\mathcal{D}$ is the Kähler covariant derivative

$$\mathcal{D}\psi^i = \partial\psi^i + \Gamma^i_{jk}(\partial\zeta^k)\psi^j, \tag{5.11}$$

and

$$\not{\mathcal{D}} = \begin{pmatrix} 0 & \mathcal{D} \\ \bar{\mathcal{D}} & 0 \end{pmatrix} \tag{5.12}$$

$$\begin{aligned} p_i &= \partial_i p = \frac{\partial p}{\partial \zeta^i} \\ p_{ij} &= D_j p_i = \partial_j p_i - \Gamma^k_{ji} p_k. \end{aligned} \tag{5.13}$$

Γ^i_{jk} and $R^{FS}_{ikj\bar{l}}$ are the Christoffel symbol and Riemann curvature tensor of the ambient space (with the ambient metric $G^{FS}_{i\bar{j}}$). As promised, the Lagrange multiplier λ confines the fields ζ^i to the hypersurface $p = 0$, while χ forces the ψ^i to be tangent to this space. This in turn projects the tensor K^{FS}_{ij}, defined by

$$K^{FS}_{ij} \psi^i \psi^j = (p_k p^k)^{-\frac{1}{2}} p_{ij} \psi^i \psi^j, \tag{5.14}$$

onto the tangent space of $p = 0$, and it is in fact the extrinsic curvature of the hypersurface in the ambient metric $G^{FS}_{i\bar{j}}$ [8]. Hence every term in the sigma-model action has a simple geometrical interpretation, as they must.

Landau-Ginzburg Models

We claimed above that the minimal models describe statistical systems in two dimensions close to a phase transition. We shall now exploit this connection to write down actions for the 2-dimensional field theories that interest us here [72][50][59][67][46][54][55].

Recall first that near a critical point the 'classical' (i.e. not including fluctuations) information about the phase transition is contained in the Landau-Ginzburg potential. This is a polynomial in the order parameter, which we can take to be a scalar field ϕ. It can e.g. be the average value of the magnetization in an Ising model, which near the critical point is described by the simplest of the minimal models (N=0, n=1). In the oridnary theory of phase transitions (N=0) these potentials arise in a mean field description, which is the classical contribution to the full partition function. The basic assumption of Landau-Ginzburg theory is that the thermodynamic potentials are *analytic* near the critical point. At least for low dimensional systems this is not true, and the classical critical exponents derived from this assumption are quite different from the correct ones given in (3.76). When the fluctuations are strong enough to overpower the classical piece, there are contributions to the partition function from all orders in the loop expansion. The renormalization group (RG) technique is an approximate way of summing up these contributions. Nevertheless, many of the qualitative features of the transition are captured by the classical theory, and the physical insight offered by the Landau-Ginzburg description is irreplaceable. Furthermore, when we turn to the N=2 theories, which have super-Landau-Ginzburg descriptions, a powerful non-renormalization theorem actually protects the potential from receiving any corrections [70][22][26][48][19].

Consider therefore the bosonic Landau-Ginzburg theories ($n = 1, 2, \ldots$):

$$\mathcal{L}^{(n)} = (\partial\phi)^2 + \mathcal{V}(\phi), \tag{5.15}$$

where

$$\mathcal{V} = m^2\phi^2 + \ldots + g\phi^{2(n+1)}. \tag{5.16}$$

Zamolodchikov has shown that at the multi-critical point, defined by:

$$\frac{\partial\mathcal{V}}{\partial\phi} = \frac{\partial^2\mathcal{V}}{\partial\phi^2} = \ldots = \frac{\partial^{2n+1}\mathcal{V}}{\partial\phi^{2n+1}} = 0, \tag{5.17}$$

and the non-trivial fix-point, the theory $\mathcal{L}^{(n)}$ is equivalent to the minimal model (of type A) at level n. Hence the 'bare' actions for the A-type models are given by:

$$\mathcal{L}^{*(n)} = (\partial\phi)^2 + g\phi^{2(n+1)}, \tag{5.18}$$

and the primary operators of the conformal algebra are represented by the scalar field ϕ, $\partial\phi$ and composites of these.

What about the D- and E-type minimal models, which together with the A-type models are known to exhaust the list of all possible unitary minimal models? Since we have already considered all possible polynomials of one variable, we should now allow two real order parameters ϕ_1 and ϕ_2. Thom and Arnold have shown that all possible extrema of smooth functions $\mathcal{V} = \mathcal{V}(\phi_1, \phi_2)$ are captured by the following list of 'germs' of critical points:

$$\begin{aligned} A_k^{\pm} &= \pm\phi_1^{k+1} \quad (k \geq 1) \\ D_k^{\pm} &= \pm\phi_1^{k-1} + \phi_1\phi_2^2 \quad (k \geq 4) \\ E_6^{\pm} &= \phi_1^3 \pm \phi_2^4 \\ E_7 &= \phi_1^3 + \phi_1\phi_2^3 \\ E_8 &= \phi_1^3 + \phi_2^5. \end{aligned} \tag{5.19}$$

The labels A, D and E refer to the fact that these potentials are in one-to-one correspondence with the simply laced Lie algebras. In other words, near a critical point any smooth potential $\mathcal{V} = \mathcal{V}(\phi_1, \phi_2)$ must have the same qualitative behaviour as one of these ADE-germs. The germs appear in a Taylor expansion near the extremum. The higher order terms vanish by a smooth change of variables, while the absence of lower order terms characterize the type of extremum. The remaining 'middle' terms contain all the information about how the function goes critical.

However, when fluctuations are important (as they always are in two dimensions (N=0,1)), the higher order terms in the expansion can no longer be ignored and the analyticity of the potential does not reflect the physics correctly. It is therefore not obvious in what sense the ADE-classification of germs of Landau-Ginzburg potentials 'explains' the ADE-classification of 2-dimensional quantum field theories with $c < 1$ (minimal models with N=0). Apparently the fluctuations (quantum effects) can only upset the quantitative, and not the qualitative, aspects of the Landau-Ginzburg description. In particular, the quasi-homogeneity property of the germs (which is not obvious for more than two order parameters) seems to survive the fluctuations, albeit with new (renormalized, anomalous) scaling weights. Namely, if the free energy scales like a weighted homogenous polynomial then the scaling laws (3.74) are necessarily true. This property is called Widom scaling in physics.

In a completely analogous manner, the N=2 invariant minimal models can be represented by Landau-Ginzburg potentials of chiral scalar superfields Φ. The lowest components of these potentials are polynomials of complex scalar fields (ϕ) for which an ADE-classification also exists. The germs are the same as shown in (5.19) with all negative signs omitted. The A-type N=2 invariant minimal model at level n is described by the germ $A_{n+1} = \phi^{n+2}$, and the Landau-Ginzburg action at the critical point is now:

$$S^{LG} = \int d^2z d^4\theta \Phi\bar{\Phi} + g\int d^2z d^2\theta \Phi^{n+2} + \bar{g}\int d^2z d^2\bar{\theta}\bar{\Phi}^{n+2}. \tag{5.20}$$

Notice that if we assume that the coupling constant g is dimensionless then the scaling dimension of Φ is $\frac{1}{n+2}$. Furthermore, since the chiral superfields of the minimal model at level n have quantum numbers $l = |m|$, we see from (4.1) that the exact scaling dimensions of these fields is given by:

$$2h_{l,\pm l} = \frac{l}{n+2} \geq \frac{1}{n+2}. \tag{5.21}$$

Since F-terms are not renormalized in two dimensions, we conclude that the Landau-Ginzburg superfield Φ represents the most relevant chiral operator ($l = 1$). The other chiral operators are represented by composites Φ^l $(l = 2, 3, \ldots, n)$ of the Landau-Ginzburg field. Here the product of fields should be defined by an appropriate normal ordering, which is the only field-renormalization possible in two dimensions. In the singularity theory of Thom and Arnold (the lowest components of) these fields are called perturbations of the germ, and the fact that Φ^{n+1} is not a relevant perturbation of Φ^{n+2} reflects that it can be eliminated from the Landau-Ginzburg theory by its equation of motion. Thom calls a germ together with all its relevant perturbations a 'catastrophe'.

The F-term of the Landau-Ginzburg representation of the tensor construction is now simply given by adding the Landau-Ginzburg potentials for the minimal models being tensored. Our favourite example of five copies of the N=2 model at level $n = 3$ is therefore described by the Landau-Ginzburg model:

$$S^{LG} = \int d^2zd^4\theta K(\Phi, \bar{\Phi}) + \int d^2zd^2\theta \sum_{A=1}^{5} \Phi_A^5 + h.c., \tag{5.22}$$

where the D-term $K(\Phi, \bar{\Phi})$ remains to be specified. Although it is an irrelevant operator under the RG flow, this is the term which carries information about the metric of the target space (actually the metric of the ambient space in our case, whose restriction to the hypersurface is the target space metric), so that it would seem necessary to know it close to criticality in order to conclude which fixpoint this Landau-Ginzburg action is in fact describing. As it stands, (5.22) could be describing any number of situations, including the trivial one where K is taken to be the naive Landau-Ginzburg kinetic term $K = \sum_{A=1}^{5} \Phi_A \bar{\Phi}^A$ which describes flat C^5.

The reason for this is that we have failed to take into account the important U_1 projection which is part of Gepner's construction. It is a GSO-like truncation of the spectrum of states that guarantees modular invariance and spacetime supersymmetry. Its implementation in the Landau-Ginzburg construction becomes clear by comparing the intrinsic algebraic picture, the content of which is *partially* encoded in the Landau-Ginzburg action (5.22), and the extrinsic geometric picture, which is encoded in the sigma-model action:

$$S^{CCP} = \int d^2zd^4\theta ln(1 + \sum_{i=1}^{4} X_i\bar{X}^i) + \int d^2zd^2\theta\Lambda(1 + \sum_{i=1}^{4} X_i^5) + h.c. \tag{5.23}$$

The two actions S^{LG} and S^{CCP} are apparently very different. The following argument from ref.[46] indicates how to go from one to the other by *changing* the action S^{LG} into a *completely new theory* S^{OLG} which should then be compared with S^{CCP} to see if they are in the same universality class. Since the F-term of 2-dimensional N=2 theories is not renormalized, this can only be the case if $S_F^{CCP} = S_F^{OLG}$. This immediately tells us how to construct S^{OLG} from S^{LG}. Namely, in order to turn the quintic potential in five fields into a constraint in four independent variables, a non-linear 'change of coordinates' (actually a 'change of degrees of freedom' would be a better description) suggests itself:

$$\begin{aligned} X_i &= \frac{\Phi_i}{\Phi_5} \quad (i = 1, \ldots, 4) \\ \Lambda &= \Phi_5^5. \end{aligned} \tag{5.24}$$

This is nothing but a chiral superfield extension of the usual passage from homogeneous to inhomogeneous coordinates of CP^4, as described above in Section 2.

Notice first that this map is 1-1 only if we identify homogeneous fields which differ by fifth roots of unity. In other words, the theory parametrized by the new fields is a Z_5 orbifold of the Landau-Ginzburg model we started with. This is the implementation in the Landau-Ginzburg context of Gepner's projection onto states with integer charge under the overall N=2 superconformal U_1 current $exp(2\pi i J_0)$. Since this is the diagonal part of the full U_1^5 manifest in the product of five minimal N=2 representations at level 3 ($J_0 = \sum_{A=1}^{5} J_0^{(A)}$), and since $5q_A$ is an integer for every $A = 1, \ldots, 5$, the U_1^5 invariance reduces to a Z_5^5 invariance in this case, and the GSO-projection reduces to a projection onto states invariant under the diagonal Z_5. It should now be clear that it is the Z_5 orbifold S^{OLG} of S^{LG} that should be compared with S^{CCP}.

Notice next that the Jacobian associated with this transformation is trivial (constant) precisely because we are dealing with a polynomial of the Calabi-Yau type (the degree equals the number of homogeneous variables), so that this change of coordinates does not affect the partition functions.

The discussion so far can be summarized in the following diagram:

$$\begin{aligned} \mathcal{Z}^{LG} &= \int_{C^5} \mathcal{D}\Phi_1 \ldots \mathcal{D}\Phi_5 e^{-iS^{LG}(\Phi_1,\ldots,\Phi_5)} \quad \xrightarrow{mod\ Z_5} \\ \mathcal{Z}^{OLG} &= \int_{CP^4} \mathcal{D}\Phi_1 \ldots \mathcal{D}\Phi_5 e^{-iS^{OLG}(\Phi_1,\ldots,\Phi_5)} = \int_{CP^4} \mathcal{D}\Lambda\mathcal{D}X_1 \ldots \mathcal{D}X_4 e^{-iS^{OLG}(\Lambda,X_1,\ldots,X_4)}. \end{aligned} \tag{5.25}$$

Since the F-term is linear in Λ, if we now could argue that the kinetic term (the D-term) is independent of Λ, we could integrate out Λ to obtain a delta-function in $(1+X\bar{X})$, which is indeed the hypersurface constraint given by S_F^{CCP}. If we choose to study the theory after the D-term has been scaled away[46], this is perhaps trivially true, but it requires that we make assumptions about the existence and uniqueness of D-terms that are all but impossible to check.

However, any N=2 sigma-model possesses Kähler gauge invariance which in the case of the CP^n model means that one degree of freedom can be gauged away. This is because any holomorphic function of a chiral superfield is itself a chiral superfield, whose integral over $d^4\theta$ vanishes, so we can eliminate Φ_5 from $S^{CP} = S_D^{CCP}$:

$$\int d^4\theta ln(\Phi_A\bar{\Phi}^A) = \int d^4\theta ln\Phi_5\bar{\Phi}_{\bar{5}}(1+X\bar{X}) = \int d^4\theta ln(1+X\bar{X}). \tag{5.26}$$

This shows that the beautiful argument of ref.[46] does not have to fail even in the presence of a D-term, in particular not if we use the one provided by the constrained CP^n sigma-model.

Notice finally that the F-terms are manifestly invariant under the U_1^5 expected to be present in the full action only at the conformal fixpoint. This is another expression of the non-renormalization of F-terms. Furthermore, since S_D^{CCP} only is invariant under the diagonal U_1, which must be present by N=2 invariance, S^{CCP} can not be identical to S^{OLG} which by definition/assumption should possess the full U_1^5 invariance present in Gepner's construction. This is as expected, because the target space metric used in the construction of the constrained CP^n model is the induced Fubini-Study metric, which is not conformally invariant.

Nevertheless, S^{CCP} and S^{OLG} should still describe the same physics. Presumably they are in the same universality class, which means that after S^{CCP} flows to a fixpoint S^{CCP*} under the RG, S^{OLG} and S^{CCP*} can differ at most by a marginal perturbation (the fixpoint set should essentially be the moduli space of the Calabi-Yau manifold). In fact, since the D-term is irrelevant, we expect to recover the full U_1^5 invariance present in the F-term, so that $S^{OLG} = S^{CCP*}$. If a different defining polynomial is chosen, which corresponds to a marginal deformation of the Landau-Ginzburg potential, they should differ because only the Fermat polynomial used above possesses the full $Z_5^5 \simeq U_1^5$ invariance.

Thus we have all but proven the connection between the CICY construction and Gepner's construction, in rather illuminating physical terms which should also add to our general understanding of the peculiar connection between superconformal field theories in two dimensions and algebraic geometry.

In practical terms we should now, by bringing the combined power of the extrinsic geometric and intrinsic algebraic approach to bear on the compactification problem, be able to extract reliable phenomenology from string theory. A few promising steps have already been taken in this direction. Several non-renormalization properties [70][22][26][48][19] and selection rules [57] have recently been uncovered which add up to powerful statements about flat directions in the 4-dimensional superpotential. This is of great importance when searching for a realistic mechanism for breaking the GUT group. By making a detailed comparison of the Yukawa couplings in the only known three generation model [71][3][14][62][43][34], computed in both the intrinsic and extrinsic picture, these general results were verified [52][63][45][20]. This model seems to be a promising candidate for a unified theory.

6. COUPLINGS IN SUPERCONFORMAL MODELS

Until recently attempts at extracting phenomenology from the heterotic string concentrated on deriving as much information as possible from the field theory limit where a traditional Kaluza-Klein picture seems to be valid [15]. In this naive and perhaps unrealistic approximation, the effective four dimensional (spacetime) theory is an N=1 supergravity Yang-Mills theory with at least an $E_8 \times E_6$ gauge group.

It was rather surprising how far this 'manifold' approximation could be pushed. In particular, the only known three generation model [71] turned out to admit a compactification scheme which at least roughly fits the standard model [44]. However, many low-energy parameters remained unobtainable within this framework, including the Yukawa couplings.

On the other hand, the recent progress in constructing non-trivial representations of the N=2 superconformal algebra acting on the world sheet, as described above, have paved the way for computing *all* low-energy parameters of interest in a given 'compactified' heterotic string. This is because all physical questions that can be asked of the string at tree-level now translate into questions about properties of the minimal models, which are completely understood.

We now turn to the computation of couplings in these models. In the language of conformal field theory this means that we wish to compute correlation functions of the form

$$\langle V^1_{-1} V^2_{-\frac{1}{2}} V^3_{-\frac{1}{2}} V^4_0 \cdots V^n_0 \rangle, \tag{6.1}$$

where $V_{-1(0)}$ is a vertex operator for a spacetime scalar in the $(-1)((0))$ picture and $V_{-\frac{1}{2}}$ is a vertex operator for a spacetime fermion in the $(-\frac{1}{2})$ picture. The correlation function (6.1) has the correct ghost charge (-2) for a correlator computed at string tree level [31]. The spacetime supersymmetry means that (6.1) is non-vanishing only if it arises from the F-term of a product of chiral superfields, all of the same chirality which we choose to be left-handed.

As far as the internal sector goes, the vertex operator in the (-1) picture is essentially given by the product of the r primary fields responsible for the state:

$$V_{-1}(z,\bar{z}) \sim O(z,\bar{z}) = \bigotimes_{i=1}^{r} \begin{pmatrix} l_i & q_i & s_i \\ \bar{l}_i & \bar{q}_i & \bar{s}_i \end{pmatrix}_{k_i}. \tag{6.2}$$

The transformation from the (-1) picture to the $(-\frac{1}{2})$ picture is effected by the supersymmetry generator $(S(z))$ whose action on the internal sector is given by:

$$V_{-\frac{1}{2}}(z,\bar{z}) \sim S(z)O(z,\bar{z}) = \bigotimes_{i=1}^{r} \begin{pmatrix} l_i & q_i+1 & s_i+1 \\ \bar{l}_i & \bar{q}_i & \bar{s}_i \end{pmatrix}_{k_i}. \tag{6.3}$$

The transformation from the (-1) to the (0) picture is, roughly speaking (see [26] and references therein for more details), carried out by one of the superpartners of the stress tensor:

$$\begin{aligned} V_0(z,\bar{z}) &\sim G^+_{-\frac{1}{2}}(z)O(z,\bar{z}) \\ &= \sum_{j=1}^{r} \bigotimes_{i=1}^{j-1} \begin{pmatrix} l_i & q_i & s_i \\ \bar{l}_i & \bar{q}_i & \bar{s}_i \end{pmatrix}_{k_i} \otimes \begin{pmatrix} l_j & q_j & s_j+2 \\ \bar{l}_j & \bar{q}_j & \bar{s}_j \end{pmatrix}_{k_j} \otimes \bigotimes_{i=j+1}^{r} \begin{pmatrix} l_i & q_i & s_i \\ \bar{l}_i & \bar{q}_i & \bar{s}_i \end{pmatrix}_{k_i}. \end{aligned} \tag{6.4}$$

In this section we adopt the convention that the left quantum numbers refer to the field in the (-1) picture, while the right quantum numbers refer to an appropriate choice of representation leading to a gauge invariant correlation function.

The evaluation of (6.1) is greatly fascilitated by the large number of symmetries present in the models from the unitary minimal series. As explained in detail in [57], the U_1 which is automatically present in the N=2 superconformal algebra is particularly useful, because the correlation function is allowed to exist only if all factors $(i = 1, \ldots, r)$ in the tensor product *separately* respect the U_1 symmetries. For the right movers (gauge sector) this is just conservation of the charges $\sum_{j=1}^{n} \bar{Q}_{ij}$ $(\forall i)$ (where the charge $\bar{Q}$ is given in eq.(4.4)), but the supersymmetric (left) sector is more complicated because of the picture changing, whose effect on the charges is encoded in eqs.(6.3) and (6.4). From eqs.(6.1)-(6.4) it follows that the r left U_1 charges are conserved only if there exists some choice of numbers $\{p_1, \ldots, p_n\}$, with $p_1 = p_2 = p_3 = 0$ and $p_j \in \{1, 2, \ldots, r\} (j = 4, \ldots, n)$, such that for every factor $(i = 1, \ldots, r)$ the value of the total charge $\sum_{j=1}^{n} Q_{ij}$ equals $\sum_{j=1}^{n} \delta_{i,p_j}$. In short, we can encode the U_1^r conservation in the expression:

$$\langle \prod_{j=1}^{n} V^j \rangle \sim \sum_{\{p_1, \ldots, p_n\}} \prod_{i=1}^{r} \delta(\sum_{j=1}^{n} (Q_{ij} + \delta_{i,p_j})) \prod_{i=1}^{r} \delta(\sum_{j=1}^{n} (\bar{Q}_{ij}). \tag{6.5}$$

As explained in detail in [57] other conditions may be obtained from the parafermionic correlators [73] which enter (6.1) but which were not used in deriving eq.(6.5). These conditions are just SU_2 isospin selection rules and they can be used together with U_1^r conservation to derive the Z_{k_i+2} symmetry of each tensor factor. These may be identified with the discrete symmetries of the Calabi-Yau approximation, but they will not play any rôle in our discussion of Yukawa couplings since eq.(6.5) turns out to be sufficient.

While checking for U_1 invariance is rather easy using eq.(6.5), only in the case of three-point functions shall we evaluate their strength. This is sufficient for determining most of the low energy symmetry breaking pattern.

The lowest order contribution to the four dimensional superpotential is given by the Yukawa couplings:

$$c_{ijk} = \langle V_{-1,i} V_{-\frac{1}{2},j} V_{-\frac{1}{2},k} \rangle, \tag{6.6}$$

which are simply the structure constants of the superconformal field theory. As explained in refs.[26] and [35], in the case of diagonal (A-type) affine invariants they are easily determined by exploiting the connection to SU_2 WZW theory, for which the operator product coefficients (c_A) have been computed by Fateev and Zamolodchikov

[73]. In terms of the conventional SU_2 quantum numbers j and m, which are related to the labels of the minimal models through $l = 2j$ and $q - s = 2m$, the result is:

$$\begin{aligned} &c_A\left(\begin{pmatrix} j_1 & m_1 \\ j_1 & \bar{m}_1 \end{pmatrix}, \begin{pmatrix} j_2 & m_2 \\ j_2 & \bar{m}_2 \end{pmatrix}, \begin{pmatrix} j_3 & m_3 \\ j_3 & \bar{m}_3 \end{pmatrix}\right) \\ &\quad = n_A(j_1, j_2, j_3) \begin{bmatrix} j_1 & j_2 & j_3 \\ m_1 & m_2 & m_3 \end{bmatrix} \begin{bmatrix} j_1 & j_2 & j_3 \\ \bar{m}_1 & \bar{m}_2 & \bar{m}_3 \end{bmatrix}. \end{aligned} \tag{6.7}$$

This is basically an SU_2 group invariant, which must be built from the Wigner 3j-symbols [53] denoted here by $[\ldots]$, up to a complicated normalisation function which is given by:

$$\begin{aligned} n_A(j_1, j_2, j_3) = \left[\frac{\Gamma(\frac{k+3}{k+2})}{\Gamma(\frac{k+1}{k+2})}\right]^{\frac{1}{2}} &F(j_1)F(j_2)F(j_3)P(j_1 + j_2 + j_3 + 1) \\ &P(j_1 + j_2 - j_3)P(j_1 + j_3 - j_1)P(j_2 - j_3 + j_1), \end{aligned} \tag{6.8}$$

where

$$\begin{aligned} F(j) &= \left[(2j+1)\frac{\Gamma(1 - \frac{2j+1}{k+2})}{\Gamma(1 + \frac{2j+1}{k+2})}\right]^{\frac{1}{2}} P(2j)^{-1} \\ P(j) &= \prod_{i=1}^{j} \frac{\Gamma(1 + \frac{i}{k+2})}{\Gamma(1 - \frac{i}{k+2})}. \end{aligned} \tag{6.9}$$

An analogous result holds for the exceptional invariants:

$$\begin{aligned} &c_E\left(\begin{pmatrix} j_1 & m_1 \\ \bar{j}_1 & \bar{m}_1 \end{pmatrix}, \begin{pmatrix} j_2 & m_2 \\ \bar{j}_2 & \bar{m}_2 \end{pmatrix}, \begin{pmatrix} j_3 & m_3 \\ \bar{j}_3 & \bar{m}_3 \end{pmatrix}\right) \\ &\quad = n_E(j_1, j_2, j_3; \bar{j}_1, \bar{j}_2, \bar{j}_3) \begin{bmatrix} j_1 & j_2 & j_3 \\ m_1 & m_2 & m_3 \end{bmatrix} \begin{bmatrix} \bar{j}_1 & \bar{j}_2 & \bar{j}_3 \\ \bar{m}_1 & \bar{m}_2 & \bar{m}_3 \end{bmatrix}, \end{aligned} \tag{6.10}$$

but the determination of the normalization n_E is substantially more difficult [30][21].

It is also possible to compute Yukawa couplings in the extrinsic picture [64][65], up to some unknown normalizations. They have a topological interpretation, and it is extremely satisfying that the beautiful deformation technology developed by Candelas [8][13] has an exact analogue in the Landau-Ginzburg representation of the CFT [55][36]. Furthermore, a detailed comparison of the two techniques, applied both to the $3^5 \simeq CP^4[5]$ model and the three generation model discussed below, show complete consistency and provides a strong consistency check on the whole complex of ideas developed above.

Having mastered the technology for computing low energy parameters, we should turn to some specific models in order to see if realistic and believable low energy phenomenology can be obtained in this way.

7. A THREE GENERATION MODEL

The construction of this three generation model [34] from N=2 minimal superconformal fields proceeds in several stages, each of which has an exact analogue in the algebraic construction of the Calabi-Yau manifold [62] which appears in the low energy ten dimensional field theory limit. Here we only describe the CFT construction since the algebraic geometric construction is rather complicated and not particularly illuminating, and we refer the reader who is interested in the exact correspondence to the literature [52][63][45].

To obtain the three generation model we start with the model $1 \times E16^3$. Here $E16$ denotes that we use the exceptional invariant, which in practice means that the massless modes can only have $(l, \bar{l}) = (0,0), (0,16), (4,4), (4,12), (6,6), (6,10), (8,8), (2,8)$, or 'reflections' of these given by (4.6). Since the character for $E16$ exhibits a trivial Z_2 symmetry (contained in the Z_{18} automatically present for level 16), there are further identifications on the massless fields beyond those of (4.6). For $E16$ (and also $E28$) we should in fact apply the identifications of (4.6) separately to the left and right sector.

The resulting model has 35 families, 8 anti-families and 197 massless gauge singlets. The 35 family modes are permutations of the 9 basic modes:

$$
\begin{aligned}
U_1 &= \begin{pmatrix} 0 & 0 & 0 \\ 0 & 0 & 0 \end{pmatrix}_1 \begin{pmatrix} 0 & 0 & 0 \\ 0 & 0 & 0 \end{pmatrix}_{E16} \begin{pmatrix} 6 & 6 & 0 \\ 6 & 6 & 0 \end{pmatrix}_{E16} \begin{pmatrix} 12 & 12 & 0 \\ 12 & 12 & 0 \end{pmatrix}_{E16} \\
U_2 &= \begin{pmatrix} 0 & 0 & 0 \\ 0 & 0 & 0 \end{pmatrix}_1 \begin{pmatrix} 4 & 4 & 0 \\ 4 & 4 & 0 \end{pmatrix}_{E16} \begin{pmatrix} 4 & 4 & 0 \\ 4 & 4 & 0 \end{pmatrix}_{E16} \begin{pmatrix} 10 & 10 & 0 \\ 10 & 10 & 0 \end{pmatrix}_{E16} \\
U_3 &= \begin{pmatrix} 0 & 0 & 0 \\ 0 & 0 & 0 \end{pmatrix}_1 \begin{pmatrix} 6 & 6 & 0 \\ 6 & 6 & 0 \end{pmatrix}_{E16} \begin{pmatrix} 6 & 6 & 0 \\ 6 & 6 & 0 \end{pmatrix}_{E16} \begin{pmatrix} 6 & 6 & 0 \\ 6 & 6 & 0 \end{pmatrix}_{E16} \\
U_4 &= \begin{pmatrix} 1 & 1 & 0 \\ 1 & 1 & 0 \end{pmatrix}_1 \begin{pmatrix} 0 & 0 & 0 \\ 0 & 0 & 0 \end{pmatrix}_{E16} \begin{pmatrix} 0 & 0 & 0 \\ 0 & 0 & 0 \end{pmatrix}_{E16} \begin{pmatrix} 12 & 12 & 0 \\ 12 & 12 & 0 \end{pmatrix}_{E16} \\
U_5 &= \begin{pmatrix} 1 & 1 & 0 \\ 1 & 1 & 0 \end{pmatrix}_1 \begin{pmatrix} 0 & 0 & 0 \\ 0 & 0 & 0 \end{pmatrix}_{E16} \begin{pmatrix} 6 & 6 & 0 \\ 6 & 6 & 0 \end{pmatrix}_{E16} \begin{pmatrix} 6 & 6 & 0 \\ 6 & 6 & 0 \end{pmatrix}_{E16} \\
U_6 &= \begin{pmatrix} 1 & 1 & 0 \\ 1 & 1 & 0 \end{pmatrix}_1 \begin{pmatrix} 4 & 4 & 0 \\ 4 & 4 & 0 \end{pmatrix}_{E16} \begin{pmatrix} 4 & 4 & 0 \\ 4 & 4 & 0 \end{pmatrix}_{E16} \begin{pmatrix} 4 & 4 & 0 \\ 4 & 4 & 0 \end{pmatrix}_{E16} \\
U_7 &= \begin{pmatrix} 0 & 0 & 0 \\ 0 & 0 & 0 \end{pmatrix}_1 \begin{pmatrix} 0 & 0 & 0 \\ 0 & 0 & 0 \end{pmatrix}_{E16} \begin{pmatrix} 8 & 8 & 0 \\ 8 & 8 & 0 \end{pmatrix}_{E16} \begin{pmatrix} 10 & 10 & 0 \\ 10 & 10 & 0 \end{pmatrix}_{E16} \\
U_8 &= \begin{pmatrix} 0 & 0 & 0 \\ 0 & 0 & 0 \end{pmatrix}_1 \begin{pmatrix} 4 & 4 & 0 \\ 4 & 4 & 0 \end{pmatrix}_{E16} \begin{pmatrix} 6 & 6 & 0 \\ 6 & 6 & 0 \end{pmatrix}_{E16} \begin{pmatrix} 8 & 8 & 0 \\ 8 & 8 & 0 \end{pmatrix}_{E16} \\
U_9 &= \begin{pmatrix} 1 & 1 & 0 \\ 1 & 1 & 0 \end{pmatrix}_1 \begin{pmatrix} 0 & 0 & 0 \\ 0 & 0 & 0 \end{pmatrix}_{E16} \begin{pmatrix} 4 & 4 & 0 \\ 4 & 4 & 0 \end{pmatrix}_{E16} \begin{pmatrix} 8 & 8 & 0 \\ 8 & 8 & 0 \end{pmatrix}_{E16}
\end{aligned}
\tag{7.1}
$$

The 8 antifamily modes are permutations of the 4 basic modes:

$$
\begin{aligned}
\bar{U}_1 &= \begin{pmatrix} 0 & 0 & 0 \\ 0 & 0 & 0 \end{pmatrix}_1 \begin{pmatrix} 6 & 6 & 0 \\ 6 & -6 & 0 \end{pmatrix}_{E16} \begin{pmatrix} 6 & 6 & 0 \\ 6 & -6 & 0 \end{pmatrix}_{E16} \begin{pmatrix} 6 & 6 & 0 \\ 6 & -6 & 0 \end{pmatrix}_{E16} \\
\bar{U}_2 &= \begin{pmatrix} 1 & 1 & 0 \\ 1 & -1 & 0 \end{pmatrix}_1 \begin{pmatrix} 4 & 4 & 0 \\ 4 & -4 & 0 \end{pmatrix}_{E16} \begin{pmatrix} 4 & 4 & 0 \\ 4 & -4 & 0 \end{pmatrix}_{E16} \begin{pmatrix} 4 & 4 & 0 \\ 4 & -4 & 0 \end{pmatrix}_{E16} \\
\bar{U}_3 &= \begin{pmatrix} 0 & 0 & 0 \\ 0 & 2 & 2 \end{pmatrix}_1 \begin{pmatrix} 2 & 2 & 0 \\ 8 & -8 & 0 \end{pmatrix}_{E16} \begin{pmatrix} 8 & 8 & 0 \\ 2 & -2 & 0 \end{pmatrix}_{E16} \begin{pmatrix} 8 & 8 & 0 \\ 2 & -2 & 0 \end{pmatrix}_{E16} \\
\bar{U}_4 &= \begin{pmatrix} 1 & 1 & 0 \\ 1 & 3 & 2 \end{pmatrix}_1 \begin{pmatrix} 2 & 2 & 0 \\ 8 & -8 & 0 \end{pmatrix}_{E16} \begin{pmatrix} 2 & 2 & 0 \\ 8 & -8 & 0 \end{pmatrix}_{E16} \begin{pmatrix} 8 & 8 & 0 \\ 2 & -2 & 0 \end{pmatrix}_{E16} .
\end{aligned}
\tag{7.2}
$$

Only 131 of the 197 gauge singlets are guaranteed to exists in the field theory limit by the topology of the corresponding Calabi-Yau manifold [28].

The symmetry group of this theory is $\mathcal{G} = S_3 \times G/Z_9$, where the quotient is by the Z_9 subgroup of $G = Z_3 \times Z_9^3$ generated by the group element carrying the G-charge $(1,1,1,1)$, and S_3 is the permutation symmetry of the three $E16$ factors. We can exploit this large group of discrete symmetries to lower the number of generations from 27 to 3 by 'twisting' the model by a $Z_3 \times Z_3$ subgroup of G, which amounts to imposing new boundary conditions on the toroidal mode functions that enter the construction of the partition function. This $Z_3 \times Z_3$ group consists of the Z_3 group generated by the G-charge $(0,3,6,0)$ and the Z_3 cyclic permutation subgroup of S_3.

The effect of the twisting only amounts to shifting the $\bar{q}$ quantum numbers by the G-charge when determining the massless spectrum. We list the massless spectrum resulting from modding out by the first Z_3 subgroup only. If nothing else is said we always display the scalar component of the super-multiplet, and the $(\overline{\mathbf{10}}\ )\mathbf{10} \in SO_{10} \subset E_6$ of (anti-)family multiplets. Of the original 35 untwisted (2,1) forms, the 17 permutations of $U_1, \ldots, U_6$ survive the twisting. But because the Z_3 under consideration is not freely acting we must also consider 'twisted' modes. In this case there are 6 new twisted family modes which are permutations of:

$$
V = \begin{pmatrix} 0 & 0 & 0 \\ 0 & 0 & 0 \end{pmatrix}_1 \begin{pmatrix} 2 & 2 & 0 \\ 8 & 8 & 0 \end{pmatrix}_{E16} \begin{pmatrix} 8 & 8 & 0 \\ 8 & 8 & 0 \end{pmatrix}_{E16} \begin{pmatrix} 8 & 8 & 0 \\ 2 & 2 & 0 \end{pmatrix}_{E16} . \tag{7.3}
$$

All the 8 original antifamilies survive the twisting, and in addition there are also 6 twisted anitfamily modes to match the 6 twisted family modes. They are permutations of:

$$
\bar{V} = \begin{pmatrix} 0 & 0 & 0 \\ 0 & 0 & 0 \end{pmatrix}_1 \begin{pmatrix} 0 & 0 & 0 \\ 0 & 0 & 0 \end{pmatrix}_{E16} \begin{pmatrix} 6 & 6 & 0 \\ 6 & -6 & 0 \end{pmatrix}_{E16} \begin{pmatrix} 12 & 12 & 0 \\ 12 & -12 & 0 \end{pmatrix}_{E16} . \tag{7.4}
$$

The second Z_3 factor is freely acting and its effect is discussed below.

Notice that (in the standard range) no $s = \pm 2$ states appear in these matter multiplets. It is easy to see that this always has to be the case for fields transforming as **27** or $\overline{\mathbf{27}}$, and that at most one factor with $s = \pm 2$ can appear in any massless state. These are precisely the E_6 gauge singlets, the moduli, whose existence is guaranteed in the field theory limit by the topology of the Calabi-Yau manifold. Hence acting with $G^+_{-\frac{1}{2}}(\bar{z})$ on any one factor in the family modes, $U_i \quad (i = 1, \ldots, 6)$ and V, creates 43 and 18 E_6 gauge singlets, respectively, the $h^{2,1}$ fields specifying the complex structure. (Note that $(\bar{l}, \bar{q}, \bar{s}) \simeq (0, 0, \pm 2)$ are null states.) Acting on any factor in the anti-family modes, $\bar{U}_i \quad (i = 1, \ldots, 4)$ and $\bar{V}$, with $G^-_{-\frac{1}{2}}(\bar{z})$ creates 28 and 12 E_6 gauge singlets, respectively, corresponding to the $h^{1,1}$ radii. In addition there are 24 untwisted scalars[1] which are permutations of:

$$
\begin{aligned}
S_1 &= \begin{pmatrix} 0 & 0 & 0 \\ 0 & 0 & 0 \end{pmatrix}_1 \begin{pmatrix} 2 & 2 & 0 \\ 8 & -4 & 0 \end{pmatrix}_{E16} \begin{pmatrix} 8 & 8 & 0 \\ 2 & 2 & 0 \end{pmatrix}_{E16} \begin{pmatrix} 8 & 8 & 0 \\ 2 & 2 & 0 \end{pmatrix}_{E16} \\
S_2 &= \begin{pmatrix} 0 & 0 & 0 \\ 0 & 4 & 2 \end{pmatrix}_1 \begin{pmatrix} 2 & 2 & 0 \\ 8 & -6 & 0 \end{pmatrix}_{E16} \begin{pmatrix} 8 & 8 & 0 \\ 2 & 0 & 0 \end{pmatrix}_{E16} \begin{pmatrix} 8 & 8 & 0 \\ 2 & 0 & 0 \end{pmatrix}_{E16} \\
S_3 &= \begin{pmatrix} 0 & 0 & 0 \\ 0 & 0 & 0 \end{pmatrix}_1 \begin{pmatrix} 4 & 4 & 0 \\ 4 & -2 & 0 \end{pmatrix}_{E16} \begin{pmatrix} 4 & 4 & 0 \\ 4 & -2 & 0 \end{pmatrix}_{E16} \begin{pmatrix} 10 & 10 & 0 \\ 6 & 4 & 0 \end{pmatrix}_{E16} \\
S_4 &= \begin{pmatrix} 0 & 0 & 0 \\ 0 & 4 & 2 \end{pmatrix}_1 \begin{pmatrix} 4 & 4 & 0 \\ 4 & 2 & 0 \end{pmatrix}_{E16} \begin{pmatrix} 4 & 4 & 0 \\ 4 & 2 & 0 \end{pmatrix}_{E16} \begin{pmatrix} 10 & 10 & 0 \\ 10 & -10 & 0 \end{pmatrix}_{E16} \\
S_5 &= \begin{pmatrix} 0 & 0 & 0 \\ 0 & 4 & 2 \end{pmatrix}_1 \begin{pmatrix} 4 & 4 & 0 \\ 4 & -4 & 0 \end{pmatrix}_{E16} \begin{pmatrix} 4 & 4 & 0 \\ 4 & -4 & 0 \end{pmatrix}_{E16} \begin{pmatrix} 10 & 10 & 0 \\ 6 & 2 & 0 \end{pmatrix}_{E16} \\
S_6 &= \begin{pmatrix} 0 & 0 & 0 \\ 0 & 2 & 2 \end{pmatrix}_1 \begin{pmatrix} 4 & 4 & 0 \\ 4 & 0 & 0 \end{pmatrix}_{E16} \begin{pmatrix} 4 & 4 & 0 \\ 4 & 0 & 0 \end{pmatrix}_{E16} \begin{pmatrix} 10 & 10 & 0 \\ 6 & 6 & 0 \end{pmatrix}_{E16} \\
S_7 &= \begin{pmatrix} 0 & 0 & 0 \\ 0 & 0 & 0 \end{pmatrix}_1 \begin{pmatrix} 4 & 4 & 0 \\ 4 & 4 & 0 \end{pmatrix}_{E16} \begin{pmatrix} 4 & 4 & 0 \\ 4 & 4 & 0 \end{pmatrix}_{E16} \begin{pmatrix} 10 & 10 & 0 \\ 10 & -8 & 0 \end{pmatrix}_{E16} \\
S_8 &= \begin{pmatrix} 1 & 1 & 0 \\ 1 & 1 & 0 \end{pmatrix}_1 \begin{pmatrix} 4 & 4 & 0 \\ 4 & -2 & 0 \end{pmatrix}_{E16} \begin{pmatrix} 4 & 4 & 0 \\ 4 & -2 & 0 \end{pmatrix}_{E16} \begin{pmatrix} 4 & 4 & 0 \\ 4 & -2 & 0 \end{pmatrix}_{E16} \\
S_9 &= \begin{pmatrix} 1 & 1 & 0 \\ 1 & -1 & 0 \end{pmatrix}_1 \begin{pmatrix} 4 & 4 & 0 \\ 4 & 2 & 0 \end{pmatrix}_{E16} \begin{pmatrix} 4 & 4 & 0 \\ 4 & 2 & 0 \end{pmatrix}_{E16} \begin{pmatrix} 4 & 4 & 0 \\ 4 & 2 & 0 \end{pmatrix}_{E16} \\
S_{10} &= \begin{pmatrix} 1 & 1 & 0 \\ 1 & 3 & 2 \end{pmatrix}_1 \begin{pmatrix} 4 & 4 & 0 \\ 4 & 0 & 0 \end{pmatrix}_{E16} \begin{pmatrix} 4 & 4 & 0 \\ 4 & 0 & 0 \end{pmatrix}_{E16} \begin{pmatrix} 4 & 4 & 0 \\ 4 & 0 & 0 \end{pmatrix}_{E16} ,
\end{aligned}
\tag{7.5}
$$

[1] Hence the total number of untwisted scalars in the 9 generation model is 95. The 85 quoted in [34] is presumably a misprint.

and 48 twisted scalars which are permutations of:

$$
\begin{aligned}
T_1 &= \begin{pmatrix} 0 & 0 & 0 \\ 0 & 0 & 0 \end{pmatrix}_1 \begin{pmatrix} 0 & 0 & 0 \\ 0 & 0 & 0 \end{pmatrix}_{E16} \begin{pmatrix} 6 & 6 & 0 \\ 6 & 0 & 0 \end{pmatrix}_{E16} \begin{pmatrix} 12 & 12 & 0 \\ 4 & 0 & 0 \end{pmatrix}_{E16} \\
T_2 &= \begin{pmatrix} 0 & 0 & 0 \\ 0 & 4 & 2 \end{pmatrix}_1 \begin{pmatrix} 4 & 4 & 0 \\ 4 & 2 & 0 \end{pmatrix}_{E16} \begin{pmatrix} 4 & 4 & 0 \\ 4 & -4 & 0 \end{pmatrix}_{E16} \begin{pmatrix} 10 & 10 & 0 \\ 6 & -4 & 0 \end{pmatrix}_{E16} \\
T_3 &= \begin{pmatrix} 0 & 0 & 0 \\ 0 & 2 & 2 \end{pmatrix}_1 \begin{pmatrix} 4 & 4 & 0 \\ 4 & 0 & 0 \end{pmatrix}_{E16} \begin{pmatrix} 4 & 4 & 0 \\ 12 & 12 & 0 \end{pmatrix}_{E16} \begin{pmatrix} 10 & 10 & 0 \\ 6 & -6 & 0 \end{pmatrix}_{E16} \\
T_4 &= \begin{pmatrix} 0 & 0 & 0 \\ 0 & 0 & 0 \end{pmatrix}_1 \begin{pmatrix} 4 & 4 & 0 \\ 4 & 4 & 0 \end{pmatrix}_{E16} \begin{pmatrix} 4 & 4 & 0 \\ 4 & -2 & 0 \end{pmatrix}_{E16} \begin{pmatrix} 10 & 10 & 0 \\ 6 & -2 & 0 \end{pmatrix}_{E16} \\
T_5 &= \begin{pmatrix} 0 & 0 & 0 \\ 0 & 2 & 2 \end{pmatrix}_1 \begin{pmatrix} 4 & 4 & 0 \\ 12 & 12 & 0 \end{pmatrix}_{E16} \begin{pmatrix} 4 & 4 & 0 \\ 12 & -12 & 0 \end{pmatrix}_{E16} \begin{pmatrix} 10 & 10 & 0 \\ 6 & 6 & 0 \end{pmatrix}_{E16} \\
T_6 &= \begin{pmatrix} 0 & 0 & 0 \\ 0 & 0 & 0 \end{pmatrix}_1 \begin{pmatrix} 6 & 6 & 0 \\ 6 & -6 & 0 \end{pmatrix}_{E16} \begin{pmatrix} 6 & 6 & 0 \\ 6 & 6 & 0 \end{pmatrix}_{E16} \begin{pmatrix} 6 & 6 & 0 \\ 6 & 0 & 0 \end{pmatrix}_{E16} \\
T_7 &= \begin{pmatrix} 1 & 1 & 0 \\ 1 & 1 & 0 \end{pmatrix}_1 \begin{pmatrix} 0 & 0 & 0 \\ 0 & 0 & 0 \end{pmatrix}_{E16} \begin{pmatrix} 6 & 6 & 0 \\ 6 & 0 & 0 \end{pmatrix}_{E16} \begin{pmatrix} 6 & 6 & 0 \\ 6 & -6 & 0 \end{pmatrix}_{E16} \\
T_8 &= \begin{pmatrix} 1 & 1 & 0 \\ 1 & 3 & 2 \end{pmatrix}_1 \begin{pmatrix} 4 & 4 & 0 \\ 12 & -12 & 0 \end{pmatrix}_{E16} \begin{pmatrix} 4 & 4 & 0 \\ 4 & 0 & 0 \end{pmatrix}_{E16} \begin{pmatrix} 4 & 4 & 0 \\ 12 & 12 & 0 \end{pmatrix}_{E16} .
\end{aligned}
\tag{7.6}
$$

The model described in detail above has 9 generations of massless charged particles and 173 scalars.

The final step in obtaining a 3 generation string theory is to twist again, this time with the freely acting Z_3 permutation symmetry. The resulting theory has 9 families, 6 anti-families and at least 62 singlets, all of which are untwisted. In addition there may be twisted scalars. In this paper we shall work only with the 27 and 9 generation models. As far as the charged fields are concerned the final step to obtain the 3 generation model is essentially trivial being just a projection of the 9 generation model. If there are no twisted singlets the same is true of this sector, but at the moment this question is unresolved.

Using (6.10) on the Yukawa couplings in the 9 generation model which are allowed by (6.5), we find that only very few couplings survive. In fact, of the 2300, 560 and 55706 possible couplings of the type $\mathbf{27}^3$, $\overline{\mathbf{27}}^3$ and $\mathbf{1}\cdot\mathbf{27}\cdot\overline{\mathbf{27}}$, respectively, only 22, 7 and 79 survive. They are permutations of 6, 3 and 17 independent ones listed in [45]. Of the 877975 possible singlet couplings ($\mathbf{1}^3$) 369 survive as permutations of only 71 independent ones. Notice that none of the moduli, or indeed any of the singlet modes related to the (anti-) family modes by the action of $G^{i\pm}_{-\frac{1}{2}}(\bar{z})$ ($i = 1,\ldots,r$), couple to the (anti-) family modes, as required by the non-renormalization theorems proved in [26] and [56].

This completes the first step in determining the complete structure of the effective low energy theory resulting from the 4 dimensional superconformal theory. The trilinear terms in the superpotential computed here determine all the renormalisable interactions of the zero modes of the theory and may largely determine the low energy properties of the model. These couplings correspond to a particular point in the moduli space of the corresponding Calabi-Yau theory and may receive significant corrections as the radii of compactification change. Remarkably, the trilinear couplings of the (2,1) forms do not receive such corrections so the computation performed here determines these couplings for the general Calabi-Yau model. (There may be differences as the complex structure is changed but these are calculable.) As these couplings give the structure of the quark and lepton mass matrices there is hope that there may be reliable predictions for quark and lepton masses, but before such exciting questions can be answered we must break the symmetries down to those of the standard model.

The gauge symmetries are in fact quite easily dispensed with by using the CFT version of what we used to call 'Wilson line breaking' in the extrinsic manifold picture, which can be regarded as a 'twisting' of the gauge degrees of freedom. We have already had a brief encounter with twisting in the construction of the three generation model. It is a powerful method for constructing new models, and by a simple extension also for breaking gauge symmetries. It is a natural and attractive mechanism suggested by the structure of the theory itself, whose geometric meaning is just that flux lines get trapped in the topology of the Calabi-Yau , thus allowing non-zero VEVs of the gauge potential without turning on a field strength. The twisted states in the CFT offer some exotic possibilities, but unlike the problem of supersymmetry breaking there seems to be no reason why gauge symmetry breaking should prevent progress. The reader is referred to the literature for details.

8. CONCLUDING REMARKS

Space and time limitations has forced the omission of much interesting Calabi-Yau work from this review, in particular almost all the work on algebraic geometry as well as the 4-dimensional field theory approach, but the other speakers at this school will remedy that. This was an attempt to introduce the main concepts of the compactification game. Altough the emphasis has been heavily on CFT here, because it provides the technology we need for computing low energy parameters, we shall almost certainly continue to use the spacetime picture of compactification for many purposes. Not only does it remain irresistably appealing to view the '4 dimensional' strings in this way, it has become increasingly clear that Calabi-Yau manifolds are unreasonably efficient in collating information about the full string solution, despite only being a field theory approximation to this solution. In fact, most if not all of the low energy information we are now extracting from the superconformal representations seems to be of a geometrical nature and a direct translation from the language of algebraic geometry. Furthermore, as we now move away from the highly symmetric situations enforced by the tensor construction, it is the moduli (parameter) space of Calabi-Yau manifolds that provides the intuition needed for guiding progress [10][11][12][27].

In either case the name of the game is to construct the 4-dimensional field theory that is relevant at our energies. It is quite surprising how much of the structure of these theories is fixed by spacetime supersymmetry alone, as shown in recent work by Ferrara et al. What freedom remains ('Planck slop') should be fixed by the underlying string structure, up to a discrete choice of vacuum parameters (the Calabi-Yau topology). If this is true then the string will make contact with reality if we can find a believable mechanism for breaking supersymmetry. This may be 'just' a property of the particular field theory in question, albeit probably a non-perturbative one, or it may really be a problem which requires understanding the fundamentals of string theory, since it seems to be related to the vacuum degeneracy problem. But even if this is the case it may be that the field theory contains enough 'fossils' of the event that we will be able to give an effective description of the phenomenon which is sufficiently robust to allow prediction. If past Calabi-Yau history is a reliable guide, it does indeed seem that these systems are so constrained that this is a reasonable hope. Supersymmetry breaking remains the single most challenging problem in this approach to string theory.

Perhaps concluding with a few philosophical remarks could be helpful for those uninitiated in the rites of string.

When constructing a physical theory we usually demand that it give a satisfactory account of experimental data. Occasionally we can also tolerate a new theory which is only tenuously supported by experiment, but only if it embodies some particulary attractive or powerful idea, and then only temporarily. An example of the latter was the theory of relativity. At the moment string theory fails both of these tests: it is certainly not cast in a beautiful or natural form, and it makes no falsifiable statements. In short, it is not a yet a physical theory, but we *hope* that it is the germ of one. The idea is that we by a fortuitous historical accident have stumbled onto an effective formulation which only contains the few degrees of freedom relevant at and below the compactification scale, where our traditional picture of spacetime as an arena on which physics is played out is valid.

The only reason we are willing to live with this situation is that the string paradigm may provide a consistent quantum theory of gravity and all other interactions. In order to determine whether this is actually the case we must advance in two directions: down to the scale of phenomenology, and up to the Planck scale where presumably the foundations of string theory are to be found. Here we have only described our attempts to bridge the gap up to the compactification scale. What lies beyond there is still hidden in total darkness.

Acknowledgments

My understanding of this subject has grown out of discussions and collaborations with many people, in particular P. Candelas, A.M. Dale, P. Di Vecchia, M.G. Eastwood, P. Green, B.R. Greene, T. Hübsch, J.I. Latorre, X. de la Ossa, G.G. Ross, R. Schimmrigk and T.Taylor. It is a great pleasure to thank them all here.

References

[1] M.Ademollo et al., Phys.Lett. **62B** (1976)105; Nucl.Phys. **B111**(1976)77

[2] A.D'Adda, P.Di Vecchia and and M.Lüscher, Nucl.Phys. **B146**(1978)73

[3] P.S. Aspinwall, B.R.Greene, K.H.Kirklin and P.J.Miron, Nucl.Phys. **B294**(1987)193

[4] L.Alvarez-Gaumé and D.Z.Freedman, Phys.Rev. **D22**(1980)816 ; Commun.Math.Phys.**80**(1981)443

[5] A.A.Belavin, A.M.Polyakov and A.B.Zamolodchikov, *Infinite Conformal Symmetry in Two-Dimensional Quantum Field Theory*, Nucl.Phys.**B241** (1984) 333

[6] W.Boucher, D.Friedan and A.Kent, Phys.Lett. **172B** (1986)316

[7] P.Candelas, *Lectures on Complex Manifolds*, UTTG-21-87, 1987

[8] P.Candelas, Nucl.Phys. **B298**(1988)458

[9] P.Candelas, A.M.Dale, C.A.Lütken and R.Schimmrigk, Nucl.Phys. **B298**(1988)493

[10] P.Candelas, P.S.Green and T.Hübsch, *Connected Calabi-Yau Compactifications*, UTTG-28-88 (1988)

[11] P.Candelas, P.S.Green and T.Hübsch, Phys.Rev.Lett. **62**(1989)1956

[12] P.Candelas, P.S.Green and T.Hübsch, *Rolling Among Calabi-Yau Vacua*, UTTG-10-89 (1989)

[13] P.Candelas and S.Kalara, *Yukawa Couplings for a Three Generation Superstring Compactification*, UMN-TH-601-87, UT-TH-12/87 (1987)

[14] P.Candelas, C.A.Lütken and R.Schimmrigk, Nucl.Phys. **B306**(1988)113

[15] P.Candelas, G.Horowitz, A.Strominger and E.Witten, Nucl.Phys. **B258**(1985)46

[16] A.Capelli, C.Itzykson and J.B.Zuber, Nucl.Phys. **B280**(1987)445; Commun. Math. Phys. **113**(1987)1

[17] J.Cardy, *Conformal Invariance*, in Domb and Lebowitz (eds.), *Phase Transitions*, Vol.11, Academic Press (1987)

[18] J.Cardy, Les Houces Lectures, June 28 - August 5, 1988, to appear in Les Houches, Session XLIX, 1988, *Fields, Strings and Critical Phenomena*, ed. by E.Brézin and J.Zinn-Justin, Elsevier Science Publishers B.V.(1989)

[19] S.Cecotti, L.Girardello and A.Pasquinucci, *Non-perturbative Aspects and Exact Results for the N=2 Landau-Ginzburg Models*, SISSA preprint 51/89/EP (April 1989)

[20] S.F.Cordes and Y.Kikuchi, *Correlation Functions and Selection Rules in Minimal N=2 String Compactifications*, CTP-TAMU-92/88 (December 1988)

[21] P.Di Francesco, Phys.Lett. **215B**(1988)124

[22] M.Dine and N.Seiberg, Phys.Rev.Lett.**57**(1986)2625

[23] P.Di Vecchia and S.Ferrara, Nucl.Phys. **B130**(1977)93

[24] P.Di Vecchia, J.L.Petersen and H.Zheng, Phys.Lett. **162B**(1985)32

[25] P.Di Vecchia, J.L.Petersen, M.Yu and H.B.Zeng, Phys.Lett. **174B** (1986)280

[26] J.Distler and B.R.Greene, Nucl.Phys. **B309**(1988)295

[27] L.Dixon, V.S.Kaplunovsky and J.Louis, *On Effective Field Theories Describing (2,2) Vacua of the Heterotic String*, SLAC-PUB-4959 and UTTG-19-89 (1989)

[28] M.G.Eastwood and T.Hübsch, *Endomorphism Valued Cohomology and Gauge-Neutral Matter*, University of Texas preprint, UTTG-25-89 (August 1989)

[29] S.Ferrara and A.Strominger, *N=2 Space-Time Supersymmetry and Calabi-Yau Moduli Space*, CERN-TH.5291/89 and UCLA/89/TEP/6 (February 1989), and references therein

[30] J.Fuchs and A.Klemm, *The Computation of the Operator Algebra in Non-diagonal Conformal Field Theories*, to appear in Annals of Physics,**194**(1989)

[31] D.Friedan, E.Martinec and S.Shenker,Nucl.Phys. **B271**(1986)93

[32] D.Gepner, Phys.Lett. **B199**(1987)380

[33] D.Gepner, Nucl.Phys. **B296**(1988)757

[34] D.Gepner, *String Theory on Calabi-Yau Manifolds: The Three Generation Case*, Princeton preprint, (December 1987)

[35] D.Gepner, Nucl.Phys. **B311**(1988)191

[36] D.Gepner, Nucl.Phys. **B322**(1989)65

[37] D.Gepner, Trieste Lectures, spring 1989

[38] D.Gepner and Z.Qiu, Nucl.Phys. **B285**[FS10](1987)432

[39] P.Ginsparg, *Applied Conformal Field Theory*, to appear in Les Houches, Session XLIX, 1988, *Fields, Strings and Critical Phenomena*, ed. by E.Brézin and J.Zinn-Justin, Elsevier Science Publishers B.V.(1989)

[40] P.Green and T.Hübsch, Commun.Math.Phys. **109**(1987)99

[41] P.Green and T.Hübsch and C.A.Lütken, Class.Quantum Grav.**6**(1989)105

[42] M.B.Green, J.H.Schwarz and E.Witten, *Superstring Theory*, Cambridge University Press, 1987

[43] B.R.Greene and K.H.Kirklin, Commun.Math.Phys.**113**(1987)105

[44] B.R.Greene, K.H.Kirklin, P.J.Miron and G.G.Ross, Phys.Lett. **180B**(1986)69; Nucl.Phys. **B274**(1986)574; Nucl.Phys. **B278**(1986)667

[45] B.R.Greene, C.A.Lütken and G.G.Ross, *Couplings in the Heterotic Superconformal Three Generation Model*, to appear in Nucl.Phys. **B**

[46] B.R.Greene, C.Vafa and N.Warner, *Calabi-Yau Manifolds and Renormalisation Group Flows*, HUTP-88/A047, 1988

[47] P.Griffths and J.Harris, *Principles of Algebraic Geometry* (Wiley 1978)

[48] P.S.Howe and P.C.West, *Chiral Correlators in Landau-Ginsburg Theories and N=2 Superconformal Models*, CERN preprint CERN-TH.5390/89 (May 1989), and references therein

[49] T.Hübsch, Commun.Math.Phys. **108**(1987)291

[50] D.Kastor, E.Martinec and S.H.Shenker, *RG Flow in N=1 Discrete Series*, EFI 88-31, to appear in Nucl.Phys. **B**

[51] A.Kato,*Classification of modular invariant partition functions in two dimensions*, Univ. of Tokyo preprint UT-509(May 1987)

[52] A.Kato and Y.Kitazawa, *E_7 Type Modular Invariant Wess-Zumino Theory and Gepner's String Compactification*, Univ. of Tokyo preprint UT-535(August 1988)

[53] L.Landau and E.M.Lifshitz, *Quantum Mechanics: Nonrelativistic Theory* (Pergamon 1958)

[54] J.I.Latorre and C.A.Lütken, Phys.Lett. **B222**(1989)55

[55] W.Lerche, C.Vafa and N.P.Warner, *Chiral Rings in N=2 Superconformal Theories*, HUTP-88/A065 and CALT-68-1540 (February 1989)

[56] C.A.Lütken and G.G.Ross, Phys.Lett. **B213**(1988)152

[57] C.A.Lütken and G.G.Ross, Phys.Lett. **B214**(1988)357

[58] M.Lynker and R.Schimmrigk, Phys.Lett. **B215**(1989)681

[59] E.J.Martinec, *Algebraic Geometry and Effective Lagrangians*, EFI 88-76, 1988

[60] M.E.Peskin, *Introduction to String and Superstring Theory II*, Lectures presented at the 1986 Theoretical Advanced Study Institute in Particle Physics, University of California, Santa Cruz, June 23 - July 19, 1986

[61] V.Petkova, *Two-Dimensional (Half)-Integer Spin Conformal Theories with Central Charge $c < 1$*, Sofia preprint (March 1988)

[62] R.Schimmrigk, Phys.Lett. **193B**(1987)175

[63] G.Sotkov and M.Stanishkov, Phys.Lett. **215B**(1988)674

[64] A.Strominger and E.Witten, Commun.Math.Phys.**101**(1985)341

[65] A.Strominger, Phys.Rev.Lett.**55**(1985)2547

[66] T.Taylor and T.Hübsch, private communications

[67] C.Vafa and N.Warner, *Catastrophes and the Classification of Conformal Field Theories*, HUTP-88/A03 (November 1989)

[68] J.Wess and J.Bagger, *Supersymmetry and Supergravity*, Princeton University Press, 1982

[69] E.Witten, Phys.Rev. **16**(1977)2991

[70] E.Witten, Nucl.Phys. **B268**(1986)79

[71] S.-T.Yau, with an appendix by G.Tian and S.-T.Yau,in *Proc.of Argonne Symposium on Anomalies, Geometry and Topology*, eds. W.A.Bardeen and R.A.White (World Scientific, Singapore 1985)

[72] A.B.Zamolodchikov, Sov.J.Nucl.Phys. **44**(1986)529

[73] A.B.Zamolodchikov and V.A.Fateev, Sov.Phys.JETP **62**(1984)215

[74] B.Zumino,Phys.Lett. **87B**(1979)203

Standard Model
and Beyond

ISBN 1-56072-027-1

TWO LECTURES ON RIEMANN SURFACES AND CALABI-YAU MANIFOLDS

Paul S. Green

Department of Mathematics, University of Maryland, College Park, MD USA

Introduction: Moduli of Complex Manifolds

A complex manifold of complex dimension n can be thought of as a smooth oriented manifold of real dimension $2n$ with some extra structure which identifies those (local) complex valued functions that are to be considered holomorphic (complex analytic). Two orientable smooth manifolds are diffeomorphic if there exists a pair of mutually inverse smooth mappings (diffeomorphisms) between them. A mapping from one complex manifold to another is holomorphic if holomorphic functions pull back to holomorphic functions. Two complex manifolds are considered equivalent as complex manifolds if there exists a pair of mutually inverse holomorphic mappings between them. Such mappings obviously count also as diffeomorphisms between the underlying smooth orientable manifolds. The general moduli problem is to study the set, or "moduli space", of holomorphic equivalence classes of complex manifolds with some fixed underlying real oriented manifold. Typically such a moduli space is itself a non-compact complex manifold, and often it has a natural geometric structure as well.

Two cases of particular interest in string theory are the moduli spaces of Riemann surfaces, complex manifolds of complex dimension 1, and the moduli spaces of three dimensional Kähler manifolds with holomorphic volume form, or Calabi-Yau manifolds. As we shall see, there are some strong analogies between these two cases, the first of which has been studied intensively for over a century, and the second of which has only relatively recently received systematic attention.

The first part of these notes, roughly corresponding to the first of two lectures, consists of a review of some classical and elementary results on the moduli spaces of Riemann surfaces. We shall give particular emphasis to the case of the moduli space of tori, both because it is the simplest non-trivial case and because it exhibits the strongest analogy with the theory of moduli spaces of Calabi-Yau manifolds.

The second lecture begins by showing how the period map and Weil-Petersson metric for moduli spaces of Calabi-Yau manifolds exhibit an analogy with the the moduli space of tori, and then turns to the geometry of nodal degenerations and small resolutions. We discuss the obstruction to the existence of a Kähler structure on a small resolution and also the finiteness of the Weil-Petersson distance to a nodal degeneration in the moduli space. It is only to these last results that the author can claim to have made any contribution; they are the result of joint work with Tristan Hübsch and Philip Candelas.

Like the oral lectures from which they are derived, these notes are essentially descriptive in nature; virtually no proofs are given but some references to the literature have been added. While we assume that the reader is acquainted with the de Rham cohomology of smooth oriented manifolds, we have included three appendices summarizing some prerequisite ideas concerning respectively the integral homology of manifolds, geometry on complex manifolds, and projective algebraic varieties. The textbook [4] of Griffiths and Harris is an excellent reference for all the material presented here except for those results specifically concerning Calabi-Yau manifolds.

Lecture 1: Riemann Surfaces

Three distinct but overlapping points of view have emerged with respect to the problem of understanding the moduli spaces of Riemann surfaces. They may be referred to respectively as the geometric, analytic and algebraic points of view, and we shall discuss them in that order.

The underlying oriented manifold of a Riemann surface is a smooth oriented surface. Such surfaces are completely classified by the Euler characteristic, χ or, more commonly, by the genus, $g = \frac{1}{2}(2 - \chi)$. g, which is 0 for a sphere and 1 for a torus can be interpreted as the "number of holes". In contrast to the situation in higher dimensions, every smooth oriented surface admits the structure of a Riemann surface.

If we endow an orientable surface with a Riemannian metric, we can make sense of the length of a smooth curve and also of the angle between any two curves at any point of intersection. We can also compute the Riemannian curvature κ which, in the two dimensional case, can be thought of as a scalar which may vary from point to point. The geometric classification of Riemann surfaces depends on three important facts. The first of these is that in the (complex) one dimensional case, a holomorphic structure is equivalent to a conformal structure, which consists of an orientation together with a coherent angle measurement scheme. Such a structure can be read off, as we have observed, from any Riemannian metric. However the conformal structure does not determine the metric; two metrics which differ by a variable scalar multiple determine the same conformal structure.

The second important fact, known as the Uniformization Theorem, is that every conformal structure is derivable from a metric of constant curvature which *is* unique up to an overall scale, which can be fixed by normalizing either the volume V or the curvature. By the Gauss-Bonnet theorem, we have $\kappa V = 2\pi\chi$; thus the sign of κ is determined.

The third fact is that two complete simply connected Riemannian surfaces (i.e. surfaces with Riemannian metrics) of the same constant curvature are isometric. From this it follows immediately that the only holomorphic structure on S^2, which is simply connected, is the one which can be read off from the standard metric, which has constant positive curvature.

For a non-simply connected Riemann surface, we must consider its universal covering space. This is a simply connected space on which the fundamental group of the surface acts freely, and for which the quotient of the action is identified with the surface. The constant curvature metric induces a constant curvature metric on the universal covering space, for which the covering transformations are isometries. There are two cases: the torus, for which the curvature is 0 and the covering space with its induced geometry is the Euclidean plane; and the higher genus surfaces, for which the curvature is negative and the covering space with its induced geometry is the hyperbolic plane.

From the geometric point of view, the classification of Riemann surfaces can now be reduced to the problem of classifying fixed point free actions of the fundamental group of a surface on the appropriate plane, since we can recover the surface by dividing out the action of the fundamental group. We shall consider the case of a torus T in some detail.

We recall that the fundamental group $\pi_1(T)$ is Z^2, which may also be identified with $H_1(T)$, from which it inherits the intersection pairing (see Appendix A). Choose a symplectic basis $\{a, b\}$ for $\pi_1(T)$, and coordinatize the Euclidean plane which covers T as the complex numbers. Then $\pi_1(T)$ acts as a lattice of translations, and we can rotate and rescale so that $a(0) = 1$. Let us write z for $b(0)$. From the fact that $a \cap b = 1$, it follows that z is in the upper half plane. z is otherwise arbitrary and completely determines the lattice. Thus the upper half plane parametrizes the set of holomorphic structures on T. This parametrization is redundant, however, because z depends on a choice of basis for $\pi_1(T)$. In order to obtain the true moduli space of tori, we must compensate for this by dividing out the action of $SL(2, Z)$, the group of automorphisms of $\pi_1(T)$ respecting the intersection pairing.

$SL(2, Z)$ is the group of unimodular 2×2 matrices acting on Z^2 in the obvious way. It is not difficult to see (and is left as an exercise to the reader) that the corresponding action of $SL(2, Z)$ on the upper half plane is by linear fractional transformations

$$z \longrightarrow \frac{mz+n}{pz+q}.$$

Observe that the matrix $-I$, where I is the identity matrix, acts trivially, so that we are really dealing with an action of $PSL(2,Z) = \dfrac{SL(2,Z)}{\{I, -I\}}$.

This action is by isometries if we take the hyperbolic metric on the upper half plane, and induces a metric on the non-singular part of the moduli space of tori. The moduli space is not a manifold but has a threefold orbifold point because the action of $PSL(2, Z)$ is not free. The moduli space of tori is also non-compact, but can be compactified by the addition of a single point corresponding to the degeneration into an infinitely long infinitely thin torus.

If a Riemann surface S has genus greater than 1, its universal covering space, $\tilde{S}$, as we have already observed, may be identified with the hyperbolic plane, most conveniently represented as the upper half plane. $\pi_1(S)$ has the presentation,

$$\{a_i, b_i : \prod_{i=1}^{g} a_i b_i a_i^{-1} b_i^{-1} = 1\}.$$

By the ideas we have been developing above, classifying Riemann surface structures on S is equivalent to classifying fixed-point-free actions of $\pi_1(S)$ by orientation preserving isometries of the hyperbolic upper half plane. Such isometries are fractional linear tranformations with real coefficients and the isometry group is $PSL(2,R)$, which has dimension 3. We can get an estimate of the real dimension of the moduli space by a crude count:

We are looking at homomorphisms of $\pi_1(S)$ into $PSL(2,R)$. Since there are $2g$ generators of $\pi_1(S)$, the space of maps of the generators into $PSL(2,R)$ has $6g$ real dimensions. The relation imposes 3 constraints, and we should identify homomorphisms which differ by an inner automorphisms of $PSL(2,R)$, which cuts the total dimension by 3 more to $6g-6$. Despite the crudity of this analysis, it arrives at the correct answer. We do not need to worry about the fixed point free condition for the purposes of this dimension count because it is an "open" condition. That is, any action which is sufficiently close to a free action is also free. The existence of at least some free actions is guaranteed by the existence of Riemann surfaces of all genera. To see that the moduli space of Riemann surfaces of any genus has the structure of a complex manifold on its non-singular part, we need to switch our point of view.

We return our attention to the covering map for the torus,

$$C \to T$$

where, as before, we have identified the universal covering space with the complex numbers. The complex coordinate z on C obviously does not descend to a single valued function on T. However the holomorphic differential dz descends to a holomorphic differential on T and we can evaluate its periods,

$$\int_a dz \text{ and } \int_b dz,$$

where a and b form a symplectic basis of $H_1(T)$, as before. We can normalize dz by a constant factor so that

$$\int_a dz = 1$$

and choose

$$w = \int_b dz$$

as a (multiple valued) coordinate on the moduli space. In the case of the torus, the action of $SL(2,Z)$ comes in the same way as before and we recover the same results. However it is worth paying attention to the reason, from this point of view, that w turns out to be in the upper half plane.

From the discussion in Appendix A, it follows that

$$\int_T dz \wedge d\bar{z} = \int_a dz \int_b d\bar{z} - \int_a d\bar{z} \int_b dz = \bar{w} - w.$$

On the other hand, $\frac{1}{-21} dz \wedge d\bar{z}$ is a Kähler form (see Appendix B) and therefore a volume form for T so that

$$\frac{1}{-2i} \int_T dz \wedge d\bar{z} > 0.$$

It follows that the imaginary part of w is positive.

On a Riemann surface S of genus g, the holomorphic differentials form a vector space of dimension g. We also know that $H_1(S)$ has rank $2g$.

In order to define the period matrix for a Riemann surface of positive genus, we choose a symplectic basis

$$\{a_i, b_i\}_{1 \leq i \leq g}$$

of $H_1(S)$. It turns out that we can choose a basis $\{\omega_i\}_{1 \leq i \leq g}$ for the space of holomorphic differentials, characterized by

$$\int_{a_i} \omega_j = \delta_{i,j}.$$

The period matrix is now defined to be the matrix whose (i, j)th entry is

$$\int_{b_j} \omega_i.$$

It can be shown that the period matrix is symmetric and that its imaginary part is positive definite. The space of such matrices is known as the Siegel upper half space. For $g > 2$ the dimension of the Siegel upper half space exceeds the dimension of the moduli space so that not all symmetric matrices with positive definite imaginary part are period matrices of Riemann surfaces. However the image of the period map is a complex submanifold of the Siegel upper half space. Since the period matrix depends on the choice of a symplectic basis for $H_1(S)$, the period map is really defined only on a covering space of the moduli space. One can divide out, both in the domain and in the Siegel upper half space, by an action of the integer symplectic group $Sp(2g, Z)$ which specializes to the action of $SL(2, Z)$ in the case of genus 1. The complex structure on the Siegel upper half space is respected by the action of $Sp(2g, Z)$ and so induces a complex structure (with orbifold type singularities) on the moduli space.

From the algebraic viewpoint all Riemann surfaces are non-singular projective algebraic curves. This can be seen easily in three steps. In the first place any Riemannian metric compatible with the given conformal structure can be taken as the real part of an Hermitian metric (see Appendix B). In the second place the Kähler condition is vacuous since all two-forms are closed. Finally, the hypothesis of Kodaira's embedding theorem (see Appendix C) can be satisfied by normalizing the volume to an integer. The simplest case is a hypersurface in CP^2. If we write k for the degree of the polynomial, then the genus of the hypersurface is given by the formula

$$g(k) = \frac{(k-1)(k-2)}{2}.$$

It is clear that not all Riemann surfaces can arise as hypersurfaces in CP^2. In the first place, many genera will not appear at all. In the second place, the linear space of homogeneous polynomials of degree k in 3 variables has dimension

$$\frac{(k+1)(k+2)}{2}.$$

Since a linear automorphism of C^3 induces an automorphism of CP^3 and curves which differ by such an automorphism are equivalent, we must subtract 9 from this number to obtain the dimension of the deformation space of distinct algebraic curves which arise as hypersurfaces of degree k. For $k > 4$,

$$\frac{(k+1)(k+2)}{2} - 9 < 3g(k) - 3,$$

so that even for those genera that arise as $g(k)$ for some k, not all Riemann surfaces arise as curves in CP^2. In point of fact, there are infinitely many genera for which Riemann surfaces are not complete intersections in any CP^n.

Lecture 2: Calabi-Yau Manifolds

We shall take as our definition of a Calabi-Yau manifold, a three dimensional compact Kähler manifold M which admits a nowhere vanishing holomorphic three form Ω given locally by

$$\Omega = f(z)dz_1 \wedge dz_2 \wedge dz_3; \quad f(z) \neq 0.$$

Since the ratio of two such forms is a global holomorphic function, Ω is unique up to a multiplicative constant. It is also evident that

$$\frac{1}{8i^3}\Omega \wedge \bar{\Omega}$$

is a real volume form for M. Yau has proved ([7]) that if M is a Calabi-Yau manifold, and J is a closed positive (1,1)-form, there is a unique Kähler metric g such that $J_g - J$ is exact and $\nabla\Omega = 0$.

Let us assume that M is a Calabi-Yau manifold such that $H_3(M, Z)$ has rank $2k$ and let us fix a symplectic basis $\{a_i, b_i\}_{1\leq i\leq k}$ (see Appendix A). We define the periods of M, $A_i(M)$ and $B_i(M)$, to be $\int_{a_i}\Omega$ and $\int_{b_i}\Omega$, for a holomorphic three-form Ω on M. The periods are only well defined up to a complex factor. However since

$$0 \neq \int_M \Omega \wedge \bar{\Omega} = \sum_{i=1}^{k}(A_i(M)\bar{B}_i(M) - \bar{A}_i(M)B_i(M)),$$

neither all the $A_i(M)$ nor all the $B_i(M)$ can vanish simultaneously, and there is an element of CP^{2k-1} determined by the periods which is independent of the choice of Ω.

Following an analogy with the terminology in the theory of Riemann surfaces, we will say that the Calabi-Yau manifold M and the symplectic basis of $H_3(M, Z)$ determine a point of the Teichmüller space of $|M|$, the smooth oriented manifold underlying. In this language, we have defined a mapping from the Teichmüller space of $|M|$ to CP^{2k-1}. By a theorem of Bryant and Griffiths ([**1**]), if we normalize one of the A_i to 1, the remaining A_i provide local coordinates for a holomorphic structure on the Teichmüller space, and each B_j depends holomorphically on (all) the A_i. Such a normalization determines a specific choice of Ω and we define the Weil-Petersson metric on the Teichmüller space by taking $\log(\int_M \Omega_M \wedge \bar{\Omega}_M)$ as a Kähler potential; since the argument to the log function is determined up to the square of the modulus of a holomorphic function, the metric we obtain is independent of which period we choose to normalize and is therefore defined globally on the Teichmüller space. By a theorem of Tian ([**5**]), this metric is positive definite and therefore Kähler.

In order to descend to the moduli space, we must divide out by the action of $Sp(2k, Z)$ the automorphism group of the intersection pairing on $H^3(M, Z)$. This is much more complicated than in the case of the moduli space of tori and is not yet completely understood. It is clear, however, that the moduli space has both orbifold points and boundaries corresponding to singular degenerations. The remainder of this lecture is devoted to the simplest of these.

We consider the family of hypersurfaces M_λ in C^4 consisting of the zero loci of the equations

$$\sum_{i=1}^{4} Z_i^2 = \lambda^2,$$

where λ is a complex parameter. Notice that the Z_i are affine rather than projective coordinates, so that each hypersurface is a threefold. For $\lambda \neq 0$, M_λ is a non-singular (non-compact) complex manifold. However M_0 has a node (see Appendix C) at the origin. Rescaling by λ, we can see that the structure of M_λ is independent of λ for $\lambda \neq 0$. For this reason, and since we are interested in the limiting behaviour as λ approaches 0 along rays, we will confine our attention from this point on to positive values of λ. If we write $Z_j = X_j + iY_j$, with X and Y real, we get as real equations for M_λ,

$$X \cdot Y = 0; \;\; X^2 - Y^2 = \lambda^2.$$

Setting $Y = 0$ gives us the three-sphere, $X^2 = \lambda^2$; restricting X to be a positive multiple of $(1,0,0,0)$ gives us one sheet of a two-sheeted three-dimensional hyperboloid transverse to the sphere at

$$X = (\lambda, 0, 0, 0); \;\; Y = (0, 0, 0, 0).$$

As λ approaches 0, the three-sphere shrinks down to a point, while the hyperboloid develops a cone-like singularity.

In order to resolve the node on M_0 we introduce a CP^1 with homogeneous coordinates A and B and new affine coordinates on C^4:

$$t = Z_1 + iZ_2, \; u = Z_1 - iZ_2, \; v = Z_3 + iZ_4, \; w = -(Z_3 - iZ_4).$$

We define a variety $\tilde{M}_0$ in $C^4 \times CP^1$ by the equations,

$$tA + vB = 0; \; wA + uB = 0.$$

For any point of C^4 other than the origin, these equations determine a single point of CP^1 provided $tu - vw = 0$ and cannot be satified otherwise. At the origin, they are obviously satisfied for every point of CP^1. $\tilde{M}_0$ is non-singular and maps to M_0 via the projection from $C^4 \times CP^1$ to C^4. This mapping is bijective except that the inverse image of the origin is a CP^1. $\tilde{M}_0$ is called a small resolution of M_0.

Remark: There is a second small resolution defined by the equations,

$$tA + wB = 0; \; vA + uB = 0.$$

If we lift the singular hyperboloid from M_0 to $\tilde{M}_0$, it becomes a smooth manifold with boundary. The boundary is the CP^1 which is the inverse image of the origin under the projection. Suppose we fix an orientation on the hyperboloid. Then the boundary CP^1 has two orientations: one from its complex structure and one from being the boundary of the oriented hyperboloid. These will either agree or disagree depending on which of the two small resolutions we choose.

Before passing to more global considerations, we want to consider briefly the issue of a holomorphic three-form. We write Ω_λ for the threeform given by

$$\frac{dt \wedge du \wedge dw}{w}$$

on M_λ. It is easy to verify that the singularity for $w = 0$ is only apparent, except at the origin, because Ω_λ is invariant up to a sign under permutations of the coordinates. Moreover Ω_λ is nowhere vanishing and Ω_0 extends to a non-singular nowhere vanishing holomorphic three-form on $\tilde{M}_0$.

We are now ready to globalize the discussion of the preceding paragraphs. Let p and q be two homogeneous quintic polynomials in five variables. Then the zero-locus of $q + \lambda^2 p$ is a hypersurface in CP^4 and is well known to be Calabi-Yau provided it is non-singular. We assume now that the zero locus of p is in fact non-singular and that the only singularities of the zero locus of q occur at n points where p does not vanish and near each of which $\frac{q}{p}$, in appropriate local inhomogeneous (and nonlinear) coordinates, is the sum of the squares of the local coordinates.

With these assumptions, if we write M_λ for the zero locus of $q + \lambda p$, M_λ is non-singular, and therefore Calabi-Yau, for λ sufficiently small and different from 0. M_0 has n nodes, each of which can be resolved, as above, in either of two ways, giving us a choice of 2^n distinct small resolutions of M_0.

Since the local geometry near each node of M_0 is essentially the same as in the standard node, there is a copy of S^3 near each node on each M_λ which is parametrized by taking the local coordinates, with respect to which $\frac{q}{p}$ is a sum of squares, to be real. Suppressing the dependence on λ, we will write S_i for the three-sphere near the ith node on M_λ for $\lambda \neq 0$. We will also choose a small resolution $\tilde{M}_0$ of M_0, and write P_i

for the copy of CP^1 which resolves the ith node. Clearly $[S_i] \cap [S_j] = 0$, since the S_i are contained in disjoint neighborhoods. We can choose a symplectic basis $\{a_i, b_i\}$ of $H_3(M_\lambda)$ for which each $[S_j]$ is a linear combination of $\{a_1, \ldots, a_k\}$ and the $[S_j]$ are not all contained in any sublattice of dimension less than k. Let us look at representative cycles B_j, for $\{b_j\}_{j \leq k}$, which meet the S_i transversely. In our local geometric picture, these cycles play the role of the transverse hyperbolae. Consequently, if we let λ tend to 0 with a fixed argument, we may choose the B_j in such a way that we obtain limiting chains $\tilde{B}_j$, on any small resolution $\tilde{M}_0$ of M_0, each of whose boundary components will be one of the S_i.

We can now see that there is an obstruction to the existence of a Kähler metric on $\tilde{M}_0$. If we give each P_i the orientation from the complex structure and suppose J_g is a Kähler form on $\tilde{M}_0$, we have

$$\int_{P_i} J_g > 0$$

for each i. On the other hand, corresponding to each $\tilde{B}_j$, we have a relation among the homology classes $[P_i]$. If the $[P_i]$ satisfy any non-trival linear relation with non-negative coefficients, then we have a contradiction to the existence of a Kähler metric on $\tilde{M}_0$. I believe that this is the only obstruction to the existence of a Kähler metric on $\tilde{M}_0$.

By changing the small resolution at the ith node, we change the sign of the coefficient of $[P_i]$ in every relation. It is an easy exercise in linear algebra to check that there exists at least one small resolution for which no positive linear combination of the $[P_i]$ is zero, unless some multiple of some $[P_i]$ is zero. However, if the $[S_i]$ are linearly independent, then so are the relations induced by the $\tilde{B}_j$ and then all the $[P_i]$ are 0. Thus in order for M_0 to have a Kähler small resolution, it is necessary that the $[S_i]$ satisfy at least one linear relation.

The holomorphic three-forms on the M_λ will not have precisely the form discussed above near the nodes, but will differ from that form by a holomorphic factor which is well defined and non-vanishing on a neighborhood of each node in the ambient space. Consequently it is still the case that we can choose a holomorphic three-form Ω_λ on each M_λ that depends holomorphically on λ and gives, in the limit $\lambda \to 0$, a non-vanishing holomorphic three-form on the non-singular part of M_0 which extends holomorphically to each small resolution. Thus each small resolution of M_0 is Calabi-Yau provided it is Kähler. It is important to note that these non-singular small resolutions of M_0 are not diffeomorphic either to quintics in CP^4 or, in general, to each other. This point is addressed in detail in [6]

The last thing we will discuss is the Weil-Petersson distance from M_λ to the nodal M_0, which represents a boundary point of the moduli space. We choose Ω_λ as in the previous paragraph and write $\{A_i(\lambda), B_i(\lambda)\}$ for the periods of Ω_λ associated to our chosen symplectic basis. Locally then, the periods depend holomorphically on λ, for λ sufficiently small and different from 0. In fact the $A_i(\lambda)$ for all i and the $B_i(\lambda)$ for $i > k$ are global holomorphic functions on some punctured disk. However the $B_i(\lambda)$ for $i \leq k$ are not single valued on the punctured disk. This is the case because if we fix $|\lambda|$ and let $\arg(\lambda)$ go through a complete period, we parametrize a closed curve in the moduli space, but not in the Teichmüller space. Keeping the symplectic basis of M_λ locally constant, we do not come back to the same basis we started with, a phenomenon known as monodromy.

We write μ for the symplectic automorphism connecting the initial and final bases, and by careful consideration of the local geometry at each node, we can show that

$$\mu(x) = x + \sum_{i=1}^{n} (z \cap [S_i])[S_i].$$

From this it follows that

$$\mu(a_i) = a_i; \ \mu(b_i) = b_i + \sum_{j=1}^{k} \gamma_{ij} a_j,$$

$$\text{where } [S_i] = \sum_{j=1}^{k} \beta_{ij} a_j, \text{ and } \gamma_{ij} = \sum_{\ell=1}^{n} \beta_{\ell i} \beta_{\ell j}.$$

All the periods can be shown to be bounded as $\lambda \to 0$ along any ray, and it follows that $A_i(\lambda)$ for all i and $B_i(\lambda)$ for $i > k$ are holomorphic also at $\lambda = 0$, while $B_i(\lambda)$ for $i \le k$ has the form

$$F_i(\lambda) + \frac{1}{2\pi i} \sum_{j=1}^{k} \gamma_{ij} A_j(\lambda) \log(\lambda),$$

where F_i is holomorphic on the unpunctured disk. Since the Weil-Petersson Kähler potential can be computed from the periods, we can now show that although the potential has a singularity at $\lambda = 0$, the Weil-Petersson length of a radius of the disk is finite. See [2] for details.

The restriction of the previous discussion to nodal degenerations of quintics in CP^4 was purely for notational convenience; all the results apply equally well to nodal degenerations of any family of Calabi-Yau hypersurfaces in a four dimensional non-singular projective variety.

Appendix A: Integral Homology of Manifolds

Let N be an oriented k-dimensional submanifold of the oriented n dimensional manifold, M. Then if ω is a k form on M, we can form the integral

$$\int_N \omega$$

which vanishes if either ω is exact or N is the boundary of some $k+1$-dimensional manifold with boundary contained in M. Further, if N' is another oriented submanifold of dimension k' and N and N' intersect transversely, then $N \cap N'$ is a submanifold of dimension $k + k' - n$, or empty if $k + k' < n$. Furthermore $N \cap N'$ inherits an orientation from those of M, N and N', which depends on the order of N and N' if and only if $(n-k)(n-k')$ is odd.

The real homology groups, $\{H_i(M,R)\}_{0\leq i\leq n}$ may be most conveniently thought of as the linear duals of the de Rham cohomology groups, $\{H^i(M,R)\}$. From this point of view, the integral homology groups (or more properly the integral homology groups modulo elements of finite order — but we shall refer to this complication only when absolutely necessary to prevent the dissemination of outright untruths) are lattices in the real homology groups. An element of a homology group is often referred to as a homology class.

The connection between the two preceding paragraphs is that the linear mapping from $H^k(M,R)$, the k dimensional de Rham cohomology group, to the real numbers defined by integration over N is an integral homology class. We refer to this class as $[N]$, or sometimes just as N when there is no possibility of confusion.

The intersection of submanifolds also has an analogue at the homology level. To see what this is, we recall that by Poincaré duality, $H^k(M,R)$ and $H^{n-k}(M,R)$ are dual, with the duality pairing given by integrating the wedge product over M. This allows us to identify $H_k(M,R)$ with $H^{n-k}(M,R)$ (*not* with its dual). Using this identification, we obtain, from the wedge product of forms, a product, which we write

$$\cap : H_k(M,R)\otimes H_{k'}(M,R)\rightarrow H_{k+k'-n}(M,R).$$

and refer to as the intersection product. It turns out that the intersection of integral classes is integral and, in particular,

$$[N]\cap[N'] = [N\cap N'].$$

While it is not the case in general that every integral homology class is represented by a submanifold of M, homology classes are represented by somewhat more general geometric objects called *cycles* over which integration still makes sense. Similarly, if $[N]=0$, N is not necessarily the boundary of a submanifold of M but is the boundary of a *chain*. Our notation tends, harmlessly, to obscure the distinction between a general homology class and one which is carried by a submanifold. In particular, we integrate promiscuously over integral homology classes.

In the body of these notes we are particulary interested in the case where M is a complex manifold of odd complex dimension, n. In this case, the intersection pairing on the integral homology in the middle dimension, n, is antisymmetric and takes values in $H_0(M,Z)$, which can be identified with the integers. By Poincaré duality (over the integers), this pairing is a duality pairing, and the only invariant of such a pairing is the dimension of the lattice. Thus we always can (and usually do) choose a symplectic basis $\{a_i, b_i\}$ with

$$a_i\cap a_j = b_i\cap b_j = 0;\quad a_i\cap b_j = \delta_{ij}.$$

It then follows that if ω and τ are n-forms,

$$\int_M \omega\wedge\tau = \sum_{i=1}^{k}\left(\int_{a_i}\omega\int_{b_i}\tau - \int_{a_i}\tau\int_{b_i}\omega\right),$$

where $2k$ is the rank of $H_n(M)$.

There is a large automorphism group, $Sp(k,Z)$ which reflects the ambiguity in a choice of symplectic basis. In particular $Sp(2,Z) = SL(2,Z)$, reflecting the fact that the symplectic automorphisms of a two dimensional symplectic lattice are just the unimodular automorphisms of the lattice.

Appendix B: A Review of Some Kähler Geometry

Let M be a complex manifold of complex dimension n. We recall that a (p,q)-form on M is one which is expressed locally as a finite sum of terms of the form

$$f(z,\bar{z})dz_{i_1} \wedge \cdots \wedge dz_{i_p} \wedge d\bar{z}_{j_1} \wedge \cdots \wedge d\bar{z}_{j_q}.$$

An Hermitian inner product, g, on M is given in local coordinates by

$$\sum_{i,j=1}^{n} g_{i\bar{j}} dz_i \otimes d\bar{z}_j.$$

Associated with g is a $(1,1)$-form,

$$J_g = \frac{1}{-2i} \sum_{i,j=1}^{n} g_{i\bar{j}} dz_i \wedge d\bar{z}_j.$$

Clearly g and J_g determine one another. If g is positive definite, J_g is called positive. In this case the n-fold wedge product J_g^n is a (real) volume form on M. If, in addition, J_g is closed, then g is called a Kähler metric and J_g its Kähler form. M is called a Kähler manifold if it admits at least one Kähler metric. It is shown in the second lecture that this is a non-trival restriction, even if M is a complex manifold. It is immediate from the definitions that the restriction of a Kähler metric to any complex submanifold of M is again Kähler. An important consequence is the fact that the integral of the appropriate power of the Kähler form over any compact complex submanifold of M is positive.

A function f, usually defined only locally, is called a Kähler potential for g if

$$g_{i\bar{j}} = \frac{\partial^2 f}{\partial z_i \partial \bar{z}_j}.$$

If g arises from a function f in this way, J_g is automatically closed but g is *not* automatically positive definite. A Kähler metric can always be derived locally (but *never* globally if M is compact) from a potential.

Example: Let a and b be a symplectic basis for the one dimensional homology of a torus (see Appendix A). Then it is shown in the first lecture that if ω is a holomorphic one-form normalized so that

$$\int_a \omega = 1,$$

then $\int_b \omega$ is a local holomorphic coordinate on the moduli space of tori. It turns out that $log \int_T \omega \wedge \bar{\omega}$ is a Kähler potential for the hyperbolic metric. The verification of this observation is left to the reader, and is recommended as an excellent exercise.

Remark: Kähler potentials of the form $f = \log g$ arise commonly. Note that if we replace g by $h\bar{h}g$ for some holomorphic h, then $\frac{\partial^2 f}{\partial z_i \partial \bar{z}_j}$ does not change.

We close this section with a remark on covariant derivatives: If g is an Hermitian metric (not necessarily Kähler) on M, there is a unique covariant derivative, ∇, singled out by the conditions

$$\nabla g = 0; \quad \nabla_{\partial \bar{z}_i} \partial z_j = 0.$$

The second condition means that the covariant derivative of a holomorphic vector field in anti-holomorphic directions vanishes.

Appendix C: Projective Varieties

In order to define CP^n, complex projective space of n dimensions, we begin with C^{n+1} and divide out by the multiplicative action of the non-zero complex numbers. That is we set

$$(z_0, \ldots, z_n) \sim (\lambda z_0, \ldots, \lambda z_n)$$

for all $\lambda \neq 0$. Then if $p(z_0, \ldots, z_n)$ is a homogeneous polynomial, the equation $p = 0$ is well defined on CP^n. A projective variety is a subset of CP^n which is locally the intersection of the zero loci of a finite number of homogeneous polynomials. The simplest varieties to describe are those for which the polynomials can be chosen globally. They are known as complete intersections, since the vanishing of a single polynomial defines a hypersurface, and the vanishing of several an intersection of hypersurfaces.

A projective variety may have singular points, at which the polynomials which determine it locally cannot be chosen to be transverse. In the case of a hypersurface, a singularity occurs when the defining polynomial vanishes together with all its derivatives. The simplest hypersurface singularity is a node, characterized by the condition that the symmetric matrix of second derivatives has rank n.

CP^n is a complex manifold, and in fact a Kähler manifold (see Appendix B), and therefore so is any non-singular projective variety. However not all Kähler manifolds are projective varieties. The embedding theorem of Kodaira states that a Kähler manifold is a projective variety if and only if it admits a Kähler metric with the property that the integal of the Kähler form over every integral two dimensional homology class (see Appendix B) is an integer. The best known examples of Kähler manifolds which are not projective are generic $K3$ surfaces. Projective $K3$ surfaces can be realized as quartics (i.e. the zero loci of fourth degree polynomials) in CP^3. However the moduli space of such surfaces has codimension one in the moduli space of all $K3$ surfaces.

It follows from Bertini's theorem that the generic complete intersection of hypersurfaces (of any degrees) is non-singular. However there are severe topological restrictions on which non-singular projective varieties can be realized as complete intersections in any CP^n. Since a product of complex projective spaces is a projective variety, the class of easily constructible varieties can be enlarged by considering complete intersections of hypersurfaces in products of complex projective spaces. Bertini's theorem applies in this case also, and a large number of topologically distinct Calabi-Yau manifolds arise in this way. The interested reader should consult [**3**].

References

[1] R. Bryant and P. Griffiths: in *Progress in Mathematics* **36** *pp.* 77–102 (Birkhäuser, Boston, 1983).

[2] P. Candelas, P. Green, and T. Hübsch: in *Strings '88* S.J. Gates et al (ed.), *pp.* 155–190 (World Scientific, Singapore, 1989).

[3] P. Green and T. Hübsch: *Comm. Math. Phys* **119** (1988) *pp.* 431–441.

[4] P. Griffiths and J. Harris: *Principles of Algebraic Geometry* (Wiley, New York, 1978).

[5] G. Tian: in *Mathematical Aspects of String Theory*, S.-T. Yau (ed.), *pp.* 629–646 (World Scientific, Singapore, 1987).

[6] G. Tian and S.-T. Yau: in *Mathematical Aspects of String Theory*, S.-T. Yau (ed.), *pp.* 543–559 (World Scientific, Singapore, 1987).

[7] S.-T. Yau: *Proc. Nat'l. Acad. Sci. U.S.A.* **74** (1977) *p.* 1798.

Standard Model
and Beyond

ISBN 1-56072-027-1

LECTURES ON COMPACTIFIED STRING THEORY

B.R. Greene

Lyman Laboratory of Physics, Harvard University, Cambridge, MA 02138 USA

In these lectures we describe various aspects of classical string vacua. After a brief introduction to first quantized string propagation we focus on Calabi-Yau and minimal model vacua. We describe some general characteristics of these theories, discuss the resolution of the previously mysterious correspondence between these two classes of string vacua and then specialize to the phenomenologically interesting three generation examples. Throughout our discussion we emphasize three main themes. First, the general trend from results which appeared to be lowest order unrefined calculations to the subsequent realization that many of these results are exact or at least are an accurate reflection of the exact results. Second, the ways in which string based models of particle physics give us the potential to go far beyond standard methods of model building. Finally, the interesting and fruitful interplay between abstract studies in conformal field theory and concrete studies in string phenomenology as a conceptual framework for string studies.

1. Introduction

We have learned over the past few years that although there are only a small number of distinct string theories there are a huge number of distinct classical vacua [1]. Part of the initial enthusiasm surrounding string theory was the prospect that a unified theory with no free dimensionless parameters was at hand, thus giving hope that the goal of computing many of the fundamental observables of nature was coming well within reach. Unfortunately, the 'low energy' aspects of string theory (that is, those aspects which can potentially make contact with the physics of standard model particle masses and couplings) are very sensitive to the structure of the vacuum configuration. This is not surprising as low energy physics corresponds to small fluctuations about the vacuum state. The enormous vacuum degeneracy, therefore, disrupts the noble plan of computing the many parameters in the standard model which are presently put in by hand. It is important to emphasize, though, that this is a problem with our present formulation and understanding of string theory rather than a problem inherent to the theory itself. It is quite possible that a truly quantum theory of string (as opposed to the present perturbative formulation) will lift the vacuum degeneracy and leave only a small or possibly unique choice for the ground state. In such a situation, the attractive and important uniqueness of string theory will reemerge. Lacking such a means of addressing the vacuum degeneracy problem, a great deal of effort has been spent over the last few years in trying to fully understand and classify the space of classical vacua (as we shall discuss, this is the space of two-dimensional conformal field theories). There are a number of reasons for focusing on this task. First, the space of classical solutions to a system may be viewed as the configuration space for a quantum formulation. Hence, understanding the space of classical string vacua may be an important step in the quantum string program. Second, by studying string vacua and the low energy physics associated with them, one can begin to probe some of the phenomenological properties of the string. One certainly has to be careful when employing this reasoning as the physics of a system need not be accurately reflected in the properties obtained by semiclassical analysis about classical vacua. For example, perturbative expansion about the configuration of vanishing vector potential in QCD is not a good guide to the physics of confinement. There is no compelling argument that the physics derived by studying fluctuations about classical string vacua is not similarly misleading; nonetheless, the natural way in which a host of characteristic features of standard model phenomenology (generation structure, chiral multiplets, appropriate gauge groups, etc...see below) emerge from such studies leads us to suspect that we may in fact be probing relevant physics. It would certainly take an unlikely (and unfortunate) coincidence of mathematics and nature for the physics we calculate to so closely approach that of the standard model and yet have no relevance.

In these lectures we shall focus on the space of classical string vacua. The first part of the lectures will introduce superconformal theories as heterotic string vacua and give examples of perturbative (Calabi-Yau) and nonperturbative constructions. We will indicate how one extracts the phenomenology associated with a chosen vacuum state and discuss some useful and important nonrenormalization theorems which make these phenomenological studies sensible. We will then, in the second part, describe a surprising way in which certain complicated geometrical vacua are the lowest order approximations to other naively nongeometrical exactly soluble nonperturbative vacua. In the third part of these lectures we shall apply some of these generic features to specific string vacua chosen for their promising phenomenological prospects. We shall briefly review the work that has been done on these three generation theories and then describe some recent results in this area.

I would like to thank my collaborators over the past few years P.Aspinwall, J.Distler, K. Kirklin, A. Lutken, P.Miron, R.Plesser, G.G. Ross, C. Vafa and N.P. Warner some of whose joint efforts are described in these lectures.

2. Classical String Vacua

2.1. Strings in $D = 10$

As has been emphasized many times [1], string theory has developed in a rather nontraditional fashion. Ordinarily a physical theory is fully defined and then, if its exact analysis proves intractable, a perturbative approach is followed. In string theory, though, a perturbative approach has been formulated before a complete (that is, nonperturbative) formulation of the theory has been achieved. The present formulation of string theory is composed of a set of rules defining a perturbative expansion to an as yet undefined underlying theory. Phrased slightly differently, the familiar expansion of interactions as a sum over processes on two-dimensional world sheets of increasing genus is expected to be the Feynman diagram expansion to some second quantized theory, but as yet such a theory has not been truly formulated.

Lets concentrate for the moment on the first term in this perturbative expansion - that is, the tree level or classical approximation to string theory. In its initial and simplest formulation a (first quantized) string propagating in flat spacetime is described by the action

$$\int_{S^2} d\sigma d\tau \sqrt{h} \eta^{\mu\nu} \partial_a X_\mu \partial_b X_\nu h^{ab} \tag{2.1}$$

where h and η are the metrics on the sphere and spacetime respectively. (The action for a heterotic string, which shall be our interest, also includes terms for the left λ and right ψ moving world sheet fermions. We shall not explicitly write these down.) Tree level string theory can thus be described by a two dimensional quantum field theory on the sphere. A well known and important property of this action is its invariance under rescalings of the two dimensional metric, that is Weyl rescalings. This leads to conformal invariance (in two dimensions) and plays an important role in the consistent quantization of this two dimensional theory. In this flat space setting, the maintanence of two dimensional conformal invariance at the quantum level places severe restrictions on the theory: the dimension of the spacetime is uniquely restricted to be ten (for the superstring). In this sense then, ten dimensional flat spacetime is a consistent ***vacuum state*** for the string. There is an especially useful way to think about the relation between conformal invariance and string vacua [2]. First note that the emission and absorption of on shell physical states in string theory are described by vertex operators V which have scaling dimension $(1,1)$ (that is, in complex coordinates $V(\lambda z, \lambda \bar{z}) = \lambda^{-2} V(z, \bar{z})$). Now, how can we characterize string vacua? To gain some intuition regarding this question, lets momentarily return to field theory. In quantum field theory we minimize an effective potential to determine the vacuum expectation values of fields Φ in our theory and then write the fluctuating quantum field as $\Phi_Q = \Phi - < \Phi >$. If we find, however, that Φ_Q has a nonvanishing one point function, then this signals that we have not identified the true vacuum value for this field, and hence we must shift (if we can) to the nearby correct value. The signal that we have identified the vacuum values, therefore, is that the one point functions of our quantum fields all vanish. In string theory the analogous statement is that in a correctly identified vacuum state the one point function of the vertex operators $< V(z, \bar{z}) >$ must vanish. On the sphere (more precisely, the complex plane) we can use translation invariance to move the operator to the origin, and then scale invariance to write

$$< V(0) > = \lambda^{-2} < V(0) > \tag{2.2}$$

and hence the one point function of V must vanish. Thus we see that conformal invariance may be used (on the sphere) to prove that we do have a (classical) string vacua. This is a useful way to think about conformal field theories as string vacua, i.e. as solutions to the 'string equations of motion' (although we do not know what these equations are). The sigma model defined on flat ten space above (along with fermionic terms we have not written) is a simple example of such a conformal theory. In fact, it is actually a superconformal theory in that the action respects two dimensional world sheet supersymmetry. This high degree of simplicity, though, also brings a correspondingly low degree of realism. Our universe does not appear to be well described by a flat ten space background, so lets try to find string vacuum configurations which are closer to observation.

In order to do this, it is worthwhile to capture the essential features of our flat ten space solution and rephrase these characteristics in a way that will allow for a robust generalization of what we mean by a vacuum configuration.

As emphasized above, the central underlying feature of flat ten dimensional spacetime as a string vacuum is that viewed as a two dimensional field theory it respects conformal invariance. So, we now turn to a brief discussion of conformal field theory [3]. A conformal transformation in $\mathbf{R}^d$ is the subgroup of coordinate transformations which leave the metric unchanged up to a (coordinate dependent) scale factor. In two dimensions this notion is especially powerful as any analytic coordinate transformation

$$z \to f(z), \bar{z} \to f(\bar{z}) \tag{2.3}$$

maps

$$dz \otimes d\bar{z} \to |\frac{\partial f}{\partial z}|^2 dz \otimes d\bar{z} \tag{2.4}$$

and hence is a conformal transformation. The two dimensional conformal group is thus infinite dimensional.

This infinite dimensional symmetry places many constraints on a two dimensional conformal field theory. To understand these constraints, it is useful to use the conformal symmetry to select a special subset of fields in a given theory - the 'primary fields'. A primary field Φ is characterized by its simple transformation properties under conformal transformations: if $z \rightarrow f(z)$ then

$$\Phi(z,\overline{z}) \rightarrow (\frac{\partial f}{\partial z})^{h} (\frac{\partial \overline{f}}{\partial \overline{z}})^{\overline{h}} \Phi(f(z), \overline{f}(\overline{z})). \tag{2.5}$$

In this expression h and $\overline{h}$ are real valued and are called the conformal weights of the field Φ. The primary fields play a key role in conformal field theory as their correlation functions embody (via conformal Ward identities) essentially all of the data necessary to evaluate any correlation function in the theory. Furthermore, the correlation functions of these primary fields are highly constrained by the conformal symmetry. For example, without even specifying an underlying theory, the three point function of chosen primary fields Φ_i, Φ_j, Φ_k is *determined* up to a constant c_{ijk} (which depends on the specific theory being studied). We shall return to this point later. Another important operator is the energy momentum tensor T (that is T_{zz} as well as $\overline{T} = T_{\overline{z}\overline{z}}$). By Noether's theorem, associated with any continuous symmetry there is a conserved charge, and furthermore, this conserved charge plays the role of generator of the symmetry from which it arose. For conformal transformations, say $z \rightarrow z + \epsilon(z)$, the conserved charge is

$$Q = \frac{1}{2\pi i} \oint dz T(z)\epsilon(z) \tag{2.6}$$

(more precisely, we should include the antiholomorphic part as well) where the contour integral is about a circle in the complex plane which corresponds to an equal time surface in radial quantization. Understanding conformal transformations thus boils down to specifying the operator product expansions of fields with T as this is all that is required to compute the action of the operator Q. One operator product expansion of particular interest is that of T with itself:

$$T(z)T(w) \equiv \frac{c/2}{(z-w)^4} + \frac{2T(w)}{(z-w)^2} + \frac{\partial T(w)}{(z-w)} + \ldots \tag{2.7}$$

The constant c in the first term is known as the central charge and depends upon the specific theory being studied.

Having made this brief excursion into some of the terminology of conformal field theory, lets return to our main point: rephrasing the essential characteristics of the flat ten space vacuum configuration in the language of conformal field theory . The theory (2.1) is especially easy to analyze as it is simply a sum of free bosonic and free fermionic theories. A free boson is described by a $c = 1$ conformal field theory , while that of a free fermion has $c = 1/2$. Thus, the full theory (2.1)may be characterized as a $c = 10(1 + 1/2) = 15$ conformal field theory . More precisely, the theory not only respects conformal symmetry, but superconformal symmetry as well. (One can understand the $c = 15$ above by noting that with the proper inclusion of reparametrization ghosts (the $b-c$ system with c = -26) and the superconformal ghosts ($\beta - \gamma$ system with c = + 11) the full theory will have $c_{total} = 0$ only if the 'matter' fields contribute $c = 15$.) In fact, our interest will focus on vacua which not only have these superconformal properties, but which actually respect an $N = 2$ superconformal symmetry. (We caution the reader that this second supersymmetry is not a local symmetry of our original action. If it were the critical dimension would be changed to $D = 2$.) We require this enhanced symmetry because it turns out that this is necessary to have theories with spacetime supersymmetry [4]. As we shall see, the latter property is crucial to showing that certain perturbative vacua are true ground states of the string.

Thus, our study of (2.1) leads us to extract that the important features which a vacuum state must posses are $c = 15$ and $N = 2$. To this point we have concentrated only on the right moving modes, as no mention of $\bar{c}$ has been made. For the superstring we can simply take left-right symmetry. For the heterotic string the left movers come from a bosonic string and hence the conformal theory is characterized by $\bar{c} = 26$ and $\overline{N} = 0$.

Finally, we emphasize that it is most convenient when analyzing the physical properties of such string configurations to pass to the light cone gauge. Then we can rephrase the above by saying that the physical degrees of freedom of the string are described by a ($c = 12$,$\bar{c} = 24$) ($N = 2$, $\overline{N} = 0$) conformal theory. We will now exploit this rephrasing.

In order to carry through on our goal to construct more realistic string vacua it is worthwhile at this point to divide the ten bosonic fields X^{μ} in (2.1)into a set of four $X^{1,2,3,4}$ which describe a four dimensional subspacetime (which we shall later identify with our usual spacetime) and the remaining six $X^{5,6,7,8,9,10}$ which we shall call 'internal degrees of freedom' (although we hasten to add that at this point in the discussion there is nothing internal about these six fields - they are on equal footing with the first four). We similarly divide the ten rightmoving fermions (which have not been explicitly displayed in (2.1)) into two groups. (For an analysis of the superstring, we would similarly treat the left moving fermions. Our interest is the heterotic string for which we take the 32 left moving fermions to all be of an 'internal' nature. Nonetheless, we shall continue our analysis as if we were concerned with superstrings and hence only discuss, say, the right movers. Any such superstring vacuum can generally be converted to a heterotic vacuum without much difficulty, as we shall discuss [5].) From this point of view, then, the theory (2.1)is composed of an 'external part' (which we describe in lightcone gauge) with $N = 2$ and $c = 3$ and an 'internal part' with $N = 2$ and $c = 9$. Having rephrased our ten dimensional solution in this language, it is now straightforward to generalize our program.

2.2. Strings in $D = 4$

We construct string vacua which have the interpretation of (first quantized) strings propagating in a *four* dimensional spacetime by combining an the $N = 2$ $c = 3$ theory of above (the four dimensional spacetime part of the theory in lightcone gauge) with *any* $N = 2$ $c = 9$ theory. The case described above in which the latter theory is composed of six free bosons and fermions is just the simplest possibility. The internal part of the theory need not, in general, have an interpretation in terms of spacetime dimensions. We have emphasized the word 'any' above to stress the generality of this construction which allows us to move away from a flat ten dimensional universe to a more realistic formulation. There are some nontrivial constraints such as unitarity and modular invariance which such a theory must satisfy, so 'any' must be taken with such provisos. Thus, when we speak of four spacetime dimensional classical string vacua we are referring to the space of unitary modular invariant $N = 2$ $c = 9$ superconformal field theories.

2.3. Examples of String Vacua

2.3.1. Geometrical Constructions

The earliest and most intuitive constructions of string vacua are those which do have a geometrical spacetime interpretation in the spirit of Kaluza-Klein theory. In such scenarios, an effective four dimensional string theory is constructed by 'curling' up the excess six dimensions which are present in the flat ten space formulation. That is, we modify the action (2.1) for the fields $X^5, ...X^{10}$ to the form

$$\int_{S^2} d\sigma d\tau \sqrt{h} g^{\mu\nu}(X) \partial_a X_\mu \partial_b X_\nu h^{ab} \tag{2.8}$$

where g is the metric on the curled up (compact) space K. The universe is then of the form $M_4 \times K$ where M_4 are the four dimensions of common experience.

This approach to constructing string vacua is called 'compactification' (as we imagine compactifying the extra six dimensions, although this picture is of no cosmological significance), and the action (2.8) is known as a (two dimensional) nonlinear sigma model with target manifold K. In the last section we stated the constraints which any would-be internal theory must satisfy: it must have $N = 2$ and $c = 9$. These requirements place severe restrictions on the allowed choices for the curled up space K. Lets first consider conformal invariance [6]. An efficient way to study the conditions under which (2.8) is conformally invariant is to view the metric $g^{\mu\nu}$ as a coupling constant (more precisely, a coupling function) in the two dimensional theory. Conformal invariance implies scale invariance, which in turn implies vanishing of the β function (more precisely 'β functional'). This is because the β function tells us how a coupling parameter changes with scale and hence it vanishes in a scale invariant theory. The β function for g can be computed perturbatively without much difficulty [6]. Before giving the result, though, we should define the perturbation expansion parameter. We have been a bit sloppy in our presentation of the action in (2.1) and (2.8). For these actions to be dimensionless (in units where h =1), they must be multiplied by a parameter which is usually called $1/\alpha'$ where α' has the dimension of area. Now notice that a change in the value of α' can be absorbed in a change in the scale of the metric g on K. That is, small α' is equivalent to a large rescaling of g. Thus, the dimensionless expansion parameter is $\frac{\sqrt{\alpha'}}{R}$ where R is the characteristic size of K. (We will make this more precise later.) If we chose K to have a large radius (in units of α'), we have a small expansion parameter with which, for example, we can compute the one loop β function. Thus result of such a computation is that

$$\beta(g_{\mu\nu}) = R_{\mu\nu} \tag{2.9}$$

where $R_{\mu\nu}$ is the Ricci curvature tensor of the manifold K. Thus, to lowest nontrivial order we find that the condition for conformal invariance requires that $R_{\mu\nu} = 0$, i.e. K must be ***Ricci flat.*** Certain nonrenormalization theorems which we shall discuss ensure that higher order corrections pose no problems for perturbatively adjusting these lowest order solutions [7]. (For a more direct argument see [8].) Moving on to the next requirement of $N = 2$ superconformal invariance, it turn out that this requires that K be a complex ***Kahler*** manifold. This means that K can be described with local complex coordinates for which the transition functions from patch to patch are holomorphic, and that the metric $g_{\mu\nu}$ satisfies $\partial_i g_{j\bar{k}} = \partial_j g_{i\bar{k}}$ where we have written g in the complex coordinate system on K. Finally, it can also be shown that with no other nonvanishing background fields, the central charge of such a nonlinear sigma model is not corrected from its free field value of 9. Thus, we have found that K must be a three complex dimensional (six real dimensions) Kahler manifold with vanishing Ricci tensor.

This is the result originally found in the classic paper of Candelas, Horowitz, Strominger and Witten [9].

We have already given the simplest such example in our flat ten space solution. In that case, K is the manifold C^3, which meets all of the requirements. The next example, in order of increasing complexity, is closely related to C^3, namely C^3/Γ where Γ is a lattice. More explicitly C^3/Γ is a complex three dimensional torus, T^3 in which the global identifications on coordinates are specified by Γ. The torus, like C^3, is not only Ricci flat but completely flat as well. It is somewhat hard to picture a manifold which is Ricci flat but not flat since in two and three real dimensions these conditions are the same. In four dimensions (that is, two complex dimensions) there is one such manifold and it is called $K3$. In three complex dimensions, on the other hand, such manifolds abound and are called Calabi-Yau manifolds. The simplest Calabi-Yau spaces are constructed by taking the quotient of T^3 by a finite group action G which, when expressed as a matrix acting on the coordinates of T^3, has determinant equal to one. Such spaces, which are called toroidal orbifolds, are not manifolds in general because they have singularities at the points which are left invariant by the action of G. The requirement of unit determinant ensures that these singularities can be repaired, thus giving rise to a smooth Calabi-Yau manifold. In fact, as far as string theory is concerned, one can construct consistent theories without actually performing the desingularization. For such spaces, the curvature tensor vanishes everywhere except at the singular points. Finally, one can construct Calabi-Yau manifolds directly, without making use of actions on T^3. A robust class of such manifolds are those which can be described by the vanishing locus of homogeneous equations in projective space - the 'complete intersection Calabi-Yau manifolds' or CICY's for short. To make this concrete, lets describe the simplest CICY: it is a submanifold of CP^4 (which we recall is the equivalence classes of five complex coordinates $z_1, ..., z_5$ under the equivalence relation $(z_1, ..., z_5) \equiv \lambda(z_1, ..., z_5)$ where λ is a complex number) given by the vanishing locus of

$$z_1^5 + z_2^5 + z_3^5 + z_4^5 + z_5^5 = 0. \tag{2.10}$$

We have chosen to write this quintic polynomial in a symmetric form, however, any quintic equation (whose derivative is nonvanishing on the vanishing locus of the equation) is equally acceptable.

A couple of remarks are in order. First, although it is not difficult to construct Calabi-Yau manifolds, it is difficult to determine their Ricci flat metric. In fact, there is only a numerical procedure which is burdensome even for a supercomputer. Thus, although we can formally write down a supersymmetric nonlinear sigma model with Calabi-Yau target space, in reality we can not write the action in closed form. Thus, nonlinear sigma models with Calabi-Yau target spaces are extremely complicated objects. Nonetheless, we shall see that they are amenable to much detailed analysis, and in fact some give rise to *exactly* soluble conformal field theories. Second, unlike the free theories built on flat ten space, the flat torus, or any of its consistent orbifolds, the nonlinear sigma models with target space being these smooth Calabi-Yau manifolds which are Ricci flat but not flat have higher order corrections to the lowest order β function equation (2.9). These corrections modify (2.9) order by order in α' and hence we must correspondingly modify the metric to make sure the equation is satisfied. So in addition to being unable to write down the tree level nonlinear sigma model (due to our lack of knowledge of the metric) this action must be modified order by order in perturbation theory to maintain conformal invariance. As mentioned, there is no obstacle to performing such a modificaition.

We shall return to the physics associated with these vacua shortly. For now, lets move on to some vacua which do not appear to have an interpretation as a curled up six manifold but nonetheless are viable examples.

2.3.2. (Naively) Nongeometrical vacua

As we emphasized earlier, there is no need for the $N = 2$ $c = 9$ superconformal theory to have any geometrical interpretation in terms of spacetime dimensions. Rather, it should more generally be viewed as an internal theory whose existence is necessitated by the structure of string theory. Thus it is worthwhile to construct string vacua out of the some of the simplest and best understood nonfree $N = 2$ conformal theories - the minimal models [10]. We first describe the minimal models and then indicate how they are used in the construction of string vacua. Our discussion is a bit detailed and assumes some familiarity with relevant background material. The reader not concerned with such details may skip this section; the essential idea is that what follows is one particular method of constructing string vacua which are *exactly soluble* conformal field theories. We also note that our discussion [11]rephrases the analysis of [10]so that the resulting construction is explicitly seen to simply amount to a standard orbifolding as expected from the work of [12][13]and as generalized by [14].

The minimal $N = 2$ models [15]are so called because of the fact that these theories are characterized by having a *finite* number of superconformal primary fields. They are also completely degenerate thus giving rise to a sufficiently rich null vector structure to allow for the *computation* (through the solution of differential equations) of all correlation functions. In more detail, the minimal models form a discrete series of unitary theories with central charge $c = 3P/(P+2) < 3$.

The idea behind using these discrete series to construct string vacua is quite simple [5]. We take a tensor product of r such theories, say at levels $P_1, ..., P_r$ and ensure that the total central charge is nine. That is, we choose the P_i such that

$$\Sigma_i^r \frac{3P_i}{P_i+2} = 9. \tag{2.11}$$

Actually, although this is the basic idea it is not the entire story [5]. In order to have a spacetime supersymmetric vacua the $U(1)$ charges of all fields in the theory must be integral. To satisfy this constraint we *orbifold* this tensor product to project out all states without integral $U(1)$ charges, and add in twisted sectors as dictated by modular invariance. In this way we arrive at exactly soluble string vacua which *a priori* have no geometrical interpretation. Later on, though, we shall come back to this point and establish a surprising geometrical connection.

In the next few pages we spell this minimal model construction out in more detail. For the reader only interested in focusing on the main ideas of these lectures, he/she can safely skip to the next section.

In [5] the characters of the N=2 algebra and their modular transformation properties were computed, using the connection between the level P superconformal model and a $\mathbf{Z}_p$ parafermionic theory . The primary fields of the level P model are labeled $\Phi^{l;\bar{l}}_{q,s;\bar{q},\bar{s}}(z,\bar{z})$, where $l, \bar{l} = 0, 1, \ldots, P$; $q := q \bmod 2(P+2)$; $s := s \bmod 4$; $l+q+s = 0 \pmod 2 \; = \bar{l}+\bar{q}+\bar{s}$, and the following identification holds

$$\Phi^{l;\bar{l}}_{q,s;\bar{q},\bar{s}} \sim \Phi^{P-l;P-\bar{l}}_{q+P+2,s+2;\bar{q}+P+2,\bar{s}+2} \tag{2.12}$$

. Fields with even (odd) s create states in the NS (resp. R) sector. The characters of the N=2 algebra are defined as

$$\chi^l_{q,s}(\tau, z, u) = e^{-2\pi i u} \operatorname{Tr}_{\mathcal{H}^{l;\bar{l}}_{q,s;q,\bar{s}}} e^{2\pi i z J_0} \, e^{2\pi i \tau (L_0 - c/24)} \tag{2.13}$$

where the trace is taken over a projection $\mathcal{H}^{l,\bar{l}}_{q,s;\bar{q},\bar{s}}$ to definite 'fermion' number (mod 2) of a highest weight representation of the (left-moving) N=2 algebra with highest weight vector $\Phi^{l;\bar{l}}_{q,s;\bar{q},\bar{s}}(0)$, where fermion number 1 is assigned to the superpartners $G^{\pm}$ of the energy-momentum tensor [1]. For some values of l, q, s the

[1] For $s = 2$ we have in fact $\Phi^{l;\bar{l}}_{q,2;\bar{q},\bar{s}} \sim G^{+}_{-1/2}\Phi^{l;\bar{l}}_{q,0;\bar{q},\bar{s}}$, but for our purposes it will be treated as a primary field, creating the highest weight vector of $\mathcal{H}$. The projection we make breaks the left-moving supersymmetry, and in assembling representations to form our model we will have to ensure that one unbroken N=2 algebra is preserved.

field $\Phi^l_{q,s}$ creates a null state; we set $\chi^l_{q,s} = 0$ for these values. The conformal weights and $U(1)$ charges of all the states in $\mathcal{H}^{l,\bar{l}}_{q,s;\bar{q},\bar{s}}$ satisfy

$$h = \frac{l(l+2) - q^2}{4(P+2)} + \frac{s^2}{8} + \text{integer} \tag{2.14a}$$

$$Q = -\frac{q}{P+2} + \frac{s}{2} + 2(\text{integer}) \tag{2.14b}$$

As shown in [5] the modular transformation matrices factorize; the index l transforms under the representation of the modular group carried by level P affine $SU(2)$ characters, while the indices q and s transform under the representation carried by level $-(P+2)$ and level 2 $SU(2)$ theta functions. The modular invariant combinations for the affine characters have been classified in [16], and those for the theta system in [17]. The latter comprise the left-right symmetric solution and quotients of this by subgroups of the $\mathbf{Z}_k$ symmetry present at level k. These may be used to construct modular invariant partition functions for the minimal N=2 models.

A tensor product of r such models such that $\sum_{j=1}^r 3P_j/(P_j+2) = 9$, with the addition of two free superfields (representing flat four-dimensional Minkowski space in the light-cone gauge), forms a theory with $c = \bar{c} = 12$ and (2,2) world-sheet supersymmetry. This may be converted into a consistent theory of superstrings in four dimensions if the constraints of modular invariance and spacetime supersymmetry can be simultaneously imposed. The solution to this requirement [5], can be phrased as an orbifold-like construction, in which we project out states that do not satisfy the conditions, and add twisted sectors to the Hilbert space. We first introduce some notation. Let

$$\chi^{\vec{l}}_{\vec{\mu}} = \frac{\vartheta_{\mu_0,2}}{\eta^3} \prod_{j=1}^{r} \chi^{l_j}_{\mu_j,\mu_{j+r}} \tag{2.15}$$

where ϑ is a classical theta function at level 2 and η is the Dedekind function. We have grouped the q and s indices for all subtheories as well as the $SO(2)$ representation label for the 'external' fermions into $\vec{\mu}$ which is a $(2r+1)$-vector with $\mu_j = q_j$, $\mu_{j+r} = s_j$ for $j = 1, \ldots, r$, and $\mu_0 = 0,1,2,3 \pmod 4$ for the $SO(2)$ representations $o, s, v, \bar{s}$ respectively. For two such vectors we define

$$\vec{\mu}\vec{\lambda} = \frac{\mu_0\lambda_0}{4} + \sum_{j=1}^{r} \left(-\frac{\mu_j\lambda_j}{2(P_j+2)} + \frac{\mu_{j+r}\lambda_{j+r}}{4} \right) \tag{2.16}$$

The partition function of our initial model takes the form

$$Z_0 = \sum_{\vec{l},\vec{\bar{l}},\vec{\mu}} A_{\vec{l},\vec{\bar{l}}} \chi^{\vec{l}}_{\vec{\mu}} \chi^{\vec{\bar{l}}*}_{\vec{\mu}} \tag{2.17}$$

where $A_{\vec{\bar{l}},\vec{l}} = \prod_{j=1}^{r} A_{l_j,\bar{l}_j}$ is a product of affine invariants, and the sum includes all states such that $l_j = \mu_j + \mu_{j+r} = \bar{l}_j \pmod 2$. This model possesses a group of discrete symmetries that includes $G = \mathbf{Z}_2 \times \prod_{j=1}^{r} \left(\mathbf{Z}_{P_j+2} \times \mathbf{Z}_2\right)$. The modular transformations may now be written

$$\chi_{\vec{\mu}}^{\vec{l}}(\tau+1) = T_{\vec{l}}\, e^{\pi i \vec{\mu}\vec{\mu}}\, \chi_{\vec{\mu}}^{\vec{l}}(\tau) \tag{2.18a}$$

$$\chi_{\vec{\mu}}^{\vec{l}}(-1/\tau) = \prod_{j=1}^{r} 2(P_j+2)^{-1/2} \sum_{l'_j+\mu'_j+\mu'_{j+r}=0 \pmod 2} S_{\vec{l}\vec{l'}}\, e^{-2\pi i \vec{\mu}\vec{\mu'}}\, \chi_{\vec{\mu'}}^{\vec{l'}}(\tau) \tag{2.18b}$$

Here $T_{\vec{l}}$ and $S_{\vec{l}\vec{l'}}$ are products of affine character transformation matrices, which commute with $A_{l_j,\bar{l}_j}$. The dependence on u, z which are set to 0 is suppressed. Let $\vec{\beta} = 2\vec{\gamma}$, where $\vec{\gamma} = [\gamma_0, \ldots, \gamma_{2r}]$ generates a $\mathbf{Z}_k$ subgroup of G. Then for $x, y \in \mathbf{Z}_k$ we can define the twisted partition function

$$Z(x,y) = \sum_{\vec{l},\vec{\bar{l}},\vec{\mu}} A_{\vec{l},\vec{\bar{l}}}\, e^{-\pi i x y \vec{\beta}\vec{\beta}}\, e^{-2\pi i x \vec{\beta}\vec{\mu}}\, \chi_{\vec{\mu}}^{\vec{l}} \chi_{\vec{\mu}+y\vec{\beta}}^{\vec{\bar{l}}*} \tag{2.19}$$

It is easy to check that these transform correctly under modular transformations, *i.e.*under $\tau \to (a\tau+b)/(c\tau+d)$, $Z(x,y) \to Z(ax+by, cx+dy)$, and that $Z(x,y)$ depends on x, y only $\pmod k$. It is thus possible to mod out the $\mathbf{Z}_k$ symmetry to obtain a new modular invariant partition function

$$Z_{\vec{\beta}} = \frac{1}{k} \sum_{x,y \in \mathbf{Z}_k} Z(x,y) = \sum_{\vec{l},\vec{\bar{l}},\vec{\mu},n} A_{\vec{l},\vec{\bar{l}}}\, \chi_{\vec{\mu}}^{\vec{l}} \chi_{\vec{\mu}+n\vec{\beta}}^{\vec{\bar{l}}*} \tag{2.20}$$

The sum here extends over $\vec{\mu}, n \in \mathbf{Z}_k$ satisfying $\vec{\beta}(\vec{\mu} + \frac{n}{2}\vec{\beta}) \in \mathbf{Z}$ as well as the usual (mod 2) condition. If $\vec{\beta}\vec{\beta} \in 2\mathbf{Z}$ the projections in all sectors are identical, $\vec{\beta}\vec{\mu} \in \mathbf{Z}$. This method may be extended to several such vectors, *i.e.*we an apply the same methods to $Z_{\vec{\beta}}$ to mod out by $\vec{\beta}'$. The conditions determining the spectrum will now become:

$$\vec{\mu} - \vec{\bar{\mu}} = x\vec{\beta} + x'\vec{\beta}' \tag{2.21a}$$

$$\vec{\beta}(\vec{\mu} + \vec{\bar{\mu}}) \in 2\mathbf{Z} \tag{2.21b}$$

$$\vec{\beta}'(\vec{\mu} + \vec{\bar{\mu}}) \in 2\mathbf{Z} \tag{2.21c}$$

If $\vec{\beta}\vec{\beta}' \in \mathbf{Z}$, the projections determined by $\vec{\beta}$ remain unchanged.

We can now apply this formalism to the problem of converting the tensored theory into a string vacuum. The first requirement to be met is the $N = 1$ superconformal gauge condition. This requires that our Hilbert space contain only states which are either in the NS sector of all subtheories or in the R sector. This requirement is met by moding out as above using the r vectors $\vec{\beta}_{(j)}$ which have all zero entries except $\vec{\beta}_{(j)0} = \vec{\beta}_{(j)j+r} = 2$. Since $\vec{\beta}_{(j)}\vec{\beta}_{(j)} = 2$, this projection leaves states with $\vec{\mu}\vec{\beta}_{(j)} = \frac{1}{2}(\mu_0 + \mu_{j+r}) \in \mathbf{Z}$, and together with $\vec{\beta}_{(j)}\vec{\beta}_{(k)} = 1$ this ensures the s indices are all odd or all even. The second condition to be imposed is spacetime supersymmetry. As shown in [4] the requirement of spacetime supersymmetry (after a GSO projection) is equivalent to the existence (before GSO projection) of an unbroken $N = 2$ worldsheet superconformal symmetry and the integrality of the charges of all fields under the $U(1)$ of the $N = 2$ algebra. As pointed out in [14]charge integrality under the total $U(1)$ is accomplished by moding out by $g = e^{2\pi i J_0}$. Using (2.14b) this is the same as using $\vec{\beta}_{(0)} = (2, \ldots, 2)$ in our construction, since $\vec{\beta}_{(0)}\vec{\mu} = Q + 2(\text{integer})$. Once more, $\vec{\beta}_{(0)}\vec{\beta}_{(0)} = 4$ and $\vec{\beta}_{(0)}\vec{\beta}_{(j)} = 2$ ensure this is a consistent projection onto integral Q. This projection also ensures that the 'diagonal' $N = 2$ superconformal algebra (generated by the total energy-momentum tensor and its superpartners) is not broken, and that the internal $U(1)$ augments the external $SO(10)$ current algebra to an E_6 gauge symmetry. The last requirement is a GSO projection onto definite fermion number. The operator $(-1)^{F_L} = e^{\pi i J_0}$ [18]is local with respect to all fields surviving the previous projections, commutes with all bosonic operators and anticommutes with all fermionic ones. It can thus serve as a fermion number projection. To eliminate the tachyon yielding a supersymmetric spectrum the projection we need is onto $(-1)^{F_L} = e^{\pi i \vec{\beta}_{(0)}\vec{\mu}} = -1$. The $\mathbf{Z}_2$ symmetry by which we are moding out here is not contained in the group G defined above, and is a manifestation of the fact that we are now dealing with a supersymmetric theory. Once more one can check that this projection is consistent with the previous ones. The projections in both sectors are again identical, yielding odd integer values of Q. The final step in the construction is the conversion of the superstring model we have obtained to a heterotic one. This can be accomplished by using the related modular transformation properties of $SO(2)$ and $E_8 \times SO(10)$ characters [5], and involves exchanging the o and v representations and changing the sign of the spinor representations. The partition function of the resulting heterotic model is finally

$$Z = \sum_{\substack{\vec{l},\vec{\bar{l}},\vec{\mu},\vec{\bar{\mu}} \\ \vec{\mu}-\vec{\bar{\mu}}\in\Lambda}} \pm A_{\vec{l},\vec{\bar{l}}}\chi^{\vec{l}}_{\vec{\mu}}\chi^{\vec{\bar{l}}*}_{\vec{\bar{\mu}}+\vec{\mu}_{(0)}} \tag{2.22}$$

where the sum extends over $\vec{\mu}$ such that $\vec{\mu}\vec{\beta}_0(0) \in 2\mathbf{Z}+1$ and $\vec{\mu}\vec{\beta}_{(j)} \in \mathbf{Z}$ $\quad j = 1, \ldots, r$, $\vec{\bar{\mu}}$ satisfying the same conditions and Λ is the lattice generated over $\mathbf{Z}$ by $\frac{1}{2}\vec{\beta}_{(0)}, \vec{\beta}_{(j)}$. The vector $\vec{\mu}_{(0)}$, with zero entries except $\vec{\mu}_{(0)0} = 2$ implements the heterotic construction, and the sign of each contributions is consistent with spacetime spin-statistics. We thus explicitly see that Gepner's construction amounts to a simple orbifolding of the original tensor product theory and provides us with a rich class of (naively) nongeometrical string vacua.

Having discussed string vacua in general and given examples above, we now turn to the issues involved in extracting the four dimensional physics which follows from a given choice of the vacuum state.

2.4. Four Dimensional String Phenomenology:

We have emphasized earlier that the physics which the string gives rise to in the four extended dimensions of ordinary spacetime is intimately dependent upon the structure of the vacuum state. We shall now see this explicitly as we discuss the phenomenology which follows from the two classes of string vacua described in the previous section.

2.4.1. Phenomenology from Geometrical String Compactification:

There are a number of ways of determining the four dimensional phenomenology which follows from a chosen Calabi-Yau compactification. The method whose flavour is most stringy involves truncating the nonlinear sigma model field theory to a quantum mechanical theory by keeping only a finite number of modes with nontrivial σ dependence. This Born-Oppenheimer approach is a direct (albeit approximate) generalization of the standard flat space mode analysis of the string. We shall follow here a more intuitive approach which is based upon Kaluza-Klein type reasoning. The strategy here is to use a spacetime approach in which we examine what happens to the (well known) flat ten space spectrum of the string when six of the dimensions curl up into a small compact space. Before examining this question, it is worthwhile to consider a toy example which captures all of the essential features of the analysis while avoiding the mathematical details which can obscure the main point.

Consider a flat five dimensional spacetime R^5 with a free massless scalar field Φ. Since Φ is massless, it satisfies a klein-Gordon equation of the form

$$D_5\Phi = 0 \tag{2.23}$$

where D_5 is the five dimensional wave operator. Now imagine that we curl the fifth dimension of R^5 up into a circle of radius R. How does the field Φ manifest itself to an observer which lives in the remaining R^4? To answer this question, let $x_1, ..., x_4$ be coordinates on the R^4 and let y be the coordinate on the circle. The equation (2.23)in these coordinates is just

$$(\sum_{i=1}^{4} \frac{\partial^2}{\partial x_i^2} + \frac{\partial^2}{\partial y^2})\Phi(x,y) = 0. \tag{2.24}$$

It is now worthwhile to Fourier expand $\Phi(x,y)$ about the circle:

$$\Phi(x,y) = \sum_n \phi_n(x) e^{2\pi i n y/R} \tag{2.25}$$

Plugging (2.25)into (2.24)we find that the mth mode $\phi_m(x)$ satisfies the equation

$$(\sum_{i=1}^{4} \frac{\partial^2}{\partial x_i^2} + 4(\frac{\pi m}{R})^2)\phi_m(x) = 0. \tag{2.26}$$

The mode with $m = 0$ appears massless while all other modes appear as *massive* fields whose masses are quantized in units of the inverse of the radius of the circle. Notice that the zero mode $\phi_0(x)$ is that mode whose internal counterpart satisfies the equation

$$\frac{\partial^2}{\partial y^2}\chi(y) = 0 \tag{2.27}$$

(which in this situation has the solution $\chi(y) = e^{2\pi i 0} = 1$). This toy examples illustrates all of the important points: a massless field in higher dimensions gives rise to both massless and massive fields in lower dimensions. The massless fields are associated with solutions of harmonic equations (which generalize (2.27)) on the internal space and the massive modes form an infinite tower of fields whose masses are proportional to the inverse of the radius of the curled up space. If this radius is small these masses are huge and hence these fields may be integrated out of our low energy theory leaving an effective theory for the massless modes.

With this understanding of the toy examples, lets now turn to our real interest: string vacua of the form $M_4 \times K$. Following the reasoning in the toy example, we first must determine the massless field content in the ten dimensional theory. As is well known, these massless fields comprise an $N = 1$ $D = 10$ Super-Yang-Mills theory coupled to supergravity with gauge group $E_8 \times E_8$. When we pass to $M_4 \times K$, these fields will give rise to a finite set of massless fields and an infinite tower of massive ones. The massless fields will be associated with 'zero modes' on K, that is, with the harmonic forms on K. These forms satisfy an equation analogous to (2.27) but substantially more complicated. Luckily, it is not too difficult to count the number of solutions to the relevant equations (without explicitly solving them) and hence determine the massless spectrum.

More precisely, when we pass to $M_4 \times K$, one of the E_8 factors is broken to E_6. The reason for this is that in addition to the geometric Calabi-Yau condition given earlier, one must ensure anomaly freedom when we convert our superstring vacuum into a heterotic vacuum. This requirement takes the form $TrF \wedge F = trR \wedge R$ (more precisely, this need only be true in cohomology and hence we demand $c_2(V) = c_2(T_K)$ where V is the vacuum gauge bundle and T_K is the tangent bundle of K). For Ricci flat but not flat Calabi-Yau manifolds, the easiest way to solve this [9] is to identify an $SU(3)$ subbundle of the E_8 gauge bundle with T_K which breaks E_8 to E_6. It is worthwhile to decompose the massless E_8 super-Yang-Mills multiplet (we ignore the supergravity multiplet and one E_8 factor) under E_6. We recall that this is a gauge supermultiplet and hence fills out the 248 dimensional adjoint representation of E_8. Under $E_6 \times SU(3)$ we have

$$248 = (78, 1) \oplus (27, 3) \oplus (\overline{27}, \overline{3}) \oplus (1, 8). \tag{2.28}$$

The first term gives rise to the E_6 gauge bosons in four dimensions, while the last term gives rise to E_6 singlet fields which shall not be of much concern to us here. The middle two terms give rise to 27's and $\overline{27}'s$ of E_6, representations which are familiar from more conventional attempts at building Grand Unified Theories based on E_6. An analysis of the equations which generalize (2.27) to this context [9] shows that the 27's are associated with harmonic forms with two holomorphic and one antiholomorphic index, i.e. $(2, 1)$ forms. The space of such forms is denoted by $H^{2,1}(K)$ and has dimension $h^{2,1}$. Similarly, the $\overline{27}$'s are associated with $H^{1,1}(K)$ with dimension $h^{1,1}$. Since a standard model family fits inside a 27 and we anticipate that the 27 and $\overline{27}$'s will join up and become massive, the four dimensional theory has $|h^{1,1} - h^{2,1}|$ generations of elementary particles (where a generation here is a 27 of E_6). This difference has an important topological interpretation:

$$h^{1,1} - h^{2,1} = \frac{1}{2}\chi(K) \tag{2.29}$$

where $\chi(K)$ is the Euler characteristic of K. Thus, one of the most basic topological invariants of the curled up component of spacetime determines the number of elementary particle multiplets [9]! We see here that our string based models are already taking us beyond conventional model building by giving us a possible geometric understanding of chirality and the gauge multiplet structure. As we now discuss, we can in fact go much further.

One of the most striking features of superstring inspired models, and an aspect which gives such models the potential to go way beyond conventional grand unification, is the fact that the Yukawa couplings are determined by the topology (and complex structure) of the chosen compactification manifold. One way of understanding this is to note that all couplings in the theory originate from the supersymmetric gauge couplings of the underlying ten dimensional E_8 super-Yang-Mills theory. On passing to an effective four dimensional theory, we essentially integrate out the dependence on the extra six dimensions. As the relevant modes which live on the chosen six dimensional space are the harmonic forms (as discussed), four dimensional couplings arise from integrating various combinations of these forms over the chosen Calabi-Yau manifold. As shown in [19]the Yukawa coupling, at sigma model and string tree level, between three 27's, say $27_i, 27_j, 27_k$ is given by

$$g_{ijk} = \int \omega \wedge A_i^m \wedge A_j^n \wedge A_k^p \omega_{mnp} \tag{2.30}$$

where A_i is an element of $H^{2,1}(K)$ (the superscript explicitly written on A_i is a holomorphic tangent space index), and ω is the holomorphic three form which is always present on a Calabi-Yau manifold (ω is the global holomorphic section of the canonical bundle whose existence is guaranteed since the canonical bundle is trivial). This coupling (2.30) is not quite topological as it depends upon the complex structure of K. A general computational method has been given in [20]which reduces the calculation of (2.30) to an elegant algebraic procedure.

As shown in [19] [21]the Yukawa coupling between three $\overline{27}$'s, again to lowest order in sigma model and string perturbation theory, is given by

$$g_{ijk} = \int A_i \wedge A_j \wedge A_k \tag{2.31}$$

where A_i is an element of $H^1(T_K^*)$.

The latter formula (2.31) is purely topological; it is simply the triple intersection form on the manifold K. This has a simple interpretation in terms of the homological duals of the $(1,1)$ forms A_i which are real four surfaces in K. In terms of these duals, the Yukawa coupling is essentially the number of mutual points of intersection of the corresponding four surfaces. There is no general prescription for computing these couplings, but as we shall mention later, they have been computed for a phenomenologically promising compactification.

Of course, these formulae are only lowest order results for these Yukawa couplings. However, as shall be discussed below, some powerful nonrenormalization theorems greatly enhance their validity.

We should emphasize at this point that in reality there is less information in these lists of couplings than one might think. This is because we have not addressed the normalizations of the fields in our Lagrangian. Notice, though, that the number independent field normalizations is in general much less than the number of independent Yukawa couplings. So even if we left the normalizations completely unspecified we would still have a nonzero number of predictions. In practice, the choice of manifold often leads to a large group of holomorphic automorphisms which substantially reduces the number of *a priori* independent couplings and normalizations. In addition to greatly reducing the number of unspecified parameters from field normalizations, discrete symmetries play a number of other important roles in superstring phenomenology, so we now turn to a brief description of these symmetries.

Another striking improvement offered by string based models is that the phenomenologically essential discrete symmetries which are put in by hand in conventional model building arise ***automatically*** through the process of compactification. The basic idea is that although Calabi-Yau manifolds do not admit any continuous symmetries, they often admit large groups of discrete symmetries. The matter fields in the theory (which are associated to harmonic forms on the manifold) transform amongst themselves under the action of these discrete symmetries, thus placing severe restrictions on couplings in the Lagrangian [22]. It is easy to see from the integral expressions for the renormalizable couplings given in the last section that the only nonzero couplings will be those which respect these discrete symmetries. For the higher point couplings (and field normalizations) which we do not know how to explicitly compute at present, the discrete symmetries provide a powerful analytic tool.

2.4.1.1. Some Nonrenormalization Theorems:

Having discussed the essential ingredients and characteristics of Calabi-Yau based model building, a natural and in fact crucial question to ask is to what extent the parameters we calculate are renormalized and possibly overwhelmed by higher order corrections. That is, are the calculations we have just outlined at all relevant?

The corrections we envision come from four sources: perturbative and nonperturbative sigma model corrections and perturbative and nonperturbative string effects. About the latter there is not much one can presently say. However, a number of powerful statements can be made regarding the previous three.

The first result is a nonrenormalization theorem due to Witten [7] which relies on the fact that our compactifications preserve an $N = 1$ spacetime supersymmetry. This affords us the luxury of invoking macroscopic analysis to solve what would otherwise be complicated microphysics [23]. Since our effective four dimensional theory is supersymmetric one knows that the potential terms in the Lagrangian can be derived from a superpotential which is a *holomorphic* function of the chiral superfields in the theory. This holomorphicity is a stringent constraint and, as shown in [7] is sufficient to prove that the superpotential is not renormalized in sigma model perturbation theory. We will return to this nonrenormalization theorem towards the end of the talk, but for now we simply note that it proves that perturbative sigma models effects can not destabilize a Calabi-Yau compactification, and it also shows that the sigma model tree level computations of Yukawa couplings are actually valid to all orders in sigma model perturbation theory.

Under mild assumptions about the consistency of string theory at higher loops, a similar nonrenormalization theorem for the topological string loop expansion was proven in [24]. Thus, the remarks made above regarding the insensitivity of the superpotential to sigma model loops, extends to the effects of string loops as well.

Turning to the final source of corrections, we consider nonperturbative sigma model effects as manifested in world sheet instantons [2]. As shown in [2], generically one expects world sheet instantons to renormalize the Yukawa couplings in the superpotential (but not to destabilize $(2,2)$ compactifications). On closer inspection of fermion zero modes in an instanton background, however, one can see [25]that although the $\overline{27}^3$ couplings are renormalized [2], the 27^3 couplings are not. The proof is straightforward and is most lucidly presented by comparison with the same calculation for the $\overline{27}^3$ couplings. Let us first recall that a world sheet instanton is a topologically nontrivial map of the string world sheet into K. As shown in [2], the relevant nontrivial maps in searching for corrections to the superpotential are the antiholomorphic (or holomorphic) maps whose image is a sphere (a CP^1) in K. Removing a point from this sphere and putting local holomorphic coordinates X^1, X^2, X^3 on K, we can think of the instanton as lying, say, in the X^3 plane. Explicitly then, the map from the world sheet to K is of the form $X^3(\bar{z}) = \frac{a\bar{z}+b}{c\bar{z}+d}$ with $ad - bc = 1$. There are four left moving fermion zero modes in such a background: two λ^3 , one $\lambda^{\bar{1}}$ and one $\lambda^{\bar{2}}$ zero modes. Whether or not a chosen coupling is renormalized by world sheet instantons essentially comes down to the question of whether these zero modes can be absorbed in the associated correlation function. For a $\overline{27}^3$ coupling we can consider a three point function of the form $< \phi^\alpha \phi F^\beta >$ where ϕ (F) denotes a scalar (F axillary) superfield component and α and β are indices for the 10 in the $\overline{27}$ of $SO(10)$. If we denote our $(1,1)$ forms by $b_{i\bar{j}}$, this is explicitly

$$< (e^{-\phi} b^1_{i\bar{k}} \psi^{\bar{k}} \lambda^i \lambda^\alpha)(e^{-\phi} \epsilon^{k}_{\bar{i}\bar{j}} b^2_{k\bar{k}} \psi^{\bar{k}} \lambda^{\bar{i}} \lambda^{\bar{j}})(\epsilon^{\bar{k}}_{lm} b^3_{k\bar{k}} \psi^l \psi^m \lambda^k \lambda^\beta) > \tag{2.32}$$

Now, in an instanton background, we can absorb the two λ^3 zero modes in the first and third vertex operators, and the $\lambda^{\bar{1}}$ and $\lambda^{\bar{2}}$ zero modes in the second operator. Doing so yields the instanton correction to the formula (2.31)

$$\delta^{\alpha\beta}\int X^*(b^1)\int X^*(b^2)\int X^*(b^3) \tag{2.33}$$

as computed in [2]. What happens if we perform the same calculation for the 27^3 couplings? All of the $\lambda^{\bar{i}}$'s in the above vertex operators must be replaced by λ^{i}'s (and vice versa). When we try to absorb the λ^3 zero modes, we find that the only possibility is to use the second operator in the correlation function. But this is of the form $\epsilon_{ijk}\lambda^j\lambda^k$ and hence cannot absorb both of the λ^3 zero modes. Thus, the correlation function vanishes in an instanton background. One can even go beyond this sigma model based instanton expansion by making use of the superconformal invariance of the compactification. That is, by use of the $(0,2)$ superconformal Ward identities [26]it is not hard to prove that couplings of the form $27^3 R^n$ vanish for positive n [25]. Thus, the 27^3 coupling is independent of the radius, and hence is not renormalized from its tree level value (2.30).

For Calabi-Yau manifolds with negative Euler characteristic (which includes the large class of complete intersection varieties in products of projective spaces) the 27's (that is, the fields which to lowest order in the sigma model are built out of the $(2,1)$ forms) correspond to the matter generations (while the $\overline{27}$'s correspond to antigenerations). Thus, the phenomenologically relevant sigma model tree level couplings (2.30)between matter generations are not renormalized in the sigma model!

Using these techniques and certain well motivated choices for K, detailed low energy models have been constructed. We shall discuss some of this work in the last section. Before moving on to discuss the ways in which one extracts phenomenology from the other apparently non-geometrical class of vacua, the minimal models, we first present some interesting results on a kind of 'dual' to the Calabi-Yau manifolds discussed above.

2.4.1.2. Mirror Pairs of Manifolds:

In our discussion of nonperturbative corrections, we have noted that the 27^3 couplings are independent of the radial (Kahler) Calabi-Yau moduli. We have similarly noted that the $\overline{27}^3$ couplings are topological (even with instanton corrections) and hence are independent of the complex strucutre Calabi-Yau moduli. We thus see a nice symmetry in these two statements. In retrospect, this is what we expect since from a conformal field theory point of view, the distinction between 27's and $\overline{27}$'s and between Kahler moduli and complex structure moduli is essentially one of *convention.* To understand this statement, bear in mind that the vertex operators we have invoked for these fields in the last section differ only in their $U(1)$ charge (where the $U(1)$ here is fermion number). That is, we replace λ's by $\overline{\lambda}$'s to pass from a $\overline{27}$ to a 27. Since the distinction between these fields is characterized by a convention (the arbitrary choice of the reference sign of the $U(1)$ charge) we expect their properties to display a manifest symmetry. This is in fact the case if we analyze their properties using nonpertabive techniques such as Ward identities which are only sensitive to the conformal properties of the fields. The reader is refered to [25] for explicit examples of this. However, if we treat the fields in a perturbative manner, our expansion in powers of α'/R^2 clearly disrupts the symmetry as we have singled out a Kahler radial mode. This is why, for example, the symmetry between 27's and $\overline{27}$'s is obscured in the instanton expansion: $\overline{27}^3$ couplings are renormalized while 27^3 couplings are not. (The essential symmetry which is being obscured here is that, as mentioned, $\overline{27}^3$ couplings are independent of the complex strucutre Calabi-Yau moduli while 27^3 couplings are independent of the Kahler moduli.) One could restore 'perturbative symmetry' if there were, for example, a means of doing perturbation theory in powers of complex structure moduli. It is a bit difficult to really understand what such a statement means if taken at face value. However, if it were the case that for every Calabi-Yau manifold K there were a 'mirror pair' K' which gives rise to the same conformal field theory and has Hodge numbers which are *interchanged* (i.e. $h^{1,1}_{K'} = h^{2,1}_{K}$ and $h^{2,1}_{K'} = h^{1,1}_{K}$) we would have a way of understanding perturbation theory in complex structure moduli: simply pass to the mirror pair and do ordinary radial perturbation on that new sigma model.

The reader may well wonder why we even care to restore symmetry at the level of perturbation theory if we already have complete symmetry in the exact theory. For one thing it would give us a means of computing both of the 27^3 and $\overline{27}^3$ exactly via geometrical means. Sometimes this is the most direct approach, especially when one does not have a good handle on the exact CFT which the Calabi-Yau 's are approximating. Furthermore, even from the point of view of the exact CFT theory it would be useful and satisfying to have the conventional ambiguity in the sign of a $U(1)$ charge reflect itself in the approximating geometry. (In fact, as discussed in [11], there *is* such a geometrical ambiguity - namely T_K vs. T^*_K as the vacuum gauge bundle. See [11]for details.) Finally, there are a number of other interesting applications of such a duality, as discussed in [11].

This possibility of mirror manifolds has recently been confirmed for the class Calabi-Yau manifolds for which there exists one point in their moduli space which has a minimal model realization [11]. In particular, in [11]a particular operation of orbifolding on minimal model vacua is shown to yield an isomorphic conformal field theory as it results in reversing the sign of the left moving $U(1)$ charges of the states. At the level of the underlying geometry, though, this operation produces a new and topologically inequivalent manifold. In fact, this new manifold has hodge numbers which are reversed relative to the initial manifold, i.e. it is the mirror pair. Thus, we have, for this class of vacua, found the duality operation which shows that two topologically inequivalent Calabi-Yau manifold give rise to the same conformal theory. Some examples of this phenomenon are given in the accompanying tables.

In all of these tables, the column 'symmetries' refers to the symmetry of the theory by which we orbifold. The numbers in the given row vector correspond to the powers of the n_i^{th} roots of unity acting on the coordinates z_i. The value of n_i is simply the degree of z_i in the defining equation. For example, in the first table, $(1,1,1,3,4)$ means that we are orbifolding by $(z_1, ..., z_5) \rightarrow (\alpha z_1, \alpha z_2, \alpha z_3, \alpha^3 z_4, \alpha^4 z_5)$ where α is a nontrivial fifth root of unity.

Theory	Symmetries	$h^{2,1}$	$h^{1,1}$	χ
$\vec{P} = (3,3,3,3,3)$ or $Z_1^5 + \cdots + Z_5^5 = 0$ in $\mathbf{C}P^4$		101	1	-200
	(1,1,1,3,4)	49	5	-88
	(0,2,2,3,3) (0,1,2,3,4)	21	17	-8
	(0,0,0,2,3)	17	21	8
	(1,1,1,3,4) (0,2,2,3,3)	5	49	88
	(0,1,2,3,4) (0,2,2,3,3) (1,1,1,3,4)	1	101	200

Theory	Symmetries	$h^{2,1}$	$h^{1,1}$	χ
$\vec{P}=(3,8,8,8)$ or $Z_1^5+Z_2^{10}+\cdots+Z_4^{10}+Z_5^2=0$ in $WCP^4_{2,1,1,1,5}$		145	1	-288
	(1,1,1,6)	99	3	-192
	(0,4,8,8)	47	11	-72
	(0,0,3,7)	39	15	-48
	(1,0,2,6)	37	13	-48
	(0,4,7,9)	29	17	-24
	(0,5,7,8)	17	29	24
	(0,3,3,4)	15	39	48
	(0,4,7,9) (1,1,1,6)	13	37	48
	(0,5,7,8) (0,0,3,7)	11	47	72
	(0,4,7,9) (1,0,2,6)	3	99	192
	(0,5,7,8) (0,4,7,9)	1	145	288

Theory	Symmetries	$h^{2,1}$	$h^{1,1}$	χ
$\vec{P}=(1,\overline{16},\overline{16},\overline{16})$ or $Z_0^3+\cdots+Z_3^3=0$ $Z_1X_1^3+\cdots+Z_3X_3^3=0$ in $\mathbf{C}P^3\times\mathbf{C}P^2$		35	8	-54
	(1,7,7,16)	41	12	-58
	(1,9,9,12)	32	9	-46
	(1,5,8,17)	26	9	-34
	(1,0,6,6)	22	7	-30
	(1,1,13,16)	29	18	-22
	(0,0,6,12)	23	14	-18
	(1,0,6,6) (0,0,6,12)	14	23	18
	(1,0,6,6) (1,1,13,16)	18	29	22
	(1,0,4,8)	7	22	30
	(1,1,12,17)	9	26	34
	(2,2,7,15)	9	32	46
	(2,2,7,15) (1,0,6,6)	12	41	58
	(1,0,4,8) (1,0,6,6)	8	35	54

2.4.2. Phenomenology from Minimal Model String Vacua:

The phenomenology which follows from the construction of string vacua along the lines discussed in section 2.3.2 can be extracted much more simply than the analysis discussed for geometrical compactifications in the last section.

We begin with the massless spectrum. As outlined in the first section of these lectures, the emission and absorption of spacetime states is accomplished by two dimensional operators called vertex operators. These operators have conformal dimension $(1,1)$. For example, in the (-1) picture, a class of vertex operators can be written as $e^{-\phi}O(z,\bar{z})\lambda^{\alpha}e^{ikX}$. In this expression λ is a free left moving fermion, ϕ is the bosonized superconformal ghost field and O is a field contributed by the minimal model structure. In this case, the vertex operator we have written corresponds to a massless particle ($k^2 = 0$) if the conformal dimensions of O are $(1/2, 1/2)$. Such an operator would create states in the 10 of $SO(10)$ (recall that α is a vector index of $SO(10)$). We can similarly identify all other massless spacetime particle with particular operators in the minimal model construction. Determining the spectrum then boils down to finding all of these required minimal model states. This latter task is easily done by a (computer) search through the summands in the partition function given earlier [27]. Notice that this procedure is quite different from the geometrical methods of the last section which involved finding harmonic forms on a Calabi-Yau manifold.

After obtaining the set of massless spacetime fields, the next step is find the effective spacetime Lagrangian which describes their dynamics. The approach here is again appears to be very different from the geometrical methods above. The strategy is to compute scattering amplitudes in the two dimensional theory describing the string and then construct an effective spacetime Lagrangian which reproduces those scattering amplitudes. For example, let $V_i^{\psi_i}, V_j^{\psi_j}$ and $V_k^{\phi_k}$ be the vertex operators for two massless spacetime fermions and a massless spacetime boson. Our aim is to determine the magnitude of the Yukawa coupling h_{ijk} in the effective four dimensional Lagrangian which contains the term $h_{ijk}\psi_i\psi_j\phi_k$. To accomplish this we compute the three point function $< V_i^{\psi_i} V_j^{\psi_j} V_k^{\phi_k} >$ and pick out the momentum independent part (since we are trying to fix a potential term in the Lagrangian). By setting $h_{ijk} =< V_i^{\psi_i} V_j^{\psi_j} V_k^{\phi_k} > |_{zero\ momentum}$ we ensure that the tree level amplitude computed in the effective Lagrangian matches that of the underlying string theory. The three point functions are all calculable in these minimal model theories so this procedure can be carried out explicitly in any chosen theory. We will show an example of this shortly.

It is important to stress that the procedure just described is good deal more powerful than the geometrical methods that must be employed in the Calabi-Yau compactification program. The couplings we compute here are properly normalized as we choose our fields to have canonical two point functions. This is something which goes beyond the power of the geometrical approach.

The distinctions between the geometrical and minimal model constructions and modes of analysis are manifest. Nevertheless, we shall shortly see that these constructions are not as distant from each other as they might seem to be at this point.

3. The Calabi-Yau Geometry of Minimal Model Constructions

3.1. Gepner's Observation:

By studying the massless spectra and discrete symmetries of specific minimal model constructions, Gepner [10] came to the surprising realization that the data so derived matches that found for certain Calabi-Yau compactifications. For example, in the minimal model theory $(3,3,3,3,3)$ (this refers to taking five copies of the level 3 theory) Gepner found 101 massless 27's of E_6 and 1 massless $\overline{27}$ of E_6. This theory also respects the 75000 element discrete symmetry group $Z_5^4 \times S_5$. On the geometrical side, the Calabi-Yau manifold $Y_{4;5}$ introduced earlier has 101 massless 27's of E_6 and 1 massless $\overline{27}$ of E_6. Furthermore, this theory also respects the symmetry group $Z_5^4 \times S_5$ (and the massless modes in each construction transform in identical representations of this group). This appears to be strong, albeit circumstantial evidence that these theories might be identical. A sensitive test of this conjectured correspondence is to see if they agree at the level of interactions. Due to the nonrenormalization theorem for the 27^3 couplings [25], we can in fact compare the geometrical coupling calculation with that of the exact minimal model theory. As shown in [25] these couplings agree perfectly, (with the unknown geometrical normalizations scaled out of the comparison). Two more correspondences based upon comparison of spectra, symmetries and couplings were also made in [28].

We emphasize that this apparent correspondence is far from expected. As discussed earlier, the Calabi-Yau compactifications correspond to very complicated nonlinear sigma model whose origin is purely geometrical. The minimal models on the other hand are the simplest $N = 2$ theories and there is nothing geometrical in the construction of string vacua with these theories as building blocks.

We now present an argument which uncovers the fundamental link between these classes of vacua. For other discussions of possible links see [29][30]

3.2. The Geometry of Some 2-D Field Theories:

The action for a general 2-dimensional $N = 2$ supersymmetric quantum field theory takes the form

$$\int d^2zd^4\theta K(\overline{\Phi_i}, \Phi_i) + (\int d^2zd^2\theta \quad W(\Phi_i) + c.c) \tag{3.1}$$

where W is a holomorphic function of the chiral superfields Φ_i. It is important to bear in mind that the renormalization group flow for such a theory is driven by the superpotential, W, as the D terms supply irrelevant perturbations. In fact, W is not renormalized and hence provides us with an invariant of the renormalization group flow with which to characterize such 2-dimensional theories [31]. For example, the Landau-Ginzburg superpotential $W(\Phi) = \Phi^{P+2}$ corresponds to the $A-$series modular invariant $N = 2$ minimal theory of level P and central charge $c = \frac{3P}{P+2}$. Thus, for a tensor product of minimal models $(P_1, ..., P_r)$, we have a superpotential $W(\Phi_1, ..., \Phi_r) = \Phi_1^{P_1+2} + \ldots \Phi_r^{P_r+2}$.

Before plunging into the general analysis, it is worthwhile to emphasize the essential points of our argument through the simplest example. As we mentioned above, the Calabi-Yau manifold $Y_{4;5}$ defined by $z_1^5 + z_2^5 + z_3^5 + z_4^5 + z_5^5 = 0$ in CP^4 has been identified with the theory obtained by tensoring together five copies of the third member of the $N = 2$ discrete series. One motivation for the argument which follows is the observation made in [32]that since the Landau-Ginzburg potential associated with a level P theory in the $N = 2$ discrete series is Φ^{P+2}, five copies of the $P = 3$ series have a potential $\Sigma_{i=1}^5 \Phi_i^5$ which is suggestively close to the defining equation for $Y_{4;5}$ given above. We will now show that this is not an accident.

Although there is a formal similarity between the quintic potential and the defining equation for $Y_{4;5}$, we are faced with two obstacles: $Y_{4;5}$ is a submanifold of CP^4 whereas the potential is a function on C^5 (suppressing superspace dependence) and, moreover, there is no constraint forcing the quintic potential to vanish. As we now show, each of these problems is the others cure: consider the path integral

$$\int [d\Phi_1] \ldots [d\Phi_5] e^{i \int d^2zd^2\theta(\Phi_1^5 + \ldots \Phi_5^5)} \tag{3.2}$$

where, we have (to a first order approximation) neglected the D terms in the action, assuming they are very small. Of course we can take any D terms we wish since they constitute the irrelevant operators and for our analysis we have taken them to be very small. The basic concept of universality tells us the structure of the flow is unaffected by such choices. Changing variables to $\xi_1 = \Phi_1^5$, $\xi_i = \frac{\Phi_i}{\Phi_1}$ for $i \geq 2$ we can rewrite this path integral as

$$\int [d\xi_1] \ldots [d\xi_5] e^{i \int d^2 z d^2 \theta \xi_1 (1 + \xi_2^5 + \ldots \xi_5^5)} \tag{3.3}$$

Performing the ξ_1 integration we thus find a delta function constraint

$$\delta(1 + \xi_2^5 + \ldots + \xi_5^5)$$

which is precisely the defining equation of $Y_{4;5}$ in inhomogeneous coordinates!*

Crucial to this argument is the fact that the Jacobian for the given change of variables is a constant, thus allowing simple evaluation of the ξ_1 integration. As we shall see shortly, the analogue of this change of variables for other minimal superconformal constructions corresponds precisely to the topological constraint of vanishing first Chern class - the hallmark of Calabi-Yau manifolds. Moreover, we will also show that a large class of Calabi-Yau manifolds can be realized by a similar manipulation of a uniquely determined LG potential. Hence, Gepner's construction may be thought of as a special case of this result in which the relevant LG potentials correspond to those of the $N = 2$ minimal models.

Even though we have not specified what the fixed point form of the D terms is, by RG flow there is bound to be one and it is in some sense not a primary object for the conformal theory. In particular all the information about the metric of the Calabi-Yau manifold that is encoded in the D terms (of course this is the final form of the metric which makes the string propagation on C-Y have a superconformal symmetry) is fixed by the RG flow characterized by the superpotential.

Thus, modulo technical details and generalizations presented below, this analysis provides the underlying connection between Gepner's construction and Calabi-Yau compactifications.

The example of the quintic in CP^4, as mentioned, provides the simplest illustration of our arguments. A generic complexity we encounter in other examples is that most of Gepner's models involve the tensor product of minimal models at *different* levels for which the above prescription naively yields a superpotential of nonconstant degree of homogeneity. For example, the theory $(1, 2, 2, 10, 10)$ (the tensor product of one level 1 theory, two level 2 theories and two level 10 theories) gives rise to the superpotential $z_1^3 + z_2^4 + z_3^4 + z_4^{12} + z_5^{12}$ which cannot be interpreted

* More precisely, we can perform the same calculation in the other four affine patches to obtain the projective variety $Y_{4;5}$. Furthermore, note that even though the integral over ξ_1 excludes the origin, the result is a delta function.

as a defining equation in CP^4. The reason for this is that points in C^5 which differ by an overall scaling of the coordinates are identified in CP^4 and hence any well defined algebraic constraint in the latter space must be homogeneous. However, a simple generalization of projective space allows us to make sense of such naively inhomogeneous expressions: rather than identifying points whose coordinates differ by an overall scaling, we identify points whose coordinates differ by a fixed but not necessarily uniform power of a scaling parameter. Spaces defined in this way are called weighted projective spaces [33]. For example, the potential for the theory $(1,2,2,10,10)$ is 'quasi-homogeneous' of degree twelve if the respective powers of the coordinate scalings are in the ratio $4:3:3:1:1$, that is, we identify points via $[z_1, z_2, z_3, z_4, z_5] \sim [\lambda^4 z_1, \lambda^3 z_2, \lambda^3 z_3, \lambda z_4, \lambda z_5]$. We denote such a space by $WCP^4_{4,3,3,1,1}$. As we shall shortly see, all of the Calabi-Yau manifolds associated with the minimal model constructions can be described as varieties in weighted projective spaces.

To this end, consider a LG theory of the form (ignoring the D terms)

$$\int [d\Phi_1] \ldots [d\Phi_5] e^{i \int d^2 z d^2 \theta (\Phi_1^{l_1} + \ldots \Phi_5^{l_5})} \tag{3.4}$$

where for definiteness we have restricted our attention to five complex variables. This is the LG theory corresponding to the minimal model theory $(l_1 - 2, l_2 - 2, l_3 - 2, l_4 - 2, l_5 - 2)$. Let d be the degree of homogeneity of this equation - that is, d is the least common multiple of the l_i. Now we can rewrite the superpotential as

$$W = \Phi_1^{l_1} (1 + \frac{\Phi_2^{l_2}}{\Phi_1^{l_1}} + \ldots \frac{\Phi_5^{l_5}}{\Phi_1^{l_1}}) \tag{3.5}$$

(Essentially we are assuming that we are working in a patch in superspace in which Φ_1 is nonvanishing.) We now change variables from $(\Phi_1, \ldots, \Phi_5)$ to $(\xi_1, \ldots, \xi_5)$ where

$$\xi_1 = \Phi_1^{l_1} \quad \xi_i^{l_i} = \frac{\Phi_i^{l_i}}{\Phi_1^{l_1}}. \tag{3.6}$$

In terms of these new variables, we can write the path integral (3.4)as

$$\int [d\xi_1] \ldots [d\xi_5] \Omega e^{i \int d^2 z d^2 \theta \xi_1 (1 + \xi_2^{l_2} + \ldots \xi_5^{l_5})} \tag{3.7}$$

where Ω is the Jacobian for this coordinate transformation. In particular, this Jacobian is easily seen to be proportional to Φ_1^j where $j = 1 - l_1 + l_1 (\Sigma_{i=2}^5 \frac{1}{l_i})$. The crucial point is that j vanishes precisely if

$$\sum_{i=1}^{5} \frac{1}{l_i} = 1. \tag{3.8}$$

When this condition is met, we can perform the ξ_1 integration in (3.7)to yield a multiplicand

$$\delta(1+\xi_2^{l_2}+\ldots\xi_5^{l_5}) \tag{3.9}$$

Interpreting the scalar component of the superfields as background spatial coordinates, we thus see that the above delta function constrains classical motion of the string to lie on the hypersurface defined by the vanishing locus of the argument of the delta function†. As shown in [12], (3.8) is precisely the requirement that this hypersurface has vanishing first Chern class and hence is the Calabi-Yau manifold $\Sigma_{i=1}^5 z_i^{l_i}=0$ in $WCP^4_{\frac{d}{l_1},\ldots,\frac{d}{l_5}}$.

There is one additional subtlety in the change of variables (3.6). It is *not* a single valued transformation of coordinates. It is a single valued change of coordinates only if we demand the identification

$$(\Phi_1,\ldots,\Phi_5)\sim(e^{2\pi i/l_1}\Phi_1,\ldots,e^{2\pi i/l_5}\Phi_5). \tag{3.10}$$

This means that if we wish to obtain the Calabi-Yau manifold, i.e., if we demand ξ_i are the complete fields, then we cannot simply take the tensor product of the minimal discrete models. We have, in addition, to consider an orbifold of that model by dividing by the action (3.10). This is in fact the same as the '$U(1)$ projection' needed in [5] to make the theory space-time supersymmetric. It is gratifying to find a simple interpretation of this projection from the point of view of LG theory.

Essential to Gepner's construction is the requirement that the minimal models are chosen so that the sum of their central charges is nine, corresponding to a compactification of six dimensions. Where does this constraint manifest itself in our analysis? To answer this, we compute the central charge of the theory in (3.4). Since the central charge of the level P minimal model is $\frac{3P}{P+2}$, we have total central charge

$$c=\sum_{i=1}^{5}\frac{3(l_i-2)}{l_i}. \tag{3.11}$$

Setting $c=9$ implies $\Sigma_{i=1}^5\frac{1}{l_i}=1$ which is the same condition we found in (3.8). This explains why requiring the sum of the central charges to be nine yields Calabi-Yau compactification. The condition of the vanishing of the first Chern class is identical to $\sum c_i=9$!

The essential point of [12] , then, is the equivalence of the following three statements:

† The θ component of the delta function constrains the fermionic superfield components to be tangent to this manifold.

1. The central charge $c = \sum c_i$ of the combined minimal models equals 9.

2. Viewed as the defining equation for an algebraic hypersurface in a uniquely determined weighted projective space, $W = 0$ cuts out a manifold (possibly after resolution) of vanishing first Chern class - i.e. a Calabi-Yau manifold.

3. The path integral, after a suitable change of variables, gives rise to a multiplicand $\delta(W)$, which can be interpreted as restricting the (scalar components) of the Φ_i to the above mentioned Calabi-Yau manifold.

Thus, Gepner's construction, which is characterized by the first statement above, manifestly gives rise to string propagation on Calabi-Yau manifolds. Furthermore, the 'generalized GSO projection' employed by Gepner to ensure a spacetime supersymmetric spectrum corresponds to orbifoldizing the initial Landau-Ginzburg theory [12] [13]. From the point of view in [12], this quotienting arises via the requirement that the change of variables mentioned in the third statement be well defined. In our discussion of Gepner's construction we have repharsed things so that it equivalence to orbifolding is manifest.

This applies to supersymmetric compactifications in other dimensions as well: to obtain an $N-1$ (complex) dimensional theory, we can choose $N+1$ minimal models at level P_i with central charge $c_i = 3P_i/(P_i+2)$ such that

$$\sum_{i=1}^{N+1} \frac{P_i}{P_i+2} = N-1. \tag{3.12}$$

The corresponding Calabi-Yau is determined by the zero locus of $z_1^{P_1+2} + \ldots z_{N+1}^{P_{N+1}+2}$, in weighted projective space. The condition (3.12) can be rewritten in the form $\sum 1/(P_i+2) = 1$ which is the generalization of the condition (3.8) arising from the requirement that the Jacobian of the transformation discussed above be constant, and it is the same condition as the vanishing of first Chern class [12].

The foregoing analysis shows that when $N+1$ copies of a minimal model are tensored together to give a $c = 3(N-1)$ theory, it automatically corresponds to a compactification on an $(N-1)$ (complex) dimensional Calabi-Yau. Of course we can take more or less copies of minimal models to obtain a theory with this central charge. That is, we can take a tensor product of r minimal models so long as $\Sigma_{i=1}^{r} \frac{P_i}{P_1+2} = N-1$. How do we relate these to Calabi-Yau compactifications? When we take fewer variables, we can always add new fields with quadratic superpotential without changing the theory [31]. In this way we can always increase the number of fields to $N+1$ and repeat the analysis above. If the number of fields is more than $N+1$, we group together various fields as dictated by the requirement of vanishing first Chern class. Another way of saying this is that we group the fields in such a way that there is a constant Jacobian change of variables such that the path integral over the LG potential gives rise to a product of delta functions. For the case of $N = 3$ there is one model with $r > 4$ (in addition to 17 with $r \leq 4$) and for $N = 4$ there are 21 models with $r > 5$ (in addition to 147 with $r \leq 5$). As shown in [12], these are realized by grouping the fields so as to correspond to (orbifolds of) various products of tori and $K3$. A similar analysis holds for $N \geq 5$.

We also note that a single choice of minimal models $(P_1, P_2, ..., P_r)$ gives rise to more than one superconformal theory since we have a choice of which modular invariants to use from the $A - D - E$ series. The identification with Gepner's constructions corresponds to the A series; the precise Calabi-Yau counterparts for the other modular invariants is discussed in [12][13].

We should point out that this argument applies well beyond the minimal model construction as the latter simply corresponds to a particular choice of the monomials in the Landau-Ginzburg potential. Other Calabi-Yau manifolds can similarly be expressed with different monomial summands [34][35]. The fixed points of the RG flows for these theories are equally good conformal theories and they allow us to go beyond the restricted class of minimal model building blocks.

We thus have a clear understanding of the fundamental link underlying the observed correspondence between the exactly soluble minimal model constructions and Calabi-Yau compactifications. It is gratifying to see the close connection between notions in conformal field theory and algebraic geometry.

4. Application to Three Generation Models

In the previous sections of these lectures we have described some general properties of certain heterotic string vacua. In this final section we shall focus ourselves on some explicit vacua selected for their promising phenomenology. In particular, we shall now review work that has been done on three generation heterotic string models and describe recent results in this area of investigation.

The point of any such phenomenological study is not to build *the* model; our understanding of string theory is, at present, far too limited for such a task. Rather, the philosophy is to push as far as we can within the context of concrete and well motivated choices for the vacuum state in order to possibly extract relevant physics as well as to guide further study by exposing both the phenomenological virtues and deficiencies in our present formulation of string theory.

We begin our discussion of the work which has been done on three generation string phenomenology from a geometrical viewpoint. We shall then turn to some aspects of three generation minimal model theories. After this we shall briefly discuss some recent work on obtaining three generation $(2,0)$ theories which have $SO(10)$, $SU(5)$ or the standard model gauge group at the compactification scale.

4.1. Three Generation Geometrical String Models:

Although it is believed that there are a finite number of topologically distinct Calabi-Yau manifolds, there is at present no complete catalog of these manifolds. A large and tractable subspace is provided by those Calabi-Yau manifolds which are realizable as complete intersections in products of complex projective spaces (CICY's). Specifically, these are smooth manifolds which can be expressed as the vanishing locus of n homogeneous polynomials in an ambient $n+3$ dimensional product of complex projective spaces. This subspace of the set of all Calabi-Yau manifolds has been studied by a number of authors [9][36][37][38][39]and it appears that there are in the neighborhood of 8000 manifolds of this type [37](although there may well be nontrivial identifications between some of these manifolds) [37]. Selecting one manifold from this subset of the Calabi-Yau manifolds, though, still requires, *a priori*, a completely arbitrary choice. However, recall that the number of generations in the low energy theory is given by one half of the magnitude of the Euler characteristic of K. The Euler characteristic of a complete intersection variety is easily calculated [40]and hence it is, at first sight, a simple exercise to determine which CICY's will give rise to the favoured number of three generations. The nontriviality of this exercise is apparent when one realizes that a given Calabi-Yau manifold can give rise to a number of 'descendent' manifolds: if a Calabi-Yau manifold K_0 admits a freely acting group G of holomorphic automorphisms, then the quotient manifold K_0/G is also Calabi-Yau. The latter manifold has an Euler characteristic which is reduced from that of K_0 by a factor of the order of G. Thus, to search for CICY's which can give rise to three generations we can not simply search the space of simply connected CICY's, but most also address the multiply connected descendents. The question of whether a given symmetry acts freely on a particular complex manifold depends on the complex structure of the manifold, not only on its topological type. Thus, it would appear that topological reasoning would not be sufficiently powerful to determine whether, say, a simply connected CICY with $\chi = -6k$ admits a free group action of order k, thus giving rise to a three generation compactification. In [39],[38], however, it was shown that if one restricts attention to projectively inherited symmetries (that is, those symmetries which not only act on the CICY manifold, but also on the ambient projective space into which it is embedded), then simple topological reasoning can be used, and is remarkably effective. Of the space of all CICY manifolds, it was argued that apparently only *one* three generation example can be constructed [38][39].

This manifold, R , was originally constructed by Tian and Yau [40]and takes the form R_0/Z_3 where R_0 is a subvariety of $CP^3 \times CP^3$ given by the vanishing locus of $\Sigma_{i=0}^3 x_i^3$, $\Sigma_{i=0}^3 y_i^3$, and $\Sigma_i x_i y_i$ with the x's and y's being homogeneous $CP^3 \times CP^3$ coordinates. The free Z_3 action is generated by $g : (x_0, x_1, x_2, x_3, y_0, y_1, y_2, y_3) \rightarrow (x_0, \alpha^2 x_1, \alpha x_2, \alpha x_3, y_0, \alpha y_1, \alpha^2 y_2, \alpha^2 y_3)$ with α a nontrivial cube root of unity. This manifold has $h^{2,1} = 9$, $h^{1,1} = 6$ [41]and, clearly, $\pi_1 = Z_3$. We can thus apply the general methodology outlined in section (B) above to find that the compactification scale theory is characterized by [41]:

Gauge Group

The nontriviality of the fundamental group allows us to break the E_6 symmetry to $SU(3)^3$ or $SU(6) \times U(1)$. The former is phenomenologically favoured and shall be chosen.

Four Dimensional Matter Fields

Under $SU(3)^3$ the 27 of E_6 decomposes as $27 = (3,\overline{3},1) \oplus (\overline{3},1,3) \oplus (1,3,\overline{3})$. Identifying the first $SU(3)$ factor with color, we call these representations quarks (Q), antiquarks (q) and leptons (λ). Similarly for the $\overline{27}$ representation with bars placed over our definitions. As discussed, the cohomology of R determines the four dimensional spectrum of massless particles. With the Hodge numbers given above, the spectrum contains the representations $3 \times (\lambda, Q, q)$, $6 \times (\lambda, \overline{\lambda})$, $4 \times (Q, \overline{Q}, q, \overline{q})$. Thus, we have three net generations (due to the Euler characteristic of R) with the given number of extra 'particle,antiparticle' pairs due to the values of the Hodge numbers of R. In addition to these fields, as discussed earlier, there are also some E_6 singlet fields arising from the cohomology group $H^1(EndT_R)$. Unfortunately, we have not been able to compute the dimension of this group and only have a lower bound [36]of $h^{2,1}$.

Yukawa Couplings

Using the algebraic method [20]of computing (2.30), the 27^3 Yukawa couplings of this model have been computed [42], [43]. We will not list these numbers but simply note that by a minor adjustment of the complex structure for R given above, the ratio of these Yukawas can be quite large. There is no general procedure for computing the $\overline{27}^3$ couplings. However, by adapting some well known properties of the cubic surface in CP^3 these couplings have been computed in [44]. As these couplings are topological they are independent of the complex structure of R and turn out to be of order unity. We emphasize that these couplings are not the true properly normalized Yukawa couplings. We will return to this point later.

Discrete Symmetries

The projectively inherited discrete symmetry group of R is easily determined [41]to be the group $S_3 \times Z_3 \times Z_2 \times Z_2$ †. These symmetries greatly reduce the number of independent undetermined field normalizations. In fact, for the complex structure of R given above, these symmetries imply that there are only three independent normalizations. (For more phenomenologically desirable complex structures, there are eight such independent normalizations [43].) Furthermore, the effective four dimensional Lagrangian contains nonrenormalizable terms for which there is no simple means like (2.30)or (2.31)to determine the strength of the couplings. For these terms, the discrete symmetries provide a powerful means of determining which couplings must vanish.

The above data provides us with an effective Lagrangian for the massless modes, at the compactification scale. We emphasize that this data is computed from the choice of the manifold R (a choice which we hope to have convinced the reader is quite natural) and not simply introduced in an *ad hoc* fashion as is familiar from conventional grand unified model building.

Having determined the compactification scale theory, we now turn to its low energy implications. It is this stage of the analysis which requires that we make certain assumptions, which we now clearly identify [41]:

Assumption 1 : Our compactification scale gauge group has rank 6 thereby necessitating an intermediate scale Higgs mechanism. Before supersymmetry is broken, however, the minimum of the scalar potential is achieved when the VEV's of all fields vanish. Thus, precisely which fields acquire VEV's and hence realize the Higgs mechanism is determined, in large part, by the supersymmetry breaking mechanism. At present, no completely satisfactory mechanism has been found. Furthermore, even if one did have such a mechanism, determining which fields are destabilized first requires knowledge of the normalized Yukawa couplings - something which we cannot compute geometrically. (We will discuss this further below.) Thus, our incomplete understanding forces us to *select*, from a number of

† We thank P.Frampton for identifying the order eighteen nonabelian group found in [41]as $S_3 \times Z_3$.

possibilities, which fields will be triggered by supersymmetry breaking to acquire nonzero VEV's. The symmetry breaking pattern

$$SU(3)^3 \rightarrow SU(3) \times SU(2)^2 \times U(1) \rightarrow SU(3) \times SU(2) \times U(1) \qquad (4.1)$$

then follows. The scale of such symmetry breaking is determined by the couplings in the superpotential allowed by the discrete symmetries, and hence is not set by *ad hoc* considerations.

Assumption 2 : In our discussion of the four dimensional spectrum above, we noted that we only have a lower bound of $h^{2,1}$ on the number of singlets arising from $H^1(EndT)$. We include these identifiable singlets in our analysis and assume that the other singlets which we are unable to identify do not substantially effect our analysis. We will return to this assumption later.

Assumption 3 : As mentioned, although we can calculate the value of the renormalizable Yukawa couplings in the superpotential, our only handle on the nonrenormalizable terms are the discrete symmetries of the manifold. We thus make the fairly mild assumption that any nonrenormalizable coupling allowed by the symmetries actually does occur.

With these assumptions and the compactification scale data provided by our choice of the manifold R, we find [41]that the low energy theory has a spectrum which, up to a couple of TEV mass particles, is that of the supersymmetric standard model with an appropriate spectrum of Higgs doublets for weak gauge symmetry breaking. The discrete symmetries and magnitude of the intermediate scale Higgs mechanism ensure an acceptably small rate of proton decay. The value of $sin^2\theta$ is within the experimentally determined range, and neutrinos have acceptably small Majorana masses. Finally, the structure of the quark and lepton mass matrices are remarkably close to observation and we find a phenomenologically viable Kobayashi-Maskawa matrix. These results, modulo the assumptions stated above, are quite encouraging. However, we hasten to add that these results are very sensitive to these assumptions. If one chooses alternative symmetry breaking Higgs fields, the phenomenology generally is unacceptable [45][46]. Furthermore, our second assumption is fairly strong as singlets in these theories can sometimes have important consequences. Finally, the *values* of the higher point couplings addressed in the third assumption are important for detailed low energy predictions, as are the undetermined field normalizations.

So where do we stand? Well, it is encouraging that if we calculate everything we can (gauge group, massless spectrum, value of renormalizable couplings, zeroes of nonrenormalizable couplings) and are optimistic about those things which we can not compute (supersymmetry breaking, singlet spectrum, values of nonrenormalizable couplings, field normalizations) that the model essentially yields the standard model at low energy. However, it is important to see if we can go beyond some of these assumptions. The relation between Calabi-Yau compactifications and exactly soluble theories offers us such a possibility as we now discuss. (For some important recent developments in the phenomenology of the model discussed above the reader should consult [47]

4.2. Three Generation Minimal-Model String Models:

As mentioned earlier, the minimal model approach to string compactification does give us the added power of computing properly normalized Yukawa couplings as well as the values of nonrenormalizable couplings. An essential question then is whether we can find a three generation exactly soluble theory, and in particular, whether there exists such a counterpart to the geometrical compactification discussed above. As shown in [10]the answer to these questions is yes.

Gepner found that an orbifold of the theory obtained from the minimal models $(1, 16, 16, 16)$ with the exceptional invariant used in the level 16 theories corresponds to a three generation geometrical construction due to Schimmrigk [48]. This construction involves an orbifolding of a variety in $CP^3 \times CP^2$ and hence at first sight seems to have nothing to do with the three generation manifold we discussed above. However, as shown in [49]these two manifolds are diffeomorphic. That is, they lie in the same topological class differing only by their respective complex structures. Nonetheless, physics is dependent on the choice of complex structure and it is not known whether one can smoothly deform the complex structures from one of these manifolds to the other. Thus, we can certainly expect to get models with different characteristics.

The properties and phenomenology of this exactly soluble version have been studied in some detail recently [50], and work is still underway [51]. The basic strategy employed by [51]is to deform away from the exactly soluble point as geometrically this corresponds to a radius of the compactified space which is of the order of the Planck scale. Such a small radius makes it difficult to generate a natural hierarchy of scales one needs a source of small couplings. Such a source is provided by choosing the compactification scale to be an order of magnitude or so larger than the Planck scale thus introducing the small factor R_K/R_{Planck}. Since the 27^3 couplings are independent of the radius, the values computed in the minimal model construction are still valid in such a deformed theory. The same is not true for the $\overline{27}^3$ couplings, however at such a radius the geometrically determined couplings should be very accurate. The reason for this is that the instanton corrections to these couplings are suppressed by the factor $e^{-R_K/R_{Planck}}$.

Thus, a lot of information can be determined at this favoured point in moduli space.

As was also the case in the geometric construction, the real uncertainty which remains is that of supersymmetry breaking. It appears that the most promising possibility for supersymmetry breaking is that of gaugino condensation in the hidden sector [52]. Using Witten's [53]no-scale truncation as a guide, Ross has studied the possibilities raised in [52]in the context of the three generation minimal model construction. He has found that the large number of gauge singlet fields in the theory allows for the generation of a considerable hierarchy of scale between the Planck mass and the supersymmetry breaking scale. This SUSY breaking is then communicated to the observable sector by gravitational effects which are modeled by soft supersymmetry breaking terms. By being optimistic about the detailed form of these effects, one can then press on with the analysis, making use of the calculational power of the minimal model approach.

Following analysis similar to the consideration described in the last section, and aided by having the properly normalized Yukawa couplings, Ross has investigated this theory in some detail. In addition to finding the same phenomenological successes we mentioned in the last section, Ross has been able to go further and obtain a prediction for the mass of the top quark:

$$m_{top} = (1.05)\sqrt{2}M_W \tag{4.2}$$

Of course, our lack of full understanding of string theory has forced some important assumptions to underly this result. Nonetheless, it makes manifest the potential power of string theory to predict parameters which would otherwise remain experimentally determined mysteries.

3.3. Three Generation Models with $SO(10)$, $SU(5)$ and $SU(3) \times SU(2) \times U(1)$ Gauge Groups:

The phenomenological studies described in the last two sections are based on string vacua which have an E_6 gauge group at the compactification scale. We have seen that such theories can give rise to realistic physics, yet it is important to bear in mind that this 'large' gauge group (compared with the standard model) with its fundamental 27-dimensional representations (compared with the 15 dimensional standard model families) makes model building rather delicate and somewhat cumbersome. One must ensure that the extra coloured particles and gauge bosons which can mediate proton decay become sufficiently heavy. Furthermore, extra colour neutral fields must also become sufficiently heavy to not ruin the value of $sin^2\theta$. Model building would certainly be simpler if the unifying group were one of the far more economical GUT groups like $SO(10)$ or $SU(5)$, or even the standard model group itself. A general means of obtaining such theories was presented in [7] and a simple method for realizing these ideas was presented in [24]. A simple way to describe these ideas follows our discussion in section 3.1.. In our description of the three generation geometrical analysis we spoke of an intermediate scale scenario in which fields whose potential is flat to a high order obtain large VEV's after a SUSY breaking trigger is turned on. Imagine now a situation in which some fields have *exactly* flat potentials. This means that at any value of their VEV's we have a good string vacuum. There is no need to break SUSY to turn on their VEV's since *any* value of their VEV's is equally valid from an energetics point of view. In particular, these fields can be given Planck scale VEV's. In other words, we can naturally push the intermediate scale scenario all the way to the Planck scale under such circumstances. Should these fields be the standard model invariant fields in the 27 and $\overline{27}$ (they must turn on simultaneously to maintain D-flatness) one exact flat direction will lead to an $SO(10)$ theory, two will lead to $SU(5)$. In combination with Wilson loops we can then in fact have $SU(3) \times SU(2) \times U(1)$.

Of course, an exactly flat direction means that an infinite number of couplings of arbitrarily high order must be shown to vanish. As shown by [24], certain R-symmetries (discrete symmetries which do not commute with supersymmetry) can be sufficiently powerful to accomplish this task. For the simple case of $Y_{4;5}$ they were able to use R-symmetries to construct an $SO(10)$ theory.

Can such a program be carried out in the three generation case? At first sight it seems that the answer to this is no because the three generation manifold presented in 3.1. does not admit any R-symmetries. Happily, though, if we modify the defining equation by changing the hyperplane H to

$$x_0 y_0 + x_1 y_1 + x_2 y_2 = 0 \tag{4.3}$$

then the manifold (which is still smooth) does admit R-symmetries [54]. These are the symmetries which *independently* scale x_3 and y_3 by cube roots of unity. In fact, using these R-symmetries one can find exactly flat directions which give rise to three generation models with $SO(10)$ or $SU(5)$ or the standard model gauge group at the compactification scale. Of course, a detailed study of such models must be carried out to assess their phenomenological viability. They are certainly a more promising starting point than any of the models discussed above.

5. Concluding Remarks

In these lectures we have focused on the space of classical string vacua. We have discussed some general properties of these vacua and illustrated our remarks with the particularly interesting three generation examples. We hope that these lecture have conveyed three important themes. First, we have seen a general trend over the last few years in which results that appeared to be lowest order perturbative computations (and hence subject to a wealth of potentially overwhelming higher order corrections) turn out to either exact or at least accurate indicators for the exact result. For example, Calabi-Yau compactification were originally introduced as lowest order string and sigma model vacuum solutions. As such, it was not at all clear that physics derived from these solutions would accurately reflect the physics of conformal theory to which the geometry is a lowest order approximation. In fact, it was not clear that one could necessarily correct the Calabi-Yau sigma model order by order to maintain conformal invariance. Subsequent nonrenormalization theorems (as discussed), showed that the geometrical analysis captures many essential features of the exact conformal theory. The exactly soluble minimal model constructions allow us to explicitly verify these results. This is not to say, of course, that the lowest order geometry fully reflects the exact theory, but it is surprising how far such an *a priori* crude approximation takes us. The second theme has been concerned with the many ways in which string theory offers us a means of going way beyond the conventional methods of model building. We have explicitly illustrated this notion through our discussion of string based phenomenology. In particular, the top quark mass 'prediction', although shrouded in many unknowns of string theory, serves to emphasize the possible phenomenological steps forward offered by this theory. Finally, we are also tried to convey the fruitful interplay between comparatively abstract studies in conformal field theory with more concrete studies in string phenomenology. The uniting principle between the two being a concern with and a drive towards realism. We feel that this interplay is one interesting and useful means of steering a path through the wealth of details and possibilities raised by string theory towards the ultimate goal of converging upon a final verdict regarding this rich proposal for a truly unified theory.

References

[1] For a Review of String Theory and an extensive list of references see M.Green, J.Schwarz and E. Witten , *Superstring Theory*, Vol. I and II, Cambridge University Press (1987).

[2] M. Dine, N. Seiberg, X. G. Wen, and E. Witten, Nucl. Phys. **B278** (1987) 769; **B289** (1987) 319.

[3] For detailed discussion of conformal field theory the reader can consult A.A. Belavin, A.M. Polyakov, and A.B. Zamolodchikov, Nucl. Phys. **B241** (1984) 333; D.Friedan, E. Martinec and S. Shenkar, Nucl. Phys. **B271** (1985) 93; M. Peskin, TASI lecture notes; P.Ginsparg, Les Houches lecture notes, HUTP-88/A054.

[4] D. Friedan, A. Kent, S. Shenker and E. Witten, unpublished; A. Sen Nucl. Phys. B 278 (1986) 289; Nucl. Phys. B 284 (1987) 423; T. Banks, L.J. Dixon, D. Friedan and E. Martinec, Nucl. Phys. B 299 (1988) 613.

[5] D.Gepner, Nucl. Phys. **B296** (1987) 757.

[6] C.Callan, D. Friedan,E. Martinec and M. Perry, Nucl. Phys. **B262** (1985) 593; A.Sen, Phys. Rev. Lett **55** (1985) 1846.

[7] E. Witten, Nucl. Phys. **B268** (1986) 79.

[8] P.Candelas, M.D. Freeman, C. Pope, M.F. Sohnius and K.S.Stelle, Phys. Lett. **177B** (1986) 341 ;
D.Nemeschansky and A. Sen, Phys. Lett. **178B** (1986) 365.

[9] P.Candelas, G.Horowitz, A.Strominger and E. Witten, Nucl. Phys. **B258** (1985) 46.

[10] D. Gepner, Phys. Lett. B199 (1987) 380.

[11] B.R.Greene and M.R.Plesser, HUTMP-89/B241.

[12] B.R. Greene, C. Vafa and N.P. Warner, Harvard preprint HUTP-88/A047, Nucl. Phys. B, to appear.

[13] E.Martinec, EFI preprint 88-76.

[14] C. Vafa,*String Vacua and Orbifoldized LG models*,Harvard preprint HUTP-89/A018.

[15] P. DiVecchia, J. L. Petersen and H. B. Zheng, Phys. Lett. **174B** (1986) 280;
W. Boucher, D. Friedan and A. Kent, Phys. Lett. **172B** (1986) 316;
A. B. Zamalodchikov and V. A. Fateev, Sov. Phys. JETP **63** (1986) 913;
Z. Qiu, Phys. Lett. **188B** (1987) 207.

[16] A. Cappelli, C. Itzykson and J. B. Zuber, Nucl. Phys. **B280** [FS16](1987) 445.

[17] D. Gepner and Z. Qiu, Nucl. Phys. **B285** [FS19](1987) 423.

[18] W. Lerche, C.Vafa and N .P. Warner,*Chiral Rings in N=2 Superconformal Theories*, Harvard preprint HUTP-89/A065.

[19] A. Strominger and E. Witten, Comm. Math. Phys. **101** (1985) 341.

[20] P.Candelas, Nucl. Phys **B298** (1988) 458.

[21] A. Strominger, Phys. Rev. Lett. **55** (1985) 2547.
[22] E.Witten, Nucl.Phys. **B258** (1985) 75.
[23] M.Dine and N.Seiberg IASSNS/HEP-87/50.
[24] M. Dine and N. Seiberg, Phys. Rev. Lett. **57** (1986) 2625.
[25] J. Distler and B. Greene, Cornell preprint CLNS 88/834 and Harvard preprint HUTP-88/A020 , to appear in Nucl. Phys.**B**.
[26] L. Dixon, lectures at the 1987 ICTP Summer Workshop in High Energy Physics and Cosmology.
[27] A.Lutken and G.G.Ross, Phys. Lett. **213B** (1987) 152;
M.Lynker and R.Schimmrigk, Univ. of Texas preprint, UTTG-17-88, (1988).
[28] D. Gepner, Princeton preprint 1988.
[29] E.Martinec in *V.G. Knizhnik memorial volume*, L.Brink editor.
[30] J.I. Latorre and C.A. Lutken, Nordita preprint 88/42.
[31] C.Vafa and N.P.Warner, HUTP-88/A037, 1988
[32] D.Kastor, E.Martinec and S.Shenker, EFI preprint 88-31.
[33] S-T. Yau in the *Proceedings of the Argonne Symposium on Anomalies, Geometry and Topology*, (1985)
[34] B.Greene, HUTMP-88/B235.
[35] S.Kalara and J.Lykken, Minnesota preprint, 1989.
[36] T.Hubsch, Comm. Math. Phys. **108** (1987) 291, P.Green and T.Hubsch, Comm. Math. Phys **109** (1987) 99.
[37] P.Candelas, A.Dale, C.A.Lutken and R.Schimmrigk, Nucl.Phys **B298** (1988) 493.
[38] P.Candelas, C.A. Lutken and R. Schimmrigk, Nucl. Phys. **B306** (1988) 113.
[39] P.Aspinwall,B.Greene,K.Kirklin and P.Miron, Nucl. Phys. **B294** (1987) 193.
[40] S-T. Yau in the *Proceedings of the Argonne Symposium on Anomalies, Geometry and Topology*, (1985).
[41] B.R.Greene,K.H.Kirklin,P.J.Miron and G.G.Ross, Nucl. Phys. B278 (1986) 667, Nucl.Phys. B292 (1987) 602, Phys. Letts. 180B (1986) 69.
[42] P. Candelas and S.Kalara, Minnesota Preprint UMN-TH-601-87.
[43] B.Greene,K.Kirklin, P.Miron and G.Ross, Phys.Lett. **192B** (1987) 111.
[44] J.Distler, B.Greene,K.Kirklin and P.Miron, Phys. Lett. **195B** (1987) 41.
[45] J.Ellis,K.Enqvist, D.V.Nanopoulos and K.A.Olive,CERN preprint TH.4713/87 (1987).
[46] S.Kalara and R.Mohapatra, Phys.Rev. **D36** (1987) 3474.
[47] R.Arnowitt and P.Nath, Northeastern and Texas A & M preprints 1988,1989.
[48] R.Schimmrigk, Phys. Letts. 193B (1987) 175.

[49] B.Greene and K.Kirklin, Comm. Math. Phys. **113** (1987) 105.

[50] B.Greene, C. Lutken and G.G. Ross, Nucl. Phys. B to appear.

[51] G.G.Ross, in preparation

[52] S.Ferrara, L. Cirardello, H.P. Nilles, Phys. Lett. **125B** (1983) 457 ;
J.-P. Derendinger, L.E. Ibanez and H.P. Nilles, Phys. Lett. **155B** (1885) 65.

[53] M.Dine, R. Rohm and E. Witten Phys. Lett. **156B** (1985) 55; ;
P.Binetruy and M.K. Gaillard, Phys. Letts. **(168B** (1986) 347.

[54] J.D.Breit, B.A. Ovrut and G. Segre, Phys. Lett. **162B** (1985)303; ;
M.Mangano, Z. Phys. **C28** (1985) 613.

[55] E.Witten, Phys. Lett. **155B** (1985) 151.

[56] B.R.Greene, Phys. Rev. D, to appear.

Standard Model
and Beyond

ISBN 1-56072-027-1

SPECTRAL GEOMETRY AND COMPACTIFICATION

J. Sladkowski and R. Mánka
Department of Theoretical Physics, Silesian University, Uniwersytecka 4, 40-007 Katowice, Poland

The fundamental problem of multidimensional field theories of both Kaluza-Klein [1] and superstring inspired type [2] is to answer the question: "What mechanism chooses the internal manifold ?" The answer is conclusive in the following sense. The internal manifold decides which field configuration is physical, defines the gauge group and the number of generations [1,3]. Obviously the criterion of judgments should be a physical one. The present situation could be described as geometrical phenomenology: we analyze the appropriate manifolds one by one to find the one we hope Nature has chosen (cf. lectures given by P. Candelas, B. Greene and T. Hübsch). We would like to point out that this problem is strictly connected with spectral geometry [4]. To raise the problem let us discuss the compactification of D =10 pure electrodynamics on an arbitrary internal manifold K. Let us suppose that $M^{(10)} = M^{(4)} \times K$. $M^{(4)}$ is the flat D = 4 Minkowski space. We suppose that K is chosen so that the classical Einstein equations are fulfilled. The product structure of $M^{(10)}$ means that:

$$g_{MN} = \begin{pmatrix} \eta_{\mu\nu} & 0 \\ 0 & -g_{mn} \end{pmatrix} \tag{1}$$

with the obvious notation [1,5]. The Lagrange function has the form:

$$L = \frac{1}{2k^2} R - \frac{1}{4} F_{MN}F^{MN} - \partial_M BA^M + \frac{1}{2} B^2 + i\ \partial_M \bar{c}\ \partial^M c + \Lambda\ , \tag{2}$$

where $F_{MN} = \partial_M A_N - \partial_N A_M$ and Λ denotes the cosmological constant. L is BRS invariant [5]. For simplicity we will treat the gravitational field as a classical one. The Lagrange function (2) leads to the following equation of motion:

$$- \frac{1}{\sqrt{|g|}} \partial_M \left(\sqrt{|g|}\quad F^{MN} \right) = - \partial^N B\ , \tag{3a}$$

$$- \frac{1}{\sqrt{|g|}} \partial_M \left(\sqrt{|g|}\quad A^M \right) = B\ , \tag{3b}$$

$$- \frac{1}{\sqrt{|g|}} \partial_M \left(\sqrt{|g|}\quad \partial^M c \right) = 0\ , \tag{3c}$$

$$- \frac{1}{\sqrt{|g|}} \partial_M \left(\sqrt{|g|}\quad \partial^M c \right) = 0\ . \tag{3d}$$

The Einstein equations have the form:

$$R_{MN} - \frac{1}{2} \left(R + \Lambda \right) g_{MN} = k^2 T_{MN}\ , \tag{4}$$

where

$$T_{MN} = - F_{MK}F^{K}{}_{N} - \frac{1}{4} F_{CD}F^{CD}g_{MN} + T_{MN}^{unphys} \quad . \tag{4b}$$

If $M^{(10)} = M^{(4)} \times K$ then

$$- \frac{1}{\sqrt{|g|}} \partial_M \left(\sqrt{|g|} \, \partial^M \right) = []^{(4)} - \Delta^{(6)} \tag{5a}$$

and

$$\Delta^{(6)} = - \frac{1}{\sqrt{|g|}} \partial_m \left(\sqrt{|g|} \, \partial^m \right) \quad . \tag{5b}$$

The fields $A_\mu(x,y)$ and $A_m(x,y)$ can be expanded in terms of the harmonics $\phi_r(y)$ and $A^r_m(y)$. They fulfil the following equations

$$- \frac{1}{\sqrt{|g|}} \partial_m \left(\sqrt{|g|} \, \partial^m \phi_r \right) = \lambda_r \phi_r \, , \tag{6a}$$

$$- \frac{1}{\sqrt{|g|}} \partial_m \left(\sqrt{|g|} \, \partial^m F^{mn}_r \right) = \lambda' A^n_r \tag{6b}$$

$$- \frac{1}{\sqrt{|g|}} \partial_m \left(\sqrt{|g|} \, \partial^m A^m_r \right) = 0 \tag{6c}$$

Usually we are interested in the zero modes of (6) because we interpret them as low energy physical fields. Explicitly

$$A_\mu(x,y) = \sum_r A^r_\mu(x) \phi_r(y) \quad , \tag{7a}$$

$$A_m(x,y) = \sum_r \varphi^r(x) A^r_m(y) + \alpha_m(y) \quad , \tag{7b}$$

$$c(x,y) = \sum_r c^r(x)\phi_r(y) \qquad \text{and} \qquad \bar{c}(x,y) = \sum_r \bar{c}(x)^r \phi_r(y) \quad , \tag{7c,d}$$

$$B(x,y) = \sum_r B^r(x)\phi_r(y) \quad . \tag{7e}$$

$\alpha_m(y)$ fulfills the equation $\sqrt{|g|}\; F^{mn}_r(y) = \text{const}$. For example if $K = S^2$, $\alpha_m(\varphi,\vartheta) = \{0, C\cos\vartheta\}$ and $F_{12} = -C\sin\vartheta$. This means that $T_{\mu\nu} = 0$ and $T_{mn} \neq 0$ on the classical level. The cosmological constant can be chosen so that Einstein equations are fulfilled. Now we are preparred to apply the variational method proposed in [5,8,9] to analyze the process of choosing the internal manifold. The Bogolubov inequality for free energy [6] has the form

$$F \approx F_1 \leq F_0 + \langle H - H_0 \rangle_0 \quad , \tag{8}$$

where H_o is the Hamiltonian of the trial system. We take the following form of the trial system [5, 7-9]:

$$\begin{aligned} L_o = & -\frac{1}{4} F^r_{\mu\nu} F^{r\mu\nu} + \frac{1}{2} M^2_r A_{r\mu} A^\mu_r - \partial_\mu B_r A^{r\mu} \\ & + \frac{1}{2} B^2_r - i\partial_\mu \bar{c}^r \partial^\mu c^r + i\, m^{ghost\,2}_r \bar{c}_r c_r \\ & + \frac{1}{2} \partial_\mu \varphi_r \partial^\mu \varphi_r - \frac{1}{2} m^2_r \varphi_r \varphi_r \quad . \end{aligned} \tag{9}$$

This leads [5,7-9] (in one loop approximation) to:

$$F_1 = F_0(M^2_{r}, m^2_r) + \int d^{(6)}y \sqrt{|g|} \left(\frac{1}{2k^2} R + \frac{1}{4} F_{mn}F^{mn} - \Lambda \right) \tag{10}$$

When temperature $T \to 0$, $F_0 \to \varepsilon_0$ where ε_0 denotes the vacuum state energy [5,7-9]. Each degree of freedom gives the following contribution:

$$\varepsilon_0 = \int \frac{d^4k}{(2\pi)^2} \ln (k^2 + \lambda_r) \quad . \tag{11}$$

In a supersymmetric theory the fermion sector contribution cancels that of the boson one and ε_0 vanishes [10]. In general (nonsupersymmetric) case ε_0 must be regularized [5,7-9]. For $T = 0$ we obtain

$$f_1 = \frac{F_1}{vol} = \sum_{f} \sum_{p=1}^{n_f} \sum_{r} \int \frac{d^4k}{(2\pi)^2} \ln (k^2 + \lambda_{rp}) + \int d^{(6)}y \sqrt{|g|} \left(\frac{1}{2k^2} R + \frac{1}{4} F_{mn}F^{mn} - \Lambda \right) \tag{12}$$

$n_f = 2$ for $f = \varphi_r(x)$ and $n_f = 6$ for $f = A_{r\mu}$. The regularization of ε_0 introduces some constant C_r. Using standard methods [5,7-9] we obtain:

$$\varepsilon^f_0 = \sum_r \frac{h\lambda^2_{rf}}{64\pi^2} \ln \left(\frac{\lambda_{rf}}{C^2_r} - \frac{25}{6} \right) . \tag{13a}$$

It is convenient to introduce dimensionless proper values defined by the relation

$$\lambda_{r,f} = \nu^2 e^{25/6} \tilde{\lambda}_{r,f} \tag{13b}$$

The constant ν^2 is related to the internal space. In the Kaluza-Klein theory we have $\nu^2 = R^{-2}$, where R denotes the radius of the internal space [12]. Now

$$\varepsilon_o^f = \sum_r \frac{h\, e^{25/3}}{64\pi^2} \nu^4 \tilde{\lambda}_{r,f}^2 \ln\left(\lambda_{r,f} \frac{\nu^2}{\Lambda^2}\right) \tag{13c}$$

The total vacuum energy ε_o is obtained by summing contributions of all degrees of freedom f:

$$\varepsilon_o = \sum_f n_f\left(C_r\ \lambda_{rp}^2 \ln\lambda_{rp} + C_r'\ \lambda_{rp}^2\right) \tag{14}$$

ε_o is infinite if we take all the massive modes into consideration. It may be regularized by the ζ-function method [13]. For the Laplace operator Δ we define

$$\zeta_\Delta(s) = \sum_r \lambda_r^{-s} \ . \tag{15}$$

Then [5]

$$\varepsilon_o = \sum_{r,f} n_f \varepsilon_{orf} = \sum_f n_f \left[C\zeta_{\Delta_f}(-2) + C'\zeta'_{\Delta_f}(-2)\right] \tag{16}$$

ε_o can be interpreted as a quantum correction to the vacuum energy. In D = 4 it is related to the Casimir energy [11]. ζ_Δ is closely related to the internal manifold. It can be expressed as:

$$\zeta_\Delta(s) = \frac{1}{\Gamma(s)} \int_0^1 \left(\nu_\Delta(t) - 1\right) t^{s-1} dt \ , \tag{17}$$

where ν_Δ is nothing else than the statistical sum of a system whose Hamiltonian is equal to the Laplace operator Δ:

$$\nu_\Delta(t) = \sum_r e^{-t\lambda_r} \ . \tag{18}$$

In the high temperature limit (t->0) [13,4]

$$\upsilon_{\Delta}(t\to 0) = \frac{1}{(4\pi t)^{n/2}} \left(a_o + a_1 t + \ldots \right) , \tag{19}$$

where a_k are the spectral invariants [4,13].

Let us look for a minimum of ε_o as function of ν^2. If $\zeta_R(-2)=0$ then $\nu^2=0$ (or $\zeta_R'(-2)=0$). The case $\nu^2=0$ corresponds to decompactification. The function ζ vanishes at negative integers in odd-dimensional spaces [12]. This explain the instability of compactification of a Kaluza-Klein theory on an odd-dimensional internal space (e.g. on S^1). If $\zeta(-2)\neq 0$ we have

$$\nu^2_{min} = \Lambda^2 \exp \left(-\frac{1}{2} - \frac{\zeta'(-2)}{\zeta(-2)} \right) \tag{20a}$$

and

$$\varepsilon_{omin} = -\frac{1}{2} \Lambda^4 \zeta(-2) \exp \left(-1 - \frac{2\zeta'(-2)}{\zeta(-2)} \right) , \tag{20b}$$

where Λ is the renormalization group parameter. The analytical continuation of $\zeta(s)$ leads to the following representation [13]:

$$\zeta_{\Delta}(s) = \frac{1}{(4\pi)^2} \sum_{k=o}^{2} \frac{a_k}{\Gamma(3-k)(s-3+k)} + R(s) = \zeta_R(s) + R(s) \tag{21}$$

where R(s) is a regular function. This means that ε_o gives direct information about the internal manifold via a_o, a_1, a_2 (in ten dimensions). The spectrum of the Laplace operator contains important information about the manifold [4] e.g. it says when the manifold is Kähler or when the Ricci curvature vanishes. We can define the regularization procedure then $\nu^2=0$ (or $\zeta_R'(-2)=0$). The case $\nu^2=0$ corresponds to by discarding R(s) in (21). Thus, in a systematic way, we have expressed ε_o in terms of a_0, a_1, and a_2:

$$\varepsilon_0 = \frac{h\, e^{25/3}}{64\, \pi^2} \left(\nu^4\, \zeta_R'(-2) + \nu^2\, \zeta_R(-2)\, \ln\left(\frac{\nu^2}{\Lambda^2}\right)\right) . \tag{22}$$

In the regularization method we use ζ_R and ζ_R' can be calculated at least for simple manifolds [13]. For S^6 and $RP^6=S^6/\mathbb{Z}_2$ we have $a_0= 16(4\pi)^3/375$, $a_1= 2(4\pi)^3/15$, $a_2= (4\pi)^3/5$ and $a_0= 8(4\pi)^3/375$, $a_1= 19(4\pi)^3/240$, $a_2 = 10009(4\pi)^3/(120x512)$, respectively. The important fact is that for a space of the form K/Γ, where Γ is a discrete symmetry group, the spectral invariants are in general smaller then for K. This can be visualized by noticing that only a part of the spectrum survives. This means that K/Γ should be preferred in the compactification process because it has lower vacuum energy ε_0. In general, a_0 is proportional to the volume of the space and this means that the compactification process prefers collapsing. Analogous analysis can be performed in the string theory compactification. If we adapt this approach to the Calabi-Yau compactificationwe can draw the following conclusions. As far as the supersymmetry is unbroken vacuum energy vanishes because the contributions coming from the bosonic and the fermionic sectors cancel themselves. All Calabi-Yau spaces, that is Calabi-Yau vacuum solutions are physically equivalent and there is no preferred generation number. All we can do is to choose ad hoc the one we want to. At this stage we can use discrete symmetries to get relations between some parameters of the model. At the very moment of supersymmetry breaking the effective potential gets non zero contribution. All Calabi-Yau spaces cease to be equivalent from the physical point of view: the one with the lowest vacuum energy is preferred. If Calabi-Yau space K possess a (discrete) symmetry group G that acts freely it collapses into the quotient space K/G which is also a Calabi-Yau space but has Euler characteristic $X_{K/G}$ given by the formula

$$X_{K/G} = X_k /\, n(G) \quad , \tag{23}$$

where n(G) is the number of elements in G. The collapsing occurs because the quotient space has lower vacuum energy [14] (remember that for a

Calabi-Yau space we have $a_1 = a_2 = 0$) . Generally speaking the lowest "net" generation number N_g is preferred. This provides theoretical arguments in favour of the $N_g = 2$ models discussed in [15]. This analysis suggests also that the generation number N_g has physical meaning only after supersymmetry breaking. We should also be careful in using discrete symmetries to obtain low-energy predictions. In general, Calabi-Yau spaces with big Euler characteristic are possible internal spaces for superstring compactification as far as they possess discrete symmetry groups that after supersymmetry breaking lead to quotient spaces with $N_g = 2,3$ or 4.

Generally speaking, Nature directs the compactification process so that the configuration with lower vacuum energy (cosmological constant) is obtained. The Hosotani mechanism emerges in a natural way. This is a physical criterion of choosing one of the known Calabi- Yau spaces (in a given class).

Let us notice that $f_1(K)$ is a functional defined on the set of manifolds fulfilling Einstein equations. In some cases this set might be given a topology [16]. $f_1(K)$ should have a minimum on the physical internal manifold. The spectrum of the scalar Laplace operator alone may be insufficient to describe the internal manifold in an unambiguous way. This is because there might exist isospectral deformations of the internal space although no physically interesing examples are known.

References

[1] See e.g. D. Bailin and A. Love, Rep. Prog. Phys. 50 (1987) 1087.

[2] See e.g. M. Green, J.H. Schwarz and E. Witten, Superstring theory (Cambridge U.P., Cambridge, 1986) or J. Sładkowski, Fortschr. der Phys. in press.

[3] P. Candelas, G. Horowitz, A. Strominger and E. Witten, Nucl. Phys. B258 (1985) 46.

[4] P.B. Gilkey, Invent. Math. 26 (1974) 231, V.K. Patodi, J. Indian Math. Soc. 34 (1970) 131, J.J. Duistermaat and V.W. Guillemin, Invent. Math. 29 (1975) 39.

[5] R. Mańka, Class. and Quantum Grav. 4 (1987) 1785.

[6] R. Feynman, Statistical Mechanics (W.A. Benjamin, Menlo Park, 1972).

[7] R. Mańka, J. Kuczyński and G. Vitiello, Nucl. Phys. B276 (1986) 533.

[8] R. Mańka, Annals of Phys. 171 (1986) 1.

[9] R. Mańka and J. Sładkowski, Can. J. Phys. 65 (1987) 395.

[10] R. Mańka, J. Sładkowski and G. Vitiello, Phys. Lett. B222 (1989) 31.

[11] S. Blau, M. Visser and A. Wipf, Los Alamos preprint LA-UR-88-1098 (1988).

[12] R. Mańka and J. Sładkowski, Acta Phys. Pol., in press.

[13] N.E. Hurt, Geometric Quantization in Action (D. Reidel, Dordrecht, 1983) R.S. Cahn and J.A. Wolf, Comment. Math. Helv. 51 (1976) 1.

[14] R. Mańka and J. Sładkowski, Phys. Lett. B224 (1989) 97.

[15] J. Sładkowski, J. Phys. G, in press.

[16] K. Fukaya, Invent. Math. 87 (1987) 517.

Standard Model
and Beyond

ISBN 1-56072-027-1

A PROBABILISTIC APPROACH OF CONSTRUCTING THE GENERATING FUNCTIONAL IN THE STOCHASTIC QUANTIZATION METHOD

Ion Porchescu

Institute of Applied Physics, 277028, Grosul str. 5, Kishinev, Russia

INTRODUCTION

In recent years, there has been a considerable intrest in a new method propozed by Parisi and Wu [1] for quantizing of Euclidean field theory. This is due to the fact that stochastic quantization (SQ) scheme seems, first, to offer some advantages [2-8] over conventional ones, particularly in dealing with constrained systems, and second ,to give a new vision of the problem of quantum behaviour. The basic idea of SQ is to consider the Euclidean path integral "measure" exp(-S[x])dx (S[x]-is the Euclidean action) as the stationary distribution of a stochastic process (SP) governed by Langevin equation (LE). The system relaxes to stationary state for large fictious "time" τ . The prescription of the SQ method formulated by Parisi and Wu [1] is the following:
I. One supplements the field $\varphi(x)$ with an additional fictious axis of "time" τ .
II. Starting from the corresponding classical field equation

$$\frac{\delta S}{\delta \varphi}(x\tau) = \int dx' d\tau' \left[\frac{\delta \mathcal{L}}{\delta \varphi(x\tau)} - \partial_\mu \frac{\delta \mathcal{L}}{\delta(\partial_\mu \varphi(x\tau))} \right] = 0 \qquad (1)$$

one constracts a LE

$$\frac{\partial}{\partial \tau}\varphi(x\tau) + \frac{\delta S}{\delta \varphi}(x\tau) = \eta(x\tau) \qquad (2)$$

for a fictious SP modulated by two new forces: a Gaussian white noise $\eta(x\tau)$ and a "friction" force $\frac{\partial}{\partial\tau}\varphi(x\ )$. Here $\mathcal{L}(\varphi(x\tau)\ ,\ \varphi_{,}(x\tau))$ is the Lagrangian density. The white noise

$$\langle \eta(x\tau) \rangle_\eta = 0$$

$$\langle \eta(x\tau)\eta(x'\tau') \rangle_\eta = 2h\delta(\tau-\tau')\,\delta(x-x') \qquad (3)$$

modulates nonequilibrium quantum fluctuations while the "friction" force is responsible, first, for evolution to equilibrium distribution, and, second, for the fact that non-stationary fluctuations are a result of stochastic forces at all preciding "time" superposition [4].

III. Given some initial conditions, one has to solve LE (2). To emphasize the noise dependence explicitly the solution is indicated by $\varphi_\eta(x\tau)$. Many "time" stochastical correlation functions of $\varphi_\eta(x\tau)$ are defined by performing Gaussian average over noise η :

$$\langle \varphi_\eta(x_1\tau_1)\dots\varphi_\eta(x_n\tau_n) \rangle_\eta \qquad (4)$$

IV. The central assertion in the SQ method is that the Euclidean Green functions may be obtained from large "time" limit of single "time" average (4) :

$$\lim_{\tau\to\infty} \langle \varphi_\eta(x_1\tau)\dots\varphi_\eta(x_n\tau) \rangle = \frac{\int d[\varphi]\exp(-S[\varphi])\,\varphi(x_1)\dots\varphi(x_n)}{\int d[\varphi]\exp(-S[\varphi])} \qquad (5)$$

Let us consider the single "time" stochastical correlation functions. They are formally obtained with respect to functional integration over a single "time" probability distribution density (PDD) $P(\varphi,\tau)$ [13].

$$\langle \varphi_\eta(x_1\tau)\dots\varphi_\eta(x_n\tau) \rangle = \int d[\varphi]P(\varphi,\tau)\,\varphi(x_1)\dots\varphi(x_n), \qquad (6)$$

where

$$P(\varphi,\tau) = \langle \delta[\varphi-\varphi_\eta(\tau)] \rangle_\eta . \qquad (7)$$

The single "time" functional δ-function implies, for example, lattice space approximation at "time" τ :

$$\delta[\varphi-\varphi_\eta(\tau)] = \prod_x \delta(\varphi(x)-\varphi_\eta(x\tau)). \qquad (8)$$

Single "time" PDD satisfies the Fokker-Plank equation (FPE):

$$\frac{\partial}{\partial\tau}P(\varphi,\tau) = \int dx\, \frac{\delta}{\delta\varphi(x)}\left\{ \frac{\delta}{\delta\varphi(x)} + \frac{\delta S}{\delta\varphi(x)} \right\}P(\varphi,\tau). \qquad (9)$$

Additionally, we have to give some initial conditions

$$P(\varphi,0) = \delta[\varphi - c], \qquad (10)$$

where c are constants. It is obvious that there are two equivalent formulations of the SQ method based on the LE(2) and on the FPE (9). We may reformulate in the FPE language the main assertion of the SQ method implying consideration of the Euclidean path measure as the stationary distributi- of the SP (3) [2]:

$$\lim_{\tau\to\infty} P(\varphi,\tau) = N \exp(-S[\varphi]) \quad . \tag{11}$$

Let us consider in detail the "ergodicity" problem (11) [12,14]. Equation (9) may be viewed as the Schrödinger equation in imaginary "time" with non-Hermitian Hamiltonian:

$$(\frac{\partial}{\partial\tau} + \int dx\hat{H})P(\varphi,\tau) = 0, \tag{12}$$

where

$$\hat{H} = \hat{p}^2 - i\hat{p}\frac{\delta S}{\delta\varphi} \quad , \quad \hat{p} = - i\frac{\delta}{\delta\varphi} \quad . \tag{13}$$

If we set:

$$P(\varphi,\tau) = \Psi(\varphi,\tau) \exp(-\frac{1}{2} S[\varphi]) \tag{14}$$

a simple calculation leads to Schrödinger equation in imaginary "time" with the Hermitian Hamiltonian :

$$(-\frac{\partial}{\partial\tau} + \int dx H)\Psi(\varphi,\tau)=0 \tag{15}$$

where

$$H = \hat{p}^2 + U(\varphi)$$

$$U(\varphi) = \frac{1}{4}(\frac{\delta S}{\delta\varphi})^2 - \frac{1}{2}\frac{\delta^2 S}{\delta\varphi^2} \tag{16}$$

by "gauge" transformation (14) we have associated Hermitian Hamiltonian (16) with the known energetical band with non--Hermitian Hamiltonian (13). More detailed analysis of the correlation between "gauge" transformation (14) and energetical bands of operators (13,16) is given in works [12,14]. If we denote the complete set of eigenstates of H by Ψ_n :

$$H\ \Psi_n(\varphi) = E_n\Psi_n(\varphi)$$

$$\Psi_n(\varphi\tau) = \Psi_n(\varphi) \exp(-\tau E_n) \tag{17}$$

$$\Psi_0(\varphi) = \exp(-\frac{1}{2} S[\varphi])$$

we may write for the solution of (15):

$$\Psi(\varphi,\tau)=a_0\exp(-\frac{1}{2}S[\varphi])+\sum_{n=1}^{\infty}a_n\varphi_n(\varphi)\exp(-\tau E_n) \tag{18}$$

If we suppose that $E_n > 0$ for $n > I$ then for large "time" limits of $\Psi(\varphi,\tau)$ and $P(\varphi,\tau)$ we get :

$$\lim_{\tau\to\infty}\Psi(\varphi,\tau) = a\exp(-\frac{I}{2}S[\varphi]) \tag{I9}$$

$$\lim_{\tau\to\infty}P(\varphi,\tau) = a\exp(-S[\varphi]) \tag{20}$$

We fix the normalization by:

$$\lim_{\tau\to\infty}\int d[\varphi]P(\varphi,\tau) = I \tag{21}$$

and arrive finally at:

$$\lim_{\tau\to\infty}P(\varphi,\tau) = \frac{\exp(-S[\varphi])}{\int d[\varphi]\exp(-S[\varphi])} \tag{22}$$

as has been shown in [2]. Here we have intentionally considered the simplest case of application of SQ method- scalar field theory with nondegenerate vacuum to show the leading part of PDD in solving the "ergodicity" problem in the SQ . The essence of the probabilistic approach consists in isolating single "time" PDD from the expression of dynamical generating fuctional. But , first , let us cosider the Gozzi [9] approach in constructing this expression. To evaluate the many "time" noise averages let us consider , following [9], the generating functional:

$$Z[J] =\int D[\varphi]D[\eta]P(\varphi,0)\Delta[\varphi-\varphi_\eta]\exp(J\varphi), \tag{23}$$

where $P(\varphi,0)$ is some initial probability, $\Delta[\varphi-\varphi_\eta]$ and $D[\varphi]$ imply a lattice approximation of space supplemented by "time" axis:

$$D[\varphi] = \prod_{\tau}\prod_{x} d\varphi(x\tau) \tag{24}$$

$$\Delta[\varphi-\varphi_\eta] = \prod_{\tau}\prod_{x}\delta(\varphi(x\tau) - \varphi_\eta(x\tau)) . \tag{25}$$

After formal transformation [11]

$$\Delta[\varphi-\varphi_\eta] = \Delta[\dot{\varphi}+\frac{\delta S}{\delta\varphi}-\eta]\left|\left|\frac{\delta\eta}{\delta\varphi}\right|\right| \tag{26}$$

and the η integration we have the following expression:

$$Z[J]=\int D[\varphi]P(\varphi,0)\exp\{-\int_0^{\tau}dt(-J\varphi+\frac{1}{4}(\dot{\varphi}+\frac{\delta S}{\delta\varphi})^2-\frac{1}{2}\frac{\delta^2 S}{\delta\varphi^2})\} \tag{27}$$

Performing the integration of the term

$$\int_0^{\tau} \dot{\varphi} \frac{\delta S}{\delta \varphi} dt = S[\varphi(\ )] - S[\varphi(0)]$$

we get for Z[J] :

$$Z[J]=\int d[\varphi(0)]d[\varphi(\tau)]\ D''[\varphi])\exp\{-\frac{(S[\varphi(0)]-S[\varphi(\tau)]\ )}{2}+$$
$$+\ dt(J\varphi-\mathcal{L}^{FP}(t))\}P(\varphi,0) \qquad (28)$$

To understand the physical meaning of (28) we consider a simplest one-dimensional quantum mechanical system[3] .The kernel of the evolution operator is given by a path integral over all trajectories x(t) from the initial state $x_0=x(t_0)$ to the final state $x_f= x(t_f)$ of the exponential S[x] where S is the classical action functional :

$$\langle x_f t_f | x_0 t_0\rangle = \int d[x]\ \exp\{-\frac{S[x]}{h}\} \quad . \qquad (29)$$

If we consider vacuum expectation values of the position operators in the Heisenberg picture

$$\hat{Q}(t) = \exp\{-\frac{t\hat{H}}{h}\}\ \hat{Q}\ \ \exp\{\ .\frac{t\hat{H}}{h}\}$$

where $\hat{H}$ is the Hamiltonian of a given quantum system, it is possible to obtain the following expression:

$$\langle 0|T\ \hat{Q}(t_1)\ldots\hat{Q}(t_n)|0\rangle = \int dx_0\, dx_f \Psi_0^*(x_f t_f)\ \Psi_0(x_0 t_0)$$
$$d''[x]\exp(-\frac{S[x]}{h})\ x(t_1)\ldots x(t_n)\ , \qquad (30)$$

where

$$\Psi_0(x_0 t_0) = \langle \Psi_0 | x_0 t_0\rangle$$
$$\Psi_0^*(x_f t_f) = \langle x_f t_f | \Psi_0\rangle \qquad (31)$$

Comparing (30) with (28) and taking into account (17) we see an analogy between the Fokker-Plank dynamical ensemble and the quantim system. We can also derive the second conclusion: to get the Euclidean measure $\Psi_0(\Psi)$ we have choose the initial probability density in the from :

$$P(\varphi,0) = N\exp\ (-S[\varphi]). \qquad (32)$$

The physical meaning of (32) is clear: we put the system from the begining in the stationary (equilibrium) state and the presence of the Langevin dynamics does not modify it [9]. Implying that vacuum of H (16) is $\Psi_0(\varphi)=\exp(-I/2\ S[\varphi])$ Gozzi named (28) the vacuum-vacuum generating functional. Though this approach is elegant, it seems for us that the factor $P(\varphi,0)$ is introduced "by hand" in expression (23) for the generating functional.

Generating Functional. Probabilistic approach.

It is the our aim to build the generating functional of the SQ method from the assumption that large "time" limit of noise average is equivalent to the quantum one without of additional "by hand" introduction of the factor $P(\varphi,0)$ [9] . Due to the leading role of the Markov proprety we would like to do a brief survey of the Markov process theory [15-17]. By definition a SP is markovian under the condition that if a present state is known, any additional information about the history of a process is nonessential for future prognostication of the process behaviour. Thus, if the present is known, than past and future are relatively independent [17]. In particular, for a process governed by the Langevin equation to be markovian the correlations of noise η have to be -correlated at "time" τ . The Markov process is described by only two probability functionals: single "time" PDD and two "time" conditional PDD . By definition:

$$P(x_1;...;x_k|x_{k+1};...;x_n)=P(x_1;...;x_n)/P(x_{k+1};...;x_n) \quad (33)$$

For the Markov SP we have

$$P(x_1|x_2;...;x_n)= P(x_1|x_2) \quad (34)$$

where $P(x|x')$ and $P(x;...;x')$ are respectively the conditional and joint PDD,

$$P(x_1| x_3) = \int dx_2 P(x_1|x_2)P(x_2|x_3) \quad (35)$$

$$P(x_1) = \int dx_2 P(x_1|x_2) \; P(x_2) \quad (36)$$

This implies that $\tau_1 > \tau_2 > \; ... \; > \tau_n$
Equations (35-36) are named the Chapman-Kolmogorov equations respectively for the two "time" conditional PDD and single "time" PDD. We have to note that (36) is not the consequence of the Markov property but is a general property of the joint PDD :

$$\int dx_2 P(x_1;x_2;x_3)= P(x_1;x_3) \quad (37)$$

Let us go back to building the generating functional. A starting point will be the following expression for the generating functional :

$$Z[J] = \int \mathbf{D}[\varphi]P[\varphi,\tau]\exp(J\varphi), \quad (38)$$

where probability density functional:

$$P[\varphi,\tau] = \langle \Delta [\varphi - \varphi_\eta] \rangle_\eta \tag{39}$$

can be written in the following form [14]

$$P[\varphi,\tau] = \lim_{N\to\infty} P_N(\varphi_1\tau_1;\ldots;\varphi_N\tau_N) \tag{40}$$

$$\varphi_i = \varphi(x_i), \qquad \tau_0 \le \tau_i \le \tau$$

and $P_N(\ldots)$ is the N-dimensional joint PDD. Using the defition (33-37) for (40) we have

$$P_N(\varphi_1\tau_1;\ldots;\varphi_N\tau_N) = P(\varphi_1\tau_1/\varphi_2\tau_2)\ldots P(\varphi_{N-1}/\varphi_N)P(\varphi_N,\tau_N), \tag{41}$$

where

$$P(\varphi,\tau) = \langle \prod_x \delta(\varphi(x\tau) - \varphi_\eta(x\tau)) \rangle_\eta \tag{42}$$

$$P(\varphi\tau/\varphi_1\tau_1) = \frac{\langle \prod_{x,x_1} \delta(\varphi(x\tau) - \varphi_\eta(x\tau))\,\delta(\varphi(x_1\tau_1) - \varphi_\eta(x_1\tau_1)) \rangle_\eta}{P(\varphi_1,\tau_1)}. \tag{43}$$

For the functional (43) we have an obvious boundary condition:

$$P(\varphi\tau/\varphi_1\tau) = \delta[\varphi-\varphi_1]. \tag{44}$$

The boundary condition for the joint PDD (41) is the following:

$$P_N(\varphi_1\tau;\ldots;\varphi_N\tau) = \delta[\varphi_1-\varphi_2]\ldots[\varphi_{N-1}-\varphi_N]P(\varphi_N,\tau). \tag{45}$$

The SQ prescription implies two operations over the many "time" stochastical averages $\langle \varphi_\eta(x_1\tau_1)\ldots\varphi_\eta(x_n\tau_n) \rangle_\eta$; namely

1) $\lim\limits_{\tau_i=\tau} \langle \ldots \rangle$

2) $\lim\limits_{\tau\to\infty} \langle \ldots \rangle$

By means of (45) the first operation automatically isolates the single "time" PDD while the second one is responsible for the relaxation to equilibrium state. To express the conditional PDD (43) through the path integral we have to note that the "Hamiltonian" operator $\hat{H}$ (12) associated with $P(\varphi\tau/\varphi_1\tau_1)$ is non-Hermitian. After the "gauge" transformation

$$P(\varphi\tau/\varphi_1\tau_1) = W(\varphi\tau/\varphi_1\tau_1)\exp\{-\frac{1}{2}(S[\varphi(\tau)] - S[\varphi(\tau_1)])\} \tag{46}$$

we get the Fokker-Plank equation with the Hermitian "Hamiltonian" (16). The boundary condition for W is the same as for P (44)

$$\mathcal{W}(\varphi\tilde{\tau}/\varphi_1\tilde{\tau}) = \delta[\varphi-\varphi_1]. \tag{47}$$

Associating with the Hermitian "Hamiltonian" a Lagrangian $\mathcal{L}^{FP}$ according to the well-known prescription [11]:

$$-\mathcal{L}^{FP} = p\dot{\varphi} - H \;,\; i\dot{\varphi} = \frac{\partial H}{\partial p} = -2p \;, \tag{48}$$

where H is the classical Hamiltonian function, we get for following path integral expression

$$\mathcal{W}(\varphi\tilde{\tau};\varphi_1\tilde{\tau}_1) = \int d[\varphi]\exp\{-\int_{\tau_1}^{\tilde{\tau}} dt\,\mathcal{L}^{FP}(t)\} \tag{49}$$

where

$$\mathcal{L}^{FP}(t) = \frac{1}{4}(\dot{\varphi})^2 - \frac{1}{2}\frac{\delta^2 S}{\delta\varphi^2} + \frac{1}{4}\left(\frac{\delta S}{\delta\varphi}\right)^2 \tag{50}$$

Using (49-50), we arrive finally at the following path integral expression for the generating functional of the SQ method wich coincides with the vacuum-vacuum form of the generating functional [9].

$$Z[J]=\int d[\varphi(0)]d[\varphi(\tilde{\tau})]D''[\varphi]P(\varphi,0)\exp\{-\frac{1}{2}S[\varphi(\tilde{\tau})]+\frac{1}{2}S[\varphi(0)]\}$$
$$\exp\{-\int_0^{\tilde{\tau}} dt[\mathcal{L}^{FP}(t)-J\varphi]\} \tag{51}$$

We have considered the simplest case [19], scalar field theory SQ, to build the generating functional. We think that the essential moment of the procedure is isolating of single "time" probability. This is true also in Zwanzinger's schems [18]; hence this approach may by regarded also as an alternative to Alvarez-Estrada and Munoz-sudupe [10] functional approach to the stochastic quantization of the Euclidean Yang-Mills field theory.

Acknowledgement

I would like to thank Professor V.A.Moskalenko for useful discussions and for his support, Dr. G.D'Agostinio for useful discussions.

REFERENCES

[1] G.Parisi and Y.Wu ,Scientia Sinica 24, (1981) 483
[2] P.H.Damgaard and H.Huffel ,Phys.Rep. 152, (1987) 227
[3] A.Gonzalez-Arroyo ,Lect.Notes Phys.,vol.216 (eds. L. Garrido),Springer-Verlag,Berlin,(1985)
[4] A.A.Migdal ,Uspehi Phys. Nauk 149, (1986) 3

[5] S.Chaturvedi, A.K.Kapoor ,V.Srinivasan ,Preprint Hyderabad,HUTP-83-3 ,Hyderabad ,(1985) p.5
[6] V.Bujek ,Preprint JINR ,P2-84-419,Dubna, (1984) p.6
[7] V.Bujek and S.Olejnik, Czech.J.Phys. A 38, (1988), 31
[8] M.Namiki, In Concept of Probability,(Eds. E.I.Bitsakis and C.A.Nicolaides),Kluer Academic Publ.,London, (1989)
[9] E.Gozzi, Phys.Rev.D 28, (1983), 1922
[10] R.F.Alvares-Estrada and A.Munoz-Sudupe ,J.Phys. A 20 , (1987), 733
[11] F.Langouche, D.Rockaerts, E.Tirapegui ,Physica A 95 , (1979) ,252
[12] B.Muhlschlegel, In Path Integrals and their applications in Quantum ,Statistical and Solid State Physics, (Eds.G.Papandopoulos and J.T.Devreese),Plenum Press, N.Y. and London , (1978) p.40
[13] E.Floratos and J.Ilioupolos, Nucl.Phys. B 214, (1983), 392.
[14] C.M.Bender, F.Cooper, B.Freedman Nucl.Phys. B 219, (1983), 61
[15] K.B.Kliatskin, Stochastic equations,and wave in random nonhomogeneous media, Nauka, Moskow, (1979), p.11
[16] C.W.Gardiner, Handbook of Stochastic Methods, Springer -Verlag, Berlin, (1983), p.40
[17] W.Horsthemke and R.Lefever, Noise-induced transitions, Springer-Verlag, Berlin, (1984), p.51
[18] D.Zwanzinger, In Fundamental problems of gauge theory, (Eds. G. Velo and A.Wightman), Plenum Press,N.Y., (1985), p.39
[19] I.Porchescu, Kh. Namsrai, Preprint JINR, EZ-89-359, Dubna, (1989), p.5

Subject Index